Functional nanofibers and their applications

The Textile Institute and Woodhead Publishing

The Textile Institute is a unique organisation in textiles, clothing and footwear. Incorporated in England by a Royal Charter granted in 1925, the Institute has individual and corporate members in over 90 countries. The aim of the Institute is to facilitate learning, recognise achievement, reward excellence and disseminate information within the global textiles, clothing and footwear industries.

Historically, The Textile Institute has published books of interest to its members and the textile industry. To maintain this policy, the Institute has entered into partnership with Woodhead Publishing Limited to ensure that Institute members and the textile industry continue to have access to high calibre titles on textile science and technology.

Most Woodhead titles on textiles are now published in collaboration with The Textile Institute. Through this arrangement, the Institute provides an Editorial Board which advises Woodhead on appropriate titles for future publication and suggests possible editors and authors for these books. Each book published under this arrangement carries the Institute's logo.

Woodhead books published in collaboration with The Textile Institute are offered to Textile Institute members at a substantial discount. These books, together with those published by The Textile Institute that are still in print, are offered on the Woodhead web site at: www.woodheadpublishing.com. Textile Institute books still in print are also available directly from the Institute's web site at: www.textileinstitutebooks.com

A list of Woodhead books on textiles science and technology, most of which have been published in collaboration with the Textile Institute, can be found towards the end of the contents pages.

Woodhead Publishing Series in Textiles: Number 134

Functional nanofibers and their applications

Edited by
Qufu Wei

The Textile Institute

WOODHEAD PUBLISHING

Oxford Cambridge Philadelphia New Delhi

Published by Woodhead Publishing Limited in association with The Textile Institute
Woodhead Publishing Limited, 80 High Street, Sawston,
Cambridge CB22 3HJ, UK
www.woodheadpublishing.com
www.woodheadpublishingonline.com

Woodhead Publishing, 1518 Walnut Street, Suite 1100, Philadelphia,
PA 19102-3406, USA

Woodhead Publishing India Private Limited, G-2, Vardaan House,
7/28 Ansari Road, Daryaganj, New Delhi – 110002, India
www.woodheadpublishingindia.com

First published 2012, Woodhead Publishing Limited

British Library Cataloguing in Publication Data
A catalogue record for this book is available from the British Library.

Library of Congress Control Number: 2012936173

ISBN 978-0-08-101649-7 (print)
ISBN 978-0-85709-564-0 (online)
ISSN 2042-0803 Woodhead Publishing Series in Textiles (print)
ISSN 2042-0811 Woodhead Publishing Series in Textiles (online)

The publisher's policy is to use permanent paper from mills that operate a sustainable forestry policy, and which has been manufactured from pulp which is processed using acid-free and elemental chlorine-free practices. Furthermore, the publisher ensures that the text paper and cover board used have met acceptable environmental accreditation standards.

Typeset by Newgen Publishing and Data Services, India

Contents

Contributor contact details *xi*
Woodhead Publishing Series in Textiles *xv*

Part I Types and processing

1 Nanofibers: principles and manufacture 3
Q. Wei, D. Tao and Y. Xu, Jiangnan University, China

1.1 Introduction 3
1.2 Principles of electrospinning 5
1.3 Solution electrospinning 10
1.4 Melt electrospinning 15
1.5 Future trends 18
1.6 References 19

2 Types and processing of structured functional nanofibers: core-shell, aligned, porous and gradient nanofibers 22
Q. Wei, Jiangnan University, China and X. Xia, Xinjiang University, China

2.1 Introduction 22
2.2 Core-shell nanofibers 23
2.3 Aligned nanofibers 27
2.4 Porous nanofibers 30
2.5 Gradient nanofibers 31
2.6 Applications of structured functional nanofibers 33
2.7 Future trends 34
2.8 References 36

3 Processing of composite functional nanofibers 38
Y. Cai, Q. Wei and F. Huang, Jiangnan University, China

3.1 Introduction 38
3.2 Formation of polymer and polymer composite nanofibers 38
3.3 Formation of polymer and nanoparticle composite nanofibers 40
3.4 Formation of polymer and inorganic salt composite nanofibers 43
3.5 Examples and applications of composite functional nanofibers 43
3.6 Future trends 49
3.7 Acknowledgments 49
3.8 References 49

4 Carbon and polymer nanofiber reinforcements in polymer matrix composites: processing and applications 55
J. Zhang, The Royal Melbourne Institute of Technology (RMIT) University, Australia and T. Lin and X. Wang, Deakin University, Australia

4.1 Introduction 55
4.2 Composite formation with nanofibers 57
4.3 Strengths and weaknesses of nanofibers in composites 60
4.4 Applications of nanofiber composites 61
4.5 Future trends 67
4.6 References 68

5 Inorganic functional nanofibers: processing and applications 71
N. Wu, Tianjin Polytechnic University, China and Q. Wei, Jiangnan University, China

5.1 Introduction 71
5.2 Processing of inorganic nanofibers 72
5.3 Case study: preparation, structure and properties of TiO_2 inorganic nanofibers 75
5.4 Doping inorganic nanofibers 82
5.5 Applications of inorganic functional nanofibers 84
5.6 Future trends 87
5.7 References 88

6 Surface functionalization of polymer nanofibers 92
F. Huang, Q. Wei and Y. Cai, Jiangnan University, China

6.1 Introduction 92
6.2 Surface functionalization using physical technologies 93
6.3 Surface functionalization using chemical technologies 101
6.4 Surface functionalization using nanotechnologies 106
6.5 Surface functionalization using biotechnology 112
6.6 Future trends 115
6.7 References 116

Part II Applications

7 Functional nanofibers for filtration applications 121
G. G. Chase, The University of Akron, USA,
S. Swaminathan, Hollingsworth and Vose Company, USA
and B. Raghavan, Parker Hannifin Corporation, USA

7.1 Introduction 121
7.2 Key principles of functional nanofibers for filtration 125
7.3 Filtration of nanoparticles from gas streams 131
7.4 Nanofiber catalyst support structures 142
7.5 Future trends and conclusions 147
7.6 Acknowledgments 149
7.7 References 149
7.8 Appendix: notation 151

8 Functional nanofibers for drug delivery applications 153
Q. Wei, Jiangnan University, China and A. Wei, Anhui Polytechnic University, China

8.1 Introduction 153
8.2 Drug delivery mechanisms 155
8.3 Applications of functional nanofibers in drug delivery: a case study 160
8.4 Future trends 167
8.5 References 168

9 Functional nanofibers for tissue engineering applications 171
I. A. Rodriguez, J. M. McCool and G. L. Bowlin, Virginia Commonwealth University, USA

9.1 Introduction 171

9.2 Electrospinning of nanofibers 172
9.3 Applications of nanofibers in tissue engineering 172
9.4 Conclusions 186
9.5 References 188

10 Functional nanofibers in lithium-ion batteries 197
H. Qiao and Q. Wei, Jiangnan University, China

10.1 Introduction 197
10.2 Components and materials in lithium-ion batteries 198
10.3 Applications of nanofibers in lithium-ion batteries 200
10.4 Examples of applications of nanofibers 202
10.5 Future trends 205
10.6 References 206

11 Functional nanofibers in sensor applications 209
X. F. Wang, B. Ding and J. Y. Yu, Donghua University, China

11.1 Introduction: modern electrospinning technology 209
11.2 Formation of nanostructured sensing materials 215
11.3 Sensing mechanisms 219
11.4 Future trends and conclusions 229
11.5 Acknowledgments 229
11.6 References 230

12 Functional nanofibers in clothing for protection against chemical and biological hazards 236
A. H. Nurfaizey, University of Canterbury, New Zealand and Universiti Teknikal Malaysia Melaka, N. Tucker, New Zealand Institute for Plant and Food Research Limited, New Zealand, J. Stanger and M. P. Staiger, University of Canterbury, New Zealand

12.1 Introduction 236
12.2 Chemical and biological warfare background 237
12.3 Electrospun nanofibers 244
12.4 The advantages of using electrospun nanofiber in protective clothing 248
12.5 Issues with development of electrospun fibers 255
12.6 Future trends 256
12.7 References 258

13 Functional nanofibers in food processing 262
J. A. Lopes da Silva, University of Aveiro, Portugal

13.1 Introduction 262
13.2 Electrospun fibrous mats in food-related separation applications 265
13.3 Electrospun microfibers and nanofibers for enzyme immobilization 271
13.4 Electrospun nanofibers in food sensors 280
13.5 Electrospun micro- and nanofibers used for encapsulation and controlled release of bioactive compounds 286
13.6 Future trends and conclusions 290
13.7 Acknowledgments 292
13.8 Sources of further information 292
13.9 References 294

14 Functional nanofibers in sound absorption, electromagnetic wave attenuation and bioreactor application 305
Q. Wei and D. Gao, Jiangnan University, China

14.1 Introduction 305
14.2 Applications in sound absorption 306
14.3 Applications of functional nanofibers as electromagnetic wave attenuating material 306
14.4 Applications as a bioreactor: filtration membranes 315
14.5 Applications as a bioreactor: immobilization of enzymes 317
14.6 Future trends 325
14.7 References 327

15 Functional nanofibers for water purification 331
H. Ma, B. Chu and B. S. Hsiao, Stony Brook University, USA

15.1 Introduction 331
15.2 A new class of thin-film nanofibrous composite (TFNC) membranes 335
15.3 TFNC membranes for microfiltration 339
15.4 TFNC membranes for ultrafiltration (UF) 348
15.5 TFNC membranes for nanofiltration and reverse osmosis 357
15.6 Conclusions 361
15.7 Acknowledgment 362
15.8 References 362

16 Functional nanofibers in microelectronics applications 371
T. S. Natarajan, Indian Institute of Technology Madras, India

16.1 Introduction: polymers in electronics 371
16.2 Electrospinning of nanofibers 373
16.3 Nanofibers in microelectronics 377
16.4 Conclusions 401
16.5 Acknowledgments 401
16.6 References 401

17 Future trends in the processing of functional nanofibers 411
Q. Wei, Jiangnan University, China

17.1 References 413

Index *415*

Contributor contact details

(* = main contact)

Editor, Chapters 1, 2, 5, 6, 8, 10, 14 and 17

Qufu Wei
School of Textiles & Clothing
Jiangnan University
1800 Lihu Avenue
Wuxi 214122
Jiangsu Province
China

Email: qfwei@jiangnan.edu.cn

Chapter 1

Dan Tao and Yang Xu
Jiangnan University
1800 Lihu Avenue
Wuxi 214122
Jiangsu Province
China

Email: dandan8791@163.com; zh3212@vip.sina.com

Chapter 2

Xin Xia
Xinjiang University
Xinjiang
Wulumuqi 830046
China

Email: xjxiaxin@163.com

Chapters 3 and 6

Yibing Cai and Fenglin Huang
Key Laboratory of Eco-textiles
Ministry of Education
School of Textiles and Clothing
Jiangnan University
Wuxi 214122
China

Email: yibingcai@jiangnan.edu.cn; flhuang@jiangnan.edu.cn

Chapter 4

Jin Zhang
Sir Lawrence Wackett Aerospace Research Centre
School of Aerospace, Mechanical and Manufacturing Engineering
The Royal Melbourne Institute of Technology (RMIT) University
Melbourne
VIC 3083
Australia

Tong Lin and Xungai Wang*
Australian Future Fibres Research and Innovation Centre (AFFRIC)
Deakin University
Geelong
VIC 3217
Australia

Email: xwang@deakin.edu.au

Chapter 5

Ning Wu
Key Laboratory for Advanced Textile Composite Materials of Ministry of Education
Institute of Textile Composites
Tianjin Polytechnic University
Tianjin 300160
China

Email: wuning@tjpu.edu.cn; wuning2006@hotmail.com

Chapter 7

George G. Chase
Department of Chemical and Biomolecular Engineering
The University of Akron
185 E Mill Street
Akron
Ohio 44325-3906, USA

Email: gchase@uakron.edu

Sneha Swaminathan
Hollingsworth and Vose Company
219 Townsend Road
West Groton, MA 01472

Email:Sneha.Swaminathan@hovo.com

Bharath Raghavan
Parker Hannifin Corporation
Filtration Group
Filtration & Separation
260 Neck Road
Haverhill, MA 01835, USA

Email: bharath.raghavan@parker.com

Chapter 8

Anfang Wei
Anhui Polytechnic University, Wuhu 241000
Anhui Province
China

Email: wanfang2564yahoo.com.cn

Chapter 9

Gary L. Bowlin
Virginia Commonwealth University
401 W Main Street, Room E1254
Engineering East
P.O. Box 843067
Richmond, VA 23284-3067
USA

Email: glbowlin@vcu.edu

Isaac A. Rodriguez and Jennifer M. McCool
Virginia Commonwealth University
401 West Main .
Engineering East
P.O. Box 843067
Richmond, VA 23284-3067
USA

Email: rodriguezia@vcu.edu; mccooljm2@vcu.edu

Chapter 10

Hui Qiao
Jiangnan University
1800 Lihu Avenue
Wuxi 214122
Jiangsu Province
China

Email: huiqiao@jiangnan.edu.cn

Chapter 11

Xianfeng Wang, Bin Ding* and Jianyong Yu
Nanomaterials Research Center
Modern Textile Institute
Donghua University
Shanghai 200051
China

Email: xfwang81@yahoo.com; binding@dhu.edu.cn; yujy@dhu.edu.cn

Chapter 12

Abdul Hamid Nurfaizey,* Jonathan Stanger and Mark P. Staiger
Mechanical Engineering Department
University of Canterbury
Christchurch 8140
New Zealand

Email: nurfaizey.abdulhamid@pg.canterbury.ac.nz; jonathan.stanger@pg.canterbury.ac.nz; mark.staiger@canterbury.ac.nz

Abdul Hamid Nurfaizey
Faculty of Mechanical Engineering
Universiti Teknikal Malaysia Melaka
Melaka 76100
Malaysia

Email: nurfaizey@utem.edu.my

Nick Tucker
The New Zealand Institute for Plant & Food Research Limited
Gerald Street
Lincoln 7608
New Zealand

Email: Nick.Tucker@plantandfood.co.nz

Chapter 13

J. A. Lopes da Silva
QOPNA Research Unit
Department of Chemistry
University of Aveiro
3810-193 Aveiro
Portugal

Email: jals@ua.pt

Chapter 14

Dawei Gao
Jiangnan University
1800 Lihu Avenue
Wuxi 214122
Jiangsu Province
China

Email: gdw8668@163.com

Chapter 15

Hongyang Ma, Benjamin Chu and Benjamin S. Hsiao*
Department of Chemistry
Stony Brook University
Stony Brook
NY 11794-3400
USA

Email: bhsiao@notes.cc.sunysb.edu; bchu@notes.cc.sunysb.edu

Chapter 16

T. S. Natarajan
Department of Physics
Indian Institute of Technology Madras
Chennai 600036
India

Email: tsn@physics.iitm.ac.in; tsniit@gmail.com

Woodhead Publishing Series in Textiles

1 **Watson's textile design and colour Seventh edition** *Edited by Z. Grosicki*
2 **Watson's advanced textile design** *Edited by Z. Grosicki*
3 **Weaving Second edition** *P. R. Lord and M. H. Mohamed*
4 **Handbook of textile fibres Vol 1: Natural fibres** *J. Gordon Cook*
5 **Handbook of textile fibres Vol 2: Man-made fibres** *J. Gordon Cook*
6 **Recycling textile and plastic waste** *Edited by A. R. Horrocks*
7 **New fibers Second edition** *T. Hongu and G. O. Phillips*
8 **Atlas of fibre fracture and damage to textiles Second edition** *J. W. S. Hearle, B. Lomas and W. D. Cooke*
9 **Ecotextile '98** *Edited by A. R. Horrocks*
10 **Physical testing of textiles** *B. P. Saville*
11 **Geometric symmetry in patterns and tilings** *C. E. Horne*
12 **Handbook of technical textiles** *Edited by A. R. Horrocks and S. C. Anand*
13 **Textiles in automotive engineering** *W. Fung and J. M. Hardcastle*
14 **Handbook of textile design** *J. Wilson*
15 **High-performance fibres** *Edited by J. W. S. Hearle*
16 **Knitting technology Third edition** *D. J. Spencer*
17 **Medical textiles** *Edited by S. C. Anand*
18 **Regenerated cellulose fibres** *Edited by C. Woodings*
19 **Silk, mohair, cashmere and other luxury fibres** *Edited by R. R. Franck*
20 **Smart fibres, fabrics and clothing** *Edited by X. M. Tao*
21 **Yarn texturing technology** *J. W. S. Hearle, L. Hollick and D. K. Wilson*
22 **Encyclopedia of textile finishing** *H.-K. Rouette*
23 **Coated and laminated textiles** *W. Fung*
24 **Fancy yarns** *R. H. Gong and R. M. Wright*
25 **Wool: Science and technology** *Edited by W. S. Simpson and G. Crawshaw*
26 **Dictionary of textile finishing** *H.-K. Rouette*
27 **Environmental impact of textiles** *K. Slater*
28 **Handbook of yarn production** *P. R. Lord*

29 **Textile processing with enzymes** *Edited by A. Cavaco-Paulo and G. Gübitz*
30 **The China and Hong Kong denim industry** *Y. Li, L. Yao and K. W. Yeung*
31 **The World Trade Organization and international denim trading** *Y. Li, Y. Shen, L. Yao and E. Newton*
32 **Chemical finishing of textiles** *W. D. Schindler and P. J. Hauser*
33 **Clothing appearance and fit** *J. Fan, W. Yu and L. Hunter*
34 **Handbook of fibre rope technology** *H. A. McKenna, J. W. S. Hearle and N. O'Hear*
35 **Structure and mechanics of woven fabrics** *J. Hu*
36 **Synthetic fibres: Nylon, polyester, acrylic, polyolefin** *Edited by J. E. McIntyre*
37 **Woollen and worsted woven fabric design** *E. G. Gilligan*
38 **Analytical electrochemistry in textiles** *P. Westbroek, G. Priniotakis and P. Kiekens*
39 **Bast and other plant fibres** *R. R. Franck*
40 **Chemical testing of textiles** *Edited by Q. Fan*
41 **Design and manufacture of textile composites** *Edited by A. C. Long*
42 **Effect of mechanical and physical properties on fabric hand** *Edited by Hassan M. Behery*
43 **New millennium fibers** *T. Hongu, M. Takigami and G. O. Phillips*
44 **Textiles for protection** *Edited by R. A. Scott*
45 **Textiles in sport** *Edited by R. Shishoo*
46 **Wearable electronics and photonics** *Edited by X. M. Tao*
47 **Biodegradable and sustainable fibres** *Edited by R. S. Blackburn*
48 **Medical textiles and biomaterials for healthcare** *Edited by S. C. Anand, M. Miraftab, S. Rajendran and J. F. Kennedy*
49 **Total colour management in textiles** *Edited by J. Xin*
50 **Recycling in textiles** *Edited by Y. Wang*
51 **Clothing biosensory engineering** *Y. Li and A. S. W. Wong*
52 **Biomechanical engineering of textiles and clothing** *Edited by Y. Li and D. X.-Q. Dai*
53 **Digital printing of textiles** *Edited by H. Ujiie*
54 **Intelligent textiles and clothing** *Edited by H. R. Mattila*
55 **Innovation and technology of women's intimate apparel** *W. Yu, J. Fan, S. C. Harlock and S. P. Ng*
56 **Thermal and moisture transport in fibrous materials** *Edited by N. Pan and P. Gibson*
57 **Geosynthetics in civil engineering** *Edited by R. W. Sarsby*
58 **Handbook of nonwovens** *Edited by S. Russell*

59 **Cotton: Science and technology** *Edited by S. Gordon and Y.-L. Hsieh*
60 **Ecotextiles** *Edited by M. Miraftab and A. R. Horrocks*
61 **Composite forming technologies** *Edited by A. C. Long*
62 **Plasma technology for textiles** *Edited by R. Shishoo*
63 **Smart textiles for medicine and healthcare** *Edited by L. Van Langenhove*
64 **Sizing in clothing** *Edited by S. Ashdown*
65 **Shape memory polymers and textiles** *J. Hu*
66 **Environmental aspects of textile dyeing** *Edited by R. Christie*
67 **Nanofibers and nanotechnology in textiles** *Edited by P. Brown and K. Stevens*
68 **Physical properties of textile fibres Fourth edition** *W. E. Morton and J. W. S. Hearle*
69 **Advances in apparel production** *Edited by C. Fairhurst*
70 **Advances in fire retardant materials** *Edited by A. R. Horrocks and D. Price*
71 **Polyesters and polyamides** *Edited by B. L. Deopura, R. Alagirusamy, M. Joshi and B. S. Gupta*
72 **Advances in wool technology** *Edited by N. A. G. Johnson and I. Russell*
73 **Military textiles** *Edited by E. Wilusz*
74 **3D fibrous assemblies: Properties, applications and modelling of three-dimensional textile structures** *J. Hu*
75 **Medical and healthcare textiles** *Edited by S. C. Anand, J. F. Kennedy, M. Miraftab and S. Rajendran*
76 **Fabric testing** *Edited by J. Hu*
77 **Biologically inspired textiles** *Edited by A. Abbott and M. Ellison*
78 **Friction in textile materials** *Edited by B. S. Gupta*
79 **Textile advances in the automotive industry** *Edited by R. Shishoo*
80 **Structure and mechanics of textile fibre assemblies** *Edited by P. Schwartz*
81 **Engineering textiles: Integrating the design and manufacture of textile products** *Edited by Y. E. El-Mogahzy*
82 **Polyolefin fibres: Industrial and medical applications** *Edited by S. C. O. Ugbolue*
83 **Smart clothes and wearable technology** *Edited by J. McCann and D. Bryson*
84 **Identification of textile fibres** *Edited by M. Houck*
85 **Advanced textiles for wound care** *Edited by S. Rajendran*
86 **Fatigue failure of textile fibres** *Edited by M. Miraftab*
87 **Advances in carpet technology** *Edited by K. Goswami*

88 **Handbook of textile fibre structure Volume 1 and Volume 2** *Edited by S. J. Eichhorn, J. W. S. Hearle, M. Jaffe and T. Kikutani*
89 **Advances in knitting technology** *Edited by K.-F. Au*
90 **Smart textile coatings and laminates** *Edited by W. C. Smith*
91 **Handbook of tensile properties of textile and technical fibres** *Edited by A. R. Bunsell*
92 **Interior textiles: Design and developments** *Edited by T. Rowe*
93 **Textiles for cold weather apparel** *Edited by J. T. Williams*
94 **Modelling and predicting textile behaviour** *Edited by X. Chen*
95 **Textiles, polymers and composites for buildings** *Edited by G. Pohl*
96 **Engineering apparel fabrics and garments** *J. Fan and L. Hunter*
97 **Surface modification of textiles** *Edited by Q. Wei*
98 **Sustainable textiles** *Edited by R. S. Blackburn*
99 **Advances in yarn spinning technology** *Edited by C. A. Lawrence*
100 **Handbook of medical textiles** *Edited by V. T. Bartels*
101 **Technical textile yarns** *Edited by R. Alagirusamy and A. Das*
102 **Applications of nonwovens in technical textiles** *Edited by R. A. Chapman*
103 **Colour measurement: Principles, advances and industrial applications** *Edited by M. L. Gulrajani*
104 **Fibrous and composite materials for civil engineering applications** *Edited by R. Fangueiro*
105 **New product development in textiles: Innovation and production** *Edited by L. Horne*
106 **Improving comfort in clothing** *Edited by G. Song*
107 **Advances in textile biotechnology** *Edited by V. A. Nierstrasz and A. Cavaco-Paulo*
108 **Textiles for hygiene and infection control** *Edited by B. McCarthy*
109 **Nanofunctional textiles** *Edited by Y. Li*
110 **Joining textiles: Principles and applications** *Edited by I. Jones and G. Stylios*
111 **Soft computing in textile engineering** *Edited by A. Majumdar*
112 **Textile design** *Edited by A. Briggs-Goode and K. Townsend*
113 **Biotextiles as medical implants** *Edited by M. King and B. Gupta*
114 **Textile thermal bioengineering** *Edited by Y. Li*
115 **Woven textile structure** *B. K. Behera and P. K. Hari*
116 **Handbook of textile and industrial dyeing. Volume 1: Principles, processes and types of dyes** *Edited by M. Clark*
117 **Handbook of textile and industrial dyeing. Volume 2: Applications of dyes** *Edited by M. Clark*

118 **Handbook of natural fibres. Volume 1: Types, properties and factors affecting breeding and cultivation** *Edited by R. Kozlowski*
119 **Handbook of natural fibres. Volume 2: Processing and applications** *Edited by R. Kozlowski*
120 **Functional textiles for improved performance, protection and health** *Edited by N. Pan and G.Sun*
121 **Computer technology for textiles and apparel** *Edited by J. Hu*
122 **Advances in military textiles and personal equipment** *Edited by E. Sparks*
123 **Specialist yarn and fabric structures** *Edited by R. H. Gong*
124 **Handbook of sustainable textile production** *M. I. Tobler-Rohr*
125 **Handbook of natural fibres Volume 1: Types, properties and factors affecting breeding and cultivation** *Edited by R. Kozlowski*
126 **Handbook of natural fibres Volume 2: Processing and applications** *Edited by R. Kozlowski*
127 **Industrial cutting of textile materials** *I. Viļumsone-Nemes*
128 **Colour design: Theories and applications** *Edited by J. Best*
129 **False twist textured yarns** *C. Atkinson*
130 **Woven textiles: Principles, developments and applications** *K. Gandhi*
131 **Textiles and fashion: Materials design and technology** *R. Sinclair*
132 **Modelling, simulation and control of the dyeing process** *R. Shamey and X. Zhao*
133 **Advanced process control in textile manufacturing** *A. Majumdar, A. Das, R. Alagirusamy and V. K. Kothari*
134 **Functional nanofibers and applications** *Q. Wei*
135 **The global textile and clothing industry** *R. Shishoo*
136 **Simulation in textile technology: Theory and applications** *D. Veit*

Part I
Types and processing

1
Nanofibers: principles and manufacture

Q. WEI, D. TAO and Y. XU,
Jiangnan University, China

Abstract: Nanofibers possess many unique properties, which enable them to be used for many different applications. Various techniques have been explored to produce nanofibers. Electrospinning has been recognized as one of the most efficient techniques for producing polymer nanofibers. This chapter briefly reviews the techniques used to make nanofibers, including the key principles of electrospinning. The use of solution electrospinning and melt electrospinning to fabricate polymer nanofibers is also discussed.

Key words: electrospinning, nanofibers, solution, melt.

1.1 Introduction

Nanofibers offer small pore sizes and large surface areas with novel and significantly improved physical, chemical and biological properties as the fiber diameter decreases to nanolevel. Hence, nanofibers have attracted a tremendous amount of attention over the past few decades.

Various techniques have been tried to produce polymer nanofibers. Ondarcuhu and Joachim (1998) demonstrated that nanofibers could be drawn with ultimate dimensions comparable with those of single-wall carbon nanotubes by a process similar to dry-spinning. It was also reported that the nanofiber could be positioned precisely on a surface during fabrication, and could then be cut or manipulated by an atomic force microscope (AFM) tip. Feng *et al.* (2002) reported that nanofibers were obtained by simply extruding a polyacrylonitrile (PAN) solution through an anodic alumina template into a solidifying solution. It was found that a water contact angle exceeding 170° was achieved by the surface of aligned PAN nanofibers without any surface treatment. It has also been reported by Ma and Zhang (1999) that a novel, highly porous structure, which was a three-dimensional interconnected fibrous network with a fiber diameter ranging from 50 to 500 nm, was created from biodegradable aliphatic polyesters. The nanofibrous matrices were prepared from the polymer solutions by a procedure involving thermally induced gelation, solvent exchange and freeze-drying

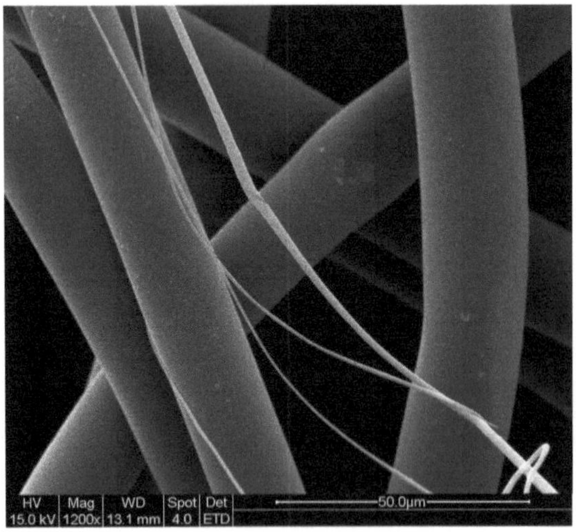

1.1 SEM image of PLA fibers and nanofibers.

to mimic natural extracellular matrix architecture. Polymer nanofibers can also be formed by electrospinning. Shao *et al.* (2008) reported the formation of PAN nanofibers by electrospinning. The electrospun PAN nanofibers could be used as substrates for depositing the functional layer of zinc oxide (ZnO) on the PAN nanofiber surfaces by sol-gel technique. The functional carbon nanofibers were obtained by the carbonization of ZnO coated PAN nanofibers.

The utilization of electrostatic force to produce microdiameter fibers was first put forward by Anton Formhals's series of patents published from 1934 to 1944. However, until the 1990s, electrospinning derived from 'electrostatic spinning' has been considered the primary technique for producing nanofibers as the interest in these fibers expanded. Since then, electrospinning has been adopted in the fabrication of various polymer nanofibers.

In terms of textile fibers, nanofibers (Ramakrishna *et al.*, 2005; Brown and Stevens, 2007) are generally defined as one-dimensional flexible solid-state nanomaterials characterized by a diameter equal to or less than 0.1 μm (100 nm) and an aspect ratio greater than 100:1. However, this strict definition is impractical for textile processes; nanofibers are usually defined as fibers with a diameter of less than 1 μm (1000 nm) or less than 0.5 μm (500 nm). Figure 1.1 indicates the significant difference in diameter between normal polylactic acid (PLA) fibers and electrospun polylactic acid nanofibers. The normal fibers in Fig. 1.1 have diameters of about 20 μm, but the electrospun PLA nanofibers have diameters of about 200 nm.

1.2 Principles of electrospinning

1.2.1 Electrospinning devices

Electrospinning has been most widely adopted to produce fibers with diameters ranging from a few nanometers to several micrometers, either in the form of solution electrospinning (S-ES) or melt electrospinning (M-ES). Basically, an electrospinning apparatus consists of the following four major components:

1. High-voltage power supply: A high-voltage power supply is needed to provide an electrostatic field for electrospinning. The supply needed could provide high voltage from a few thousand volts to a hundred thousand volts, depending on the materials being electrospun. The term 'electrospinning' comes from the use of electrostatic force to draw fibers.
2. Flow control pump: A flow control pump is usually used to adjust the flow rate of spinning fluid in order to produce the fiber with a controllable diameter. Flow rate affects the fiber diameter and productivity.
3. Spinneret: A spinneret is used to form small tubes for the polymer solution or melt to flow through. A spinneret can be arranged in different configurations depending on the formation forms of fibers.
   (a) Single spinneret: As shown in Fig. 1.2, different shapes of single spinneret can be developed either for the fabrication of electrospun pure fibers or for electrospinning component fibers depending on the configuration of the spinnerets.
   (b) Multispinnerets: Multispinnerets can be arranged either in linear array or in two-dimensional arrays and the number of nozzles can be made as required, as illustrated in Fig. 1.3. Multispinnerets are used to increase the productivity of the electrospinning.
4. Collector: To obtain different fiber assemblies, a collector can be designed in various forms, as required. A plate, a rotating drum, a rotating disc, rings or any other shapes, as shown in Fig. 1.4, are not unusual.

A typical schematic of the electrospinning process is presented in Fig. 1.5. The apparatus as well as the components can be set up and adjusted in accordance with the actual needs. With high-voltage potential applied, polymer solution or melt inside the spinneret is forced through a flow control pump to form a pendant drop at the tip of the spinneret. Free charges are induced into the polymer solution as a result of the applied high-voltage potential. In response to the applied electric field, these charged ions move toward the electrode of opposite polarity, transferring stretching forces generated by the electrical field to the polymer liquid. As the applied electric field increases, the hemispherical surface of the fluid at the tip of

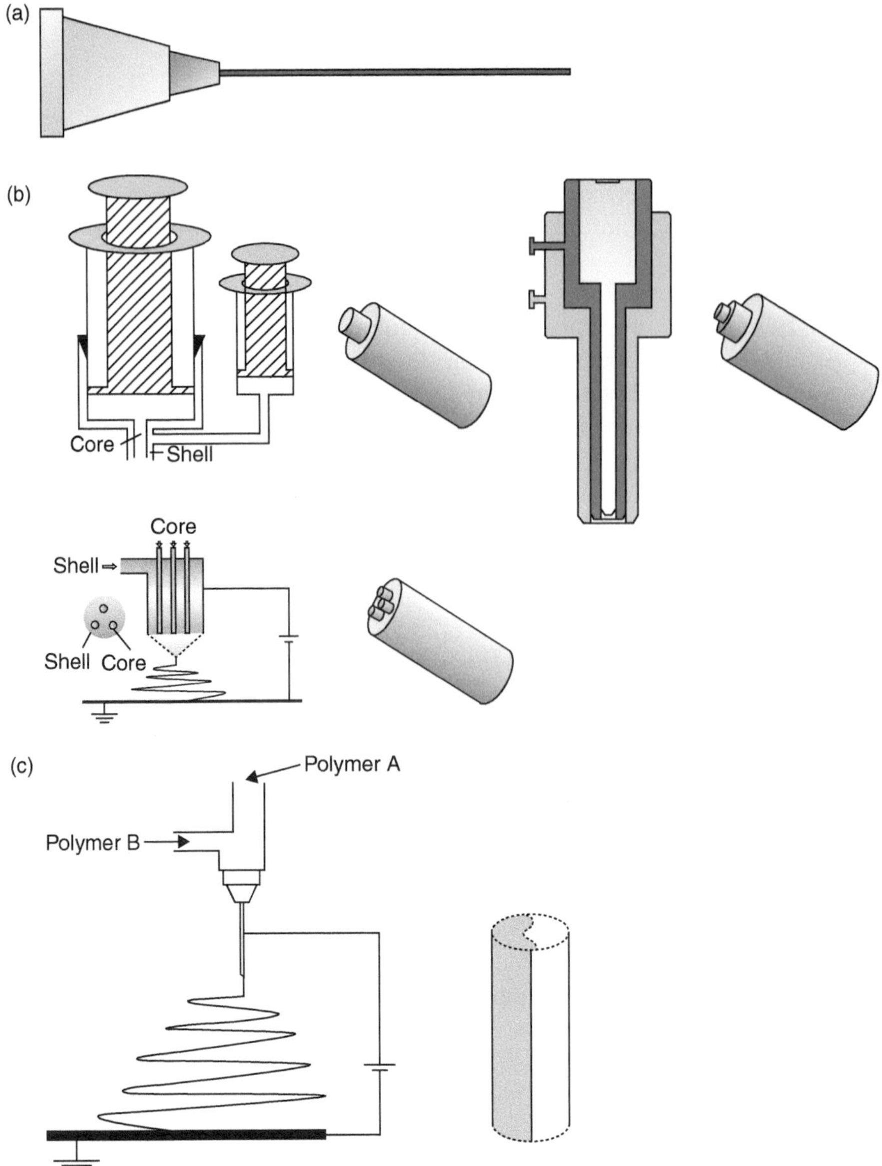

1.2 Different shapes of single spinneret: (a) normal spinneret, (b) coaxial spinnerets and (c) side by side spinneret.

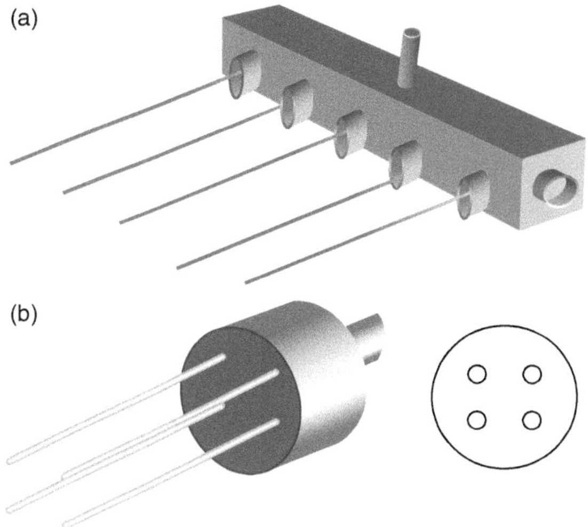

1.3 Multispinnerets with different arrangements: (a) linear array and (b) two-dimensional array.

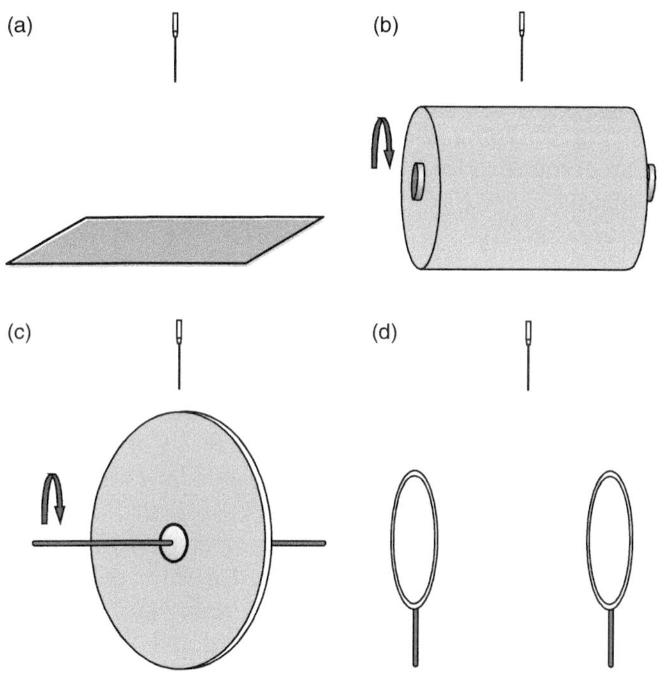

1.4 Schematic diagram of main electrospinning collectors: (a) plate collector, (b) rotating drum collector, (c) disc collector and (d) ring collector placed in parallel.

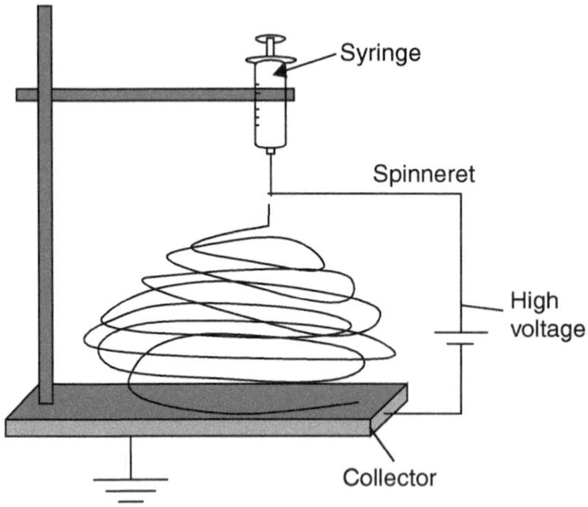

1.5 Schematic diagrams of electrospinning apparatus.

the spinneret is stretched and conic shape called Taylor cone is formed. When the applied high-voltage potential reaches a critical value, the electrostatic forces overcome the surface tension and a charged jet of the solution ejected from the tip of the Taylor cone is rapidly accelerated to the grounded collecting target. Due to instabilities in the jet, the charged jet undergoes violent whipping and elongation, allowing evaporation of solvent or cooling of melts to form solid fibers on the collector (Huang *et al.*, 2003; Ramakrishna *et al.*, 2005; Kulkarni *et al.*, 2010; Brown and Stevens, 2007; Subbiah *et al.*, 2005).

1.2.2 Fiber formation mechanism of electrospinning

The electrospinning process may be considered as several operational sections: (i) charging of the fluid, (ii) formation of the cone-jet, (iii) thinning of the jet in the presence of an electric field, (iv) instability of the jet and (v) collection of the jet on an appropriate collector (Rutledge and Fridrikh, 2007).

Numerous studies on electrospinning have been dedicated to modeling the formation mechanisms of electrospun fibers (Byusse, 2008). Taylor (1969) described the viscous jets at the outer edge of the capillary tip (spinneret) breaking into conical drops and then forming jets at the top of the cone (Taylor cone) under electric force in great detail. He proposed a formula predicting the critical value (V_c) of potential needed for fiber formation based on the assumption that drops at the top of Taylor cone are surrounded by air and form steady state at the capillary tip:

$$V_c^2 = \frac{4H^2}{L^2}\left(\ln\frac{2L}{R} - \frac{3}{2}\right)(0.0117\pi\gamma R) \quad [1.1]$$

where H is the distance between the capillary tip and the grounded collector, L is the capillary regime length, R is radius of the capillary and γ is the surface tension of the spinning solution. Taylor's approximate expression established a theoretical foundation for understanding the electrospinning process.

It can be concluded from current references that most electrospinning processes involve two modes, the straight jet portion and the whipping jet portion. A great deal of attention has been given to the jet formation modes, including jet initiation from the droplet (Senador *et al.*, 2001; Yarin *et al.*, 2001b), straight jet portion (Feng, 2002, 2003; Spivak and Dzenis, 1998; Spivak *et al.*, 2000) and the entire jet (Hohman *et al.*, 2001a, 2001b; Yarin *et al.*, 2001a). These research works further reveal the formation of nanofibers in the electrospinning process.

Spivak *et al.* (2000) studied the fluid jet by taking into account inertial, hydrostatic, viscous, electric and surface tension forces. Considering the jet diameter as a function of surface tension, flow rate and current, they suggested the governing equations of mass balance and linear momentum balance, as well as electric charge balance. Assuming that the jet flow is extensional and axisymmetric in the direction of the external electric field, the general three-dimensional problem can be reduced to a one-dimensional problem. By choosing the dimensionless parameter N_{w} (which describes the ratio of inertia force to surface tension in the jet) and the effective Reynolds number N_{R}, a differential equation of jet radius R to coordinate axes Z was derived:

$$\frac{\mathrm{d}}{\mathrm{d}z}\left[R^{-4} + (N_{\mathrm{w}}R)^{-1} - N_{\mathrm{R}}^{-1}\left(\frac{\mathrm{d}R^{-2}}{\mathrm{d}Z}\right)^m\right] = 1 \quad [1.2]$$

Shin *et al.* (2001) studied the phenomenon of electrospinning, inducing the relationship between the radius of capillary and the actual potential. They obtained the amplification factor $\Gamma(E_\infty, Q)$ for a perturbation connecting a distance d downstream:

$$\Gamma(E_\infty, Q) = \ln\left[\frac{A(d)}{A(0)}\right] = \int_0^d \frac{\omega(h, E, \sigma)\pi h^2}{Q} \quad [1.3]$$

Where E_∞ is the applied electric field, Q is the volumetric flow rate, h is the jet radius and σ is the surface charge distribution.

The instability equations cover a wider range of materials and process parameters than previous studies and are expressed in terms of well-known fluid properties.

1.3 Solution electrospinning

In S-ES, polymers are usually dissolved in some solvents. The solution properties, electrospinning parameters and environmental conditions may have some effects on the formation of nanofibers.

1.3.1 Solution properties

Polymer molecular weight

Polymer molecular weight reflects the number of entanglements of polymer chains in solution, thus solution viscosity. It has been found that a low molecular weight solution tends to form beads rather than fibers and a high molecular weight solution helps to form fibers with larger diameters (Huang *et al.*, 2008). Therefore, even when the polymer concentration is low, an adequate number of entanglements of the polymer chains can ensure a sufficient level of solution viscosity to fabricate a uniform fiber web during electrospinning and restrain effects of surface tension. In other words, high molecular weights are not always essential for the electrospinning process if intermolecular interactions are sufficient for the inter-chain connectivity obtained through chain entanglements.

Concentration

Various research works have revealed that a mixture of beads and fibers might form at low solution concentration. As the solution concentration increases, the shapes of the beads change from spherical to spindle-like and, finally, uniform fibers with increased fiber diameters (Deitzel *et al.*, 2001). As the concentration further increases, formation of continuous fibers are prohibited because of the inability to maintain the flow of the solution at the tip of the spinneret.

Viscosity and surface tension

Viscosity, polymer concentration and molecular weight are strongly correlated to each other. Generally, no continuous fibers could be formed with solutions of very low viscosity or very high viscosity, since there is difficulty in forming stable jets ejected from polymer solutions. As the viscosity increases, polymer solution exhibits longer stress relaxation times, preventing the ejected jets from fracturing during electrospinning. Thus, it is necessary to choose an optimal viscosity for electrospinning and different polymer solutions may contribute different viscosity ranges. It has been reported that the maximum spinning viscosities should be in the range from 1 to 215 poise (Baumgarten, 1971; Deitzel *et al.*, 2002).

Generally, a solution with a high surface tension has a negative effect on the electrospinning process because of instability of the jets and the formation of sprayed droplets. The formation of droplets, beads and fibers depends on the interaction between the surface tension of a solution and electrostatic forces. A lower surface tension of the spinning solution facilitates electrospinning to occur at a lower electric field.

Conductivity/surface charge density

Solution conductivity is mainly determined by the polymer type, solvent used and the availability of ion sable salts. Generally, there is a significant decrease in fiber diameters with an increase of solution conductivity, whereas low solution conductivity results in insufficient jet elongation acted by electrical force. However, high solution conductivity is extremely unstable in the presence of strong electric fields, resulting in a dramatic bending instability and a broad diameter distribution. Hence, suitable solution conductivity plays an important role in fiber formation during the electrospinning process (Bhardwaj and Kundu, 2010).

1.3.2 Electrospinning process conditions

Applied voltage

The increase in the electrospinning voltage enhances the electrostatic repulsive force on the solution jet which ultimately favors the stretching of fibers, leading to a decrease in the fiber diameter and also rapid evaporation of solvent from the fibers. However, as the applied voltage increases, more polymer jets are ejected from tip of the spinneret, facilitating the formation of larger fiber diameters in some cases. Meanwhile, the probability of bead formation increases as a higher voltage is applied, since the large polymer jets are not sufficiently stretched during the electrospinning. The effect of applied voltage on fiber diameter is also affected by the concentration of a polymer solution and the distance between the tip and the collector (Yordem *et al.*, 2008).

Flow rate

Flow rate influences the jet velocity and material transfer rate. A lower flow rate is more desirable due to the fact that it helps to extend the solvent evaporation time. Meanwhile, a flow rate that is too high results in bead formation due to insufficient drying time before reaching the collector.

Distance between nozzle and collector

Research work has proved that there is a minimum distance between nozzle and collector required to allow sufficient time for solvent evaporation before fibers reach the collector; if the distance is too short or too long beads may be formed. It has been found that flatter fibers or smaller fibers can be fabricated at closer distances while rounder fibers can be formed with an increase in distance. Other processing parameters, including inner diameter of nozzle, collector shape and moving speed, also have influence on fiber formation (Bhardwaj and Kundu, 2010).

1.3.3 Environmental conditions

Temperature

It has been found that the increase in solution temperature causes a decrease in solution viscosity, which favors the formation of finer fibers during electrospinning. Huang *et al.* (2008) investigated the effect of solution temperature on morphology, structure and crystallinity of poly(vinylidene fluoride) (PVDF) nanofibers by evaluating the surface tension and viscosity of electrospinning solutions at various ambient temperatures. It was found that polymer droplets were collected only if the solution temperature maintained a low temperature of 5°C, as shown in Fig. 1.6a. The mixture of beads and fibers were observed at an ambient temperature of around 15°C. As the temperature increased to 15°C, the surface tension of the PVDF solution dropped, which facilitated the electrostatic force to overcome the cohesive force of the solution. It was clearly observed that fibers with few beads were formed when the temperature was increased to 35°C, as illustrated in Fig. 1.6b. The average diameter of the PVDF nanofibers formed at 15°C was over 850 nm, but the average diameter of the PVDF nanofibers gradually reduced to about 373.4 nm as the temperature increased to 45°C, as presented in Fig. 1.7. These changes in fiber formation could be well explained by Table 1.1, from which it can be found that the decrease both in surface tension and viscosity contributed to the formation of finer nanofibers as the temperature increased. When the temperature was too low, the electrostatic force was not strong enough to overcome the surface tension of the solution and stabilize the jet, leading to the formation of beads or beaded fibers.

It was revealed from the X-ray diffraction (XRD) analysis that all the electrospun PVDF nanofibers showed clear diffraction peaks at 20.0° of 2θ (β phase), and the nanofibers electrospun at 25°C , 35°C and 45°C also

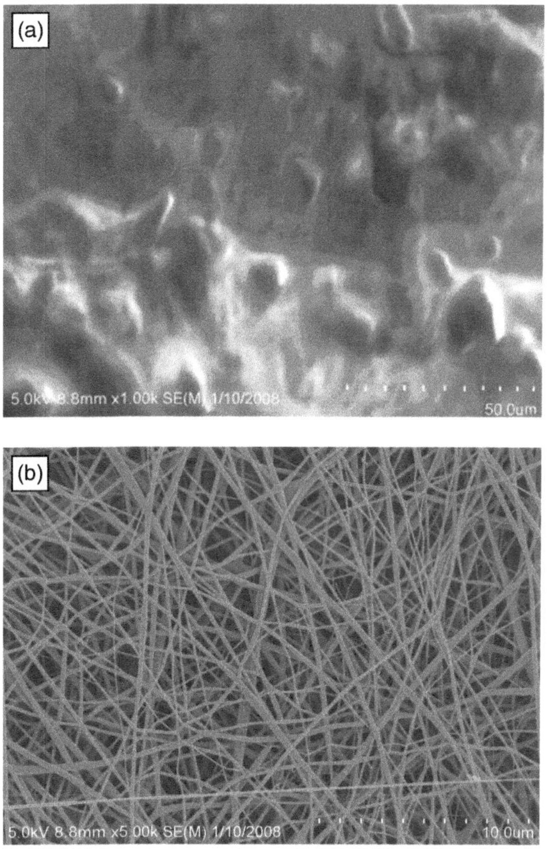

1.6 SEM images of electrospun PVDF nanofibers prepared at different ambient temperatures: (a) 5°C and (b) 35°C.

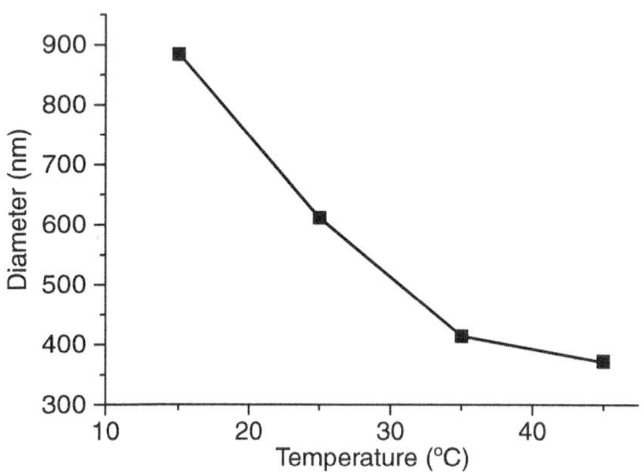

1.7 Effect of temperature on fiber diameters.

Table 1.1 Effect of temperature on the surface tension and viscosity of electrospun solution

Temperature (°C)	Surface tension (mN/m)	Viscosity (cP)
5	41.23	241.8
15	38.78	205.8
25	32.26	180.9
35	29.23	150.2
45	27.05	112.6

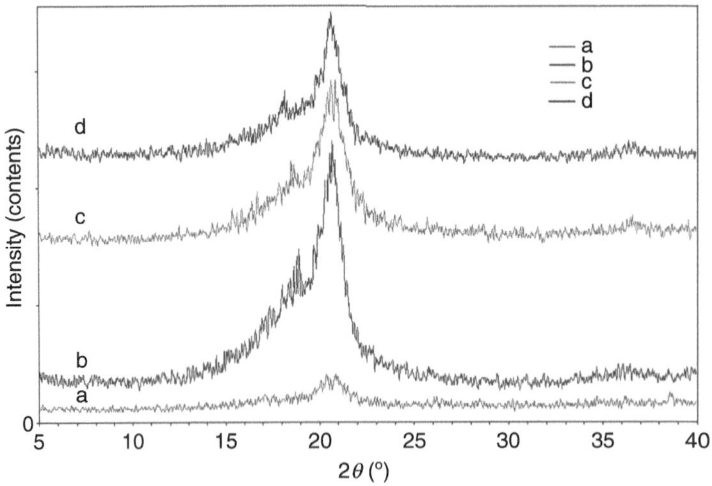

1.8 XRD patterns of PVDF nanofibers electrospun with ambient temperatures of (a) 15°C, (b) 25°C, (c) 35°C and (d) 45°C.

exhibited apparent diffraction peaks at 18.4°of 2θ (α phase), as indicated in Fig. 1.8. The XRD patterns suggested that the crystalline structure could be altered from a single structure, β, to a multiple structure, α and β, through an ambient temperature changing. This phenomenon could be attributed to strong intermolecular interactions in the PVDF molecular chains and the sensitivity of chains movement to the ambient temperature.

Humidity

It has been suggested that a volatile solvent may dry rapidly as the evaporation of the solvent is faster at a very low humidity. However, when the solvent evaporates too fast, it may result in insufficient time for the ejected jet to reach the collector, and the spinneret may easily be clogged. It is believed that a high humidity may help discharge of the electrospun fibers.

Besides, oxygen content, atmosphere (vacuum, N_2, ...) and magnetic field may also have some influence on the formation of electrospun fibers.

1.4 Melt electrospinning

Melt electrospinning is a much simpler and safe method to fabricate ultra-fine fibers compared to solution electrospinning, since some toxic solvents may have to be used in solution electrospinning. A typical schematic of a melt electrospinning device is illustrated in Fig. 1.9. With much greater solution viscosity compared to solution electrospinning, uniform fibers with larger diameters can be fabricated. Wang *et al.* (2011) reported the formation of PLA microfibers by melt electrospinning. It was observed by a high-speed charge-coupled device (CCD) camera that the melt jet was accelerated in the form of a straight line by the electrostatic force in the first motion stage. After moving for a certain distance, the melted polymer jet started to experience stress relaxation. The accelerated jets then whipped violently under the electrostatic repulsion.

Zhmayev *et al.* (2011) investigated the electrohydrodynamic (EHD) quenching in polymer melt electrospinning using infrared thermal measurements on polymer melt jets in electrospinning; the experimental results revealed rapid quenching by ambient air, of an order of magnitude faster than predicted by the classical Kase and Matsuo correlation. This drastic heat transfer enhancement could be attributed to EHD effects. The EHD-driven air flow was included in a comprehensive model for polymer melt electrospinning. The experimental results for electrospinning of Nylon-6 (N6), polypropylene (PP) and PLA melts validated the excellent agreement of both predicted jet radius and temperature profiles. Based on this study, several methods can be used to inhibit or enhance the quenching in melt electrospinning.

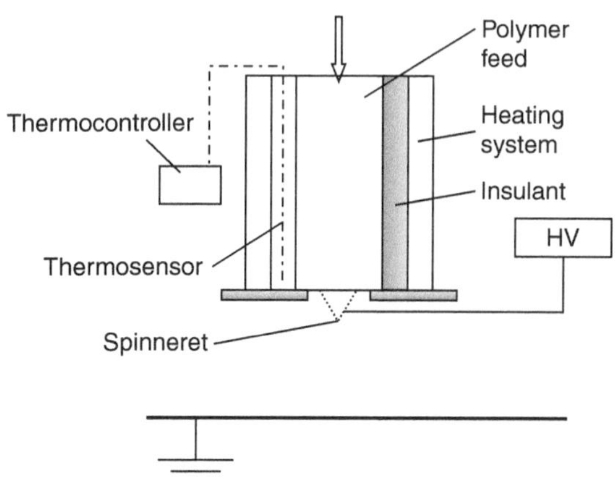

1.9 A typical schematic diagram of melt electrospinning apparatus.

1.4.1 Heating in melt electrospinning

One important consideration on setting up a melt electrospinning apparatus is to choose a heating source for the system. Various heating techniques have been explored in melt electrospinning systems. These heating devices can be classified into four types (Hutmacher and Dalton, 2011).

1. Electrically heating: electrical heating used to melt polymers in melt electrospinning.
2. Circulating fluids: high-temperature fluid used to melt polymers for melt electrospinning.
3. Heat gun: flame heating used to melt polymers for melt electrospinning, as illustrated in Fig. 1.10.
4. Laser heating: laser beam used to heat polymer in melt electrospinning.

Shimada *et al.* (2010) reported the use of a new melt electrospinning system equipped with a line-like laser beam melting device developed for the mass production of nanofibers. The system enabled fibers to be produced from locally melted polymer sheets in the presence of an electric field. The study also investigated the fiber formation mechanism with a video camera and a thermographic camera. It was found from the experimental observations that (i) many Taylor cones were formed on the molten end surface of the sheet at almost regular intervals; (ii) a single fiber was produced from the apex of each Taylor cone; and (iii) the temperature in the Taylor cone changed dramatically during the fiber formation.

There are advantages and disadvantages of using these heating techniques. Yang *et al.* (2009) reported a new design of melt electrospinning. The temperature of every part in the setup, including injector, nozzle, spinning space and collection target could be measured accurately by thermocouple

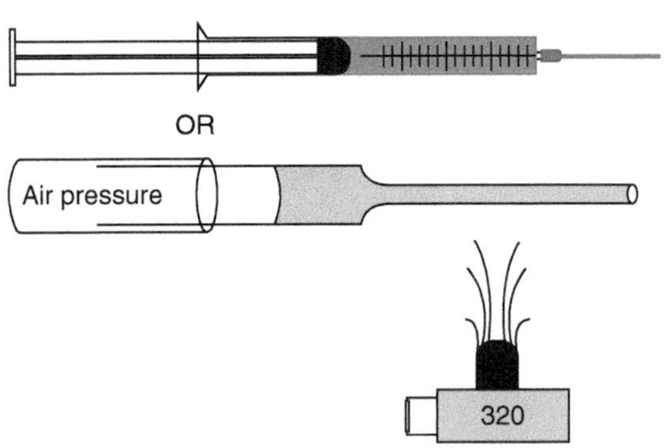

1.10 Heating for melt electrospinning: heat gun.

or infrared radiation thermometers. PP was used in this work. The experimental results revealed that the average fiber diameters decreased with the increase of nozzle temperature (T_2) and spinning space temperature (T_3). Zhmayev *et al.* (2010) reported the concept of a gas-assisted polymer melt electrospinning process. The technique allowed reduced quenching of the melt jet in the spinning region, and thus an increase in the jet attenuation rate, resulting in production of submicron scale fibers. It was found that air drag produced an additional 10% thinning compared to the unassisted melt electrospinning process under the investigated conditions in electrospinning of PLA melt, and the heating provided by the air stream resulted in an additional 20-fold jet thinning.

1.4.2 Processing parameters on fiber diameters

In most melt electrospinning, the viscosity of a molten polymer is usually much higher than that in solution, which leads to larger fiber diameters (mostly in the micrometer domain). It is also well known that the equipment required for melt electrospinning is much more complicated than that suitable for solution electrospinning because a heating and temperature-control system is needed to melt the polymer. Fiber diameter can also be controlled by adjusting the processing parameters such as electric field strength, polymeric viscosity and flow rate in melt electrospinning.

Lyons *et al.* (2004) investigated the effects of various melt electrospinning parameters on the morphology and fiber diameter of PP of different tacticities. The effect of the electric field strength at various melt flow indexes of PP on fiber uniformity, morphology and diameter was measured in this study. The experimental results revealed that the molecular weight was the predominant factor in determining the fiber diameter of the electrospun fibers. Liu *et al.* (2010) used orthogonal design to examine melt temperature, spinning voltage, spinning distance and melt flow rate (MFR) of polymer. The experimental results revealed that MFR at the reported three levels had the most important impact both on the average diameters and standard deviations of fiber diameters.

Deng *et al.* (2009) reported the use of a novel melt spinning device which incorporated a reverse of the normal polarity, with the capillary grounded and the collector grid at positive potential. Low-density polyethylene with a low-melt flow index of 2 g/10 min, which is not suitable for melt spinning using current commercial methods, was used to fabricate polymer fibers by melt electrospinning. The effects of the varying electrospinning parameters such as temperature, electrostatic field, spinning distance and capillary inner diameter were investigated. It was observed that the use of a temperature in melt electrospinning that was higher than temperatures used in normal melt spinning had to be employed to reduce the viscosity of the polymer melt sufficiently. Electrospun fibers with smooth and even surfaces, most of which had diameters smaller than 15 μm, were obtained successfully.

Dalton *et al.* (2007) demonstrated that PP fibers could be reduced from 35 ± 8 μm in diameter to 840 ± 190 nm with a viscosity-reducing additive. Melt electrospun blends of poly(ethylene glycol)-block-poly(3-caprolactone) (PEG47-b-PCL95) and poly(3-caprolactone) were lowered to 270 ± 100 nm by using the gap method of alignment for collection. The collected melt electrospun fibers could fuse together where they touched, forming relatively thick nonwoven felts. It was found that the frequent fusion points between melt electrospun fibers and a reduction in diameter for the gap method of alignment indicated that the melt electrospun fibers were still slightly molten at collection.

1.5 Future trends

Since the 1990s, a large amount of research has been focused on preparation of nanofibers by electrospinning. Considerable progress has been made over the past two decades. In the early days, most published papers just reported how to choose a polymer to electrospin fibers and containing, perhaps, some simple discussions on the effects of processing parameters on fiber diameter and morphology. However, more attention has been directed at understanding the fiber formation mechanism in the electrospinning process in recent years and many models have been proposed to explain the experimental phenomena observed. At the same time, the applications of the electrospun nanofibers have also been extensively explored, including filtration, affinity membranes and recovery of metal ions, tissue engineering scaffolds, wound healing, drug release, catalyst and enzyme carriers, sensors, energy storage, etc. (Fang *et al.*, 2008). However, there are some obvious constraints that limit the further development of electrospinning.

1.5.1 Melt electrospinning

Since solvent electrospinning offers a quick approach to producing fibers with diameters less than 100 nm, it has attracted much more attention than melt electrospinning. However, for some extensively used polymers such as polyethylene and PP, no solvents have been found for solution electrospinning at room temperature at present (Lyons and Ko, 2005). In addition, in spite of the comprehensive potential applications, environmental and health limitations, as well as productivity complications, do exist in solution electrospinning systems (Lyons *et al.*, 2004).

Therefore, it is of vital importance to explore alternative ways to prepare such kinds of polymer nanofibers. Under this circumstance, melt electrospinning has extraordinary superiority over solution electrospinning, including the absence of toxic solvents as well as high productivity. Melt electrospinning seems to have great potential to replace solution electrospinning, but a lot of effort is needed to reduce fiber diameters while simultaneously establishing a high degree of orientation in the structure to ultimately obtain the theoretically predicted strengths. When fibers with such desired properties become continuous and reproducible, melt electrospinning will become a primary technology for nanofiber preparation, as more and more areas demand a safer, more cost-effective, higher yielding and environmentally friendly alternative.

1.5.2 Large-scale production

Although electrospinning has taken precedence over other nanofiber preparation methods, the production of electrospun nanofibers is far below the requirements of large-scale production. Thus, in the long term, it is necessary to improve the technology to meet production demands.

An alternative solution might be multijet electrospinning. It has been demonstrated that the initial stream line and jets of nanofibers from the nozzles could have stable jet-motion without interruption near charged jets by using a cylindrical electrode connected with multiple nozzles (Kim *et al.*, 2006). The conical jet from the spinneret was at least as stable as jets in single needle. At present, modified single-needle, multineedle and needleless systems have been introduced to achieve multiple polymer jets and thus improve the output of nanofibers (Zhou *et al.*, 2009).

1.6 References

Baumgarten, P. K. (1971), 'Electrostatic spinning of acrylic microfibres', *Journal of Colloid and Interface Science*, **36**, 71–9.

Bhardwaj, N. and Kundu, S. C. (2010), 'Electrospinning: A fascinating fibre fabrication technique', *Biotechnology Advances*, **28**, 325–47.

Brown, P., Stevens, K. (2007), *Nanofibres and nanotechnology in textiles*, Cambridge, Woodhead and CRC Press LLC.

Byusse, W. M. (2008), *A 2D model for the electrospinning process*, Eindhoven University of Technology.

Dalton, P. D., Grafahrend, D., Klinkhammer, K., *et al.* (2007), 'Electrospinning of polymer melts: Phenomenological observations', *Polymer*, **48**, 6823–33.

Deitzel, J. M., Kleinmeyer, J. and Harris, D. (2001), 'The effect of processing variables on the morphology of electrospun nanofibres and textiles', *Polymer*, **42**, 261–72.

Deitzel, J. M., Kosik, W., McKnight, S. H., *et al.* (2002), 'Electrospinning of polymer nanofibres with specific surface chemistry', *Polymer*, **43**, 1025–9.

Deng, R. J., Liu, Y., Ding, Y. M., *et al.* (2009), 'Melt electrospinning of low-density polyethylene having a low-melt flow index', *Journal of Applied Polymer Science*, **114**, 166–75.

Fang, J., Niu, H. T., Lin, T., *et al.* (2008), 'Applications of electrospun nanofibres', *Chinese Science Bulletin*, **53**(15), 2265–86.

Feng, J. J. (2002), 'The stretching of an electrified non-Newtonian jet: A model for electrospinning', *Physics of Fluids*, **14**(11), 3912–26.

Feng, J. J. (2003), 'Stretching of a straight electrically charged viscoelastic jet', *Journal of Non-Newtonian Fluid Mechanics*, **116**(1), 55–70.

Feng, L., Li, S., Li, H., *et al.* (2002), 'Super-hydrophobic surface of aligned polyacrylonitrile nanofibres', *Angewandte Chemie International Edition*, **41**(7), 1221–3.

Hohman, M. M., Shin, M., Rutledge, G., *et al.* (2001a), 'Electrospinning and electrically forced jets. I. Stability theory', *Physics of Fluids*, **13**(8), 2201–20.

Hohman, M. M., Shin, M., Rutledge, G., *et al.* (2001b), 'Electrospinning and electrically forced jets. I I . Applications', *Physics of Fluids*, **13**(8), 2221–36.

Huang, F. L., Wei, Q. F., Wang, J. X., *et al.* (2008), 'Effect of temperature on structure, morphology and crystallinity of PVDF nanofibres via electrospinning', *e-polymers*.

Huang, Z. M., Zhang, Y. Z., Kotakic, M., *et al.* (2003), 'A review on polymer nanofibres by electrospinning and their applications in nanocomposites', *Composites Science and Technology*, **63**, 2223–53.

Hutmacher, D. W. and Dalton, P. D. (2011), 'Melt electrospinning', *Chemistry –An Asian Journal*, **6**, 44–56.

Kim, G. H., Cho, Y. S. and Kim, W. D. (2006), 'Stability analysis for multi-jets electrospinning process modified with a cylindrical electrode', *European Polymer Journal*, **42**, 2031–8.

Kulkarni, A., Bambole, V. A. and Mahanwar, P. A. (2010), 'Electrospinning of polymers, their modeling and applications', *Polymer-Plastics Technology and Engineering*, **49**, 427–41.

Liu, Y., Deng, R. J., Hao, M. F., *et al.* (2010), 'Orthogonal design study on factors effecting on fibers diameter of melt electrospinning', *Polymer Engineering & Science*, **50**, 2074–8.

Lyons, J. and Ko, F. (2005), 'Melt electrospinning of polymers: A review', *Polymer News*, **30**, 1–9.

Lyons, J., Li, C. and Ko, F. (2004), 'Melt-electrospinning part I: processing parameters and geometric properties', *Polymer*, **45**, 7597–603.

Ma, P. X. and Zhang, R. (1999), 'Synthetic nano-scale fibrous extracellular matrix', *Journal of Biomedical Materials Research*, **46**, 60–72.

Ondarcuhu, T. and Joachim, C. (1998), 'Drawing a single nanofibre over hundreds of microns', *Europhysics Letters*, **42**(2), 215–20.

Ramakrishna, S., Fujihara, K., Teo, W. E., *et al.* (2005), *An introduction to electrospinning and nanofibres Applications of electrospun nanofibres*, World Scientific.

Rutledge, G. C. and Fridrikh, S. V. (2007), 'Formation of fibres by electrospinning', *Advanced Drug Delivery Reviews*, **59**, 1384–91.

Senador Jr, A. E., Shaw, M. T. and Mather, P. T. (2001), 'Electrospinning of polymeric nanofibres: Analysis of jet formation', *Materials Research Society Symposium Proceedings*, **661**, 591–6.

Shao, D. F., Wei, Q. F., Zhang, L. W., *et al.* (2008), 'Surface functionalization of carbon nanofibres by sol gel coating of zinc oxide', *Applied Surface Science*, **254**(20), 6543–6.

Shimada, N., Tsutsumi, H., Nakane, K., *et al.* (2010), 'Poly(ethylene-co-vinyl alcohol) and nylon 6/12 nanofibers produced by melt electrospinning system equipped with a line-like laser beam melting device ', *Journal of Applied Polymer Science*, **116**(5), 2998–3004.

Shin, Y. M., Hohman, M. M., Brenner, M. P. and Rutledge, G. C. (2001), 'Electrospinning: A whipping fluid jet generates submicron polymer fibres', *Applied Physics Letters*, **78**(8), 1149–51.

Spivak, A. F. and Dzenis, Y. A. (1998), 'A symptotic decay of radius of a weakly conductive viscous jet in an external electric field', *Applied Physics Letters*, **73**(21), 3067–9.

Spivak, A. F., Dzenis, Y. A. and Reneker, D. H. (2000), 'A model for steady state jet in the electrospinning process', *Mechanics Research Communications*, **27**(1), 37–42.

Subbiah, T., Bhat, G. S. and Tock, R. W. (2005), 'Electrospinning of nanofibres', *Journal of Applied Polymer Science*, **96**, 557–69.

Taylor, G. (1969), 'Electrically driven jets', *Proceedings of the Royal Society A*, **313**, 453–75.

Wang, X. N., Xu, Y., Wei, Q. F., *et al.* (2011), 'Study on technological parameters effecting on fibre diameter of melt electrospinning', *Advanced Materials Research*, **332–4**, 1550–6.

Yang, C. R., Jia, Z. D., Wang, K., *et al.* (2009). 'Exploration of melt-electrospinning based on the novel device', Properties and Applications of Dielectric Materials, ICPADM, 1223–6.

Yarin, A. L., Koombhongse, S. and Reneker, D. H. (2001a), 'Bending instability in electrospinning of nanofibers', *Journal of Applied Physics*, **89**(5), 3018–26.

Yarin, A. L., Koombhongse, S. and Reneker, D. H. (2001b), 'Taylor cone and jetting from liquid droplets in electrospinning of nanofibres', *Journal of Applied Physics*, **90**(9), 4836–46.

Yordem, O. S., Papila, M. and Menceloğlu, Y. Z. (2008), 'Effects of electrospinning parameters on polyacrylonitrile nanofibre diameter: An investigation by response surface methodology', *Materials & Design*, **29**, 34–44.

Zhmayev, E., Cho, D. and Joo, Y. (2010), 'Nanofibres from gas-assisted polymer melt electrospinning', *Polymer*, **15**(18), 4140–4.

Zhmayev, E., Cho, D. and Joo, Y. (2011), 'Electrohydrodynamic quenching in polymer melt electrospinning', *Physics of Fluids*, **23**, 073102–073109.

Zhou, F. L., Gong, R. and Porat, I. (2009), 'Mass production of nanofibre assemblies by electrostatic spinning', *Polymer International*, **58**, 331–42.

2

Types and processing of structured functional nanofibers: core-shell, aligned, porous and gradient nanofibers

Q. WEI, Jiangnan University, China and X. XIA, Xinjiang University, China

Abstract: This chapter introduces types of structured functional nanofibers for advanced applications. The key formation mechanisms of core-shell, aligned, porous and gradient nanofibers are reviewed, including detailed information on the processes involved. Their applications and future trends are also discussed in the chapter, giving due consideration to their inherent properties – strengths and weaknesses – and how these affect potential usage.

Key words: structured functional nanofibers, electrospinning, core-shell, aligned, porous, gradient nanofibers.

2.1 Introduction

Functional nanofibers are prepared from various organic or inorganic materials. They take advantage of the structural characteristics of these materials to give nanofibers many desirable properties for advanced applications. Various techniques have been developed to fabricate structured functional nanofibers. Among these, electrospinning is a relatively simple and fast process to produce structured functional nanofibers by using different materials, modifying electrospinning apparatus, changing electrospinning parameters and including treatment of electrospinning materials before or after electrospinning.

Although the first patent on electrospinning technique was published as early as in 1934, this technique has not been improved until recently. Particular attention has been given to the fabrication by electrospinning of nanofibers with different morphologies, such as those with a porous surface, or core-sheath and aligned structures (Park, 2007). Such changes in the structures of nanofibers provide some excellent properties – for example, the possibility to create a very large specific surface area about one to two orders of magnitude larger than conventional electrospun nanofibers, excellent mechanical performance and loading capacity, etc. (Fang, 2008).

2.2 Core-shell nanofibers

Generally, core-shell nanofibers are composites in which one component can be encapsulated within another, leading to some functional enhancements (Li, 2007). Various techniques have been proposed to prepare ultrafine fibers configured in a core-sheath structure, including self-assembly, laser ablation and template synthesis. In recent years the electrospinning technique has attracted much attention in the preparation of core-sheath nanofibers. As a relatively new technique which fabricates core-shell nanofibers through a one-step process, electrospinning has been applied to a wide range of drug delivery systems, tissue engineering and other biomedical applications.

Coaxial dual-capillary spinneret and emulsion electrospinning are the two major methods used in the fabrication of core-shell nanofibers. Coaxial electrospinning technology resembles a dual-capillary device, producing one capillary surrounded by a larger one. Inner and outer solutions are delivered into each respective capillary by individual syringe pumps. When a high-voltage power supply creates an electric field between a grounded collector and a positively charged dual-capillary, the composite droplet transforms into a composite Taylor cone with a core/shell jet initiating from the tip. Under the whipping force of electrostatic repulsion, the core/shell fibers are ultimately deposited on the grounded collector, as shown in Fig. 2.1a.

The coaxial device can also be arranged in an alternative configuration which involves the assembly of one syringe inside another, controlled by an individual pump, as presented in Fig. 2.1b. The purpose of this design is to minimize loss of solution in the dead space in the connecting tubes and spinnerets, thereby allowing the encapsulation of small fluid volumes (50–100 μL). The equipment design and electrohydrodynamic behaviors of the process are complex, with polymer types, concentration, miscibility and

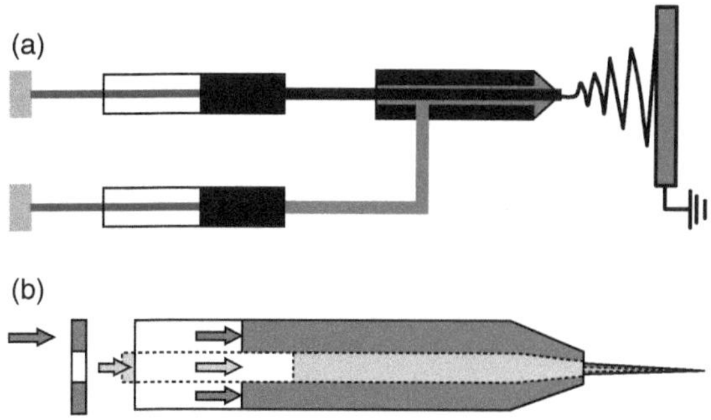

2.1 Schematic illustration of the coaxial electrospinning: (a) coaxial dual-capillary spinneret and (b) coaxial syringe.

flow rate of the outer and inner solutions all being crucial factors in coaxial electrospinning.

The miscibility of the inner and outer solutions significantly influences the process of coaxial electrospinning (Li, 2009). Interestingly, some research has shown that two immiscible components create a well-defined core/shell structure, though poor miscibility between core/shell solutions (e.g., water and chloroform) requires higher field strengths to overcome the solution surface tension. Some studies have proved that an intermediate solvent (e.g., ethanol or methanol) added to one of immiscible solutions reduces the high surface tension of components. It has further been suggested that two polymer solutions with added miscible solvents could facilitate the formation of core/shell fibers using a coaxial spinneret. If the same solvent is added to both the inner and outer solutions, a lower liquid–liquid interfacial tension is generated between them, allowing formation of the core/shell structure. Due to the rapidity of coaxial electrospinning, interaction between core and shell fluids during the process could be negligible. Meanwhile, the same solvent would also allow the use of lower applied voltage, generating reduced interfacial tension and producing thinner fibers.

The concentration of polymers in the core and shell significantly influence the stability of the compound Taylor cone. In hydrophobic/hydrophilic inner/outer solution systems, hydrophilic polymers like polyvinyl pyrrolidone (PVP) in a N,N-dimethylformamide (DMF) solution produce low liquid–liquid surface tension and high viscosity in the shell, possibly preventing the varicose breakup of the inner fluid in the coaxial jet until the coaxial electrospinning process is complete. In contrast, a polymer which has lower viscosity, producing higher surface tension in shell, will easily lead to a breakup of the typically hydrophilic inner core solution. Therefore relatively low interfacial tension, low diffusion rates of different polymer solution pairs and the miscibility of these solutions are all important factors in the production of core/shell nanofibers by coaxial electrospinning.

The flow rate ratio of the inner and outer solutions, the capillary length ratio of inner and outer tips and capillary size are also very important in controlling the core/shell fiber structures. In general, coaxial electrospinning can typically be achieved at similar field strength as monoaxial electrospinning, depending on the solvent miscibility and surface tension of the core/shell solutions. It has been proved that immiscible solutions usually require higher voltage than miscible ones. The flow ratio of the inner and outer solutions can be adjusted to keep the level of voltage necessary for efficient encapsulation of the core solution to a minimum (Zhang, 2006). Based on experimental investigations, it has been found that stable core/shell Taylor cones and a consistent yield of electrospun core-shell fibers are produced if the shell rate flow is three times higher than that in the core.

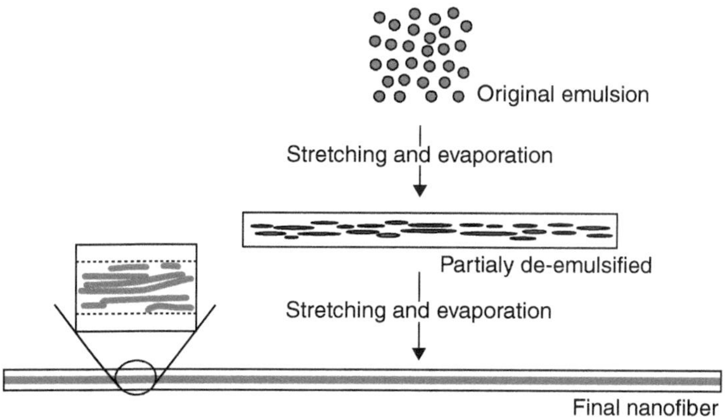

2.2 Schematic mechanism for the formation of core-shell nanofibers during emulsion electrospinning.

Emulsion electrospinning, another novel process, has also been proposed in the preparation of core-sheath fibers. It is similar to normal solution electrospinning, with the exception that the solution is replaced with an emulsion (Angeles, 2008). The electrospinning jets are generated from the emulsion liquid and stretched into ultrafine fibers. The dispersed drops in the emulsion turn into the core of the electrospun fiber, and the continuous matrix becomes the shell of the fiber, as shown in Fig. 2.2.

Water-in-oil (W/O) and oil-in-water (O/W) are the two kinds of emulsion used. Generally, emulsion electrospinning requires an adequate surfactant, a suitable solvent and the correct ratio of water to stabilize the emulsion (Xu, 2006). In the W/O case, the water phase, which usually has a lower evaporation rate, has a lower viscosity than the oil phase, so the oily matrix should be more viscous if the shell is to form. During the electrospinning process, the aqueous drops are forced to stretch and coalesce in the center to form the core-shell fibers. In the O/W case, in contrast, the oily drops should have lower viscosity, so that they can be deformed and coalesced. Both processes demonstrate the common rule that the viscosity of the drop phase and the stability of the emulsion are critical for emulsion electrospinning.

One example of this rule is the formation mechanism of poly(methyl methacrylate) (PMMA)/polyethylene oxide (PEO) (O/W). Figure 2.3 illustrates fiber formation in O/W emulsion electrospinning, using PMMA as the oil phase and PEO as the aqueous phase. Figure 2.3a shows the adsorption of PEO on an oil droplet. The sunflower droplets of O/W emulsion are illustrated in Fig. 2.3b. O'Brien's 'gel layer' model suggests that there is a dense inner layer and an outer absorbent layer, each with differing elasticity and viscosity. In the PMMA/PEO system, the droplets can be divided into

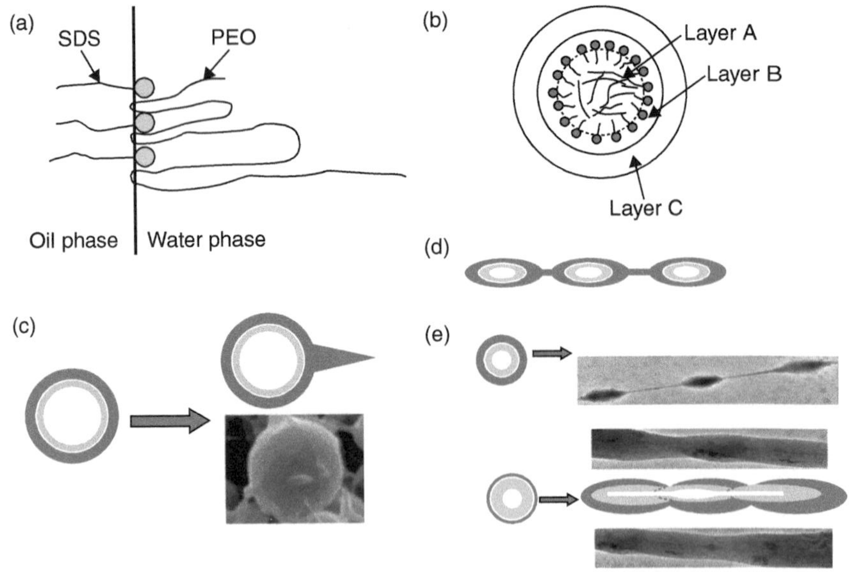

2.3 Schematic mechanisms for electrospinning O/W emulsion: (a) the interface between oil and water, (b) emulsion droplet made from PMMA/SDS/PEO, (c) 10% PEO/emulsion after electrospinning, (d) 15% PEO/emulsion after electrospinning and (e) 20% PEO/emulsion after electrospinning.

three layers: the PMMA core, the dense inner layer and the outer adsorbed layer, described as layer A, layer B and layer C, respectively.

Figures 2.3c, 2.3d and 2.3e describe the electrospinning of emulsions with different concentrations of PEO. When the PEO concentration is only 10%, the PMMA core appears very large in comparison to the coating PEO layer. During the electrospinning process, the droplets are slightly stretched in the direction of the electric field due to the thinner layer of the PEO, forming tailed beads as shown in Fig. 2.3c. Since the droplet size is only about 260.4 nm, it takes longer for two droplets to coalesce with each other in the electrospinning process. The emulsion is more likely to electrospray and form tailed beads, rather than being electrospun into nanofibers.

When the PEO content reaches 15%, the thickness of layer B and layer C both increase dramatically. Layer C is stretched further than layer A and layer B, due to the far greater elasticity of layer C. When the droplet size increases, the possibility of one droplet coalescing with another also increases, facilitating the coalescence of the droplets to form beaded nanofibers, as shown in Fig. 2.3d.

At a concentration of 20%, the surface of each PMMA/sodium dodecyl sulfate (SDS) emulsion droplet is fully surrounded by PEO. The dense inner layer becomes thicker with the increase of PEO. It appears that layer B is the most important layer in this process. Layer C stretches with layer B when it is elongated in the electric field. Using 10% PEO and 15% PEO, one droplet coalesced with another before it was fully stretched, in a much shorter period and more easily, leading to the formation of nanofibers with a helical structure. The inner beaded nanofiber made from PMMA can be seen from transmission electron microscopy (TEM) observation, as illustrated in Fig. 2.3e. From this it can be inferred that the density of the inner layer of the PMMA droplet itself is larger than the outer layer. The dark area in the nanofiber might be the broken micelles of PMMA.

In conclusion, the mechanisms of PMMA/PEO (O/W) emulsion electrospinning are based on three theoretical foundations as follows:

1. O'Brien's 'two-layer' model: inner dense layer and outer adsorbed layer.
2. When the coated layer of PEO gets thicker, the thickness of the dense inner layer increases faster than the outer adsorbent layer.
3. The droplet size grows with the increase of PEO concentration, allowing the more rapid coalescence of two droplets.

2.3 Aligned nanofibers

Aligned nanofibers with uniaxially oriented fibers possess mechanical properties much higher than that of conventional fibers, suggesting potential applications in catalysis, biomedicine, microelectronics and optics. Macroscopic alignment of the electrospun nanofibers can be successfully accomplished by using proper collection apparatus, auxiliary electrodes, or a combination of both.

The collection apparatus allows the application of various different strategies, including the use of a rotating mandrel collector, a rotating disk collector, a dual grounded collection plate and a copper wire drum, among others (as shown in Fig. 2.4). A dynamic collector such as a rotating disc as a fiber collector can deposit most of the fibers on the sharp-edged disk and collect aligned or patterned nanofibers (Zhang, 2010a). As an electrode, the sharp edge of the disc accumulates charge to attract positively charged nanofibers, and hence the charged polymer jet is pulled towards the edge of the wheel, causing the inverted conical shape of the jet at the disk edge.

Due to the tangential force acting on the fibers, produced from the rotation of the disk, the fibers are attracted to the perimeter of the disk. This force further stretches the fibers and reduces their diameter. To collect aligned nanofibers for a larger and thicker nanofibrous mat, the disc could be replaced by a drum or a shaft with a large surface area (Yogesh, 2009).

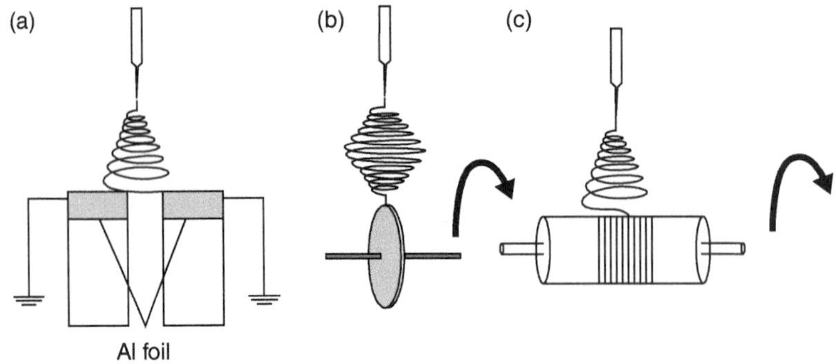

2.4 Schematics of electrospinning system for various collection apparatus: (a) a dual grounded collection plate, (b) a rotating disk collector and (c) a rotating drum collector.

Several studies have shown that nanofiber alignment is highly dependent on the rotation speed of the disc, drum or shaft. Fiber alignment increases with the rotation speed of the collecting surface. However, at much higher speeds, the take-up velocity breaks the depositing fiber jet and continuous fibers cannot be collected. Ideally, the linear rate of the rotating collector should match the evaporation rate of the solvent, thus the fibers can be deposited and taken up on the surface of the collector.

Other methods used in preparing aligned nanofibers include the use of auxiliary electrodes to create an electric field profile that influences the flight of the electrospinning jet. When using the electrodes, a fiber collector consisting of rectangular conductive electrodes is placed on a highly insulating substrate and a gap is left. Electrospun nanofibers are attracted and uniaxially deposited on the collectors. Increasing charges on the nanofibers improves fiber alignment (Chen, 2009). Altering the pattern of the electrode array could also change the orientation of uniaxial nanofibers and fabricate aligned nanofibers with a complicated pattern.

A combination of a field-controllable electrode and a rotating collector can achieve highly oriented electrospun fibers (Lee, 2009). This technique involves two key components: the use of a cylindrical electrode connected to a spinning nozzle and a field-controllable electrode generating an AC electric field fixed with a rotating collector, as shown in Fig. 2.5. To understand the simple mechanism of the electrostatic interaction generated by a field-controllable electrode connected with a rotating collector, a schematic diagram of the technique is shown in Fig. 2.6. As illustrated in Fig. 2.6a, the positively charged jet from the nozzle begins to stretch and undergoes a whipping motion toward the rotating collector, as in the normal electrospinning process, where the electrospun fibers undergo alignment and stretching by means of the high-speed rotating collector. However, in the

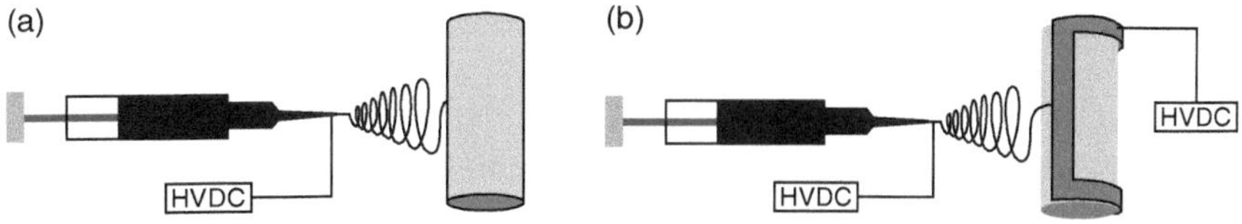

2.5 Schematic illustration of electrospinning for aligned nanofibers: (a) normal electrospinning and (b) modified electrospinning.

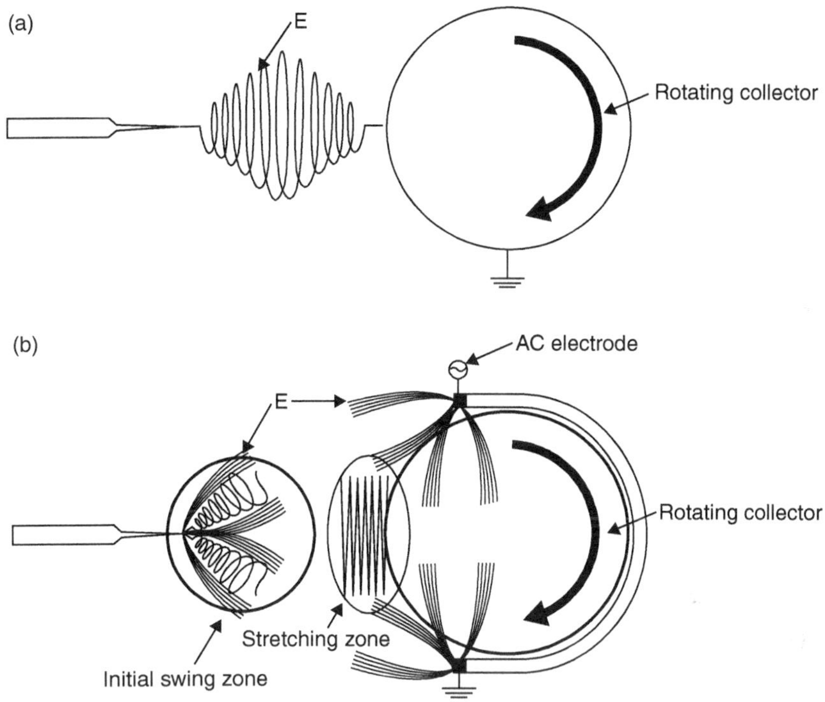

2.6 Schematic diagram of the electrospun nanofiber alignment mechanism due to a field-controllable electrode with a rotating collector: (a) normal electrospinning process and (b) modified electrospinning process with a field-controllable electrode.

field-controllable electrospinning process, the swing phenomenon of the initial solution jet may be caused by both Coulombic force between the field-controllable electrodes at the rotating collector and the positively charged spun fibers (Fig. 2.6b). Thus the whipping area can be enlarged by the movement of the initial jet solution, and the stretching and alignment can be more highly reinforced by the field-controllable electrode set up with the collector (Baji, 2010).

Aligned fibrous structures possess anisotropic mechanical properties due to the fiber orientations. Fiber orientation may enhance mechanical properties, as demonstrated by research correlating anisotropic mechanical properties with the percentage of fiber alignment (Seeram, 2006). Aligned nanofibers may possess superior mechanical properties than randomly oriented nanofibers for a wide range of applications.

2.4 Porous nanofibers

The development of porous nanofibers can facilitate the creation of fiber mats with a greatly increased surface area, which could lead to a wide range of applications in such fields as tissue engineering, sensing, filtration, separation, catalysis, superhydrophobic coatings and encapsulating materials. There are a number of techniques developed for the generation of porous nanofibers using electrospinning. Phase separation has been proposed as the main mechanism for the formation of porous nanofibers in the electrospinning process (Zou, 2011). Two commonly adopted techniques include the use of highly volatile solvents such as methylene chloride, in which solvent-rich regions facilitate formation of pores during electrospinning. Another technique in producing porous nanofibers is to employ bicomponent spinning, then selectively remove one component after fiber formation by thermal treatments or dissolution in a solvent in which the other polymer is insoluble.

Thermal treatments under different atmospheres are usually used to fabricate carbon or inorganic porous nanofibers. Electrospun porous carbon nanofibers can be produced by co-spinning polyacrylonitrile (PAN) and poly(L-lactic acid) (PLLA) combined with a removal of PLLA phase in an air environment at 280°C for 5.5 h and then carbonization at 800°C for 1 h (Ji, 2009a). Furthermore, the modified evaporative cooling technique combined with carbonization can be used to prepare porous carbon nanofibers that contain metal/metal oxide nanoparticles in the pores (Ji, 2009b). The basic idea of this method is also to use phase separation to prepare porous structures. Zou and coworkers prepared composite nanofibers by electrospinning poly(vinyl alcohol) (PVA) and $SnCl_2 \cdot H_2O$ aqueous solution. The nanofibrous membrane was first frozen in an ice-box at around 12°C for two days, then heated at 110°C for 2 h and finally at 550°C for 3 h in Ar/H_2 (Zou, 2011). Porous carbon nanofibers with Sn/SnOx nanoparticles in the pores were successfully produced in this process. The key issue in this technique is to control the solubility of the metal salt, in other words, the phase separation. The reduced metal particles could be deposited in the pores, or embedded in the carbon substrate.

Porous nanofibers fabricated by highly volatile solvents during electrospinning have several advantages. First, there is no residue left as volatilized solvent, which means the surface chemistry of electrospun fibers can be carefully regulated. Furthermore, the structural integrity of nanofiber membranes can be maintained when the pores are formed. This is because pore generation occurs during the spinning process, but not after fiber deposition and so no interconnections form among the fibers. Some other approaches can also be used in the preparation of porous nanofibers, such as porogen methods and leaching of salts from the fibers although these techniques are currently attracting less attention.

2.5 Gradient nanofibers

Gradient nanofibers which are systematically designed for their hierarchical structures possess superior fracture toughness for load-bearing applications, such as musculoskeletal tissues (tendons and ligaments), skin scaffolds and wound dressing materials, where materials are required to be more elastic than stiff (Matsuda, 2008). Furthermore, gradient nanofibers can also be used in photoelectric engineering, mechanical engineering, nuclear engineering and filtration engineering.

Electrospinning is a simple and effective technique to achieve graduated nanofibers for tissue engineering applications. The hierarchical structure of nanofibers has led to their classification into three main types. These are illustrated by the following models, which allow the fabrication of a prototype in each case, and are also shown schematically in Fig. 2.7:

Model I: Gradually graded component content (variable parameter, component content; fixed parameter, diameter and wall thickness);
Model II: Gradually graded wall thickness (variable parameter, wall thickness; fixed parameter, diameter and component content);
Model III: Gradient tube diameter (variable parameter, diameter; fixed parameter, wall thickness and component content).

The setup of the electrospinning apparatus for fabricating gradient nanofibers is relatively simple and conventional. Materials can be electrospun on a rotating/transverse mandrel (rod-type, tapered-type and some other types) to form a gradient structure. The factors influencing the average fiber

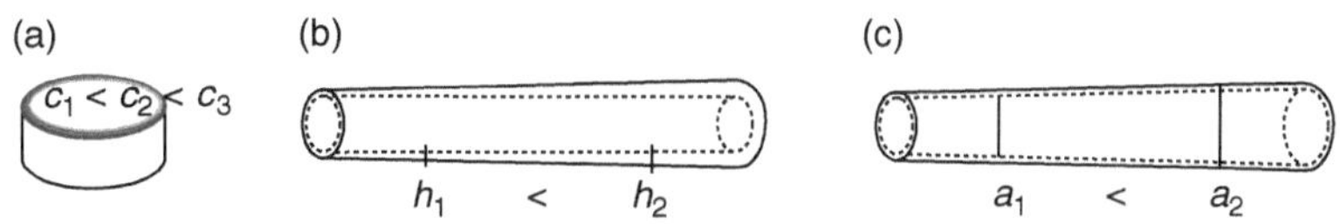

2.7 Schematic illustration of the fabrication of gradient nanofibers: (a) graded content nanofibers, (b) graded wall thickness nanofibers and (c) gradient tube diameter nanofibers (where *c* is component, *h* is the wall thickness and *a* is the diameter).

diameter and the wall thickness of electrospun meshes include choice of solvents, polymers, concentrations, air gap (distance between injection needle and collector), voltage and flow rate, among others.

Some examples of gradient nanofibers are provided as follows:

> For model I, the component gradient nanofibers are produced by increasing or decreasing the concentration of the polymer solution. Electrospinning is conducted by spinning separate gradient concentration solutions one by one (Liu, 2008).
>
> For model II, nanofibers are initially deposited along the entire length of a mandrel. Electrospinning then occurs across the whole region, excluding the required section at the end of the mandrel. A third step is fiber formation at the far edge at the end, while a final step is electrospinning on the remaining part. This process creates four different zones (Kidoaki, 2005). The gradient wall thickness is achieved by controlling the deposited areas, as shown in Fig. 2.8.
>
> For model III, the electrospun tube with gradiently reduced diameter was built by a tapered mandrel (Fig. 2.9). Its diameter decreases gradually toward the tip, as is clearly shown in cross section photographs of samples taken from different regions of the tip (Wen, 2010).

It has been realized that the design of scaffolds for tissue engineering should mimic the architectural structures formed in natural tissues, in order that the mechanical integrity of natural composites is matched. This requires a thorough understanding of the mechanical behavior of electrospun nanofibers, which will lead to the development of necessary new materials.

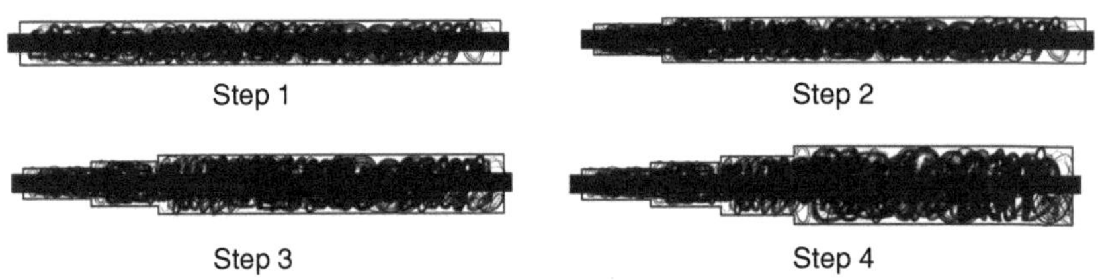

2.8 Schematic procedure of fabrication of model II, which consists of four steps.

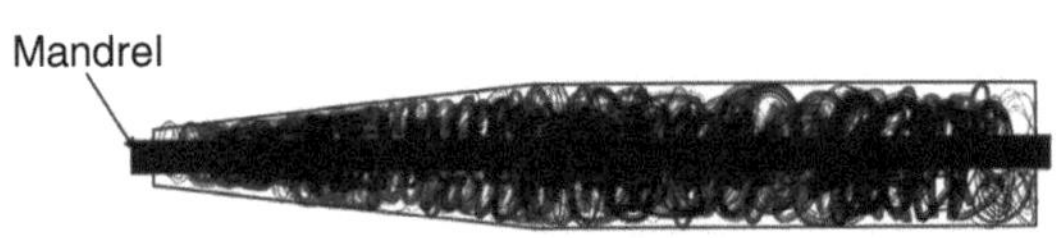

2.9 Schematics of electrospinning on a tapered mandrel.

2.6 Applications of structured functional nanofibers

The extremely fine structure and special properties of functional nanofibers make them exceedingly attractive, and possible new uses for the fibers in a wide range of advanced applications are being explored continuously. Table 2.1 lists the main application areas of structured nanofibers (Bhardwaj, 2010; Cui, 2010; Gregory, 2007; Huang, 2003; Li, 2010; Seeram, 2006; Wet, 2006). A number of other applications with specific structures, which have been fabricated in our lab, will also be briefly discussed.

C/SnO_2 core-shell nanofibers have been fabricated in our lab using coaxial dual-capillary spinneret electrospinning apparatus. Initially, PAN was used as shell and SnO_2 sol mixed with PVP was used as core, with both dissolved in DMF. The TEM image in Fig. 2.10a reveals the core-shell structure of the electrospun nanofibers. After carbonization, C/SnO_2 core-shell nanofibers were well formed, as shown in Fig. 2.10b. C/SnO_2 core-shell nanofibers can be used as an anode material for lithium ion batteries. Theoretically, a carbon matrix offers better electric conductivity and buffers the volumetric change of tin during battery cycling; therefore the cycle performance of tin anode could be improved.

PMMA/O-MMT composite porous nanofibers have also been fabricated by chloroform volatilizing during electrospinning. The morphology of porous nanofibers is presented in Fig. 2.11a. The PMMA/O-MMT composite porous nanofibers were coated with nanosize TiO_2 using radio frequency (RF) magnetron sputtering, as illustrated in Fig. 2.11b. The surface modified the PMMA/O-MMT composite porous nanofibers, producing high photodegradation efficiency in the photocatalytic degradation of Methylene Blue.

The applications of structured functional nanofibers discussed here are by no means exhaustive, and new structured nanofibers and applications are emerging continually. While the potential for nanofibers is enormous, there are also considerable challenges ahead.

Table 2.1 The application of structured functional nanofibers

The structure of nanofibers	The main application areas
Core-shell	Drug delivery, release control, catalyst and enzyme carriers, sensors, energy storage, etc.
Aligned	Muscle tissue engineering, polymeric scaffolds, wound dressings, biomedicine, high-level filtration systems, etc.
Porous	Sensors, energy storage, etc.
Gradient	Tissue engineering scaffolds, aerospace, mechanical engineering, etc.

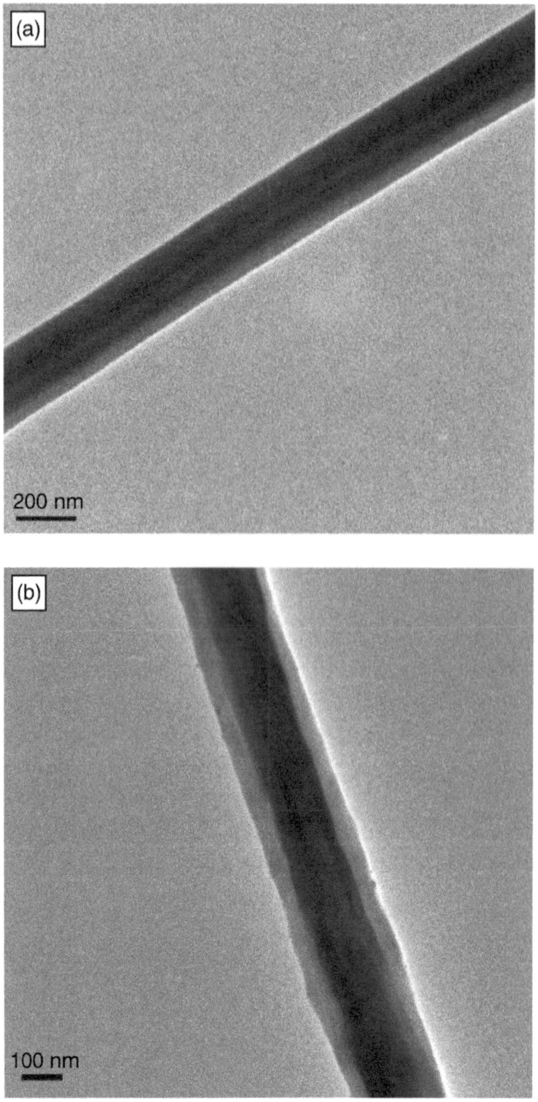

2.10 TEM images of C/SnO_2 core-shell nanofibers: (a) electrospun C/SnO_2 core-shell nanofibers and (b) carbonized C/SnO_2 core-shell nanofibers.

2.7 Future trends

With such versatility, structured functional nanofibers have been explored for use in many different applications. Currently, most effort has been focused on the preparation of combined structures, such as three-layered structures

2.11 Scanning electron microscopy (SEM) images of PMMA/O-MMT composite nanofibers: (a) electrospun PMMA/O-MMT composite porous nanofibers and (b) TiO_2 coated on PMMA/O-MMT nanofibers by RF magnetron sputtering.

based on core-shell nanofibers, triangular or square nanofiber membranes based on rebuilding collectors and, in particular, multifunctional nanofiber composites with fillers (Fragal, 2010; Xiao, 2009; Zhang, 2008, 2010b). Besides modifications of the collectors and electrodes and manipulation of the electric field, the electrospinning processes themselves have also seen several developments, such as the development of gas jacketing and multiple spinnerets. In addition, hybrid technologies are being developed for the production of structured functional nanofibers (Cai, 2010; Wu, 2010). However, future work should focus more on analysis of the shape formation mechanisms and functional properties (Zhang, 2009), in order to enhance the prospect of application.

2.8 References

Angeles, M. (2008), 'Short communication emulsion electrospinning: Composite fibers from drop breakup during electrospinning', *Polymers for Advanced Technologies*, **19**, 728–33.

Baji, A. V. (2010), 'Electrospinning of polymer nanofibers: Effects on oriented morphology, structures and tensile properties', *Composites Science and Technology*, 70, 703–18.

Bhardwaj, N. (2010), 'Electrospinning: A fascinating fiber fabrication technique', *Biotechnology Advances*, **28**, 325–47.

Cai, Y. B (2010), 'Comparison between structures and properties of ABS nanocomposites derived from two different kinds of OMT', *Journal of Materials Engineering and Performance*, **19**, 171–6.

Chen, R. (2009), 'A novel approach via combination of electrospinning and FDM for tri-leaflet heart valve scaffold fabrication', *Frontiers of Materials Science in China*, 3(4), 359–66.

Cui, W. G. (2010), 'Electrospun nanofibrous materials for tissue engineering and drug delivery', *Science and Technology of Advanced Materials*, **11**, 1–11.

Fang, J. (2008), 'Applications of electrospun nanofibers', *Chinese Science Bulletin*, **53**, 2265–86.

Fragal, M. E. (2010), 'Core–shell Zn-doped TiO2–ZnO nanofibers fabricated via a combination of electrospinning and metal–organic chemical vapor deposition', *CrystEngComm*, **12**, 3858–65.

Gregory, C. R. (2007), 'Formation of fibers by electrospinning', *Advanced Drug Delivery Reviews*, **59**, 1384–91.

Huang, Z. M. (2003), 'A review on polymer nanofibers by electrospinning and their applications in nanocomposites', *Composites Science and Technology*, **63**, 2223–53.

Ji, L. W. (2009), 'Fabrication of porous carbon nanofibers and their application as anode materials for rechargeable lithium-ion batteries', *Nanotechnology*, **20**, 155705.

Ji, L. W. (2009), 'Fabrication of porous carbon/Si composite nanofibers as high-capacity battery electrodes', *Electrochemistry Communications*, **11**, 1146–49.

Kidoaki, S. (2005), 'Mesoscopic spatial designs of nano- and microfiber meshes for tissue-engineering matrix and scaffold based on newly devised multilayering and mixing electrospinning techniques', *Biomaterials*, **26**, 37–46.

Lee, H. J. (2009), 'Highly oriented electrospun polycaprolactone micro/nanofibers prepared by a field-controllable electrode and rotating collector', *Applied Physics A*, **97**, 559–65.

Li, X. (2007), 'Core–shell structured and electro-magnetic functionalized polyaniline composites', *Synthetic Metals*, **157**, 575–9.

Li, X. Q. (2009), 'Fabrication and properties of core-shell structure P (LLA-CL) nanofibers by coaxial electrospinning', *Journal of Applied Polymer Science*, **111**, 1564–70.

Li, Y. H. (2010), 'Hierarchical structured ZnO nanorods on ZnO nanofibers and their photoresponse to UV and visible lights', *Sensors and Actuators A*, **158**, 176–82.

Liu, C. Z. (2008), 'Gradient collagen/nanohydroxyapatite composite scaffold: Development and characterization', *Acta Biomaterialia*, **21**, 1–9.
Matsuda, T. (2008), 'Electrospinning fabrication of high-trackable catheter tip with gradually graded or gradient flexibility', *Journal of Biomedical Materials Research Part B: Applied Biomaterials*, **5**, 35–41.
Park, S. A. (2007), 'Apparatus for preparing electrospun nanofibers: Designing an electrospinning process for nanofiber fabrication', *Polymer International*, **56**, 1361–6.
Seeram, R. K. (2006), 'Electrospun nanofibers: Solving global issues', *Materials Today*, **9**, 40–50.
Wen, S. P. (2010), 'Hierarchical electrospun SiO2 nanofibers containing SiO2 nanoparticles with controllable surface-roughness and/or porosity', *Materials Letters*, **64**, 1517–20.
Wet, S. R. (2006), 'A review on electrospinning design and nanofiber assemblies', *Nanotechnology*, **17**, R89–R106.
Wu, N. (2010), 'Preparation and properties of organic/inorganic hybrid nanofibres', *Fibers & Textiles in Eastern Europe*, **18**, 21–3.
Xiao, Q. Z. (2009), 'A novel sandwiched membrane as polymer electrolyte for application in lithium-ion battery', *Journal of Membrane Science*, **326**, 260–4.
Xu, X. L. (2006), 'Preparation of core-sheath composite nanofibers by emulsion electrospinning', *Macromolecular Rapid Communications*, **27**, 1637–42.
Yogesh, N. (2009), 'Nanofiber alignment on a flexible substrate: Hierarchical order from macro to nano', *Applied Materials & Interfaces*, **1**, 2093–7.
Zhang, D. M. (2008), 'Electrospinning of three-dimensional nanofibrous tubes with controllable architectures', *Nano Letters*, **10**, 3283–7.
Zhang, K. (2010a), 'Aligned and molecularly oriented semihollow ultrafine polymer fiber yarns by a facile method', *Journal of Polymer Science: Part B: Polymer Physics*, **48**, 1118–25.
Zhang, L. J. (2009), 'Nanotechnology and nanomaterials: Promises for improved tissue regeneration', *Nano Today*, **4**, 66–80.
Zhang, X. W. (2010b), 'Design and preparation of CNT@SnO2 core-shell composites with thin shell and its application for ethanol oxidation', *International Journal of Hydrogen Energy*, **35**, 8841–7.
Zhang, Y. Z. (2006), 'Coaxial electrospinning of (fluorescein isothiocyanate-conjugated bovine serum albumin)-encapsulated poly (e-caprolactone) nanofibers for sustained release', *Biomacromolecules*, **7**, 1049–57.
Zou, L. (2011), 'A film of porous carbon nanofibers that contain Sn/SnOx nanoparticles in the pores and its electrochemical performance as an anode material for lithium ion batteries', *Carbon*, **49**, 89–95.

3

Processing of composite functional nanofibers

Y. CAI, Q. WEI and F. HUANG,
Jiangnan University, China

Abstract: This chapter introduces the formation mechanisms and potential applications of composite nanofibers, including coverage of polymer/polymer composite nanofibers, polymer/nanoparticles composite nanofibers and polymer/inorganic salts composite nanofibers. The methods and principles of electrospinning and future trends of composite nanofibers are also explored. The potential uses for the different nanofibers are identified; many of these are still in the development stage, but some are already fulfilling roles in medicine, industry etc.

Key words: electrospinning, composite nanofibers, application.

3.1 Introduction

Nanofibers have attracted a great deal of attention due to their remarkable properties. Compared to conventional fibrous structures, nanofibers are lightweight with small diameters, controllable pore structures and high surface-to-volume ratio, making them ideal for use in applications as varied as filtration, sensors, protective clothing, tissue engineering, functional materials and energy storage.[1–5] However, the development of new composite nanofibers holds even greater potential and investigation of new polymer composite nanofibers aimed at further developing their multifunctional possibilities and enhancing both their physical and chemical properties has gained extensive interest. Polymer composite nanofibers generally consist of two or more composites, with a range that includes polymer/polymer composite nanofibers, polymer/nanoparticle composite nanofibers and polymer/inorganic salts composite nanofibers, among others.

3.2 Formation of polymer and polymer composite nanofibers

Polymer/polymer composite nanofibers are defined as nanofibers composed of two or more polymers. Polymer/polymer composite nanofibers can be produced using two experimental methods. The first method is a one-step electrospinning procedure. Direct electrospinning of polymer solutions, containing two kinds of polymers dissolved in homogeneous or mixed solvents, leads to the formation of composite nanofibers. Fabrication of nanofibers from biopolymers is attracting increasing interest due to the fact that biopolymers may have superior biocompatibility and biodegradability, are generally nontoxic, renewable and available at low costs, and may have additional properties such as being antioxidant, antimicrobial or effective in relation to enzymes.[6] Researchers have focused on electrospun blends of Chitosan with other compatible polymers such as poly(vinyl alcohol) (PVA) [7–10] and polyethylene oxide (PEO) [11–14] among others.[15] In blends, the fiber-forming ability of the cospinning agent is utilized to facilitate polymer entanglement and generation of a polymer jet. Electrospinning of polymer blends is also an efficient way to create composite nanofibers that have improved material properties, such as higher tensile strength.

The second method used in the production of polymer/polymer composite nanofibers is co-electrospinning or coaxial electrospinning. In this process, two kinds of polymers are dissolved in two different solvents, before being co-electrospun with the same collector. The prepared composite nanofibers may be formed with different structures, producing both hybrid structure nanofibers and core-shell structure nanofibers. The schematic views of the co-electrospinning and coaxial electrospinning setups are shown respectively in Fig. 3.1[16] and Fig. 3.2.[17] The coaxial electrospinning process is reported to be an effective and versatile process for the fabrication of core-shell structures. Using this technique, the functional component can be encapsulated as the core, or coated as the shell. Composites such as ferritin nanoparticles/poly(2-acrylamido-2-methyl-1-propanesulfonicacid),[18] poly-(N-isopropylacrylamide)/poly(methylmethacrylate),[19] polycarbonate/poly(methyl methacrylate)[20] and chitosan/poly(ethylene oxide)[21] have all been electrospun for specific applications.

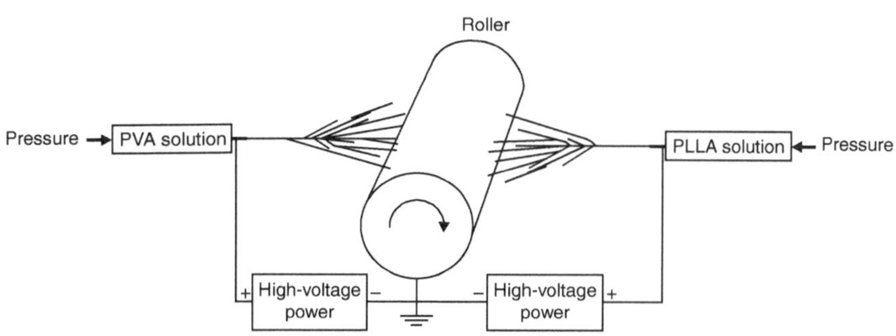

3.1 Setup of co-electrospinning.

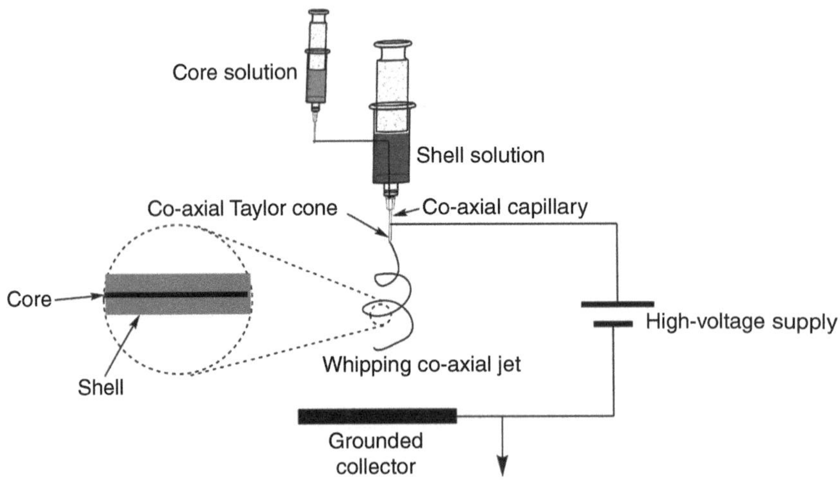

3.2 Schematic of co-axial electrospinning set-up.

3.3 Formation of polymer and nanoparticle composite nanofibers

The so-called polymer/nanoparticle composite nanofibers, consisting of nanoscale inorganic fillers and a polymer matrix, encompass the advantages of both composites. The light weight, flexibility and good moldability of the polymer materials is combined with the high strength, heat stability and chemical resistance of the inorganic materials.[22–23] As a result, these composite nanofibers have enhanced mechanical, electrical, optical, thermal and magnetic properties, with no degradation of their transparency, making them excellent candidates for many multifunctional applications including membranes, biomedical devices and energy conversion and storage systems.[23–25] Polymer/nanoparticle composite nanofibers can be obtained using several experimental methods.[23] The first method is known as the polymer template procedure, in which polymer nanofibers produced by electrospinning or other means, are immersed into a nanoparticulate containing composite solution to adsorb nanoparticles on the fiber surface. A thermal or chemical treatment step is often introduced in addition to obtain good particle–fiber adhesion.[26] The second method used in the formation of polymer composite nanofibers is a one-step electrospinning procedure, in which nanofibers are produced by the direct electrospinning of a polymer solution containing nanoparticles.[27] Comparing these two methods, the one-step electrospinning process is relatively simple and has the advantage of producing composite nanofibers at high speed and low cost. Use of the one-step electrospinning method facilitates the production of polymer/inorganic composite ultra-thin nanofibers, which combine the advantages of organic–inorganic hybrids with the large surface area of nanofibers and therefore meet the needs of new materials.[28] In addition, alternative production methods including sol-gel, sputter coating and electroless deposition are also frequently used.

Among various polymer/inorganic composite nanofibers, the most frequently used nanoparticles included montmorillonite,[1, 29–36] silver (Ag),[37–38] titanium dioxide (TiO_2),[39–41] silica (SiO_2),[42–44] magnetite (Fe_3O_4),[45] silicon nitride (Si_3N_4),[46] layered double hydroxides (LDHs),[47] zinc oxide (ZnO),[48] copper (Cu),[49] β-tricalciumphosphate (β-TCP)[50] and multiwalled carbon nanotubes.[51]

3.3.1 Two-step electrospinning method

For this method, the polymer/nanoparticle nanocomposites are initially prepared by melt blending using a twin screw extruder,[29–31, 46, 50] *in situ* emulsion polymerization[32] and *in situ* polymerization.[51] The composite nanofibers are then prepared using electrospinning once the polymer/nanoparticle nanocomposites have been dissolved in a solvent. The nanoparticles most commonly used in this process include layer silicate clays,[29–31] Si_3N_4,[46] β-TCP[50] and multiwalled carbon nanotubes.[51]

3.3.2 One-step electrospinning method

A simple one-step composite technique may also be used to prepare composite nanofibers. The nanoparticles are initially dispersed in a solvent, as is the polymer which may be dissolved in the same or an alternative solvent. The two are then compounded, before the composite solution is sonicated and electrospun to form polymer/nanoparticle composite nanofibers. The silicate clays,[33–36] TiO_2,[39] SiO_2,[42–44] Fe_3O_4[45] and LDHs[47] are most commonly used in this method. The schematic views of the electrospinning setup are shown in Fig. 3.3.[52]

3.3.3 Sol-gel method

Sol-gel technology offers many possibilities for the introduction of molecules with specific properties (thermal, electrical, optic, chromogenic, photosensitive and mechanical) into polymeric matrices. The sol-gel technique has been increasingly used to produce the functional coatings on various forms of material. The precursors used in the preparation of the 'sol' are usually inorganic metal salts or metal organic compounds, such as metal alkoxide. The precursor is subjected to a series of hydrolysis and condensation

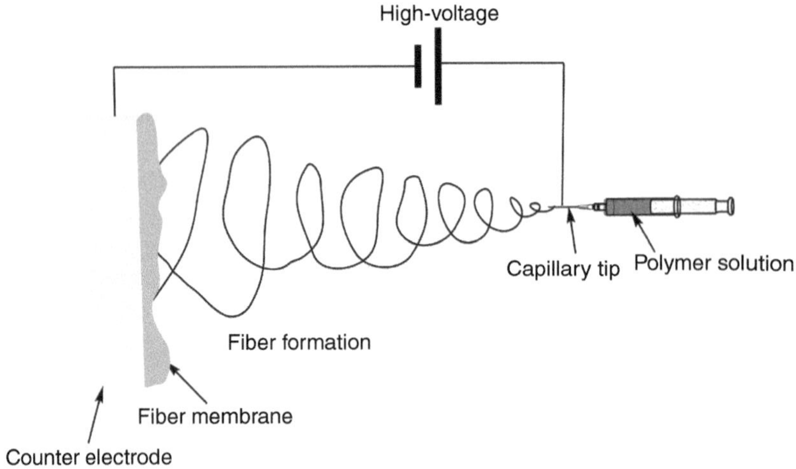

3.3 The schematic views of the electrospinning setup.

reactions for the inorganic domain formation. The use of a suitable coupling agent permits the formation of a strictly interconnected network, preventing macroscopic phase separation.[53, 54] The sol-gel technique has several advantages over other processes including its simplicity, good homogeneity, the ease of control of the film composition, its safety, its ability to produce large area coatings, its low processing temperature and the low cost of both the apparatus and raw materials.[55]

Many interesting properties of polymer nanofibers are the result of their extremely high surface to weight ratio in comparison to other conventional fibrous structures. The high pore volume and tight pore sizes make nanofibers ideal for use in such application areas as filtration, sensors, protective clothing and functional materials. The ability to deposit well-controlled coatings on nanofibers expands their application potential by changing both physical and chemical properties. Nanoparticles frequently used in such methods include TiO_2 [40, 41, 56] and ZnO.[48]

3.3.4 Electroless deposition and *in situ* reduction method

Initially the polymer/metal ion composite nanofibers are prepared by electrospinning, before the metal ions are reduced within composite nanofibers. For example, the Ag nanoparticles dispersed in polyacrylonitrile (PAN) nanofibers spun by electrospinning are prepared *in situ* by the reduction of silver ions in N_2H_5OH aqueous solution.[37] The electrospun Polyamide 6 (PA6)/Cu composite nanofibers[49] are prepared using an electroless deposition technique.

3.4 Formation of polymer and inorganic salt composite nanofibers

Polymer/transition metal salt composite nanofibers have attracted extensive attention primarily due to the fact that these fibers could easily be transformed into polymer/metal or polymer/semiconductor composite nanofibers through chemical treatment.[37,49,57–59] Various amounts of inorganic salts are initially added to the polymer electrospinning solutions. The composite solution is mixed by mechanical stirring, before being electrospun to form polymer/inorganic salts composite nanofibers. Composite nanofibers of this nature have many potential applications, such as in photographic and optical materials, catalysis and catalyst supports, ionic conductors, sensors and environmental protection materials.[57–61] Commonly used inorganic salts include $AgNO_3$,[37] $ZnCl_2$[62] and $FeCl_3$.[63]

3.5 Examples and applications of composite functional nanofibers

The application fields of electrospun polymer composite nanofibers have been steadily expanding, particularly in recent years. The relevant US patents show that some of the best representations of this can be seen in the fields of filtration systems and medical prosthesis, particularly in the areas of grafts and vessels. The high aspect ratio (l/d), large available surface area, surface morphology and potential homogeneous dispersion of nanofibers throughout matrix-made electrospun fibers mean they are of great interest as fillers in composite matrices. Other applications which have been targeted included composite reinforcement, tissue templates, electromagnetic shielding, composite delamination resistance and liquid crystal devices, among others. The mechanical properties and reinforcing behavior of continuous nanofibers differ significantly from the finite length of their conventional counterparts (engineering fibers such as carbon, glass and Kevlar fibers) that are used as reinforcements in composite developments. It should be recognized that most of these applications have not yet reached industry level, but are currently at a laboratory research and development stage. However, their promising potential is believed to be attracting attention and investment from academia, government and industry all over the world.[64,65]

3.5.1 Polymer and polymer composite nanofibers

Electrospinning of polymer blends is an efficient way to create composite nanofibers that have improved material properties, such as higher tensile strengths and wetting behavior.[14,16] The incorporation of cellulose whiskers into nanofibers of PEO by the electrospinning process can enhance the mechanical properties of the PEO.[66] Liu *et al.*[16] studied the structure and properties of poly(L-lactic acid)/poly(vinyl alcohol) composite nonwoven materials produced by co-electrospinning. The results indicate that electrospun poly(L-lactic acid) (PLLA)/PVA nonwoven materials demonstrate better wettability than pure PLLA electrospun fibrous web, and the larger the ratio of PVA, the better the wettability of the composite achieved. Chitin and chitosan have been examined and used in a wide variety of biomedical applications, including as drug delivery carriers, surgical sutures, bone-healing materials and, in particular, wound dressings.[67] Chitin and certain derivatives can accelerate the tensile strength of wounds by speeding the fibroblastic synthesis of collagen in the first few days of wound healing,[68] while chitosan can achieve hemostasis and promote normal tissue regeneration.[69] Chen *et al.*[70] prepared a collagen/chitosan nanofibrous membrane as a wound dressing by electrospinning. The membrane was found to promote wound healing and induce cell migration and proliferation. From animal studies, the collagen/chitosan nanofibrous membrane was better than gauze and commercial collagen sponge in wound healing.

A core-sheath configuration has the potential to achieve products with unique properties that are difficult to obtain from the constituent materials if spun separately. This approach can be broadly viewed as a blending of materials, except that the two materials maintain their separate identities, with the core material completely surrounded by the sheath material.[17] Zhang, Huang and coworkers[71, 72] successfully demonstrated the feasibility of preparing core-sheath nanofibers using two biodegradable materials, polycaprolactone (PCL) as the sheath and gelatin as the core, with the structures suited for use in tissue engineering and drug delivery. The authors showed that the size of the core could be controlled by simply varying the core polymer solution concentration. Gelatin and PCL, being natural and synthetic polymers respectively, degrade at different rates and, therefore, the researchers suggested that such structures will be useful in applications such as drug delivery, which may require different delivery rates at different times.

3.5.2 Polymer and nanoparticle composite nanofibers

The novel nanocomposite materials based on organic and inorganic species combined at a nanoscale have attracted great attention. These organic–inorganic hybrids have the light weight, flexibility and good moldability of organic materials along with the high strength, heat stability and chemical resistance of inorganic materials. These composites are expected to find applications in catalytic membranes, ultrafiltration, scratch- and abrasive-resistant hard coatings, nonlinear optical materials, contact lenses and reinforcement of elastomers and plastics.[65, 73, 74]

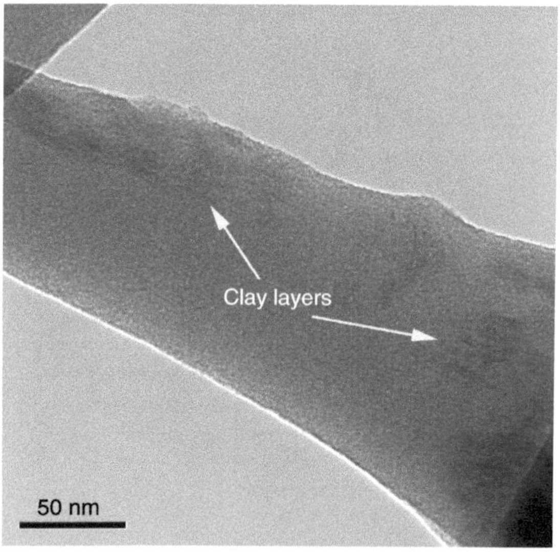

3.4 High-resolution transmission electron microscopy (HRTEM) images of electrospun PA6 composite nanofiber with 4 wt% Fe-OMT.

Cai *et al.*[34–36] prepared polymer/organic-modified Fe-montmorillonite (Fe-OMT) composite nanofibers and studied the structure, morphology and performances. The results showed the silicate clay layers to be well dispersed within the composite nanofibers as well as almost aligned along the axis of the nanofibers. The dispersion of Fe-OMT within PA6 composite nanofibers is characterized by high-resolution transmission electron microscopy images, as showed in Fig. 3.4. The loadings of Fe-OMT notably increase the thermal stability, flammability and mechanical properties of composite nanofibers. The crystalline properties of the PA6 composite nanofibers were also investigated by the wide-angle x-ray diffraction (XRD). Figure 3.5 demonstrates that the presence of nanoclays improves the γ-form crystals and induces the formation of α-form crystals of the PA6. It may be suggested from this that the nanosize clay platelets act as effective nucleating agents and promote the heterogeneous nucleation of PA6.[34] Shao *et al.*[48] prepared PAN nanofibers by electrospinning, before a functional layer of ZnO was deposited onto the surface of PAN nanofibers using sol-gel techniques. The ZnO coated PAN nanofibers were carbonized to form functional carbon nanofibers, and the surface structures of the functional carbon nanofibers were characterized by XRD and scanning electron microscopy (SEM). The SEM images of ZnO coated PAN nanofibers, preoxidized ZnO coated PAN nanofibers and carbon/ZnO nanofibers are shown in Fig. 3.6. SEM observations revealed that the ZnO was firmly and clearly distributed on the surface of carbon nanofibers as

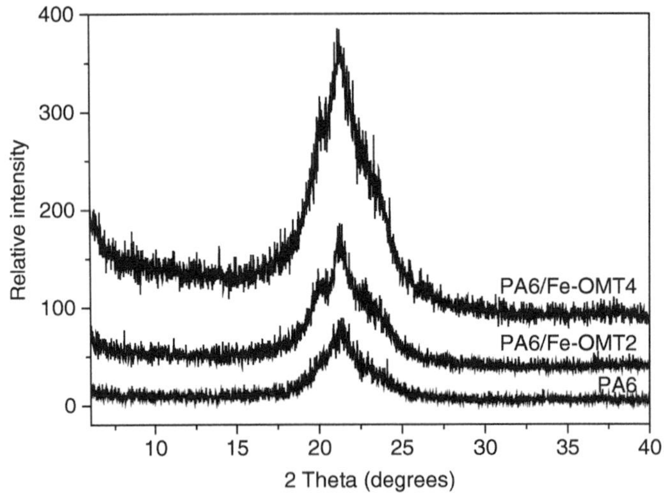

3.5 Wide-angle XRD patterns of electrospun PA6 nanofiber and PA6/Fe-OMT composite nanofibers.

nanoclusters (about 100 nm). In addition, XRD examination showed that the ZnO nanoclusters had a perfect crystalline structure, while the carbon nanofibers were of amorphous structure. The results indicate that sol-gel coating treatment has profound effects on the surface structure of carbon nanofibers.

Wu *et al.*[40, 41] studied Polyvinyl acetate (PVAc)/TiO_2 hybrid nanofibers by combining the sol-gel process with electrospinning technology, using PVAc as an organic segment and TiO_2 as an inorganic part. The SEM images indicated that the average diameter of the electrospun PVAc/TiO_2 nanofibers increased after hybridization. From the atomic force microscopy (AFM) and transmission electron microscopy (TEM) analysis, it could be concluded that the inorganic hydrolysis products lead to the formation of the particle-like surface morphology of hybrid nanofibers. It was also found from the viscosity measurement and Fourier transfer infrared (FTIR) analysis that newly associated hydrogen bonds were formed between the organic and inorganic components in hybrid nanofibers. The influences of TiO_2 content on the thermal stability and mechanical properties of PVAc/TiO_2 nanofibers are shown in Figs 3.7 and 3.8. The thermogravimetrical analysis (TGA) and tensile property analysis revealed that the thermal stability and strength were enhanced as the TiO_2 content increased. It can be reasonably concluded that hydrogen bonds or other coordinate bonds were formed in the nanofibers between the TiO_2 inorganic network and the polymeric chains.

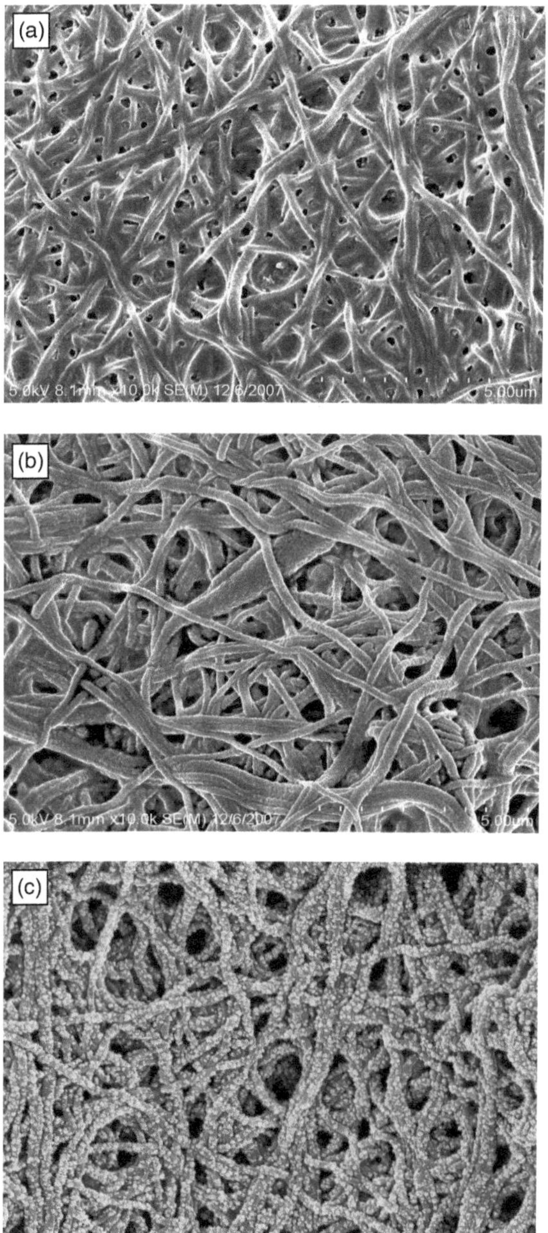

3.6 SEM images of (a) ZnO coated PAN nanofibers, (b) pre-oxidized ZnO coated PAN nanofibers and (c) carbon/ZnO nanofibers.

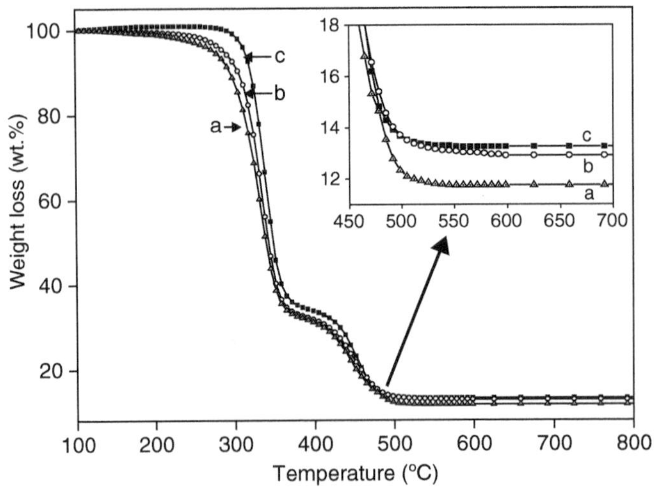

3.7 TGA curves of PVAc hybrid nanofibers with different TiO_2 content: (a) 0 wt%, (b) 0.36 wt% and (c) 0.72 wt%.

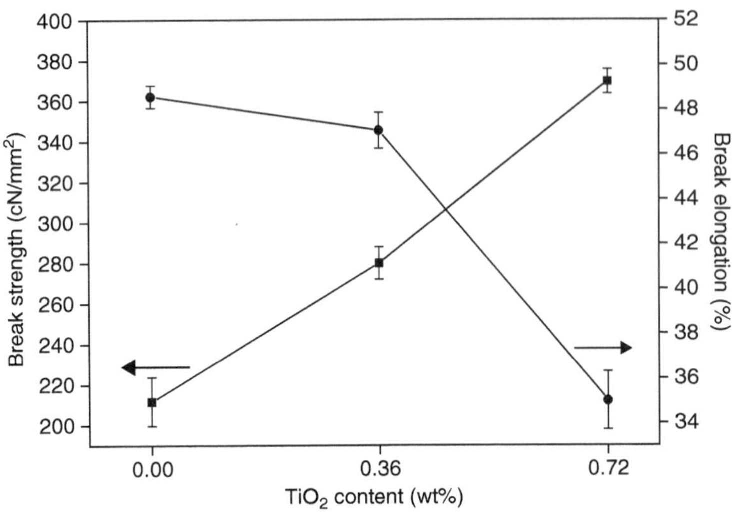

3.8 Mechanical properties of PVAc/TiO_2 nanofibers with different TiO_2 contents.

3.5.3 Polymer and inorganic salt composite nanofibers

Polymer/inorganic salts composite nanofibers have many potential applications, including in photographic and optical materials, catalysis and catalyst supports, ionic conductors, sensors and environmental protection materials.[57–61]

Ji *et al.*[62] explored the electrospinning of PAN/$ZnCl_2$ composite nanofibers and the response of these nanofibers to H_2S. The $ZnCl_2$ in composite nanofibers could react with H_2S to form ZnS crystals on the fiber surface. $ZnCl_2$-added polymer nanofibers have potential use in sensing and removing H_2S, which is colorless, flammable, heavier than air and dangerous to the environment and human health. In addition, $ZnCl_2$-added polymer nanofibers are excellent precursors for the preparation of nanofibers containing ZnS. The unique catalytic and semiconducting properties of ZnS[75–77] mean that such nanofibers have great potential for new applications in a wide variety of areas, including as catalysts and in electronic and optoelectronic nanodevices. Cai *et al.*[63] prepared PAN/$FeCl_3$ composite nanofibers using a simple compounding and electrospinning technique. The results indicated that the $FeCl_3$ loading reduced the diameter of composite nanofibers, while increasing their thermal stability properties. It was also found from shake flask tests that the reduction percentage of tested bacteria *Escherichia coli* reached 87% for the PAN composite nanofibers with 5 wt% $FeCl_3$, due to the sterilization effects of the transition metal halide $FeCl_3$.

3.6 Future trends

Electrospinning is a simple, versatile method for creating polymer-based high-functional and high-performance nanofibers that could revolutionize the world of structural materials. Other functional molecules, biomolecules or nanoparticles can easily be incorporated into electrospinning solutions to generate a combination of nanofiber functionalities.[65] However, the processing and practical application of composite nanofibers have so far been rather limited, and most have only been considered as potential applications in the future.

3.7 Acknowledgments

This chapter was financially supported by the National Natural Science Foundation of China (No. 51006046), the Natural Science Foundation of Jiangsu Province (No. BK2010140) and the Research Fund for the Doctoral Program of Higher Education of China (No. 200802951011 and No. 20090093110004).

3.8 References

1. Huang, Z.-M., Zhangb, Y.-Z., Kotakic, M. and Ramakrishna, S. (2003), A review on polymer nanofibers by electrospinning and their applications in nanocomposites, *Composites Science and Technology*, **63**, 2223–53.
2. Bognitzki, M., Czado, W., Frese, T., Schaper, A., Hellwig, M. and Steinhart, M., *et al.* (2001), Nanostructured fibers via electrospinning, *Advanced Materials*, **13**, 70–2.

3. Choi, S. W., Jo, S. M., Lee, W. S. and Kim, Y.-R. (2003), An electrospun poly(vinylidene fluoride) nanofibrous membrane and its battery applications, *Advanced Materials*, **15**, 2027–32.
4. Kim, G. M., Michler, G. H., Ania, F. and Calleja, F. J. B. (2007), Temperature dependence of polymorphism in electrospun nanofibres of PA6 and PA6/clay nanocomposite, *Polymer*, **48**, 4814–23.
5. Huang, L., McMillan, R. A., Apkarian, R. P., Pourdeyhimi, B., Conticello, V. P. and Chaikof, E. L. (2000), Generation of synthetic elastin-mimetic small diameter fibers and fiber networks, *Macromolecules*, **33**, 2989–97.
6. Bhattarai, N. and Zhang, M. Q. (2007), Controlled synthesis and structural stability of alginate-based nanofibers, *Nanotechnology*, **18**(45), 455–601(no end page)
7. Li, L. and Hsieh, Y. L. (2006), Chitosan bicomponent nanofibers and nanoporous fibers, *Carbohydrate Research*, **341**(3), 374–81.
8. Zhou, Y. S., Yang, D. Z. and Nie, J. (2006), Electrospinning of chitosan/poly(vinyl alcohol)/acrylic acid aqueous solutions, *Journal of Applied Polymer Science*, **102**(6), 5692–7.
9. Nguyen, T. T. T, Tae, B., Park, J. S. (2011), Synthesis and characterization of nanofiber webs of chitosan/poly(vinyl alcohol) blends incorporated with silver nanoparticles, *Journal of Materials Science*, **46**(20): 6528–37.
10. Lin, T., Fang, J., Wang, H. X., Cheng, T. and Wang, X. G. (2006), Using chitosan as a thickener for electrospinning dilute PVA solutions to improve fibre uniformity, *Nanotechnology*, **17**(15), 3718–23.
11. Duan, B., Dong, C. H., Yuan, X. Y. and Yao, K. D. (2004), Electrospinning of chitosan solutions in acetic acid with poly(ethylene oxide), *Journal of Biomaterials Science-Polymer Edition*, **15**(6), 797–811.
12. Subramanian, A., Lin, H.-Y., Vu, D. and Larsen, G. (2004), Synthesis and evaluation of scaffolds prepared from chitosan fibers for potential use in cartilage tissue engineering, *Biomedical Sciences Instrumentation*, **40**, 117–22.
13. Desai, K., Kit, K., Li, J. and Zivanovic, S. (2008), Morphological and surface properties of electrospun chitosan nanofibers, *Biomacromolecules*, **9**(3), 1000–6.
14. Kriegel, C., Kit, K. M., McClements, D. J. and Weiss, J. (2009), Electrospinning of chitosan-poly(ethylene oxide) blend nanofibers in the presence of micellar surfactant solutions, *Polymer*, **50**, 189–200.
15. Jung, K. H., Huh, M. W., Meng, W., Yuan, J., Hyun, S. H., Bae, J. S., *et al.* (2007), Preparation and antibacterial activity of PET/chitosan nanofibrous mats using an electrospinning technique, *Journal of Applied Polymer Science*, **105**(5), 2816–23.
16. Liu, Y., Wu, N., Wei, Q., Cai, Y. and Wei, A. (2008), Wetting behavior of electrospun poly(L-lactic acid)/poly(vinyl alcohol) composite nonwovens, *Journal of Applied Polymer Science*, **110**, 3172–7.
17. Moghe, A. K. and Gupta, B. S. (2008), Co-axial electrospinning for nanofiber structures: Preparation and applications, *Polymer Reviews*, **48**(2), 353–77.
18. Kim, M. S., Shin, K. M., Kim, S. I., Spinks, G. M. and Kim, S. J. (2008), Controlled array of ferritin in tubular nanostructure, *Macromolecular Rapid Communications*, **29**, 552–6.
19. Jo, E., Lee, S., Kim, K. T., Won, Y. S., Kim, H. S., Cho, E. C., *et al.* (2008), Core-sheath nanofibers containing colloidal arrays in the core for programmable multi-agent delivery, *Advanced Materials*, **21**(9), 968–72.
20. Kwak, G., Lee, G. H., Shim, S. and Yoon, K. B. (2008), Fabrication of light-guiding core/sheath fibers by coaxial electrospinning, *Macromolecular Rapid Communications*, **29**, 815–20.

21. Ojha, S. S., Stevens, D. R., Hoffman, T. J., Stano, K., Klossner, R., Scott, M. C., *et al.* (2008), Fabrication and characterization of electrospun chitosan nanofibers formed via templating with polyethylene oxide, *Biomacromolecules*, **9**, 2523–9.
22. Dzenis, Y. (2004), Spinning continuous fibers for nanotechnology, *Science*, **34**, 1917–9.
23. Sawicka, K. M. and Gouma, P. (2006), Electrospun composite nanofibers for functional applications, *Journal of Nanoparticle Research*, **8**, 769–81.
24. Wang, Y. Z., Li, Y. X., Yang, S. T., Zhang, G. L., An, D. M., Wang, C., *et al.* (2006), A convenient route to polyvinyl pyrrolidone/silver nanocomposite by electrospinning, *Nanotechnology*, **17**, 3304–7.
25. Li, D. and Xia, Y. N. (2004), Electrospinning of nanofibers: Reinventing the wheel? *Advanced Materials*, **16**, 1151–70.
26. Drew, C., Liu, X., Ziegler, D., Wang, X.Y., Bruno, F.F., Whitten, J., *et al.* (2003), Metal oxide-coated polymer nanofibers, *Nano Letters*, **3**(2): 143–7.
27. Lim, J. K., Moon, J. H., Yi, G. R., Heo, C. J., Yang, S. M. (2006), Fabrication of one-dimensional colloidal assemblies from electrospun nanofibers, *Langmuir*, **22**(8), 3445–9.
28. Shao, C. L., Kim, H. Y., Gong, J., Ding, B., Lee, D. R. and Park, S. J. (2003), Fiber mats of poly(vinyl alcohol)/ silica composite via electrospinning, *Materials Letters*, **57**(9–10), 1579–84.
29. Fong, H., Liu, W. D., Wang, C. S. and Vaia, R. A. (2002), Generation of electrospun fibers of nylon 6 and nylon 6-montmorillonite nanocomposite, *Polymer*, **43**, 775–80.
30. Kim, G. M., Michler, G. H., Ania, F. and Caiieja, F. J. B. (2007), Temperature dependence of polymorphism in electrospun nanofibres of PA6 and PA6-clay nanocomposite, *Polymer*, **48**, 4814–23.
31. Li, L., Bellan, L. M., Craighead, H. G. and Frey, M. W. (2006), Formation and properties of nylon-6 and nylon-6/montmorillonite composite nanofibers, *Polymer*, **47**, 6208–17.
32. Wang, M., Hsieh, A. J. and Rutledge, G. C. (2005), Electrospinning of poly(MMA-co-MAA) copolymers and their layered silicate nanocomposites for improved thermal properties, *Polymer*, **46**, 3407–18.
33. Hong, J. H., Jeong, E. H., Lee, H. S., Baik, D. H., Seo S. W. and Youk, J. H. (2005), Electrospinning of polyurethane/organically modified montmorillonite nanocomposites, *Journal of Polymer Science Part B: Polymer Physics*, **43**, 3171–7.
34. Cai, Y., Li, Q., Wei, Q., Wu, Y., Song, L., Hu, Y. (2008), Structures, thermal stability, and crystalline properties of polyamide 6/organic–modified Fe-montmorillonite composite nanofibers by electrospinning, *Journal of Materials Science*, **43**, 6132–8.
35. Cai, Y., Huang, F., Wei, Q., Song, L., Hu, Y., Ye, Y., *et al.* (2008), Structure, morphology, thermal stability and carbonization mechanism studies of electrospun PA6/ Fe-OMT nanocomposite fibers, *Polymer Degradation and Stability*, **93**, 2180–5.
36. Cai, Y., Chen, F., Wei, Q., Weng, F., Huang, F., Song, L., *et al.* (2009), Influences of organic-modified Fe-montmorillonite on structure, morphology and properties of polyacrylonitrile nanocomposite fibers, *Fibers and Polymers*, **10**, 750–5.
37. Wang, Y. Z., Yang, Q. B., Shan, G. Y., Wang, C., Du, J. S., Wang, S. G., *et al.* (2005), Preparation of silver nanoparticles dispersed in polyacrylonitrile nanofiber film spun by electrospinning, *Materials Letters*, **59**, 3046–9.
38. Chen, R., Zhao, S., Han, G. and Dong, J. (2008), Fabrication of the silver/polypyrrole/ polyacrylonitrile composite nanofibrous mats, *Materials Letters*, **62**, 4031–4.

39. Im, J. S., Kim, M. I. and Lee, Y. S. (2008), Preparation of PAN-based electrospun nanofiber webs containing TiO_2 for photocatalytic degradation, *Materials Letters*, **62**, 3652–5.
40. Wu, N., Shao, D. F., Wei, Q. F., Cai, Y. B. and Gao, W. D. (2009), Characterization of PVAc/TiO_2 hybrid nanofibers: From fibrous morphologies to molecular structures, *Journal of Applied Polymer Science*, **112**, 1481–5.
41. Wu, N., Wang, J. X., Wei, Q. F., Cai, Y. B. and Lu, B. (2009), Morphology, thermal and mechanical properties of PVAc/ TiO_2 hybrid nanofibers, *e-Polymers*, **152**, 1–8.
42. Ji, L. W. and Zhang, X. W. (2008), Ultrafine polyacrylonitrile/silica composite fibers via electrospinning, *Materials Letters*, **62**, 2161–4.
43. Jung, H. R., Ju, D. H., Lee, W. J., Zhang, X. W. and Kotek, R. (2009), Electrospun hydrophilic fumed silica/polyacrylonitrile nanofiber-based composite electrolyte membranes, *Electrochimica Acta*, **54**, 3630–7.
44. Ding, Y. H., Zhang, P., Jiang, Y., Xu, F., Yin, J. R. and Zuo, Y. D. (2009), Mechanical properties of nylon-6/SiO_2 nanofibers prepared by electrospinning, *Materials Letters*, **63**, 34–6.
45. Zhang, D., Karki, A. B., Rutman, D., Young, D. P., Wang, A., Cocke, D., *et al.* (2009), Electrospun polyacrylonitrile nanocomposite fibers reinforced with Fe_3O_4 nanoparticles: Fabrication and property analysis, *Polymer*, **50**, 4189–98.
46. Rangari, V. K., Shaik, M. Y., Mahfuz, H. and Jeelani, S. (2009), Fabrication and characterization of high strength nylon-6/Si_3N_4 polymer nanocomposite fibers, *Materials Science and Engineering A*, **500**, 92–7.
47. Zhao, N., Shi, S., Lu, G. and Wei, M. (2008), Polylactide (PLA)/layered double hydroxides composite fibers by electrospinning method, *Journal of Physics and Chemistry of Solids*, **69**, 1564–8.
48. Shao, D. F., Wei, Q. F., Zhang, L. W., Cai, Y. B. and Jiang, S. D. (2008), Surface functionalization of carbon nanofibers by sol-gel coating of zinc oxide, *Applied Surface Science*, **254**, 6543–6.
49. Tao, D., Wei, Q., Cai, Y., Xu, Q. and Sun, L. (2008), Functionalization of polyamide 6 nanofibers by electroless deposition of copper, *Journal of Coatings Technology and Research*, **5**(3), 399–403.
50. Erisken, C., Kalyon, D. M. and Wang, H. (2008), A hybrid twin screw extrusion/electrospinning method to process nanoparticle-incorporated electrospun nanofibres, *Nanotechnology*, **19**, 165302.
51. Saeed, K., Park, S.-Y., Haider, S. and Baek, J.-B. (2009), *In situ* polymerization of multi-walled carbon nanotube/nylon-6 nanocomposites and their electrospun nanofibers, *Nanoscale Research Letters*, **4**, 39–46.
52. Greiner, A. and Wendorff ,J. H. (2007), Electrospinning: A fascinating method for the preparation of ultrathin fibers, *Angewandte Chemie International Edition*, **46**, 5670–703.
53. Huang, D., Wang, Y. J., Yang, L. M. and Luo, G. S. (2006), Direct synthesis of mesoporous TiO_2 modified with phosphotungstic acid under template-free condition, *Microporous and Mesoporous Materials*, **96**, 301–6.
54. Chiang, P. C., Whang, W. T., Tsai, M. H. and Wu, S. C. (2004), Physical and mechanical properties of polyimide/titania hybrid films, *Thin Solid Films*, **59–64**, 447–8.
55. Liu, Z. F., Li, J. W., Ya, J., Xin, Y. and Jin, Z. G. (2008), Mechanism and characteristics of porous ZnO films by sol–gel method with PEG template, *Materials Letters*, **62**, 1190–3.

56. Onozuka, K., Ding, B., Tsuge, Y., Naka, T., Yamazaki, M., Sugi, S., *et al.* (2006), Electrospinning processed nanofibrous TiO_2 membranes for photovoltaic applications, *Nanotechnology*, **17**(4), 1026–31.
57. Demir, M. M., Gulgun, M. A., Menceloglu, Y. Z., Erman, B., Abramchuk, S. S., Makhaeva, E. E., *et al.* (2004), Palladium nanoparticles by electrospinning from poly(acrylonitrile-co-acrylic acid)-$PdCl_2$ solutions. relations between preparation conditions, particle size, and catalytic activity, *Macromolecules*, **37**(5), 1787–92.
58. Lu, X., Zhao, Y. and Wang, C. (2005), Fabrication of PbS nanoparticles in polymer-fiber matrices by electrospinning, *Advanced Materials*, **17**(20), 2485–8.
59. Lu, X. F., Zhao, Y. Y., Wang, C. and Wei, Y. (2005), Fabrication of CdS nanorods in PVP fiber matrices by electrospinning, *Macromolecular Rapid Communications*, **26**(16), 1325–9.
60. Bai, J., Li, Y. X., Yang, S. T., Du, J. S., Wang, S. G., Zhang, C. Q., *et al.* (2007), Synthesis of AgCl/PAN composite nanofibres using an electrospinning method, *Nanotechnology*, **18**(30), 305–601 <no end page>.
61. Thavasi, V., Singh, G. and Ramakrishna, S. (2008), Electrospun nanofibers in energy and environmental applications, *Energy & Environmental Science*, **1**, 205–21.
62. Ji, L. W., Medford, A. J. and Zhang, X. W. (2009), Electrospun polyacrylonitrile/zinc chloride composite nanofibers and their response to hydrogen sulfide, *Polymer*, **50**, 605–12.
63. Cai, Y., Wang, Q., Wei, Q., You, Q., Huang, F., Song, L., *et al.* (2010), Structure, thermal, and antibacterial properties of polyacrylonitrile/ferric chloride nanocomposite fibers by electrospinning, *International Journal of Polymer Analysis and Characterization*, **15**, 110–8.
64. Huang, Z., Zhang, Y.Z., Kotaki, M. and Ramakrishna, S. (2003), A review on polymer nanofibers by electrospinning and their applications in nanocomposites, *Composites Science and Technology*, **63**, 2223–53.
65. Chronakis, I. S. (2005), Novel nanocomposites and nanoceramics based on polymer nanofibers using electrospinning process: A review, *Journal of Materials Processing Technology*, **167**, 283–93.
66. Park, W., Kang, M., Kim, H. and Jin, H. (2007), Electrospinning of poly(ethylene oxide) with bacterial cellulose whiskers, *Macromolecular Symposia*, **249–50**, 289–94.
67. Howling, G. I., Dettmar, P. W., Goddard, P. A., Hampson, F. C., Dornish, M. and Wood, E. J. (2001), The effect of chitin and chitosan on the proliferation of human skin fibroblasts and keratinocytes in vitro, *Biomaterials*, **22**(22), 2959–66.
68. Chung, L. Y., Schmidt, R. J., Hamlyn, P. F., Sagar, B. F., Andrews, A. M. and Turner, T. D. (1994), Biocompatibility of potential wound management products: fungal mycelia as a source of chitin/chitosan and their effect on the proliferation of human F1000 fibroblasts in culture, *Journal of Biomedical Materials Research*, **28**(4), 463–9.
69. Malette, W. G. and Quigley, H. J. (1985), *US Patent*, 4, 532,134.
70. Chen, J.-P., Chang, G.-Y. and Chen, J.-K. (2008), Electrospun collagen/chitosan nanofibrous membrane as wound dressing, *Colloids and Surfaces A: Physicochemical and Engineering Aspects*, **313–4**, 183–8.
71. Zhang, Y., Huang, Z.-M., Xu, X., Lim, C. T., Ramakrishna, S. (2004), Preparation of core-shell structured PCL-r-Gelatin bi-component nanofibers by co-axial electrospinning, *Chemistry of Materials*, **16**, 3406–9.
72. Huang, Z.-M., Zhang, Y. and Ramakrishna, S. (2005), Double-layered composite nanofibers and their mechanical performance, *Journal of Polymer Science Part B: Polymer Physics*, **43**, 2852–61.

73. Shao, C., Kim, H. Y., Gong, J., Ding, B., Lee, D. R. and Park, S. J. (2003), Fiber mats of poly(vinyl alcohol)/silica composite via electrospinning, *Materials Letters*, **57**, 1579–84.
74. Larsen, G., Velarde-Ortiz, R., Minchow, K., Barrero, A. and Loscertales, I. G. (2003), A method for making inorganic and hybrid (organic/inorganic) fibers and vesicles with diameters in the submicrometer and micrometer range via sol-gel chemistry and electrically forced liquid jets, *Journal of the American Chemical Society*, **125**, 1154–5.
75. Xiong, Q., Chen, G., Acord, J. D., Liu, X., Zengel, J. J., Gutierrez, H., *et al.* (2004), Optical properties of rectangular cross-sectional ZnS nanowires, *Nano Letters*, **4**(9), 1663–8.
76. Huang, Y. and Lieber, C. M. (2004), Integrated nanoscale electronics and optoelectronics: Exploring nanoscale science and technology through semiconductor nanowires, *Pure and Applied Chemistry*, **76**(12), 2051–68.
77. Kolmakov, A. and Moskovits, M. (2004), Chemical sensing and catalysis by one-dimensional metal-oxide nanostructures, *Annual Review of Materials Research*, **34**, 151–80.

4

Carbon and polymer nanofiber reinforcements in polymer matrix composites: processing and applications

J. ZHANG, The Royal Melbourne Institute of Technology (RMIT) University, Australia and T. LIN and X. WANG, Deakin University, Australia

Abstract: Nanofibers have emerged as important fibrous materials for reinforcing or modifying polymer matrices. Carbon nanofibers (CNFs) and polymer nanofibers in composites have been critically reviewed in this chapter. The concept, methodology, potential applications, advantages and limitations of utilizing nanofibers in composite materials are discussed. The important application examples reviewed include carbon and polymer nanofibers in structural carbon fiber reinforced polymer (CFRP) matrix composites, polymer nanofiber-reinforced nanocomposites for medical use and modification of thermosetting polymer by polymer/CNF composite nanofibers. A brief account is also given of the future trends in this area.

Key words: nanofiber, composite, carbon nanofiber, electrospinning, composite nanofiber, toughening.

4.1 Introduction

The unique characteristics of nanofibers, such as large surface to volume ratio, suitability for adding surface functionalities and superior mechanical properties, have enabled them to improve the mechanical, electrical, thermal and optical properties of polymer matrix composites. Nanofibers used in composites so far include carbon nanofibers (CNFs), polymer nanofibers and polymer/nanofiller nanofibers. Among them, CNFs are the most important for incorporation into polymer matrices, which is one of the rapidly evolving areas of composite research.

CNFs have diameters varying from a few to hundreds of nanometers and lengths ranging from less than a micron to several millimeters. The internal structure of CNF is composed of stacked graphene sheets of varying shapes, commonly cones or cups.[1] A graphene layer is a hexagonal network of covalently bonded carbon atoms or a single two-dimensional layer of a

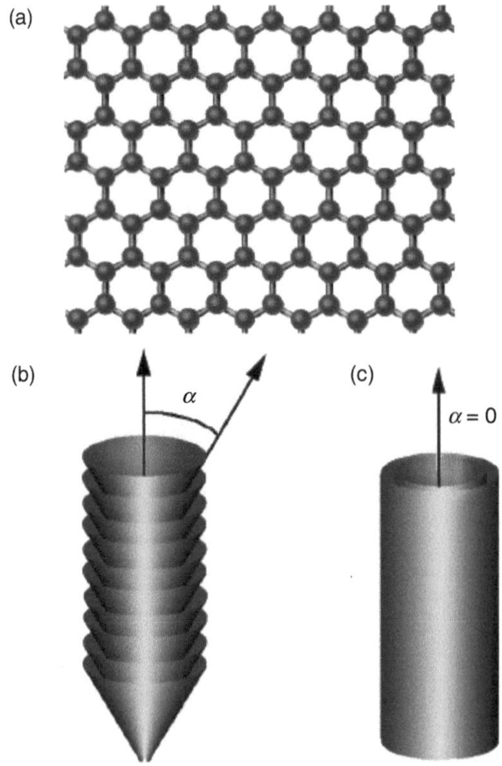

4.1 Schematic structure of carbon nanofibers and nanotubes: (a) graphene layer, (b) stacked cone (herringbone nanofiber) and (c) nanotube. (Source: Melechko *et al.*[1])

three-dimensional graphite (Fig. 4.1). In comparison with carbon nanotubes (CNTs), CNFs offer a good compromise between performance and cost. High purity CNFs can be produced at much lower cost (multiwalled CNTs are two to three times more expensive than CNFs and single-walled CNTs are even more expensive) in sufficient quantities by a catalytic chemical vapor deposition process. CNFs have diameters (70~200 nm) much lower than conventional carbon fibers (5~10 μm) but significantly larger than CNTs (1~10 nm); as a result, they are relatively easy to disperse into polymer matrices.

The introduction of CNFs to polymers promises solutions to many existing problems in composite applications. Apart from the enhancement in mechanical properties and thermal stability, CNFs can also impart composites with electrical conductivity and even provide effective shielding from electromagnetic interference (EMI). A further advantage of CNF/polymer composites is that nanoscaled CNFs improve the processability and recyclability of thermoplastic matrix nanocomposites,[2] since standard processing and recycling techniques won't break the fillers. Fiber breakage is a common problem for short-fiber reinforced thermoplastic composites. Attributed to the nanoscale of CNFs, excellent surface finish can be obtained from CNF/polymer composites. Fine structures such as fibers, films and membranes can also be reinforced by CNFs.[3, 4]

Nanofibers have been naturally used to build basic components in both animals and plants. For example, collagen nanofibers are abundant structures found in animal bodies[5] and cellulose nanofibers are commonly found in green plants. The high porosity of nanofiber structure has the advantage of allowing rapid diffusion of chemicals into the inner depths.[6] Interesting optical and mechanical properties have been demonstrated when nanofibers are used as reinforcement. Transparent composite materials prepared from nylon nanofibers with diameters of less than 200 nm and phenolic epoxy matrix resin were found by Bergshoef and Vancso.[7] When thermoplastic nanofibers were added between the carbon/epoxy lamina, significantly enhanced interlaminar fracture toughness was achieved by the structural composites.[8, 9] Core-shell nanofibers have assisted with strengthening interface adhesion between nanofiber reinforcement and dental resin matrix, resulting in improved flexural properties and significantly enhanced toughness of the advanced dental composite materials.[10] Nanofiber-reinforced composites are expanding their applications for providing unique physical and chemical properties while maintaining good mechanical performance.[11]

4.2 Composite formation with nanofibers

4.2.1 Carbon nanofiber (CNF) fabrication and preparation of CNF and polymer composites

Catalytic chemical vapor deposition technique is the dominant commercial method for the fabrication of CNFs. The procedure is usually carried out in a reactor at 500–1500°C when gas-phase molecules are decomposed and hydrocarbon or carbon monoxide is deposited on the metal (Fe, Ni, Au, Co) or metal alloy (Ni–Cu, Fe–Ni) catalysts. Separate stages such as gas decomposition, carbon deposition, fiber growth, fiber thickening, graphitization and purification are involved in the process, resulting in hollow carbon fibers with diameters dependent on the catalyst size.[12] Pyragraf® III vapor-grown carbon nanofibers (VGCNFs) manufactured by Applied Sciences Inc. (Ohio, USA) have been popularly investigated by researchers recently.[13] By introducing CNFs into polymer matrices, not only can mechanical properties of the composites be modified, but superior thermal stability, electrical conductivity and EMI shielding effectiveness to conventional conductive polymers may also be realized. Similar to CNT-filled polymer composites,

CNF distribution, dispersion, orientation and aspect ratio are important factors influencing the end properties of nanocomposites.

The main preparation methods of CNF-filled polymer composites include solution processing, melt mixing, *in situ* polymerization and heterocoagulation.[12] Dispersing CNFs in an appropriate solvent by sonication or ultrasonication, mixing the CNF suspension with the polymer in the same solvent and removing the solvent from composite are the basic steps taken for preparing CNF/polymer composites by the solution processing route. However, this is not desirable for industrial-scale application due to the large amounts of solvent involved. In contrast, melt mixing is simple and widely adopted in industrial practice. Both methods may cause reduced-aspect ratio, which consequently affects the properties of the CNF/polymer composites negatively; as a result, surface treatment and functionalization are effectively utilized to improve the dispersion quality, maintain the aspect ratio and enhance the fiber/matrix adhesion for the optimization of composite properties. Optimum tensile strength was obtained when VGCNFs were oxidized with 4% surface oxygen atoms in a CNF-filled polypropylene composite;[14] a 35% increase in tensile strength and a 140% increase in modulus was achieved in a CNF/epoxy composite by the addition of 4 wt% of highly oxidized VGCNF.[15]

In an attempt to understand the reinforcement mechanism, a number of researchers have made contributions by investigating the polymer/CNT composites. Qian *et al.* were able to observe the crazes where cracks in composite films were bridged by nanotubes; after the crazes reached the width of about 800 nm, the nanotubes fractured or became pulled out from the matrix.[16] The crazing is known as a powerful energy-absorbing and toughening mechanism and has been applied in rubber modified impact-resistant polymers. However, the crazing areas are also the weak points of the polymer composites because of the large stress concentration.[17, 18] A two-stage reinforcement mechanism was proposed by Ye *et al.*[17] The first stage is the retarded crazing extension due to the alignment of CNTs in the craze area and the slippage between CNTs, which contributes to more absorbed energy; the second stage is the CNT pull-out when the tensile stress may be fully transferred to polymer–CNT interfaces due to the superior tensile strength of CNTs over the polymer matrix. The combination of these two toughening mechanisms contributed to the higher tensile strength, modulus and toughness achieved by polymer nanocomposites.

CNFs have also been utilized to modify structural CFRP matrix composites; the resultant structure is named multiscale hybrid composites. CNFs can be introduced directly into the polymer resin and then infused into the carbon fabric, or inserted between the carbon fabrics as sprinkles or film before polymer resin is infused. The mechanical properties of the resultant multiscale composites are affected by the inclusion of CNFs. Yokozeki *et al.*

reported improved interlaminar fracture toughness as the CNF-dispersed epoxy is used.[19] The placement of CNF film between carbon/epoxy lamina showed the most significant increase in both mode I and mode II interlaminar fracture toughness. In particular, a 200% improvement in mode II interlaminar fracture toughness was achieved with the addition of CNF film. It was considered that the incorporation of CNFs into conventional CFRP composites led to increased fracture surface due to crack deflection, which may cause the enhancement of interlaminar fracture toughness. In Palmeri *et al.*'s work, only limited enhancement (10%) in tensile strength was found in CNF-modified CFRP composites[20] in comparison with unmodified composite system.

4.2.2 Polymer nanofiber preparation and formation of nanofiber composites

Polymeric fibers with diameters ranging from submicrons to nanometers (e.g., 10×10^{-3}–100×10^{-3} μm) exhibit amazing characteristics such as very high surface area to volume ratio, flexibility in surface functionalities and superior mechanical performance (e.g., stiffness and tensile strength). These unique properties made polymer nanofibers optimal candidates for many important applications such as filtration, tissue engineering scaffold, drug delivery carrier, protective clothing, electronic and semi-conductive materials and reinforced nanocomposite.[21] A number of processing techniques have been used to prepare polymer nanofibers in recent years including drawing, template synthesis, phase separation, self-assembly and electrospinning. Compared with other processing techniques, electrospinning is the only method which has potential for mass production of nanofibers from various polymers.[11] In the electrospinning process, a high voltage is applied between a negatively charged polymer fluid contained in a polymer reservoir with a capillary tip and a metallic fiber collector to generate randomly oriented or aligned nanofibers.[22] The factors influencing the production of electrospun fibers during electrospinning process include: (i) the polymer solution properties such as viscosity, elasticity, conductivity and surface tension; (ii) the process parameters such as the distance between the capillary tip and the collector, applied voltage and hydrostatic pressure in the solution reservoir and (iii) environmental effects such as temperature, humidity and air flow.[23] To fabricate polymer/nanofiller composite nanofibers, homogeneous dispersion of nanofillers in polymer solution is a prerequisite, which is usually conducted by ultrasonication.[22,24,25]

Electrospun nanofibers have been directly blended into polymer resin to reinforce matrices and also modify thermosetting polymer resin. In contrast to composites with larger reinforcements, composites reinforced with nanofibers exhibit higher toughness for equivalent volume of reinforcements.[10,26] To modify conventional CFRP composites, nanofibers have been directly electrospun onto the surface of composite prepregs or carbon fabrics, followed with laying-up, bagging and consolidation of composite components.[8,9] The thickness of the thermoplastic nanofiber interlayer can be controlled by the flow rate, electrospinning time, rotation speed of collector and applied voltage, etc., influencing the mechanical performance and thermal mechanical properties of the resultant composites.

4.3 Strengths and weaknesses of nanofibers in composites

Nanofibers have certain distinct advantages that account for their natural presence in plants and animals. The uniaxial structure can transfer forces along its longitudinal direction therefore minimizing the amount of materials required.[6] Man-made structures such as beams, cables and pillars are application examples of this concept. The additional characteristic of large surface to volume ratio provides further advantage for efficient diffusion of chemicals into nanofiber structures. As a consequence of the unique properties of nanofiber reinforcements, nanofiber composites are expected to have substantially improved mechanical properties at even a low content of reinforcement, since better load transfer at the fiber/matrix interface can be generated by virtue of the large interface area.[27] Previous research has shown enhanced tensile strength, flexural strength, elastic modulus, work-of-fracture and interlaminar fracture toughness in nanofiber-reinforced polymer matrix composites.

CNFs and CNTs are promising reinforcements and additives for use in developing nanocomposites with superior mechanical properties and additional functionalities such as conductivity, thermal stability, wear resistance and effective shielding from EMI. CNFs have been used to reinforce a variety of polymers including polypropylene, polycarbonate, nylon, polyethersulfone, polyethylene terephthalate, polyphenylene sulfide, acrylonitrile-butadiene-styrene and epoxies.[28] Conductive CNF-filled polymer composites are being used in an increasing number of applications in the aerospace, automotive and defense industries and other sectors such as construction, transportation and recreational watercraft. Some of the emerging areas in aircrafts include fire-retardant coatings and composites for fuselages, EMI shielding, lightning-strike coatings; potential applications in automotives are electrostatic painting of exterior panels, shielding of automotive electronics, improved stiffness of tires, etc.;[12] emerging defense-related applications include broadband electromagnetic shielding, conductive adhesives, enhanced thermal management and multifunctional structures.

Although an increasing number of papers have been published on nanofiber composites, the majority in current literature are related to CNF or CNT reinforcements. There are important issues that prohibit the wide adoption of polymer nanofibers as reinforcement. For instance, no sufficient quantity of unidirectional continuous nanofibers has been produced for use as reinforcement. Randomly oriented nanofibers are the most common form of polymer nanofibers collected from electrospinning. High-speed rotation of collector drum has been used to obtain aligned nanofibers; however, the degree of alignment is limited and the electrospun nanofibers are not continuous. Another factor that has to be taken into account is that the large surface to volume ratio of polymer nanofibers implies they are very light in weight. To prepare bulk composites, it is challenging to achieve sufficient weight percentage of the reinforcement. Once a moderate weight percentage of polymer nanofibers is reached, there are increased chances of getting porosity defects in nanofiber composites, attributed to the increased surface area of nanofibers.[10,26]

4.4 Applications of nanofiber composites

Conventional micro-sized synthetic fibers such as carbon, glass and aramid fibers have found their most important application as reinforcements in composites due to their high specific strength and modulus. Composite materials reinforced with these micro-sized fibers exhibit superior mechanical properties and structural performance to other engineering monolithic materials. Nanofibers prepared from suitable high-modulus polymers are promising candidates for composite reinforcement, attributing to their even higher specific strength, modulus and better molecular alignment. However, the small production scale and high equipment and maintenance costs associated with the manufacturing processes prohibit the industrial application of nanofibers in composites.

Most of the works published on nanofiber composites are related to CNF fillers. A comprehensive review of CNF-filled polymer composites was given by Tibbetts *et al.*[13] Due to the strong interfacial bonding, special attention is required when compounding the CNFs into polymer matrices. Surface treatment of CNFs such as oxidization has proven effective in preparing composites with improved tensile properties.[14,15] To disperse CNFs in thermoplastics, various methods have been used including blending in injection molder's crucible[29] and compounding in a single-screw or a twin-screw extruder. To disperse CNFs in thermosets, the high viscosity of resins usually needs to be reduced by diluent before CNF dispersion, followed by diluent removal.[30] It has been reported that, with careful preparation techniques, composite tensile strength and modulus of more than triple that of the neat resin can be realized with 15 vol% of CNF addition. Electrical conductivity can be achieved with as little as 0.5 vol% of CNFs and resistivity near 0.1 Ω cm can be possible with 15 vol% of CNFs.[31] To fabricate conductive CNF/polymer composites, there are two important objectives: first, a low percolation threshold is desirable, which is realized when a small volume fraction of nanofibers establish the conducting network; the second goal is to incorporate more CNFs to obtain more ambitious conductivity values, which is necessary for EMI shielding. Furthermore, excellent compressive strength and thermal conductivity of composites can also be realized.

Nanofibers have also been utilized to modify thermosetting resins and their carbon fiber-reinforced composites. Lin *et al.* used electrospun polyacrylonitrile (PAN) core and poly (methacrylates) (PMMA) shell nanofibers to reinforce a dental resin system, that is, 2,2-bis-[4-(methacryloxypropoxy)-phenyl]-propane (Bis-GMA).[10] Electrospun nonwoven nanofiber fabrics were dried in a vacuum oven at 60°C for 48 h before being laid up into resin to fabricate the composite specimens. The fracture surfaces of three-point flexure testing specimens showed differences among the composites reinforced with pure PMMA and PAN nanofibers and the PAN–PMMA nanofibers. Both the PAN and PAN–PMMA nanofibers kept fiber configuration in the composites; in contrast, PMMA nanofibers dissolved into the Bis-GMA resin due to the interaction between the methacryloyl groups on Bis-GMA main-chain with the methacryloyl groups on the side-chains of PMMA. For the PAN nanofiber-reinforced resins, many fiber pull-outs were observed; however, nanofibers in PAN/PMMA composites were intimately adhered to the matrix, without apparent boundaries between nanofibers and matrix. Enhanced interfacial bonding was achieved through *in situ* nanointerface in the shell structure. A schematic shown in Fig. 4.2 demonstrates the formation process of the semi-interpenetrating network (semi-IPN) in the PAN/PMMA nanofiber-reinforced composites. Consequently, the PAN–PMMA nanofiber-reinforced composites exhibited the best flexural strength, flexural modulus and work of fracture at a fiber content of 7.5 wt%. Further improvement in mechanical properties was limited by the increased number of defects in the composites with higher fiber contents.

Zhang *et al.* produced polyetherketone cardo (PEK-C) nanofibers by electrospinning and incorporated them as interlayers into carbon/epoxy composites to improve the interlaminar fracture toughness.[9] With less than 1 wt% (0.4 wt%) of nanofibers, the mode I interlaminar fracture toughness of composites increased considerably (around 70% increase). With an equal weight loading of PEK-C nanofibers on the carbon fabric, the effects of fiber diameter on the mode I delamination fracture toughness and flexure properties were examined. Finer nanofibers resulted in more stable crack propagation and better mechanical performance under flexure loading, which suggests that composites modified with finer nanofibers had better improvement in the interlaminar property without compromising

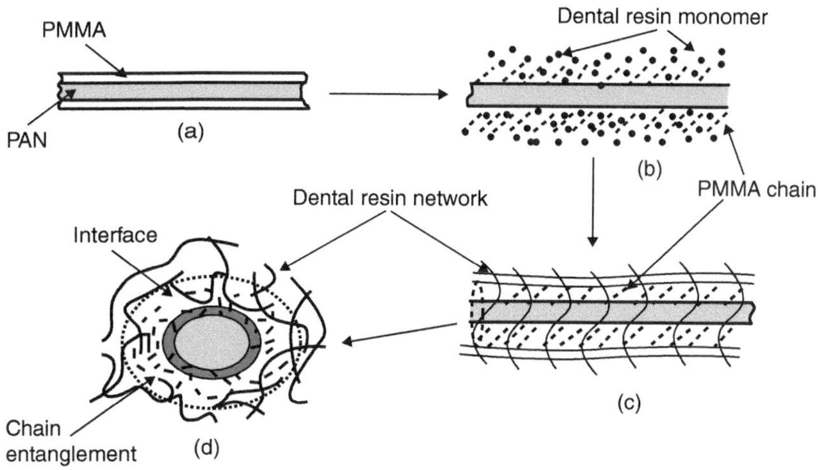

4.2 Schematic representation of the formation of semi-IPN structure in the composite: (a) PAN–PMMA nanofiber, (b) PMMA shell of nanofibers dissolved with the dental resin, (c) conform semi-IPN structure and (d) chain entanglement of the interface to form *in situ* nano-interface interaction. (Source: Lin *et al.*[10])

the in-plane performance of the toughened composites. Scanning electron microscopy (SEM) morphology on the fracture surface revealed larger particulate PEK-C–rich phases for the nanofiber interlayer with larger fiber diameter (Fig. 4.3). It was proposed that the cracks were intermittently hindered by the PEK-C-rich phases and forced to propagate between two unstable jumps. The interlayer composed of finer nanofibers gave rise to reduced distance between the ductile PEK-C-rich phases, which assisted with maintaining a more stable crack growth during delamination. The effect of interlayer thickness (with same average fiber diameter 950 nm) was demonstrated by the ascending tendency of mode I interlaminar fracture toughness and the decreasing trend of the flexural strength and elastic modulus, as the nanofiber interlayer thickened. Larger particulate PEK-C-rich phases were observed to scatter on the fracture surface of the nanofiber interlayer with larger thickness (Fig. 4.3). The glass transition temperatures of the cured epoxy did not drop until the interlayer thickness increased to 105 μm, while an average fiber diameter of 950 nm was used. The phase structure generated from phase separation was composed of fibrous PEK-C-rich macrophases and a continuous epoxy-rich macrophase, both of which consisted of microphases from secondary phase separation.

Polymer nanofiber composites also find important applications in optically functional materials. More and more transparent nanofiber-reinforced polymer composites are being produced, taking advantage of nanosized filaments that are less than one-tenth of the wavelength for light scattering. Due to its

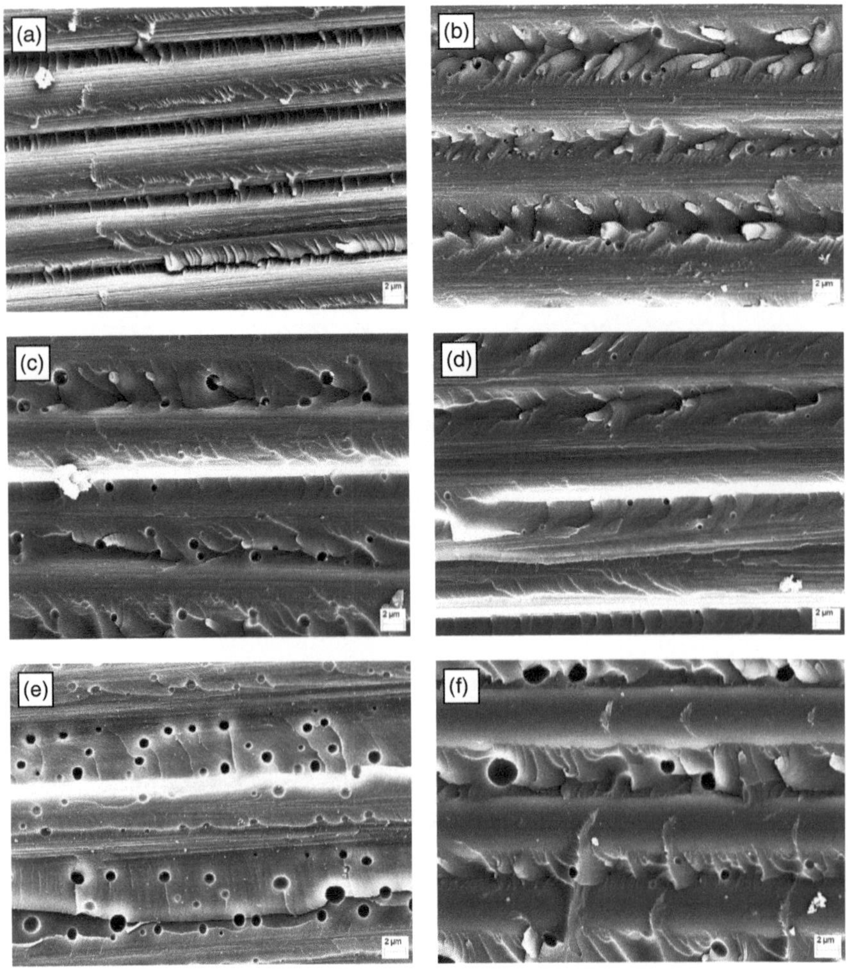

4.3 SEM images of delamination fracture surface: (a) Control composite specimen, (b) ~ (f) nanofiber modified composites, (b) Fiber diameter: 450 nm, interlayer thickness: 70 μm, (c) fiber diameter: 750 nm, interlayer thickness: 70 μm, (d) diameter: 950 nm, interlayer thickness: 40 μm, (e) fiber diameter: 950 nm, interlayer thickness: 70 μm and (f) fiber diameter: 950 nm, interlayer thickness: 105 μm. The PEK-C-rich particulate phases have been removed by solvent N,N-dimethylformamide (DMF) and are shown as dark holes in the images. (Source: Zhang *et al.*[9])

excellent mechanical performance and dimensional stability, nylon-4, 6 was electrospun to reinforce a phenolic epoxy resin by Bergshoef and Vancso.[32] Transparency was maintained with 3.9 wt% of nanofiber addition, while the strength and stiffness were substantially enhanced. Recently, Yano *et al.* have successfully prepared bacterial cellulose (BC) nanofiber-reinforced epoxy composites which are optically transparent (shown in Fig. 4.4) at a fiber content of 60 ~ 70 wt%, with a low thermal-expansion coefficient and mechanical strength five times that of engineered plastics.[33] The significant improvements are caused by the web-like network of the semi-crystalline extended chains of nanofibers.

Polymer/CNF nanofibers have also been utilized to modify a trifunctional epoxy resin by Zhang *et al.*[34] Electrospinning provides a simple but effective way to insert nanofillers, such as nanoparticles or CNTs, into polymer nanofibers and to produce continuous nanocomposite fibrils. In this work, polyetherketone cardo (PEK-C) nanofibers containing VGCNFs were electrospun (Fig. 4.5) and used for toughening and reinforcing a triglycidyl amino phenol epoxy resin. The addition of PEK-C/VGCNF nanofibers to the epoxy resin led to the distribution of VGCNFs primarily within the phase-separated PEK-C-rich domains. Synergistic effects of thermoplastic PEK-C and VGCNFs on the mechanical properties, phase morphologies and thermal stability of the resultant epoxy matrix composites were observed when the PEK-C/CNF nanofibers were blended at a low content into the epoxy resin. Strong and tough multifunctional nanocomposites were prepared with the addition of 5 wt% PEK-C/CNF nanofibers to the epoxy matrix.

Through phase separation of the PEK-C/epoxy blends at a modifier content of 5 wt%, a particulate phase structure was generated. CNFs were selectively dispersed in the particulate PEK-C-rich phases throughout the epoxy resin. Such a unique phase and CNF dispersion structure resulted in synergistic increases in flexural strength, toughness and hardness with almost negligible drop in elastic modulus (Fig. 4.6). The nanoscaled interfaces between nanofiber modifier and epoxy network led to higher flexural

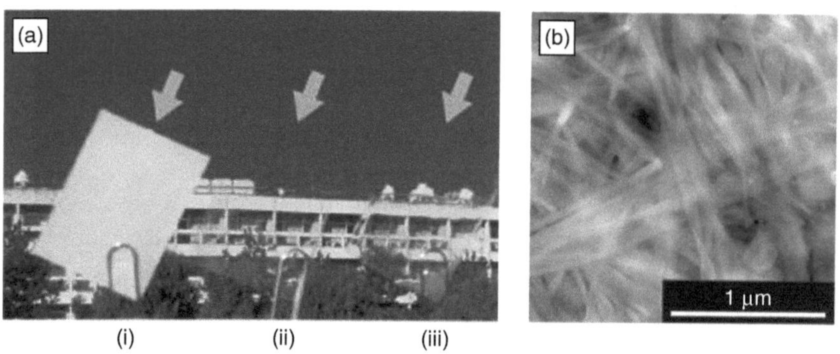

4.4 (a) Appearance of a 65 μm thick bacterial cellulose (BC) sheet: (i) with no added resin; (ii) with acrylic resin A600, 62 wt% fiber content and (iii) with epoxy resin, 65 wt% fiber content, (b) Tapping-mode atomic force microscopy (AFM) image of a BC/epoxy-resin sheet. (Source: Yano *et al.*[33])

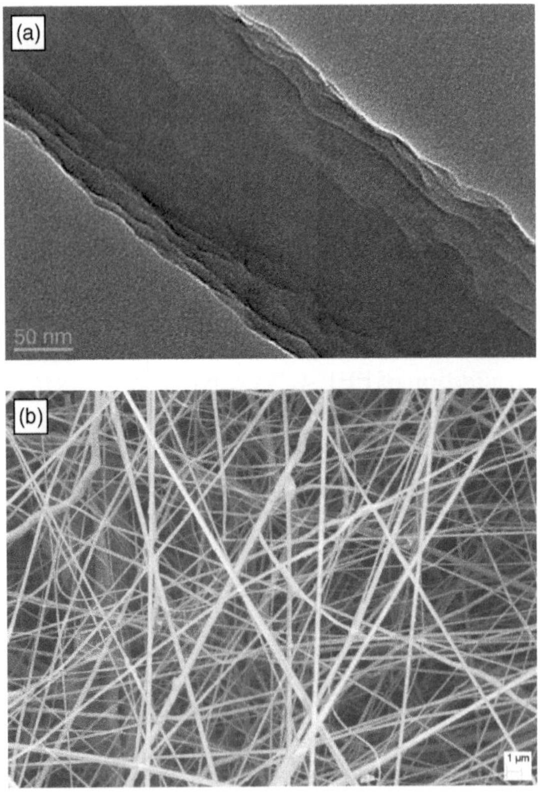

4.5 SEM and transmission electron microscopy (TEM) morphology and fiber diameter distribution of electrospun PEK-C/CNF nanofibers. (Source: Zhang *et al.*[34])

strength and glass transition temperatures of the PEK-C-rich phase, indicating stronger interfacial bonding created by using electrospun nanofibers as toughening agents and reinforcements.

Recent research efforts on producing electromagnetic absorbing composites have been demonstrated by taking advantage of the good conductivity of CNTs that increases the dielectric losses.[35] The wave absorption of materials increases as the magnetic and/or dielectric losses are promoted. $CoFe_2O_4$/multiwalled carbon nanotubes (MWCNTs) composite nanofibers have been recently successfully prepared; these are expected to exhibit improved microwave absorption due to the better match between the dielectric loss and magnetic loss, originating from the combination of paramagnetic CNT and ferromagnetic $CoFe_2O_4$ fillers incorporated.[36] Lamastra *et al.* electrospun a dispersion of MWCNTs in solution of polyvinylpyrrolidone (PVP), iron (III) nitrate nonahydrate, cobalt (II) acetate tetrahydrate, absolute ethanol and H_2O, first, before the precursor nanofibers

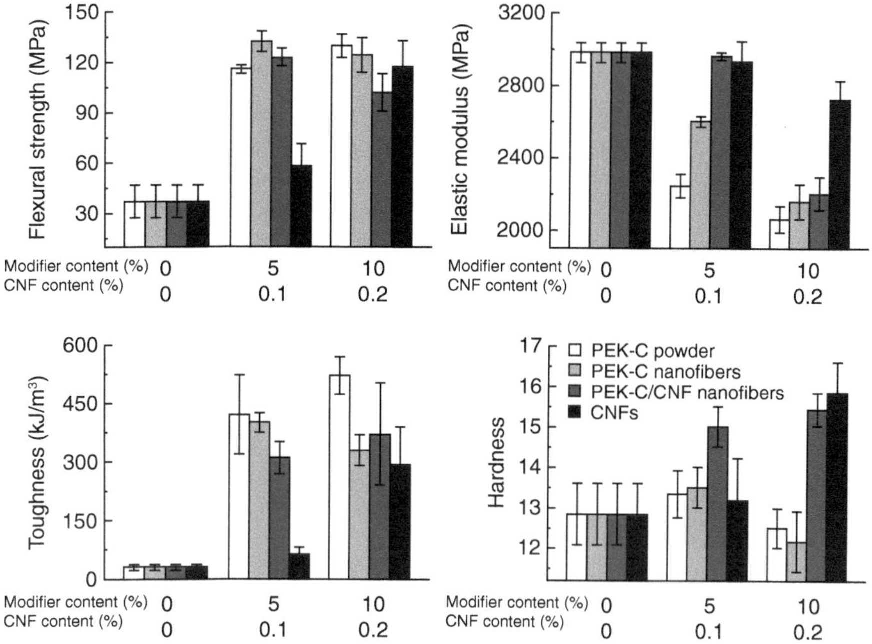

4.6 Mechanical properties of epoxies modified by PEK-C powder, PEK-C nanofibers, PEK-C/CNF nanofibers and CNFs. (Source: Zhang *et al.*[34])

were further calcined at 450°C ~ 600°C for 1 h. All the samples produced consisted of defect-free, randomly oriented fibers with uniform nanodiameter, forming a porous, highly interconnected architecture. Calcination of PVP/$CoFe_2O_4$ at 450°C and 600°C resulted in a significant reduction of fiber diameter, changing from 160 ± 30 to 60 ± 10 nm. Calcination temperature was found to affect the fiber morphology; smooth surfaces resulted from the lower calcination temperature and a necklace-like morphology was caused by a higher calcination temperature. This necklace-like structure consisted of linked crystalline particles or crystallites. The incorporation of MWCNTs into the precursor nanofibers led to their partial inclusion into the ceramic fibers. The X-ray analysis performed on calcined ceramic and composite nanofibers showed that the MWCNTs do not influence the spinel cubic phase formation of $CoFe_2O_4$.

4.5 Future trends

A number of barriers prohibit nanofibers from wide use as composite reinforcement. In the case of CNFs, surface treatment, dispersion and orientation in the matrix characterization techniques for dispersion are all challenging issues that require great attention. For polymer nanofibers, the scale-up of production, increase of nanofiber content while maintaining good mechanical properties are critical issues to consider. Nanofibers from nonsolvent-based systems will receive increasing attention and so will fiber production methods other than electrospinning. To expand the application of polymer nanofiber composites into the biomedical field, biocompatibility and long-term health hazards are among the important factors to be further examined. Theoretically, composites reinforced with nanofibers should have better mechanical performance than those reinforced by microfibers. Much work remains before the full potential of nanofiber composites can be realized.

4.6 References

1. Melechko, A. V., Merkulov, V. I., McKnight, T. E., Guillorn, M. A., Klein, K. L., Lowndes, D. H. and Simpson, M. L. (2005), Vertically aligned carbon nanofibers and related structures: Controlled synthesis and directed assembly, *Journal of Applied Physics*, **97**, 041301–39.
2. Sandler, J., Werner, P., Shaffer, M. S. P., Demchuk, V., Altstädt, V. and Windle, A. H. (2002), Carbon-nanofibre-reinforced poly(ether ether ketone) composites, *Composites Part A: Applied Science and Manufacturing*, **33**(8), 1033–9.
3. Chen, X., Yoon, K., Burger, C., Sics, I., Fang, D., Hsiao, B. S. and Chu, B. (2005), In-situ x-ray scattering studies of a unique toughening mechanism in surface-modified carbon nanofiber/UHMWPE nanocomposite films, *Macromolecules*, **38**(9), 3883–93.
4. Shi, D., Lian, J., He, P., Wang, L. M., Xiao, F., Yang, L., Schulz, M. J. and Mast, D. B. (2003), Plasma coating of carbon nanofibers for enhanced dispersion and interfacial bonding in polymer composites, *Applied Physics Letters*, **83**(25), 5301–3.
5. Chen, Z., Mo, X., He, C. and Wang, H. (2008), Intermolecular interactions in electrospun collagen-chitosan complex nanofibers, *Carbohydrate Polymers*, **72**(3), 410–8.
6. Teo, W. E. and Ramakrishna, S. (2009), Electrospun nanofibers as a platform for multifunctional, hierarchically organized nanocomposite, *Composites Science and Technology*, **69**, 1804–17.
7. Bergshoef, M. and Vancso, G. (1999), Transparent nanocomposites with ultrathin electrospun nylon – 4, 6 fiber reinforcement, *Advanced Materials*, **11**, 1362–5.
8. Li, G., Li, P., Zhang, C., Yu, Y., Liu, H., Zhang, S., Jia, X., Yang, X., Xue, Z. and Ryu, S. (2008), Inhomogeneous toughening of carbon fiber/epoxy composite using electrospun polysulfone nanofibrous membrances by in situ phase separation, *Composites Science and Technology*, **68**, 987–94.
9. Zhang, J., Lin, T. and Wang, X. (2010), Electrospun nanofibre toughened carbon/epoxy composites: Effects of polyetherketone cardo (PEK-C) nanofibre diameter and interlayer thickness, *Composites Science and Technology*, **70**, 1660–6.
10. Lin, S., Cai, Q., Ji, J., Sui, G., Yu, Y., Yang, X., Ma, Q., Wei, Y. and Deng, X. (2008), Electrospun nanofiber reinforced and toughened composites through in situ nano-interface formation, *Composites Science and Technology*, **68**(15–16), 3322–9.
11. Huang, Z. M., Zhang, Y. Z., Kotaki, M. and Ramakrishna, S. (2003), A review on polymer nanofibers by electrospinning and their applications in nanocomposites, *Composites Science and Technology*, **63**(15), 2223–53.

12. Al-Saleh, M. H. and Sundararaj, U. (2009), A review of vapour grown carbon nanofiber/polymer conductive composites, *Carbon*, **47**, 2–22.
13. Tibbetts, G. G., Lake, M. L., Strong, K. L. and Rice, B. P. (2007), A review of the fabrication and properties of vapor-grown carbon nanofiber/polymer composites, *Composites Science and Technology*, **67**, 1709–18.
14. Glasgow, D., Tibbetts, G., Matuszewski, M., Walters, K. and Lake, M. (2004), Surface treatment of carbon nanofibers for improved composite mechanical properties, 49th International SAMPE symposium proceedings, Long Beach, USA.
15. Lafdi, K. and Matzek, M. (2003), Carbon nanofibres as a nano-reinforcement for polymeric nanocomposites, 48th International SAMPE symposium proceedings, Long Beach, USA.
16. Qian, D., Dickey, E., Andrews, R. and Rantell, T. (2000), Load transfer and deformation mechanism in carbon nanotube-polystyrene composites, *Applied Physics Letters*, **76**(20), 2868–70.
17. Ye, H., Lam, H., Titchenal, N., Gogotsi, Y. and Ko, F. (2004), Reinforcement and rupture behaviour of carbon nanotubes-polymer nanofibers, *Applied Physics Letters*, **85**(10), 1775–7.
18. Coleman, J. N., Khan, U., Blau, W. J. and Gun'ko, Y. K. (2006), Small but strong: A review of the mechanical properties of carbon nanotube-polymer composites, *Carbon*, **44**, 1624–52.
19. Yokozeki, T., Iwahori, Y., Ishibashi, M., Yanagisawa, T., Imai, K., Arai, M., Takahashi, T. and Enomoto, K. (2009), Fracture toughness improvement of CFRP laminates by dispersion of cup-stacked carbon nanotubes, *Composites Science and Technology*, **69**(14), 2268–73.
20. Palmeri, M. J., Putz, K. W., Ramanathan, T. and Brinson, L. C. (2011), Multi-scale reinforcement of CFRPs using carbon nanofibers, *Composites Science and Technology*, **71**(2), 79–86.
21. Fong, H. and Reneker, D. (eds) (2001), Electrospinning and formation of nanofibers, in S. Hanser (ed.), *Structure formation in polymeric fibers*, Munich, 225–46.
22. Ko, F., Gogotsi, Y., Ali, A., Naguib, N., Ye, H., Yang, G. L., Li, C. and Willis, P. (2003), Electrospinning of continuous carbon nanotube-filled nanofiber yarns, *Advanced Materials*, **15**(14), 1161–5.
23. Aussawasathien, D. (2006), Electrospun conducting nanofiber-based materials and their characterizations: Effects of fiber characteristics on properties and applications, University of Akron, p. 14.
24. Naebe, M., Lin, T., Tian, W., Dai, L. and Wang, X. (2007), Effect of MWNT nanofillers on structures and properties of PVA electrospun nanofibres, *Nanotechnology*, **18**(225605), 1–8.
25. Naebe, M., Lin, T., Staiger, M. P., Dai, L. and Wang, X. (2008), Electrospun single-walled carbon nanotube/polyvinyl alcohol composite nanofibers: Structure – property relationships, *Nanotechnology*, **19**(305702), 1–8.
26. Fong, H. (2004), Electrospun nylon 6 nanofiber reinforced BIS-GMA/TEGDMA dental restorative composite resins, *Polymer*, **45**(7), 2427–32.
27. Schaefer, D. W. and Justice, R. S. (2007), How nano are nanocomposites?, *Macromolecules*, **40**(24), 8501–17.
28. Thostenson, E. T., Li, C. and Tsu-Wei, C. (2005), Nanocomposites in context, *Composites Science and Technology*, **65**, 491–516.
29. Tibbetts, G. and McHugh, J. (1999), Mechanical properties of vapour-grown carbon fiber composites with thermoplastic matrices, *Journal of Materials Research*, **14**(7), 1–10.

30. Rice, B., Gibson, T. and Ladfdi, K. (2004), Development of multifucntional advanced composites using a VGNF enhanced matrix, 49th International SAMPE symposium proceedings, Long Beach, USA.
31. Finegan, I. and Tibbetts, G. (2001), Electrical conductivity of vapour-grown carbon fiber/thermoplastic composites, *Journal of Materials Research*, **16**(6), 1668–74.
32. Bergshoef, M. M. and Vancso, G. J. (1999), Transparent nanocomposites with ultra-thin, electrospun nylon-4,6 fiber reinforcement, *Advanced Materials*, **11**(16), 1362–5.
33. Yano, H., Sugiyama, J., Nakagaito, A. N., Nogi, M., Matsuura, T., Hikita, M. and Handa, K. (2005), Optically transparent composites reinforced with networks of bacterial nanofibers, *Advanced Materials*, **17**(2), 153–5.
34. Zhang, J., Niu, H., Zhou, J., Wang, X. and Lin, T. (2011), Synergistic effects of PEK-C/VGCNF composite nanofibres on a trifunctional epoxy resin, *Composites Science and Technology*, **71**, 1067–71. DOI: 10.1016/j.compscitech.2011.03.008
35. Ye, Z., Li, Z., Roberts, J. A., Zhang, P., Wang, J. T. and Zhao, G. L. (2010), Electromagnetic wave absorption properties of carbon nanotubes-epoxy composites at microwave frequencies, *Journal of Applied Physics*, **108**, 054315.
36. Lamastra, F. R., Nanni, F., Camilli, L., Matassa, R., Carbone, M. and Gusmano, G. (2010), Morphology and structure of electrospun CoFe2O4/multi-wall carbon nanotubes composite nanofibers, *Chemical Engineering Journal*, **162**(1), 430–5.

5
Inorganic functional nanofibers: processing and applications

N. WU, Tianjin Polytechnic University, China and Q. WEI, Jiangnan University, China

Abstract: This chapter introduces current research into inorganic functional nanofibers including design, preparation routes, characterization methods, modifications and applications. The preparation of inorganic nanofibers with different styles and functions by combining sol-gel methods and electrospinning devices is investigated. The use of scanning electron microscopy (SEM), transmission electron microscopy (TEM), atom force microscopy (AFM), x-ray diffraction (XRD) and energy-dispersive x-ray spectroscopy (EDX) to characterize surface morphology, crystallization structure and element components are discussed. The applications of inorganic nanofibers in areas such as filtration, reaction substrates and electromagnetism are also introduced in this chapter.

Key words: inorganic nanofiber, electrospinning, preparation, applications.

5.1 Introduction

Nanostructured inorganic oxides can be used in a range of technical applications because of their stability, multifunctionality and green chemistry. The benefits of using nanostructured inorganic materials are such that they are beginning to replace conventional inorganic materials in various applications requiring nanostructures, such as catalysis, fuel cells, solar cells, hydrogen storage batteries, high mechanical strength units, biosensors, etc. When the particle size reduces to nanometer level, the large specific surface area, small size effect and quantum size effect of nanomaterials is beneficial in many ways for technological applications. For example, in the case of magnetic materials, an improved surface area allows more information to be stored in a certain area. When the dimensions of the materials become nanoscales, new properties begin to emerge, such as enhanced magnetic moments, superparamagnetism and colossal magnetoresistance. These new properties lead to the potential application of the magnetic material in permanent magnets, data storage devices, targeted drug delivery and magneto-optic memory medium.

One-dimensional (1-D) inorganic oxides in nanoscale, such as nanowires, nanorods, nanotubes, nanobelts, nanowhiskers and nanofibers have attracted much attention because of their special properties and extensive potential applications in mechanics, electronics, optics and other related areas. Researchers developed various preparation methods for 1-D inorganic nanostructures, including the preparation of nanobelts using a vapor–solid growth method,[1] vapor–liquid–solid growth for inorganic nanowires and nanorods,[2,3] solution–liquid–solid growth,[4,5] hydrothermal/solvothermal synthetic technical[6–8] and template method.[9,10] Although the research into 1-D inorganic nanostructure materials has made considerable progress using the aforementioned preparation methods, the controllability of purity, degree of uniformity, size and output have become the main stumbling blocks.

5.2 Processing of inorganic nanofibers

Electrospinning is a new method for fabricating 1-D inorganic nanofibers. Electrospinning is the process by which continuous nanofibers are produced with high surface area-to-volume ratios. In this process, a polymer solution is injected from a needle into an electric field. When the electric field overcomes the surface tension of the liquid a continuous jet is ejected, which, following subsequent solvent evaporation and bending, produces nanofibers on the collector surface. The sol-gel technique is one of the techniques that have been used to prepare ceramic nanomaterials such as ceramic nanoparticles. This technique is versatile and enables the preparation of individual as well as mixed ceramic nanoparticles at low temperatures for applications such as thin film formation, coatings, solar cells, nanoelectronics, sensors, etc.[11] The high aspect ratios achieved by electrospinning provide the ceramics with some interesting properties, making them highly suitable for specific applications. The combination of sol-gel and electrospinning techniques provides a simple, convenient and low-cost approach for fabricating inorganic nanofibers.

Generally, there are two methods of preparing inorganic nanofibers using electrospinning. First, inorganic precursors are mixed with polymers for electrospining, followed by calcination at high temperatures; second, the reaction velocity and viscosity of the inorganic precursor solution can be adjusted for direct electrospinning. It is, however, difficult to control the rheological properties of the inorganic precursor, which leads to the shape and the diameters of the inorganic nanofibers being difficult to control too. Most research therefore, is performed using the first method. The steps for the preparation of inorganic nanofibers by electrospinning are shown in Fig. 5.1.

In 2002, Shao *et al.*[12] were the first to publish the combined methods of sol-gel, electrospinning and high-temperature calcinations to fabricate

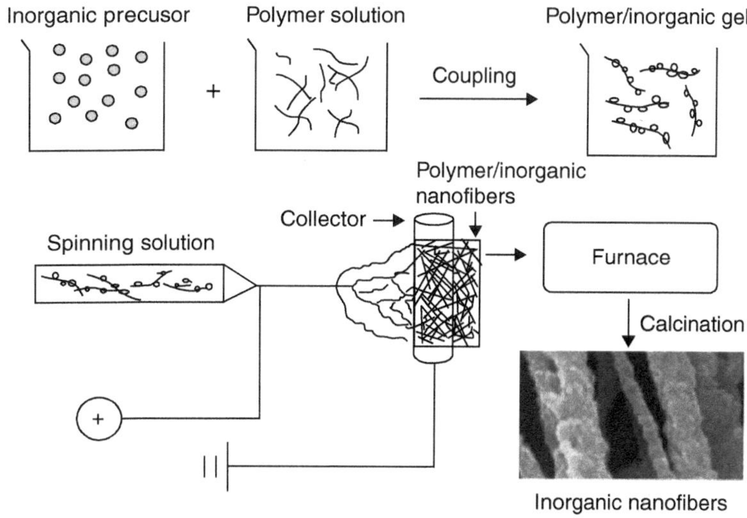

5.1 The steps in the fabrication of inorganic nanofibers by electrospinning.

inorganic nanofibers. The tetraethoxysilane sol-gel precursor was incorporated into aqueous 10 wt% polyvinylacetate (PVA) solutions and electrospun to form composite nanofibers. The as-spun composite nanofibers were subsequently calcinated at different temperatures to obtain pure SiO_2 nanofibers with different diameters and crystal forms, as illustrated in Fig. 5.2.

Larsen *et al.*[13] reported the fabrication methods of vesicles and nanofibers made from inorganic oxides and hybrid (organic/inorganic) materials in the shell and core structure and fabricated α-Al_2O_3 fibers with diameters below 200 nm (Fig. 5.3).

Li *et al.*[14] investigated the formation of ceramic hollow and porous uniaxially aligned arrays of nanofibers using electrospinning. They demonstrated that ethanol solutions containing polyvinyl pyrrolidone (PVP) and Ti $(OiPr)_4$ could be directly electrospun as uniform composite nanofibers consisting of amorphous TiO_2 and PVP. After 500°C calcination in air, these nanofibers could be converted into anatase and retain their morphology. The average diameter can be controlled under 200 nm by varying the process parameters or spinning solution parameters (Fig. 5.4).

At present, various inorganic oxide fibers have been obtained in nanoscale using high-temperature calcinations of the polymer/inorganic composite nanofibers fabricated by electrospinning. The structures of composite nanofibers and the calcination temperature have affected the crystalline phase and the surface morphology of the fibers at large. The precursor, mixed polymer and the potential applications of various inorganic nanofibers prepared

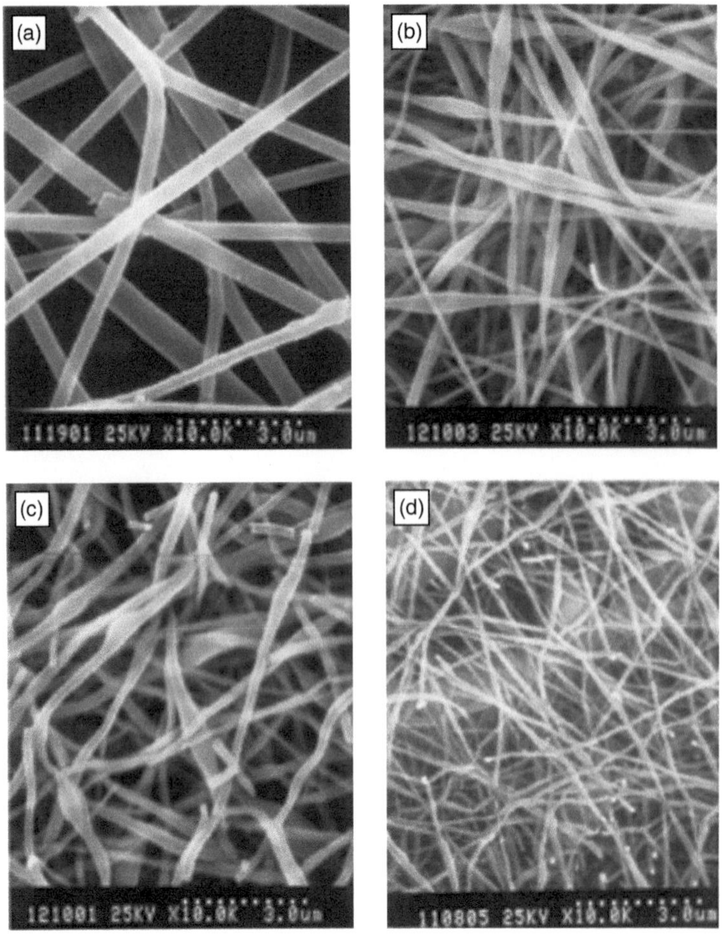

5.2 Scanning electron microscopy photographs of various fiber samples: (a) PVA/silica composite fibers with 59 wt% silica content, (b) calcination at 400°C, (c) calcination at 550°C and (d) calcination at 800°C.

by high-temperature calcinations of electrospun organic–inorganic composite nanofibers are summarized in Table 5.1.

The combination of sol-gel and electrospinning has been widely employed to fabricate inorganic nanofibers. The characterization of surface morphology, fiber structures, crystal form, components and porosity structure, etc. was achieved using various techniques including optical, electronics, microscopic, spectroscopic, thermodynamic, interface science and mechanical analyses. These characterization methods were taken from previous experience of inorganic materials analysis and some nanotechnologies. For example, the surface morphology and fiber structures were investigated

5.3 R-Al_2O_3 nanofibers. SEM image is 1.4 μm wide.

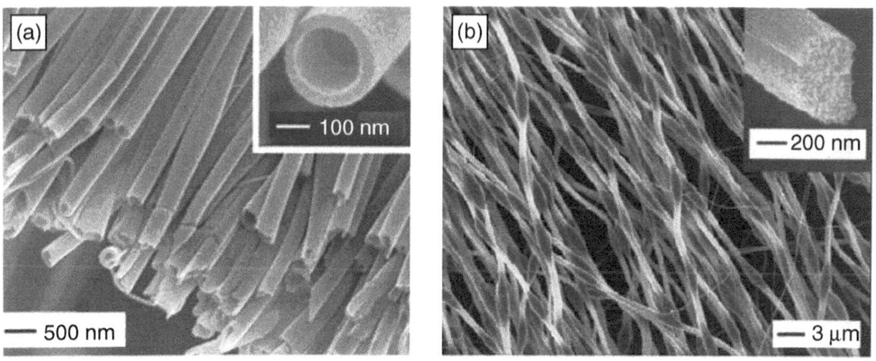

5.4 (a) Hollow and porous (b) uniaxially aligned arrays TiO_2 nanofibers by electrospinning. (Source: Li *et al.*[14])

using SEM, TEM and AFM. Fourier transform infrared spectroscopy and EDX were employed to analyze the chemical structures and components of nanofibers. The following sections discuss our own research into inorganic nanofibers as examples.

5.3 Case study: preparation, structure and properties of TiO_2 inorganic nanofibers

5.3.1 Preparation of TiO_2 nanofibers

The combination of sol-gel and electrospinning was a successful method of preparing inorganic nanofibers – for example, the preparation process of

Table 5.1 Inorganic nanofibers created using electrospinning techniques

Inorganic nanofiber	Precursor, polymer/dopant	Potential applications	References
TiO_2	1. Titanium butoxide 2. P-123 (EO-PO-EO) copolymer	Photocatalysis solar cell	15
Al_2O_3	1. Aluminum di-sec-butoxide 2. Ethanol, H_2O + HCl	Filter membranes	13
Co_3O_4	1. Cobalt acetate 2. H_2O, PVA	Biomarkers	16
Nb_2O_5	1. Niobium ethoxide 2. PVAc, ethanol, acetic acid	Superconductors	17
Fe_3O_4	1. Iron (II)chloride + iron(III) chloride 2. Graft copolymer, Polyethylene oxide (PEO) or PVA	Catalyst, drug carriers	18
ZrO_2	1. Zirconium oxychloride 2. PVA, H_2O	Electroceramics	19
GeO_2	1. Germanium isopropoxide 2. Isopropanol, H_2O,CH_3CH_2COOH 3. PVAc in acetone	Semiconductors	20
SnO_2	1. Dimethyldineodecanoate tin 2. PEO, $CHCl_3$	Sensors	21
CeO_2	1. Cerium nitrate 2. PVA, H_2O	Biomedical, catalyst	22
CuO	1. Copper acetate, PVAc, H_2O	Catalysis, semiconductors	23
ZnO	1. Zinc acetate, PVP, ethanol	Catalysis	24
SiO_2	1. Tetraethylorthosilicate, ethanol 2. HCl	Drug carriers	25
SiC	1. Poly(carbomethylsilane) (PCmS), polystyrene (PS)	Uvioresistant	26
In_2O_3	1. Indium nitrate	Gas sensors	27
$NaCo_2O_4$	1. Sodium acetate trihydrate, 2. Cobalt(III)acetate tetrahydrate, 3. PAN in DMF	Thermoelectric sensors	28

Cr_2O_3	1. Chromium chloride, PVP	Pigments	29
WO_3	1. Tungsten powder, 30% H_2O_2 2. PVP, ethanol	Field emission instruments	30
Ta_2O_5	1. Tantalum isopropoxide, ethanol 2. PVAc, DMF, acetic acid	Rectifiers, biomedical implants, dielectric	31
$NiFe_2O_4$	1. Iron(III)ethylhexanoisopropoxide 2. Nickel ethylhexanoisopropoxide 3. Acetic acid, isopropanol, PVP	Storage devices	32
$Pb(Zr0.52Ti0.48)O_3$	1. Zirconium propoxide 2. Titanium isopropoxide 3. Leadllethyl hexanoate, xylene	Piezotransducers	33
NiO/ZnO	1. Nickel acetate, zinc acetate 2. PVA, H_2O	Smart materials and semiconductors	34
$BaTiO_3$	1. Barium acetate, PVP 2. Titanium isopropoxide, ethanol 3. Acetic acid	Piezo, microwave material	35
$LiCoO_2$	1. Lithium acetylacetonate, PVP 2. Cobalt acetate, ethanol, acetic acid	Batteries, alternative energy sources	36
$LiMn_2O_4$	1. Lithium chloride, PVA 2. Manganese acetate, H_2O	Storage devices	37
$1CuAlO_2$	1. $(Cu(CH_3COO)_2{\cdot}H_2O)$, PVA 2. $(Al(NO_3)_3{\cdot}9H_2O)$,PVA	Catalysis, semiconductors	38
$Ca_{10}(PO_4)_6(OH,F)_2$	1. $Ca(NO_3)_2$, ethanol, H_2O 2. P $(C_2H_5O)_3$,	Biomedical implant artificial bone	39
$La_2Zr_2O_7$	1. Zirconium oxychloride, 2. Lanthanum nitrate, PVP	Fuel cells, catalysts	40
$CeO_2–ZrO_2$	1. Ceric nitrate, 2. Zirconyl oxy chloride, PVP	Selective membranes	41
$Al_2O_3–B_2O_3$	1. Aluminum acetate 2. Boric acid, H_2O, PVA	Reinforcement for glass fibers in electronic devices	42

(*continued*)

Table 5.1 Continued

Inorganic nanofiber	Precursor, polymer/dopant	Potential applications	References
Pd^{2+} doped TiO_2	1. $(Ti(OCH(CH_3)_2)_4$, PVP 2. $Pd(C_2H_3O_2)_2$	Gas sensors	43
Mn^{2+} doped $ZnGa_2O_4$	1. ZnO, Ga_2O_3, PVP 2. $MnCl_2{\cdot}4H_2O$	Luminescence	44
Fe^{3+} doped MnO_2	1. $MnSO_4{\cdot}H_2O$, 2. $KMnO_4$	Water filtration	45
Y^{3+} doped boehmite	1. $Al(NO_3)_3$,PEO 2. $Y(NO_3)_3$,	Catalyst	46
Eu^{3+} doped TiO_2	1. $Ti(OC_4H_9)_4)$, PVP 2. $Eu(NO_3)_3$	Photoluminescence	47
Ru doped TiO_2	1. $Ti(OCH(CH_3)_2)_4$, PVAc 2. $RuCl_3$ $3H_2O$	Solar cells	48

TiO_2 nanofibers in which the Ti $(OC_4H_9)_4$ solution was used as a molecular precursor of TiO_2. In order to control the reaction kinetics, acetylacetone was used to act as a coupling agent to moderate the reaction rate. Certain contents of acetylacetone and Ti $(OC_4H_9)_4$ were added into anhydrous ethanol and stirred (Solution A). Solution B is a mixture solution of deionized water and ethanol. Solution B was then added dropwise into Solution A and stirred at room temperature. After being mixed uniformly, stable TiO_2 sol was obtained (Solution C). Solution C was mixed with polyvinyl acetate (PVAc) solution by stirring for 6 h, to obtain PVAc/TiO_2 gel. This composite solution was placed into a syringe for electrospinning. A high voltage was applied to the spinning solution via an alligator clip attached to the syringe needle. The solution was delivered to a blunt needle via a microinfusion pump. The electrospun fibers were collected on a sheet of glass. The as-prepared composite fibers were subjected to high-temperature calcination to obtain TiO_2 nanofibers.

5.3.2 The structures of TiO_2 nanofibers

The structures of inorganic nanofibers are affected not only by the formation of composite nanofiber precursors, but also by the calcination conditions, which include calcination temperature and environment atmosphere. Generally, the effects of fiber structures on the properties of the finished materials are important for manufacturing inorganic nanofibers with desired properties. An example of inorganic nanofiber structures formed by calcination is presented in Fig. 5.5, which illustrates the observations of TiO_2 nanofibers calcinated in different temperatures. As can be seen from Fig. 5.5, the precursor nanofibers were straight and clear, as the rising of calcination temperatures from 0°C to 500°C caused the fibers to become bolder and begin to adhere together. When the calcination temperature reached 600°C, however, the nanofibers became finer and acquired more obviously porous structures.

5.3.3 The surface morphology of inorganic nanofibers

AFM has been recognized as an important analysis apparatus for characterizing the surface morphology of various materials. Wei and coworkers extensively studied various nanofiber surface morphologies using AFM, such as polymer nanofibers, organic/inorganic composite nanofibers, inorganic nanofibers, coating nanofibers, plating nanofibers and plasma-sculptured nanofibers. For instance, Fig. 5.6 revealed the surface morphologies of PVAc/TiO_2 nanofibers before and after calcinations. When the burning temperature reached 300°C, the surface of the nanofibers became even coarser.

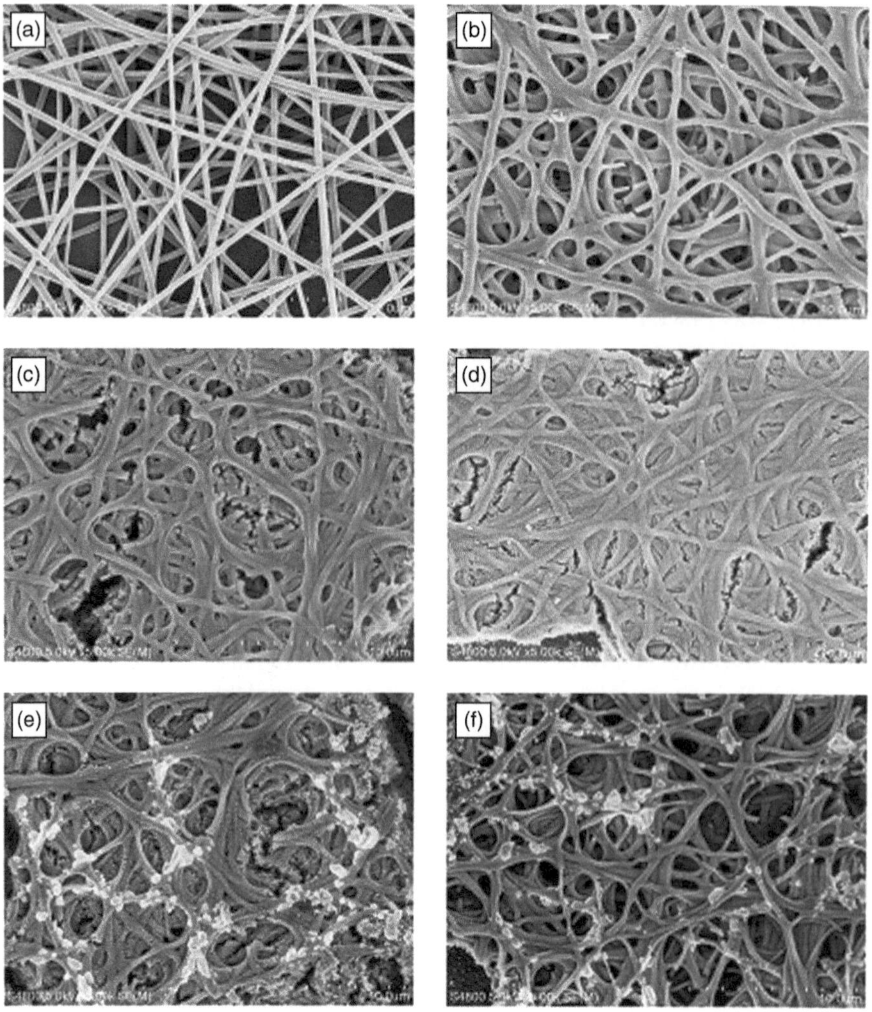

5.5 The SEM images of TiO_2 nanofiber mats calcined in different temperatures: (a) 0°C, (b) 300°C, (c) 400°C, (d) 500°C, (e) 600°C and (f) 700°C.

This may be because the nanofibers were conducting carburizing reagent at close to 300°C. After further oxidation reactions at 600°C, the surface morphology of the TiO_2 nanofiber became smoother, as illustrated in Fig. 5.6c.

5.3.4 The crystal forms of inorganic nanofibers

The crystal form is one of the most important parameters in the inorganic material research area. Some important properties – for example, optical,

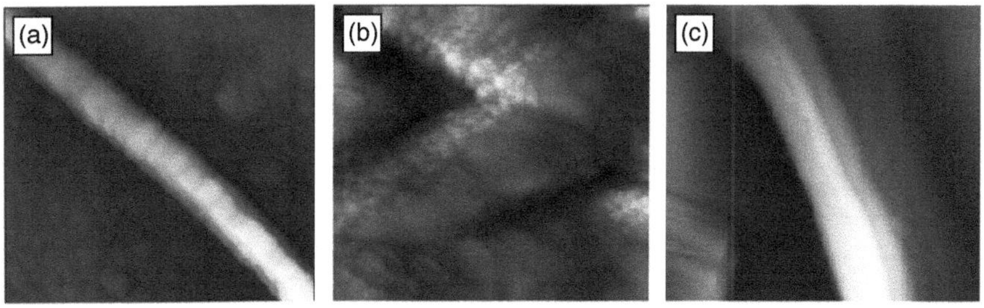

5.6 The surface morphology of PVAc/TiO_2 nanofibers calcined in different temperatures: (a) 0°C, (b) 300°C and (c) 600°C.

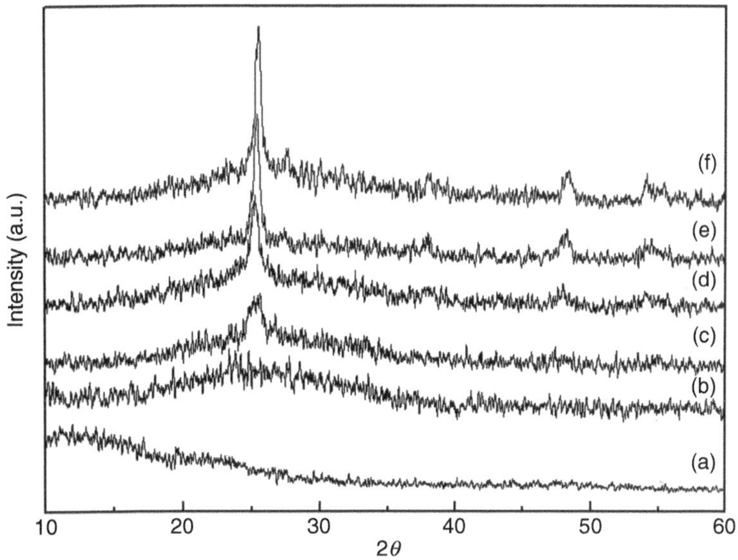

5.7 The XRD patterns of TiO_2 nanofiber mats calcined in different temperatures: (a) 0°C, (b) 300°C, (c) 400°C, (d) 500°C, (e) 600°C and (f) 700°C.

conductive electricity, photoelectric, thermal and magnetism – are significantly influenced by the crystal forms of inorganic materials. The XRD technique is usually employed to characterize the crystalline state of inorganic materials. As an example, the XRD patterns of TiO_2 nanofiber mats calcinated between 0°C and 700°C are illustrated in Fig. 5.7. Before calcination, the precursor composite nanofibers exhibited some polymer crystal peaks, the TiO_2 remained in the composite nanofiber in an amorphous state, as shown in Fig. 5.7a. At temperatures of 400°C calcination, anatase diffraction peaks mixed with some amorphous peaks and began to preliminarily form.

When temperatures reached 500°C, three anatase characteristic peaks obviously appeared in $2\theta = 25.3°, 37.8°, 48.0°$, followed by the formation of the anatase TiO_2 nanofiber. As the temperature rose to 600 and 700°C, some characteristically rutile peaks weakly appeared in 27° and 54.5°, the crystal form of TiO_2 nanofibers were then a mixture of anatase and rutile crystalline states.

5.3.5 The element components of inorganic nanofibers

During the calcination process, polymers and other organic molecules were removed from the precursor composite nanofibers. The components of nanofibers varied as the calcinate temperature increased. In order to investigate the element composition of nanofibers in different calcination phases, researchers usually use an EDX analysis method (Fig. 5.8). As in Fig. 5.7, for example, the C elements are increased first and then decreased as the temperature increases. This result is due to the composite nanofibers undergoing polymer decomposition during carbonization and creating gas in the process.

5.4 Doping inorganic nanofibers

Nanostructured metallic or semiconducting materials have attracted considerable interest due to their unique electrical conductivity, magnetism, luminescence and photocatalytic properties and the wide range of potential applications these suggest. Single metallic or semiconducting materials are, however, usually unsatisfactory in some aspect for use in industry applications. In order to improve the various application properties of these materials for better utility, researchers selected various nonmetal or metal materials to add into these semiconducting materials, such as N, C, Ag, Au, Fe^{3+}, Co^{3+}, La^{3+}, Ho^{3+} and so on. Among these dopants, metal ions doping is an extensively used way to achieve semiconductor modification. Most of the metal ion dopings are concentrated on the development of high efficiency photocatalysts, which can be formed of particles, films, fibers and so on. In recent years, metal ions doped with inorganic nanofibers have been prepared, as shown in Table 5.1.

The doping of metal ions not only changes the lattice structure, electron or hole traps, but also the surface morphology and pore structure of the inorganic nanofiber photocatalyst. As illustrated in Fig. 5.9, adding a certain amount of Fe^{3+} into the precursor solution of TiO_2 nanofibers caused filament lap joints to appear among the precursor nanofibers on the surface and, after calcination, granulate crystals formed in the interlamination of the nanofibers.

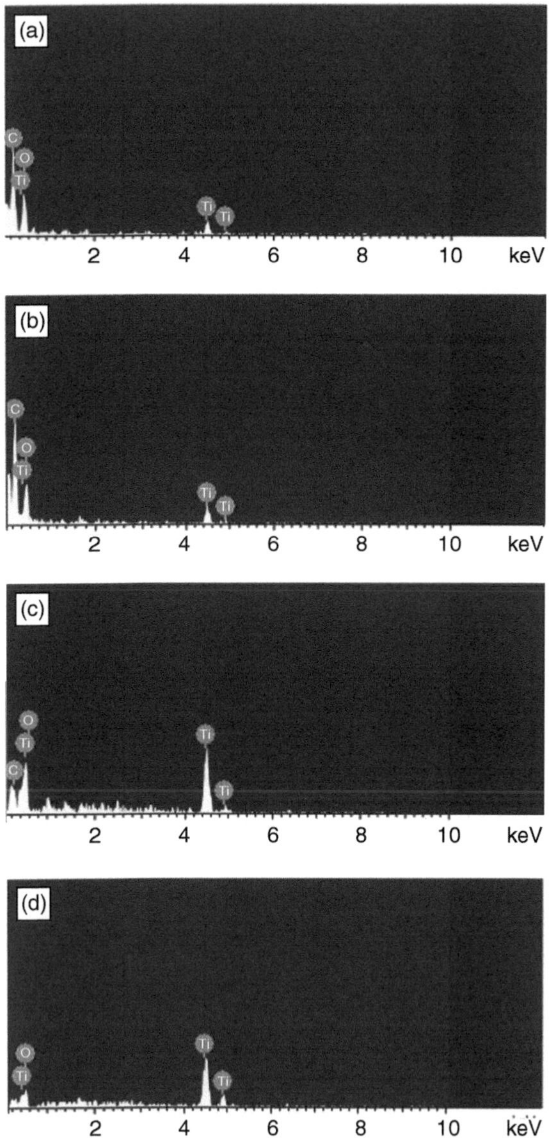

5.8 The EDX images of TiO_2 nanofiber mats calcined in different temperatures: (a) 0°C, (b) 300°C, (c) 400°C, (d) 500°C, (e) 600°C and (f) 700°C. (Source: Wu *et al.*[69])

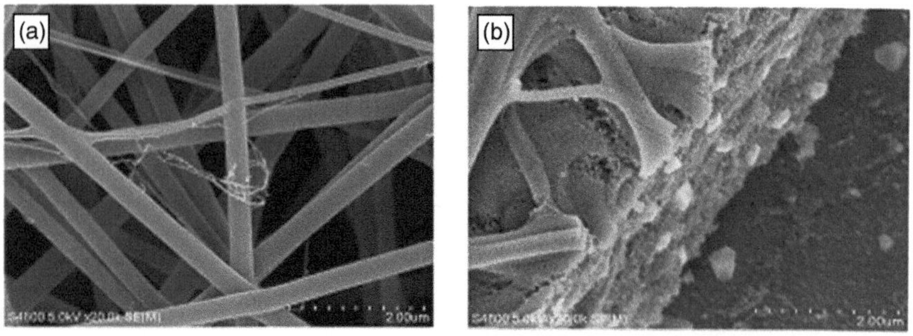

5.9 The SEM images of Fe^{3+} doped TiO_2 nanofibers (a) before and (b) after calcinations.

5.5 Applications of inorganic functional nanofibers

5.5.1 Filtration

Ceramic materials are usually used as membranes for liquid/liquid or gas/liquid separation. The nanosized membranes possess a high surface activity, which favors the selective capture of contaminants or functional parts. Due to the advantages of nanosized membranes, nanoceramic membranes could be fabricated by electrospinning and used for specialized filtration applications.

Some heavy metal contaminants could be separated from industrial effluents or rivers using ceramic membranes. Haider *et al.*[49] showed that ceramic membranes are a favorable option for the removal of heavy metal impurities such as copper and lead from waste water. This research confirmed that the nanofiber structure configuration of the filter enhanced its ability to remove heavy metal impurities.

Webster *et al.*[50] manufactured Al_2O_3 nanofibers for filtration applications. The Al_2O_3 nanofibers are used as dust attractors, viruses and bacteria killers. One of their greatest advantages is that their nanosized structure enables higher flux compared to other membranes, meaning that they do not get clogged by submicron particles. Sigmund *et al.*[51] have patented the production of a high-temperature application filter; the ceramic nanofiber mat is comprised of one or more ceramic materials selected from the group consisting of SiC, SiOC, Si3N4, SiBCN, AlN, SiAlON and SiOCN. The filter has an efficiency of greater than approximately 99.97% for 300 nm-size particles and a pressure drop of less than approximately 5.9 cm of water at an air flow velocity of approximately 5.3 cm/s, the filter has a total thickness of less than approximately 1 mm.

5.5.2 Reaction substrates

The role of inorganic nanofibers in photocatalysis and solar cell areas is to act as chemical reaction substrates. Inorganic nanofibers have attracted considerable attention in photocatalysis and solar cell applications because they possess an extraordinarily high specific surface area, which is the main factor for deciding the extent and efficiency of light-catalyzed reaction and photoelectric conversion. Some of the recent research in these areas is described here.

Dan Li's research[52] demonstrated that gold nanostructures, including particles, sheets and wires, could be selectively deposited on electrospun titania nanofibers through the photocatalytic reduction of $HAuCl_4$ in the presence of an organic capping reagent. In this research, the titania nanofiber acted as both a catalyst and a substrate at the same time. Wang *et al.*[53] fabricated BiOCl and Bi_2O_3 nanofibers using an electrospinning and calcining process. The photocatalytic test showed that BiOCl and Bi_2O_3 nanofibers possess photocatalytic properties when exposed to decomposing rhodamine B under ultraviolet light.

Im *et al.*[54] prepared a PAN nanofiber web containing TiO_2 catalysts using an electrospinning method. The prepared fibers were nanoscale-sized and contained TiO_2 catalysts on the surface of the fibers. The degradation of dye rhodamine B was more effective in the case of the suspending nanofibers when compared to the immersed nanofibers. The use of this kind of supported photocatalyst might improve the exposure to light and avoid the need for secondary operations like liquid-solid separation. In some applications, a ceramic or metal/carbon composite will often serve as a better catalyst with the carbon material because the substrate material and the metal or ceramic possess the necessary reactive function.

Li *et al.*[55] demonstrated that nickel oxides with a controlled crystalline size and fibrous shape could be prepared by the precipitation of a nickel acetate and sodium carbonate aqueous solution with the mediation of ethylene glycol. The NiO samples prepared in the solution of ethylene glycol could be spherical, fibrous or nanorod in shape, depending on the precipitation temperature. A similar metallic Ni crystalline size of about 11 nm and a fibrous structure were obtained by the reduction of these NiO samples with hydrogen at 773 K. When further exposed to the reaction of methane decomposition, the metallic Ni crystallines still exhibited similar evolutionary behavior and rapidly stabilized at the critical size of 17–18 nm. Consequently, very close catalytic activities for methane decomposition over these novel Ni catalysts were observed with carbon yields of 354–98 gC/gNi. Baker *et al.*[56] confirmed that Pt/graphite nanofiber was active in skeletal reactions of n-hexane at lower temperatures than Pt/SiO_2. The Pt/graphite nanofiber produced significantly more isomers than EUROPT-1 at a lower temperature and higher p (H_2). This research indicated that this combination may be a promising metal catalyst for low-temperature alkane isomerization reactions.

Due to its low cost and relatively high-energy conversion efficiency, the dye-sensitized solar cell (DSSC) is generally acknowledged to be the next generation of solar cell. Song *et al.*[57] managed to electrospin a TiO_2 nanofiber electrode directly onto a F doped SnO_2 (FTO) substrate from a mixture of titanium propoxide and PVAc solution. The overall conversion efficiency of the gel electrolyte system was found to be over 90% of the liquid electrolyte. ZnO nanowires created from perpendicular crystal growth on conductive glass substrates have also been suggested as a new type of DSSC electrode.[58] Li *et al.*[59] developed a vertically aligned carbon nanofiber array coated with anatase TiO_2 (CNF/TiO_2), which is an attractive possible replacement for the sintered TiO_2 nanoparticle network in the original DSSC design due to the potential for improved charge transport and reduced charge recombination. This single nanowire approach can be applied to many nanostructures involved in nanostructured DSSC and to other optoelectronic devices. Fujihara *et al.*[60] assembled DSSCs with areas of ~1 cm^2 which consisted of a thin TiO_2 nanoparticle layer and a thick TiO_2 nanorod layer as the electrode. The TiO_2 nanorods were obtained by mechanically grinding electrospun TiO_2 nanofibers. The results showed that the best-performing DSSC evaluated under AM1.5G conditions gave a current density of ~13.6 mA cm^{-2}, open circuit voltage ~0.8 V, fill factor ~51% and energy conversion efficiency ~5.8%. Zeng *et al.*[61] conducted an extensive study on poly (3-hexylthiophene) (P3HT) in combination with titanium dioxide (TiO_2) nanorods as a potential hybrid material for polymer solar cell applications. The molecular weight of P3HT, the solvent type, the hybrid compositions, the surface ligand, the film thickness and the process conditions on the TiO_2 nanorods all determined the device's performance.

5.5.3 Electromagnetism applications

Zhu *et al.*[62] described an effective method for producing superdrophilic α-Fe_2O_3 nanofibers with tunable magnetic properties. The electrospun fibers ranged from 40 ± 10 to 120 ± 20 in diameter after the decomposition of the PVA component. The a-Fe_2O_3 nanofibers can generate a superhydrophilic surface displaying the contact angle of water as 0°. By controlling the calcination temperature of electrospun composite nanofibers in the air, the magnetic property of a-Fe_2O_3 nanofibers could be tuned from superparamagnetic to ferromagnetic. Chen *et al.*[63] synthesized $(Na_{0.82}K_{0.18})_{0.5}Bi_{0.5}TiO_3$ nanofiber using sol-gel and the electrospinning process. The diameters and lengths were in the range of 150–600 nm and several hundreds of micrometers. A perovskite structure was obtained in the nanofiber. The highly effective piezoelectric coefficient was measured at 96 pm/V. This research showed that there are potential applications for $(Na_{0.82}K_{0.18})_{0.5}Bi_{0.5}TiO_3$ nanofibers in nanoscale lead-free piezoelectric devices. Wongmaneerung's research[64] demonstrated that perovskite ferroelectric $PbTiO_3$ (or PT) ceramics reinforced with SiC nanofibers could be fabricated using a simple solid-state mixed oxide and a conventional sintering method. Microstructural observations revealed that some SiC nanofibers were heterogeneously dispersed in a PT matrix and at the matrix boundaries. The grain size decreased significantly with the addition of SiC nanofiber. This research indicates that 1-D piezoelectric nanofibers may be a promising candidate for the realization of nanoscale lead-free piezoelectric devices in MEMS applications.

5.5.4 Microwave absorption

Chen *et al.*[65] fabricated the highly aligned array of Fe_3O_4/carbon nanotubes (CNT) coaxial nanofibers and doped them in poly (methyl methacrylate) matrix under a low magnetic field. The results showed that microwave absorption of the magnetically aligned composite at 8.5–12.5 GHz was evidently enhanced. Park *et al.*[66] fabricated the single-layered microwave absorbers employing both dielectric and magnetic lossy materials. CNFs were used as dielectric lossy materials and NiFe particles were used as magnetic lossy materials. An evaluation of their radar absorbing performance in the range of 2–18 GHz revealed that the mixed single-layered radar absorbing materials (RAMs) had improved absorbing characteristics with thinner matching thicknesses. The mixed RAM showed the 10 dB absorbing bandwidth of 4.0 GHz in the X-band (2.00 mm thickness) and 6.0 GHz in the Ku-band (1.49 mm thickness). Lamastra *et al.*[67] fabricated the porous, highly interconnected architecture and randomly oriented $CoFe_2O_4$/mutiwalled carbon nanotube (MWCNTs) composite nanofibers using electrospinning. The prepared nanofibrous composites are expected to exhibit improved microwave absorption due to the better match between the dielectric loss and magnetic loss. In order to achieve a higher absorbance in the 8–20 GHz range, Nanni *et al.*[68] dispersed CNFs into an epoxy resin matrix. The CNFs were chosen as a way of lowering filler content, while increasing the dielectric properties of the resulting composite. The resulting microstructures, composed of well-dispersed CNF, as well as of microaggregates, led to materials of equal filler content which displayed different dielectric properties and absorbing performances.

5.6 Future trends

This chapter discussed preparation and characterization methods of inorganic nanofibers along with some recent applications. The current research into inorganic nanofiber materials in optic, electronic and magnetism applications are inadequate and the potential for loading inorganic nanofibers on other substrates for new functional applications is in the primary phase. The potential applications of photoelectricity and electromagnetism inorganic nanostructures are still to be explored in detail and require considerable attention.

5.7 References

1. Pan, Z. W., Dai, Z. R. and Wang, Z. L. (2001), Nanobelts of semiconducting oxides, *Science*, **291**(5510), 1947–9.
2. Wagner, R. S. and Ellis, W. C. (1964), Vapor-liquid-liquid mechanism of single crystal growth, *Applied Physics Letters*, **4**(5), 89–90.
3. Johnson, J., Choi, K. P. and Yang, P. (2002), Single gallium nitride nanowire lasers, *Nature Materials*, **1**(2), 106–10.
4. Trentler, T. J., Hickman, K. M., Geol, S. C., Viano, A. M., Gibbons, P. C., and Buhro,W. E. (1995), Solution-liquid-solid growth of crystalline III-V semiconductors: An analogy to vapor-liquid-solid growth, *Science*, **270**(5243), 1791–4.
5. Rudolph, J., Reddy, K. L., Chiang, J. P., and Sharpless, K. B. (1997), Highly efficient epoxidation of olefins using aqueous H_2O_2 and catalytic methyltrioxorhenium/pyridine: Pyridine-mediated ligand acceleration, *Journal of American Chemistry Society*, **119**(26), 6189–90.
6. Dutta, P. K. and Gregg, J. R. (1992), Hydrothermal synthesis of tetragonal barium titanate, *Chemistry Materials*, **4**(4), 843–6.
7. Feng, S. H. and Xu, R. R. (2001), New materials in hydrothermal synthesis, *Accounts of Chemical Research*, **34**(3), 239–47.
8. Yao, W. T. and Yu, S. H. (2007), Recent advances in hydrothermal synthesis of low dimensional nanoarchitectures, *International Journal of Nanotechnology*, **4**(1–2), 129–62.
9. Nguyen, T. D., Mrabet, D. and Do, T. O. (2008), Controlled self-assembly of Sm_2O_3 nanoparticles into nanorods: simple and large scale synthesis using bulk Sm_2O_3 powders, *The Journal of Physics Chemistry C*, **112**(39),15226–35.
10. Jana, N. R., Gearheart, L. and Murphy, C. J. (2001), Wet chemical synthesis of silver nanorods and nanowires of controllable aspect ratio, *Chemistry Communication*, **7**, 617–8.
11. Sigmund, W., Yuh, J., Park, H. V., and Maneeratana, V. (2006), Processing and structure relationships in electrospinning of ceramic fiber systems, *Journal of American Ceramic Society*, **89**(2), 395–407.
12. Shao, C., Kim, H. and Gong, J. (2002), A novel method for making silica nanofibres by using electrospun fibres of polyvinylalcohol/silica composite as precursor, *Nanotechnology*, **13**(5), 635–7.
13. Larsen, G., Velarde-Ortiz R., Minchow, K., Barrero, A. and Loscertales, I. G. (2003), A method for making inorganic and hybrid (organic/inorganic) fibers and vesicles with diameters in the submicrometer and micrometer range via sol–gel chemistry and electrically forced liquid jets, *Journal of American Chemistry Society*, **125**(5), 1154–5.
14. Li, D., Wang, Y. and Xia, Y. (2003), Electrospinning of polymeric and ceramic nanofibers as uniaxially aligned arrays, *Nano Letters*, **3**(8), 1167–71.

15. Madhugiri, S., Sun, B., Smirniotis, P. G., Ferrarisa, J. P. and Balkus Jr., K. J. (2004), Electrospun mesoporous titanium dioxide fibers, *Microporous and Mesoporous Materials*, **69**(1–2), 77–83.
16. Guan, H. Y., Shao, C. L., Wen, S. B., Chen, B., Gong, J. and Yang, X. (2003), A novel method for preparing Co_3O_4 nanofibers by using electrospun PVA/cobalt acetate composite fibers as precursor, *Material Chemistry and Physics*, **82**(3), 1002–6.
17. Viswanathamurthi, P., Bhattarai, N., Kim, H. Y., Lee, D. R., Kim, S. R. and Morrise, M. A. (2003), Preparation and morphology of niobium oxide fibers by electrospinning, *Chemistry Physics Letters*, **374**(1–2), 79–84.
18. Wang, A., Singh, H., Hatton, T. A., Rutledge, G. C. (2004), Field-responsive superparamagnetic composite nanofibers by electrospinning, *Polymer*, **45**(16), 5505–14.
19. Shao, C. L., Guan, H. Y., Liu, Y. C., Yu, Na. and Yang, X. (2004), A novel method for making ZrO_2 nanofibres via an electrospinning technique, *Journal of Crystal Growth*, **267**(1–2), 380–4.
20. Viswanathamurthi, P., Bhattarai, N. and Kim, H. Y. (2004), GeO_2 fibers: Preparation, morphology and photoluminescence property, *Journal of Chemistry and Physics*, **121**(1), 441–5.
21. Wang, Y., Aponte, M., Leon, N., Ramos, I. Furlan, R., Evoy, S. and Santiago-Avilés, J. J. (2004), Synthesis and characterization of tin oxide microfibres electrospun from a simple precursor solution, *Semiconductor Science and Technology*, **19**(8), 1057–60.
22. Yang, X. G., Shao, C. L., Liu, Y. C., Mu, R. and Guan, H. (2005), Nanofibers of CeO_2 via an electrospinning technique, *Thin Solid Films*, **478**(1–2), 228–31.
23. Wu, H., Lin, D., and Pana, W, (2006), Fabrication, assembly, and electrical characterization of CuO nanofibers, *Applied Physics Letters*, **89**(13), 133123–5.
24. Bender, E. T., Katta, P., Lotus, A., Park, S. J., Chase, G. G. and Ramsier, R. D. (2006), Identification of CO_2 sequestered in electrospun metal oxide nanofibers, *Chemical Physics Letters*, **423**(86), 302–6.
25. Zhang, G., Kataphinan, W., Teye-Mensah, R., Katta, P., Khatria, L., Evans, E. A., Chase, G. G., Ramsier, R. D. and Reneker, D. H. (2005), Electrospun nanofibers for potential space-based applications, *Materials Science and Engineering, B*, **116**(3), 353–8.
26. Eick, B. M. and Youngblood, J. P. (2009), SiC nanofibers by pyrolysis of electrospun preceramic polymers, *Journal of Materials Science*, **44**(1),160–5.
27. Zhang, Y., Li, J., Li, Q., Zhu, L., Liu, X., Zhong, X., Meng, J. and Cao, X. (2007), Preparation of In_2O_3 ceramic nanofibers by electrospinning and their optical properties, *Scripta Materialia*, **56**(5), 409–12.
28. Maensiri, S. and Nuansing, W. (2006), Thermoelectric oxide $NaCo_2O_4$ nanofibers fabricated by electrospinning, *Materials Chemistry and Physics*, **99**(1), 104–8.
29. Hao, R., Yuan, J., and Peng, Q. (2006), Fabrication and sensing behavior of Cr_2O_3 nanofibers via in situ gelation and electrospinning, *Chemical Letters*, **35**(11), 1248–9.
30. Lu, X., Liu, X., Zhang, W., Wang, C. and Wei, Y. (2006), Large-scale synthesis of tungsten oxide nanofibers by electrospinning, *Journal of Colloid and Interface Science*, **298**(2), 996–9.
31. Dharmaraj, N., Park, H. C., Kim, C. H., Viswanathamurthic, P. and Kim, H. Y. (2006), Nanometer sized tantalum pentoxide fibers prepared by electrospinning, *Materials Research Bulletin*, **41**(3), 612–9.
32. Li, D., Herricks, T., and Xia, Y. N. (2003), Magnetic nanofibers of nickel ferrite prepared by electrospinning, *Applied Physics Letters*, **83**(22), 4586–8.

33. Wang, Y. and Santiago-Aviles, J. J. (2004), Synthesis of lead zirconate titanate nanofibres and the Fourier-transform infrared characterization of their metallo-organic decomposition process, *Nanotechnology*, **15**(1), 32–6.
34. Shao, C. L., Yang, X. H., Guan, H. Y., Liu, Y. and Gong, J. (2004), Electrospun nanofibers of NiO/ZnO composite, *Inorganic Chemistry Communication*, **7**(5), 625–7.
35. Yuh, J., Nino, J. C. and Sigmund, W. (2005), Synthesis of barium titanate ($BaTiO_3$) nanofibers via electrospinning, *Materials Letters*, **59**(28), 3645–7.
36. Fu, Z., Ma, J. and Qin, Q. (2005), Nanostructured $LiCoO_2$ and $LiMn_2O_4$ fibers fabricated by a high frequency electrospinning, *Solid State Ionics*, **176**(17–18), 1635–40.
37. Yu, N., Shao, C., Liu, Y., Guan, H. and Yang, X. (2005), Nanofibers of $LiMn_2O_4$ by electrospinning, *Journal of Colloid and Interface Science*, **285**(1), 163–6.
38. Zhao, S., Li, M., Liu, X., and Han, G. (2009), Synthesis of $CuAlO_2$ nanofibrous mats by electrospinning, *Materials Chemistry and Physics*, **116**(2–3), 615–18.
39. Kim, H. W. and Kim, H. E. (2006), Nanofiber generation of hydroxyapatite and fluorhydroxyapatite bioceramics, *Journal of Biomedical Materials Research B Applied Biomaterial*, **77**(2), 323–8.
40. Li, J. Y., Dai, H., Li, Q., Zhong, X. H., Ma, X. F., Meng, J. and Cao, X. Q. (2006), Lanthanum zirconate nanofibers with high sintering-resistance, *Materials Science and Engineering: B*, **133**(1–3), 209–12.
41. Zhang, Y., Li, J. and Li, Q., (2007), Preparation of CeO_2–ZrO_2 ceramic fibers by electrospinning, *Journal of Colloid Interface Science*, **307**(2), 567–71.
42. Dai, H. Q., Gong, J., Kim, H., and Lee, D. (2002), A novel method for preparing ultrafine alumina-borate oxide fibres via an electrospinning technique, *Nanotechnology*, **13**(5), 674–8.
43. Moon, J., Park, J., Lee, S., Zyung, T. and Kim, I. (2010), Pd-doped TiO_2 nanofiber networks for gas sensor applications, *Sensors and Actuators B*, **149**(1), 301–5.
44. Wang, L., Hou, Z. and Quan, Z. (2009), Preparation and luminescence properties of Mn^{2+}-doped $ZnGa_2O_4$ nanofibers via electrospinning process, *Materials Research Bulletin*, **44**(10), 1978–83.
45. Chen, H., Jin, L., Wang, M., Xiao, D. and Reisner, D. (2007). Reducing dissolution of MnO_2 nanofibers by doping with ferric ion. *Science and Technology of Advanced Materials*. **8**(1–2), 63–6.
46. Zhao, Y. and Frost, R. L. (2008), Synthesis and surface characterization of yttrium doped boehmite nanofibers, *Journal of Colloid and Interface Science*, **326**(1), 289–99.
47. Zhao, J., Jia, C., Duan, H., Sun, Z., Wang, X. and Xie, E. (2008), Structural and photoluminescence properties of europium-doped titania nanofibers prepared by electrospinning method, *Journal of Alloys and Compounds*, **455**(1–2), 497–500.
48. Viswanathamurthi, P., Bhattarai, N. and Kim, C. K. (2004), Ruthenium doped TiO_2 fibers by electrospinning, *Inorganic Chemistry Communications*, **7**(5), 679–82.
49. Haider, S., Waheed, A and Masry, A. (2010), Removing heavy metals from water, plastic research online, *Society of Plastics Engineers*. DOI: 10.1002/spepro.003042.
50. Webster, T. J., Hellenmeyer, E. L. and Price, R. L. (2005), Increased osteoblast functions on theta + delta nanofiber alumina, *Biomaterials*, **26**(9), 953–60.
51. Sigmund, W. M., Maneeratana, V. and Colombo, P. (2010), Ceramic nanofibers for liquid or gas filtration and other high temperature applications, NO: US 2010/0139226A1.
52. Li, D. and McCann, J. T. (2004), Photocatalytic deposition of gold nanoparticles on electrospun nanofibers of titania, *Chemical Physics Letters*, **394**(4–6), 387–91.

53. Wang, C., Shao, C., Liu, Y. and Zhang, L. (2008), Photocatalytic properties BiOCl and Bi_2O_3 nanofibers prepared by electrospinning, *Scripta Materialia*, **59**(3), 332–5.
54. Im, J. S., Kim, M. I. and Lee, Y. S. (2008), Preparation of PAN-based electrospun nanofiber webs containing TiO_2 for photocatalytic degradation, *Materials Letters*, **62**(21–2), 3652–5.
55. Li, Y., Zhang, B. and Xie, X. (2006), Novel Ni catalysts for methane decomposition to hydrogen and carbon nanofibers, *Journal of Catalysis*, **238**(2), 412–24.
56. Baker, R. T. K., Laubernds, K. and Wootsch, A. (2000), Pt/Graphite nanofiber catalyst in n-hexane test reaction, *Journal of Catalysis*, **193**(1), 165–7.
57. Song, M. Y., Kim, D. K. and Ihn, K. J. (2005), New application of electrospun TiO_2 electrode to solid-state dye-sensitized solar cells, *Synthetic Metals*, **153**(1–3), 77–80.
58. Law, M., Greene, L. E., Johnson, J. C., Saykally, R. and Yang, P. (2005), Nanowire dye-sensitized solar cells, *Nature Materials*, **4**, 455–9.
59. Li, Z., Rochford, C. and Baca, F. J. (2010), Investigation into photoconductivity in single CNF/TiO_2-dye core–shell nanowire devices, *Nanoscale Research Letters*, **5**(9), 1480–6.
60. Fujihara, K., Kumar, A., Jose, R., Ramakrishna, S. and Uchida, S. (2007), Spray deposition of electrospun TiO_2 nanorods for dye-sensitized solar, *Nanotechnology*, **18**, 365709–14.
61. Zeng, T., Lo, H., Chang, C., Lin, Y. Y., Chen, C. W. and Su,W. F. (2009), Hybrid poly(3-hexylthiophene)/titanium dioxide nanorods material for solar cell applications, *Solar Energy Materials & Solar Cells*, **93**, 952–7.
62. Zhu, Y., Zhang, J. and Zhai, J. (2006), Preparation of superhydrophilic a-Fe_2O_3 nanofibers with tunable magnetic properties, *Thin Solid Films*, **510**(1–2), 271–4.
63. Chen, Y. Q., Zheng, X. J. and Feng, X. (2010), Fabrication of lead-free $(Na_{0.82}K_{0.18})_{0.5}Bi_{0.5}TiO_3$ piezoelectric nanofiber by electrospinning, *Materials Research Bulletin*, **45**(6), 717–21.
64. Wongmaneerung, R., Singjai, P., Yimnirun, R. and Ananta, S. (2009), Effects of SiC nanofibers addition on microstructure and dielectric properties of lead titanate ceramics, *Journal of Alloys and Compounds*, **475**(1–2), 456–62.
65. Chen, Q., Chen, Y. and Cao, F. (2007), Alignment of Fe_3O_4-carbon coaxial nanofibres in a polymer for improving microwave absorption, *Proceedings of the 7th IEEE International Conference on Nanotechnology*, August 2–5, Hong Kong.
66. Park, K. Y., Han, J. H., Lee, S. B., Kim, J. B., Yi, J. W. and Lee, S. K. (2009), Fabrication and electromagnetic characteristics of microwave absorbers containing carbon nanofibers and NiFe particles, *Composites Science and Technology*, **69**, 1271–8.
67. Lamastra, F. R., Nannia, F., Camilli, L., Matassac, R., Carbone, M. and Gusmano, G. (2010), Morphology and structure of electrospun $CoFe_2O_4$/multi-wall carbon nanotubes composite nanofibers, *Chemical Engineering Journal*, **162**, 430–5.
68. Nanni, F., Travaglia, P. and Valentini, M. (2009), Effect of carbon nanofibres dispersion on the microwave absorbing properties of CNF/epoxy composites, *Composites Science and Technology*, **69**, 485–90.
69. Wu, N., Wei, Q., and Tao, D. (2009),The preparation, characterization and UV resistant property of TiO_2 nanofibers. *Journal of Donghua University (Natural Science)*,**135**(5), 500–5.

6

Surface functionalization of polymer nanofibers

F. HUANG, Q. WEI and Y. CAI,
Jiangnan University, China

Abstract: For a variety of applications, it is desirable to produce nanofibers with well-defined surface properties. Fibers with specific surface properties are also of interest in many technical applications as the surface features affect wettability, electrical conductivity, optical property and biocompatibility. This chapter reviews the surface functionalization of polymer nanofibers by various technologies. The technologies used for surface modification of polymer nanofibers include physical, chemical and biological, or a combination of these technologies. The ability to modify the surface of nanofibers would expand the applications of nanofibers, based on changes to both their physical and chemical properties.

Key words: surface, modification, function, nanofiber, interface.

6.1 Introduction

Polymer nanofibers are fibers with very small fiber diameters of less than 1000 nm. Polymer nanofibers posses many unique properties since these fibers have a very large surface area per unit mass, as well as small pore size.

Nanofibers have been extensively studied in recent decades (Reneker & Chun, 1996). They can be produced using various techniques. Among all these techniques, electrospinning has been widely used to produce polymer nanofibers. Electrospinning is the technique which uses a strong electric field to produce polymer nanofibers from polymer solution or polymer melt. If electrostatic forces overcome the surface tension of a polymer solution a charged jet is ejected and moves toward a grounded electrode. The electrospun nanofibers can be collected on a substrate located on the counter electrode (Baumgarten, 1971).

Polymer nanofibers with specific surface properties are of importance in many technical applications as the surface features affect wettability, adsorption, electrical conductivity, optical properties and biocompatibility. Modification of polymer nanofibers improves the surface properties of these materials in order to meet some special applications. Surface functionalization of polymer nanofibers has attracted a great deal of attention in recent decades (Renuga *et al.*, 2007). Various technologies have been employed to modify nanofiber surfaces, including physical, chemical and biological technologies.

6.2 Surface functionalization using physical technologies

6.2.1 Plasma treatment

Plasma treatment refers to the use of an ionized gas at either low pressure or atmospheric pressure to modify surfaces of various materials including textiles (Hocker, 2002). Plasma treatment provides a clean and environmentally friendly approach to modifying the surface of polymer nanofibers without affecting the bulk properties of the treated nanofibers (Wei *et al.*, 2005).

Wei *et al.* (2005) reported the use of cold gas to modify the surface of polyamide 6 (PA6) nanofibers prepared by electrospinning. Cold plasma utilizes energy in the form of an electric field to dissociate a process gas under vacuum into a variety of excited species. The free radicals and electrons created in the plasma can be used to modify the material surface physically and chemically. The atomic force microscopy (AFM) scans indicate a significant change in the morphology of the PA6 nanofiber surface before and after plasma treatment. A Philips Environmental Scanning Electron Microscopy (ESEM) was also used to study the wetting behavior of the fibers. In the ESEM, relative humidity was raised to 100% to facilitate the water condensation onto the PA6 nanofiber surfaces for wetting observation. The ESEM observation revealed that the plasma treatment significantly altered the surface wettability of the PA6 nanofibers.

Yoon *et al.* (2009) prepared nano/microfibrous cellulose triacetate (CTA) mats by electrospinning a fixed concentration of CTA with different methylene chloride (MC)/ethanol (EtOH) ratios and with various concentrations of CIA at a fixed MC/EtOH 80/20 (v/v) ratio. It was found that all of the electrospun CTA mats had a high water contact angle compared to the CTA cast films. The experimental results revealed that the CTA mat without plasma treatment had good surface roughness and electrospinning processability at a solvent composition of 80/20 (v/v) and 5 wt% CTA concentration and its water contact angle was 142°. The CTA fibrous mat electrospun from the 5 wt% solution of CTA was treated with a carbon tetrafluoride (CF_4) plasma for various times to further improve its hydrophobicity. Superhydrophobicity could be obtained after the CF_4 plasma treatment. The water contact angle of the CTA mat reached as high as 153° after plasma treatment for 60 s.

Huang *et al.* (2010) reported the plasma-induced grafting of acrylic acid onto the electrospun poly(vinylidene fluoride) (PVDF) nanofiber

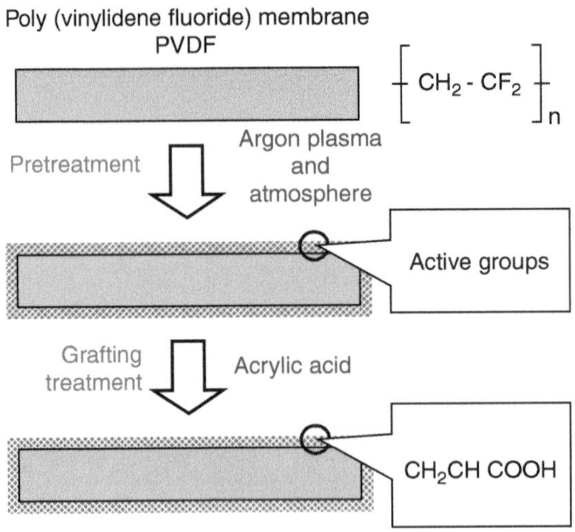

6.1 Schematic diagram of pretreatment and grafting on PVDF nanofiber membrane.

membranes to improve their wettability. The PVDF membranes with different structures were first prepared by electrospinning for the investigation into the relationship between dynamic wettability and fibrous structures. Contact angles were employed to characterize the wettability of the electrospun nanofibers. The process of plasma-induced grafting with acrylic acid on the electrospun PVDF membrane is described in Fig. 6.1. It was found that the average diameter of the PVDF nanofibers decreased with increasing applied voltage. The drawing rate of the nanofibers could be enhanced by increasing the electrostatic forces brought by the electrostatic voltage. The average diameters of the grafted PVDF nanofibers electrospun with different static voltages are illustrated in Fig. 6.2.

The wetting behavior of the electrospun PVDF nanofibers was also investigated in the work. It was observed that the surface contact angle of the electrospun PVDF nanofiber membrane was about 91.2°, in agreement with the strong hydrophobicity of PVDF material to water. A significant decrease in the contact angle on the grafted PVDF membrane was also observed, which was attributed to the grafting of hydrophilic radical, –COOH. However, it is proved from the observations that the surface chemical feature was not the only factor influencing the surface contact angle. The membranes obtained with different electrospinning voltages and grafted under the same conditions presented varied contact angles, as illustrated in Fig. 6.3. This observation revealed the effect of the specific surface area and porosity on the contact angle on the porous media.

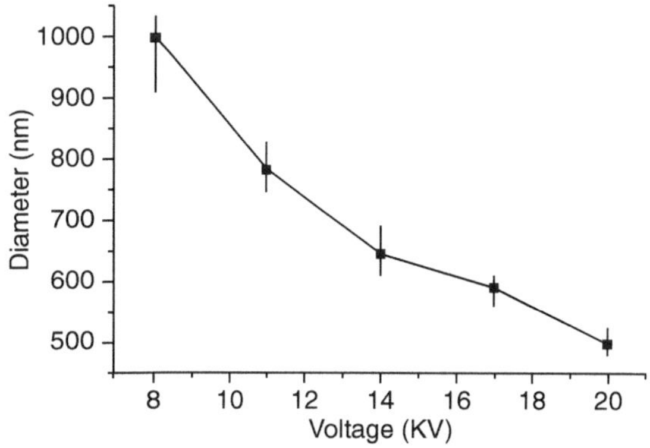

6.2 Effect of electospinning voltage on the fiber diameter.

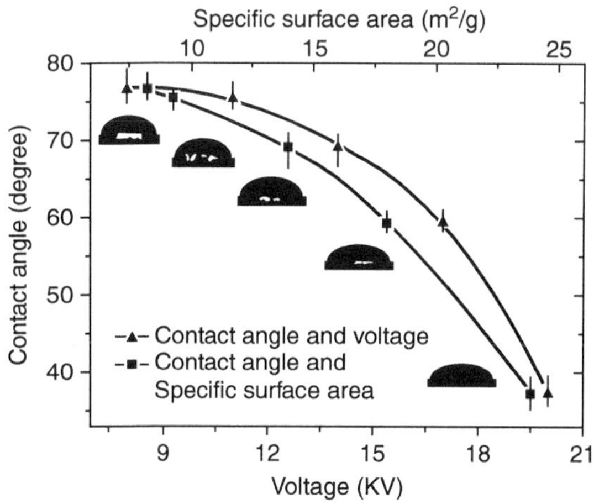

6.3 Surface contact angles of the PVDF membrane.

Figure 6.4 illustrates the dynamic wetting behavior of the PVDF nanofiber membranes before and after the acrylic acid grafting. The membrane shows a very low adsorption before the acrylic acid grafting. This is because the effect of buoyancy of water pushed the materials upwards since the surface contact angle was higher than 90°, when the sample was immersed into water. The acrylic acid grafting considerably improved the water adsorption properties of the PVDF nanofiber membrane due to the lower contact angles. It is observed from the adsorption curve that it appears to show the existence of two adsorption zones, as shown in Fig. 6.4. The first zone, A, was the initial water adsorption, which occurred over the first few seconds. Then, the water was absorbed into the fibrous webs and transferred through capillary. This could be called transferring zone, B, as shown in Fig. 6.4. It was

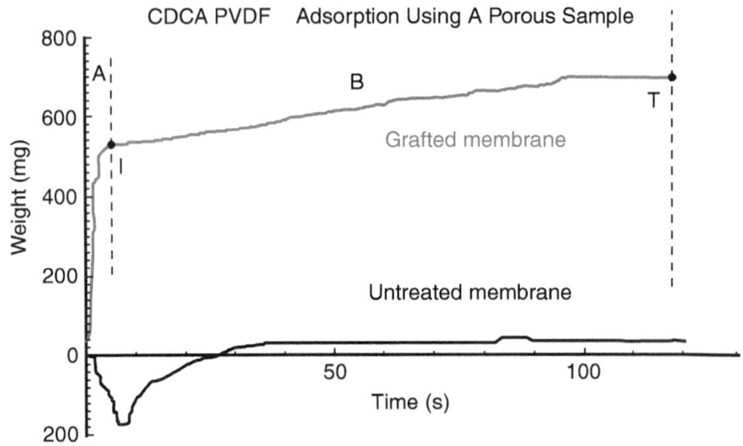

6.4 Dynamic water adsorption of PVDF membrane before and after acrylic acid grafting.

concluded that surface contact angle determined the initial adsorption and the porous structure influenced the capillary adsorption.

Plasma-induced grafting has also been used to modify biodegradable electrospun nanofibers for biomedical applications (Park *et al.*, 2007). The electrospun nanofibers with an average diameter of 340 nm were prepared by electrospinning a solution of poly(lactic-coglycolic acid) (PLGA). The electrospun PLGA nanofibers were treated with oxygen or ammonia gas plasma to modify the surface properties of the electrospun PLGA nanofibers. The hydrophilicity of the electrospun PLGA nanofibers was significantly increased by the gas plasma treatment due to the formation of polar groups on the nanofibers. It was also found that the degradation behavior of the PLGA nanofibers was accelerated by the plasma treatment, and the adhesion and proliferation of mouse fibroblasts on the plasma-treated nanofibers were significantly enhanced.

6.2.2 Physical vapor deposition

'Physical vapor deposition' (PVD) is a term used to describe a family of coating processes. The most commonly used PVD coating processes include evaporation and sputtering. These processes usually occur in a vacuum at working pressure (typically 10^{-2}–10^{-4} mbar) and generally involve bombardment of the substrate to be coated with energetic positively charged ions during the coating process to promote high density. Reactive gases such as nitrogen, acetylene or oxygen may be introduced into the vacuum chamber during deposition to create various compound coatings.

Wei *et al.* (2008) prepared PA6 nanofibers by electrospinning and the nanofibers were further sputter coated with metals. The PA6 nanofibers show three-dimensional fibrous structure, consisting of fibers with diameters ranging from less than 200 nm to over 600 nm, as displayed in Fig. 6.5a. The higher magnification of the AFM image reveals the details of the fiber surface, as shown in Fig. 6.5b. It can be seen that the fiber surface is relatively smooth.

The metallic sputter coating significantly alters the surface characteristics of the PA6 nanofibers, as illustrated in Fig. 6.5. The AFM image of 1000 × 1000 nm scan clearly shows the surface of the nanofibers sputtered with Cu and Ag. The Cu and Ag clusters are scattered on the PA6 nanofiber surface after the coating. It was observed that the size of the sputtered Cu and Ag clusters was increased when the deposition thickness was increased. This was attributed to the collision of the sputtered Cu or Ag grains. The metallic funtionalization of the nanofibers was also confirmed by energy-dispersive X-ray spectroscopy (EDX) analyses. It can be seen from Fig. 6.6a that the surface of the nanofibers dominantly consisted of C, O and N before the sputter coating. The composition of hydrogen (H) in the material was too light to be detected in the EDX analysis. The EDX spectra reveal that Cu and Ag peaks were clearly detected, as shown in Fig. 6.6b and 6.6c. A significant amount of Cu and Ag on the nanofiber surface after copper sputter coating appeared in the spectra, but the amount of C, O and N was reduced in the EDX spectra at the same time, indicating the coverage of the surface by metallic coating.

The nanofibers had a very high surface resistance, which was over 10^4 Ω cm, indicating the electrical insulation property. The metallic sputter coating, however, significantly improved the electrical properties of the nanofibers. The surface resistivity dropped to about 0.45 Ω cm for Cu coating and 0.14 Ω cm for Ag coating, as the coating had a thickness of 50 nm. The surface resistivity of the nanofibers was further reduced to about 0.08 Ω cm for Cu coating and 0.03 Ω cm for Ag coating, as the coating thickness was increased to 100 nm, indicating better surface conductivity. The study also reveals that a better surface conductivity of the nanofibers coated with silver could be obtained compared to those coated with Cu in the same coating thickness.

Transparent conductive films of tin-doped indium oxide (ITO) deposited on the PA6 nanofibers was also reported (Wei *et al.*, 2010). In this work, ITO was deposited onto the PA6 nanofiber substrates at room temperature. ITO deposition was performed on a magnetron sputter coating system. The ITO target used was a ceramic target with a diameter of 50 cm, a composition of 97 wt% In_2O_3 and 3 wt% SnO_2. Pure Ar (99.999%) was used as sputtering gas. It was observed that ITO deposition significantly altered the optical properties of the PA6 nanofibers. The PA6 nanofibers without

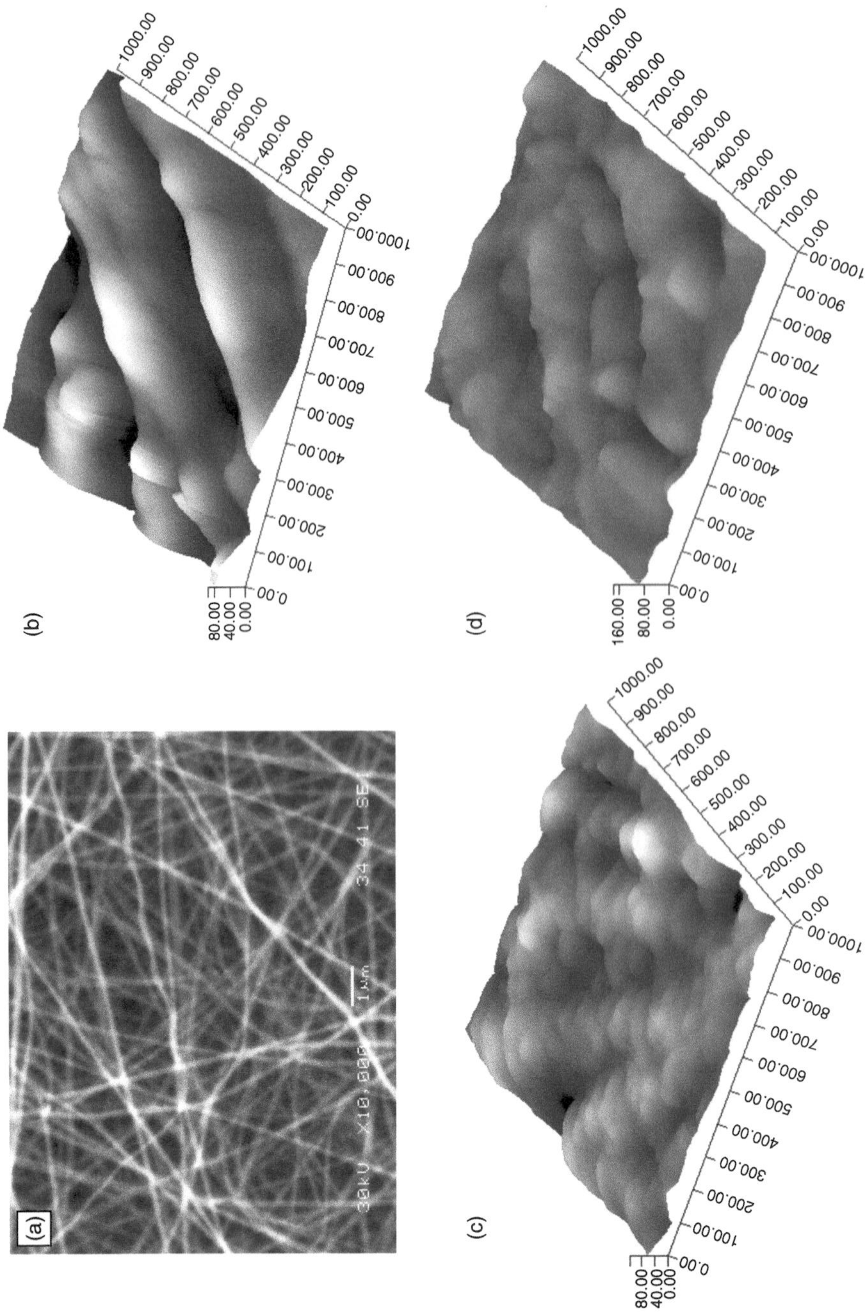

6.5 PA6 nanofibers: (a) SEM image, (b) AFM image, (c) Cu coated and (d) Ag coated.

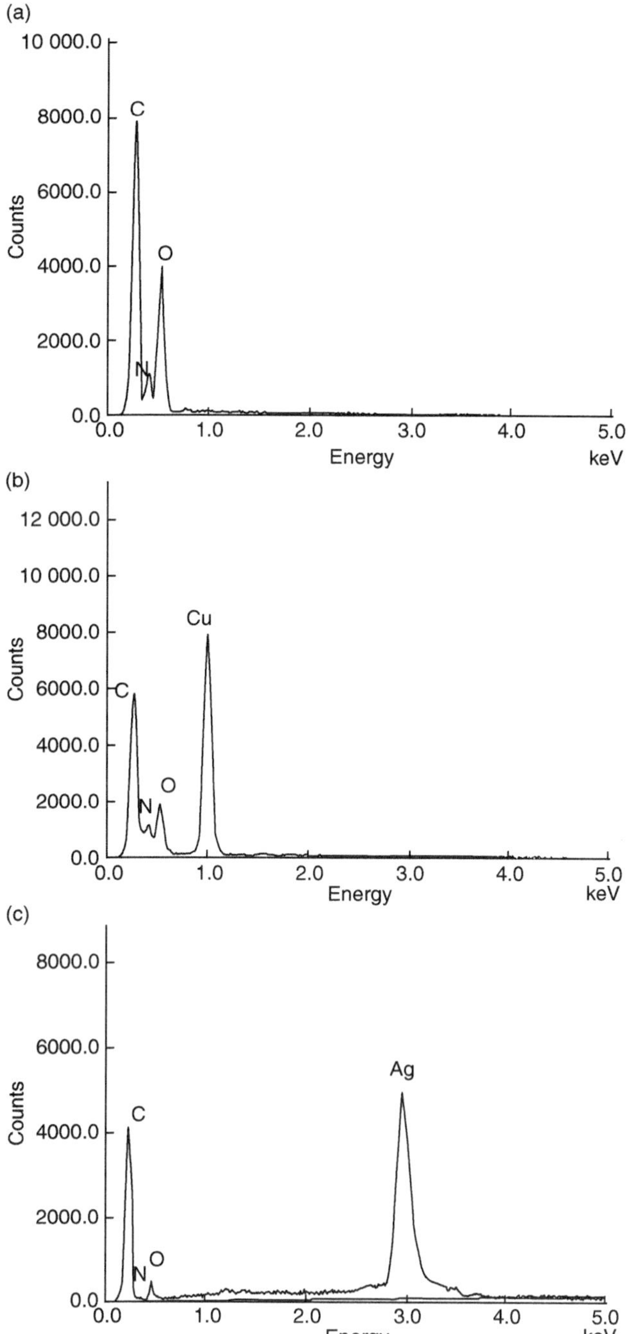

6.6 EDX spectra of the nanofibers before and after sputter coating: (a) PA6 nanofibers, (b) Cu coated and (c) Ag coated.

ITO deposition showed the transmittance of about 35% in the wavelength ranging between 400 and 600 nm, indicating a good transmittance of visible light. The transmittance dropped gradually from 35% to less than 10% in the wavelength ranging between 400 and 300 nm, indicating the UV shielding effect of the PA6 nanofibers, as presented in Fig. 6.7. The ITO-deposited PA6 nanofibers had a slightly lower transmittance than uncoated PA6 nanofiber over the wavelength ranging between 450 and 600 nm, revealing the transparent property of the ITO depositions in the visible light range. The UV shielding effect of the ITO deposition was attributed to its chemical structure.

6.2.3 Other physical technologies

Wong *et al.* (2009) have tried to use ion beam implantation in the surface modification of polymer nanofibers. The study revealed that the mechanical properties of nanosized electrospun poly(vinyl alcohol) fibers could be modified using broad-energy ion beam implantation. The elastic modulus was determined using AFM multipoint mechanical bending tests on individual fibers before and after ion beam implantation. It was observed that fiber elastic modulus was increased by 30% with a dose of 8.0×10^{15} ions/cm^2 of nitrogen ion. It was concluded that the key chemical modifications via nitrogen ion treatment were the introduction of the functional groups of amine and amide. These groups are important in promoting cell compatibility on polymer surfaces.

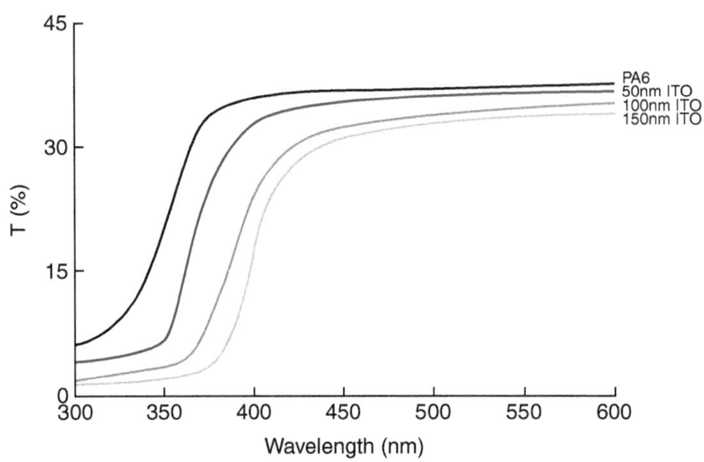

6.7 Transmittance of the PA6 nanofibers.

6.3 Surface functionalization using chemical technologies

Electrospinning forms fibers with high porosities and high surface areas, which provide good platforms for the surface functionalization by chemical technologies. Surface chemistry plays a very important role in various applications of electrospun nanofibers. Reports in the literature related to the surface modification of electrospun nanofibers by chemical technologies can be arbitrarily subdivided into several major categories: surface grafting, cross-linking, chelation and surface chemical deposition.

6.3.1 Surface grafting, cross-linking and chelation

Ma *et al.* (2005) reported the surface modification of electrospun poly(caprolactone) (PCL) nanofibers to improve their compatibility with endothelial cells (ECs) for the potential application of PCL nanofibers as a blood vessel in tissue engineering. Nonwoven PCL nanofibers (PCL NF) and aligned PCL nanofibers (APCL NF) were prepared by electrospinning. The electrospun PCL nanofibers were first treated with air plasma to introduce active groups of –COOH on the PCL nanofiber surface; this was followed by covalent grafting of gelatin using water soluble carbodiimide as the coupling agent. It was found that gelatin grafting obviously enhanced EC spreading and proliferation compared with the original material. The experimental results also revealed that gelatin-grafted APCL NF readily oriented ECs along the fibers.

It was reported that nonwoven meshes composed of polysulfone (PSU) ultra-fine fibers (diameter 1–2 μm) were fabricated via electrospinning and then modified toward development of a novel affinity membrane. In this work, the electrospun PSU fiber mesh was heat treated at 188°C to improve the mechanical strength of the fiber mesh. For surface modification, carboxyl groups were introduced onto the PSU fiber surfaces through grafting copolymerization of methacrylic acid (MAA) initiated by Ce(IV) after an air plasma treatment of the PSU fiber mesh. Toluidine Blue O, a dye, which can form stable complex with carboxyl groups, was used as a model target molecule to be captured by the polymethacrylic acid (PMAA)-grafted PSU fiber mesh. Furthermore, the carboxyl groups on the PMAA-grafted PSU fiber mesh were used as coupling sites for immobilization of other protein ligands. It was observed that the nonwoven fibrous membrane had much smaller pressure drop and higher flux compared with conventional microfiltration membranes (Ma *et al.*, 2006).

A novel high-flux filtration medium, consisting of a three-tier composite structure – that is, a nonporous hydrophilic nanocomposite coating top layer, an electrospun nanofibrous substrate midlayer and a conventional nonwoven microfibrous support, for oil/water emulsion separations – was reported for the first time by Wang *et al.* (2005a). The nanofibrous substrate was prepared by electrospinning of poly(vinyl alcohol) (PVA) followed by chemical cross-linking with glutaraldehyde (GA) in acetone. It was observed that the resulting cross-linked PVA substrates showed excellent water resistance and good mechanical properties. The tensile testing revealed that the strength at break of the electrospun nanofibrous substrate increased, while the elongation to break decreased after cross-linking. It was believed that the linear polymer chains in the noncross-linked nanofibers can easily slide by one another during tensile deformation, resulting in low tensile strength and high elongation. However, for the three-dimensionally cross-linked PVA nanofibers, the chains were tightly connected by covalent bonds, whereby the chain sliding became more difficult. Therefore, the cross-linked PVA substrates had higher tensile strength and lower elongation.

Feng *et al.* (2011) reportedly modified polyacrylonitrile (PAN) nanofibers by metal ion chelation. Amidoxime PAN (AOPAN) nanofibers were prepared by reaction with hydroxylamine hydrochloride; these were used as the matrix for metal ions chelation. The scanning electron microscopy (SEM) observations revealed that AOPAN nanofibers retained the fibrous structure and it seemed that the surface of AOPAN became a little rougher than that of elctrospun PAN nanofibers, as illustrated in Fig. 6.8a and 6.8b. This observation indicated that the structures of AOPAN nanofibers had not been distorted obviously during the reaction with hydroxylamine hydrochloride. The modified PAN nanofibers became rougher after four hours, as presented in Fig. 6.8c.

The adsorption isotherm of AOPAN nanofibers indicated that the initial metal ions adsorption increased very fast. The initial increase in metal ions adsorption might be attributed to many available amidoxime groups in AOPAN nanofibers. As more sites in the amidoxime groups were filled, it became difficult for more metal ions to find vacant sites. At the same time, adsorption amount of metal ions reached saturation gradually with the increase in concentration of metal ions, as shown in Fig. 6.9.

Figure 6.10 shows the desorption rate of metal ions from metal chelated AOPAN nanofibers. It can be clearly observed that the initial desorption rate increased very fast and then the desorption rate gradually slowed down. According to the experimental data, the desorption rate of Cu^{2+} and Cd^{2+} was 87% and 92%, respectively, within 1 h, as indicated in Fig. 6.10. The desorption test revealed that the AOPAN nanofibers could be reused multiple times as a matrix for metal ions chelation.

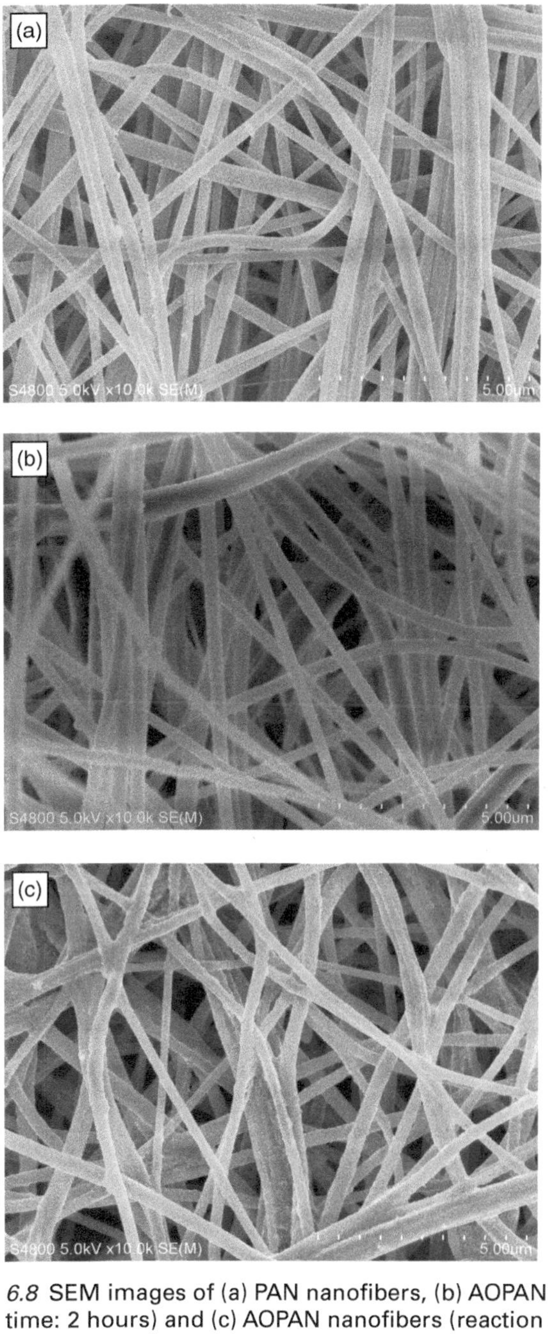

6.8 SEM images of (a) PAN nanofibers, (b) AOPAN nanofibers (reaction time: 2 hours) and (c) AOPAN nanofibers (reaction time: 4 hours).

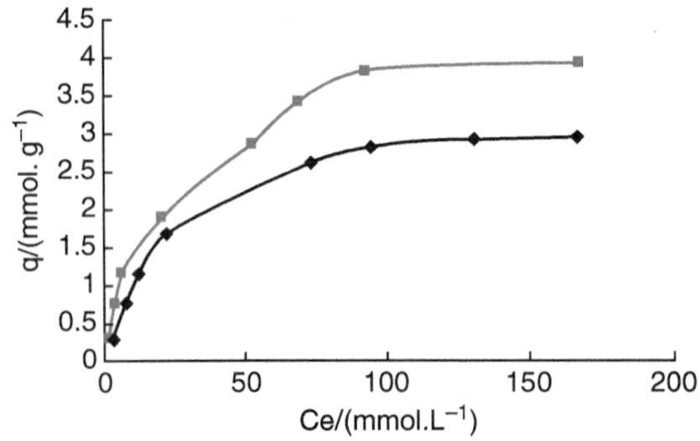

6.9 Adsorption isotherm of (■) Cd^{2+} and (◆) Cu^{2+} on the AOPAN nanofibers.

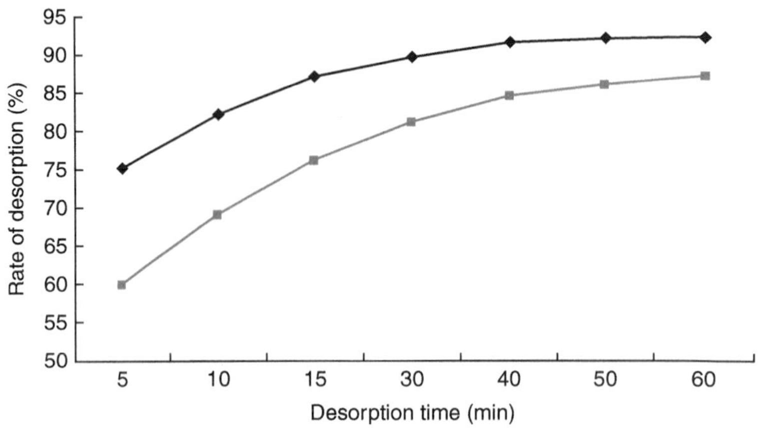

6.10 Desorption of (◆) Cu^{2+} and (■) Cd^{2+} on the metal chelated AOPAN nanofibers.

6.3.2 Electroless deposition

Tao *et al.* (2008) reported the surface functionalization of PA6 nanofibers by electroless deposition. Oxygen low temperature plasma treatment was applied to substitute the conventional roughing process using concentrated sulfuric acid-potassium dichromate. The SEM images in Fig. 6.11 show structures and surface morphology of the electrospun PA6 nanofibers without and with Cu coating. The nanofibers maintained their porous structures, but the diameters of the nanofibers were obviously increased after the Cu deposition. The copper deposition looked quite uniform on the nanofibers and the copper particles were clearly recognized. The copper deposition in this work was confirmed by EDX analysis.

6.11 SEM images of the electrospun PA6 fibers before and after Cu coating: (a) as-spun and (b) after Cu coating.

The ion source of Cu in electroless deposition is copper salt. If the concentration of Cu^{2+} is too high, the plating bath is not stable. The deposition rate becomes slow if the concentration of Cu^{2+} is too low. The effect of $CuSO_4$ concentration on the surface resistance of Cu coated nanofibers was also investigated and it was found that the surface resistance decreased from 831.35 to 24.41 ohm/sq as the concentration of $CuSO_4$ changed from 10 to 25 g/L. This was attributed to the increase in thickness with the increase of the $CuSO_4$ concentration in the deposition process.

Pinto *et al.* (2004) reported the deposition of thin films of nickel on the electrospun polyaniline fibers doped with 2-acrylamido-2-methyl-1-propanesulfonic acid. The metal-coated fibers showed a significant decrease in their resistance by three orders of magnitude while retaining the large surface area that is inherent to small-diameter fibers. Metallic tubes have also been synthesized by a polymer-based template approach using electroless deposition. Metals like gold, copper and nickel were deposited as thin films on the electrospun polymer fibers which ranged in diameter from 160 to 400 nm. The electrospun fiber template was degraded at 300 and 650°C and tubes between 450 and 730 nm were obtained with wall thicknesses of 50–150 nm (Ochanda & Jones, 2005). A new form of high surface area bio-electrode, based on nanofibers of electrospun gold with immobilized fructose dehydrogenase, was developed (Marx *et al.*, 2011). In this work, the gold nanofibers were prepared by electroless deposition of gold nanoparticles on an electrospun poly(acrylonitrile)-HAuCl(4) fiber. The study revealed that the electrochemical surface area of the gold microfibers reached 0.32 ± 0.04 m^2/g. Fructose dehydrogenase was covalently coupled to the gold surface through GA cross-links to a cystamine monolayer.

The enzyme exhibited mediated electron transfer directly to the gold electrode and catalytic current characteristics of fructose oxidation in the presence of a ferrocene methanol mediator were observed. The limit of detection of fructose was 11.7 μM and the K(M) of the immobilized enzyme was 5 mM. The microfiber electrode was stable over 20 cycles with a 3.05% standard deviation. The response time of the sensor was less than 2.2 s and reached half maximum value within 3.6 s. The sensor was proven to be accurate and precise in both serum and popular beverages sweetened with high fructose corn syrup. The addition of glucose isomerase enabled the sensor to perform with glucose, thus expanding the available analyte selection for the sensor.

6.4 Surface functionalization using nanotechnologies

6.4.1 Sol-gel method

Sol-gel technology offers many possibilities to introduce molecules with specific properties (thermal, electrical, optic, chromogenic, photosensitive and mechanical) into polymeric matrices. The sol-gel technique has been increasingly used to produce the functional coatings on various forms of materials. The precursors used in the preparation of the 'sol' are usually inorganic metal salts or metal organic compounds such as metal alkoxide. The precursor is subjected to a series of hydrolysis and condensation reactions for the inorganic domain formation. The use of a suitable coupling agent permits the attainment of a strictly interconnected network, preventing macroscopic phase separation (Huang *et al.*, 2006). The sol-gel technique has several advantages over other processes due to its simplicity, good homogeneity, easy control of the film composition, safety, large area coatings, low processing temperature, low cost of the apparatus and raw materials (Liu *et al.*, 2006).

Polymer nanofibers have interesting properties which are the result of their extremely high surface to weight ratio compared to the other conventional fibrous structures. These make nanofibers ideal for use in such application areas as filtration, sensors, protective clothing and functional materials, attributed to the high pore volume and tight pore sizes. The ability to deposit well-controlled coatings (such as sol-gel technique) on nanofibers would expand the application of nanofibers, based on changes to both the physical and chemical properties of the nanofibers. Shao *et al.* (2008) prepared PAN nanofibers by electrospinning, and then the functional layer of ZnO was deposited onto the surface of PAN nanofibers by sol-gel techniques. The ZnO coated PAN nanofibers were carbonized to form functional carbon nanofibers.

Wu *et al.* (2009) studied polyvinyl acetate (PVAc)/titanium dioxide (TiO_2) hybrid nanofibers by combining sol-gel process with electrospinning technology, which consisted of PVAc as organic segment and TiO_2 as inorganic part. The SEM images indicated that the average diameter of the electrospun PVAc/TiO_2 nanofibers was increased after hybridization. From the AFM and transmission electron microscopy (TEM) analysis, it could be concluded that the inorganic hydrolysis products led to the formation of the particle-like surface morphology of hybrid nanofibers. It was also found from the viscosity measurement and FTIR analysis that newly associated hydrogen bonds were formed between the organic and inorganic components in hybrid nanofibers. The thermogravimetric analysis (TGA) and tensile property analysis revealed that the thermal stability and strength were enhanced as an increase of TiO_2 content. It can be reasonably concluded that interaction of the hydrogen bond or other coordinate bonds took place between the nanofibers of TiO_2 inorganic network and polymeric chains. Xiao *et al.* (2009) had investigated the surface modification of nonwoven poly(ethylene terephthalate) (PET) fabrics by Al_2O_3 sol-gel coatings. The SEM observation revealed the deposition of the Al_2O_3 particles on the PET fibers, but the fabric structures remained unchanged. The dense network and porous structures of the Al_2O_3 particles formed on the fiber surfaces were observed by AFM scans. The Al_2O_3 coating on the PET fiber surface was also confirmed by FTIR. The Al_2O_3 coating contributed to the mechanical properties of the nonwoven PET fabrics. The wettability of the fabric was significantly improved by using the Al_2O_3 coating.

6.4.2 Atomic deposition

For a variety of applications, it is desirable to make such nanofibers with well-defined surface properties. Polymer nanofibers with specific surface properties are of interest in many technical applications as the surface features affect electrical conductivity, optical property and biocompatibility. Various techniques have been tried to modify the surface properties of polymer nanofibers. Wang *et al.* (2005b) presented a convenient and effective way to dope Ag nanoparticles into PAN nanofiber film by *in situ* reduction of silver ions in N_2H_5OH aqueous solution. The Ag nanoparticles with average diameter of 10 nm were dispersed homogeneously in PAN nanofibrous film. D and G peak observed in surface enhanced Raman scattering (SERS) spectrum revealed that PAN-embedded Ag nanoparticles had been partly converted to graphite structure at room temperature. It is probable that the graphitization of PAN is caused by the function of Ag nanoparticle as a catalyst for dehydrogenation of hydrocarbon compound. Demir *et al.* (2004) produced catalytic palladium (Pd) nanoparticles on electrospun copolymers of acrylonitrile and acrylic acid (PAN-AA) mats via reduction of $PdCl_2$ with aqueous hydrazine solution. Pd atoms nucleate and form small crystallites, and two to four crystallites adhere together and form agglomerates. Spherical Pd nanoparticles were dispersed homogeneously on the electrospun nanofibers. The effects of copolymer composition and amount of $PdCl_2$ on particle size were investigated. Pd particle size mainly depends on the amount of acrylic acid functional groups and $PdCl_2$ concentration in the spinning solution. Increasing acrylic acid concentration on polymer chains and $PdCl_2$ concentration leads to larger Pd nanoparticles. Therefore, it is possible to tune the number density and the size of metal nanoparticles. The catalytic activity of the Pd nanoparticles in electrospun mats was determined by selective hydrogenation of dehydrolinalool (3,7-dimethyloct-6-ene-1-yne-3-ol) in toluene at 90°C. Electrospun fibers with Pd particles have 4.5 times higher catalytic activity than the current Pd/Al_2O_3 catalyst.

6.4.3 Layer-by-layer (LbL) deposition

Besides the many efforts to develop new nanostructured and functional nanofibers, tailoring the surface chemistry and functionalization of electrospun nanofibers is also important for application of these materials. For example, modification of surface properties such as chemical composition, functionality and charge, as well as biocompatibility, is essential for utilization of the resulting nanofibers in specific applications.

Park *et al.* (2008) reported that the mechanical properties of polyelectrolyte multilayer-coated nylon 6 fibrous mats were enhanced by surface nanocoating of polyelectrolyte. Moreover, it was observed that the surface morphology of nanofiber could be finely tuned by regulating the amount of polyelectrolyte nanocoating, thereby giving rise to a controlled surface morphology, hydrophilicity and porosity of the nanofibrous mats. These are crucial properties as polymeric scaffolds for cell culture. The multilayer coating formed by LbL deposition (with a biological polymer as the outmost layer) seems to offer many attractive advantages over the monolayer coating which can be prepared using many other methods such as simple dip coating. Meanwhile, Park *et al.* (2009) also reported the controlled surface morphologies and the cell culture of polyelectrolyte multilayer-coated nylon 6 fibrous mats with different numbers of layers. Polyelectrolyte multilayer-coated nylon 6 fibers were successfully prepared by an alternative deposition of alginic acid sodium salt and chitosan via an LbL electrostatic self-assembly. The surface morphology, stiffness and hydrophilicity of polyelectrolyte multilayer-coated nylon 6 fibrous mats could be finely tuned by regulating the amount of polyelectrolyte nanocoating. It was observed that the morphology of polyelectrolyte multilayer-coated nylon 6 fibers was uniform and smooth, indicating a dense and harder nanocoating of polyelectrolytes onto nylon 6 fibers. Compared to pure nylon 6 fibrous mat (tensile strength 10.6 ± 1 MPa), the tensile strength of polyelectrolyte-coated nylon 6 fibrous mats was increased significantly to 35.2 ± 2 MPa for five bilayer-coated fiber mats. In addition, it was found that at an initial stage, after one day of cell culturing, the electrospun nylon 6 fibrous mats coated with five bilayers of alginic acid and chitosan show the highest cell affinity (good adhesion), while the electrospun nylon 6 fibrous mats coated with ten bilayers show the lowest cell affinity. After cell seeding for three days, it was observed that rate of proliferation is enhanced by increasing the number of bilayers to three (good proliferation), and then drastically decreased with further increasing the number of bilayer.

Hyun *et al.* (2009) report on the fabrication and characterization of LbL stacked MnOx-RuO_2 nanofiber mats via coelectrospinning and their potential application for electrochemical capacitors. The morphology of multistacked MnOx-RuO_2 fiber mats provides high porosity for the efficient permeability of electrolyte into the inner structure while maintaining a high surface area for enhanced surface activity. In an electrochemical capacitor using MnOx-RuO_2 fiber mats, MnOx offers a simple ion pathway and RuO_2 provides a fast electron pathway. The electrochemical capacitors utilizing coelectrospun MnOx-RuO_2 fiber mats exhibited a high specific capacitance of 208.7 F/g at 10 mV/s. Pan *et al.* (2009) prepared hybrid multilayered hollow nanofibers (HMHNFs) composed of TiO_2/polyelectrolyte (PE) by a combination of electrospinning method and LbL technology.

The results showed that both the average diameter and the wall thickness of the HMHNFs can be well controlled by the template, as well as the number of coating layers. The dried morphology of the obtained HMHNFs is dependent on the inner-deposited numbers of the polyelectrolyte layers. When compared with other nanostructured TiO_2 materials, such as commercial TiO_2 nanoparticles and TiO_2 films, the hollow TiO_2/PE hybrid nanofibers exhibited higher photocatalytic activities. Ge *et al.* (2007) fabricated hollow multilayered PE fibers by the combination of the electrostatic LbL and electrospinning methods. The influence of the deposition conditions on the structure of the obtained hollow multilayered PE fibers was studied systemically. The results showed that the formation of multilayered PE films on polystyrene (PS) fiber surface was strongly influenced by the number of repetitions of deposition and the pH value. The spaces among the adjacent fibers in fibrous mats were blocked by 15 poly(allylamine hydrochloride)/poly(styrenesulfonate) bilayers. The film growth speed was the highest in a strongly acidic solution. The theoretical Young's modulus of such a hollow multilayered PE nanofiber was found to be around 21.6 GPa, which is much larger than that of normal synthetic organic fibers but is similar to that of human bone fibers.

Murphy *et al.* (2005) produced homogeneous and ultrathin films of carbon nanofibers with a chitosan binder with controlled thickness based on layer-by-layer deposition process. Films with a thickness similar to the nanofiber diameter (50–200 nm) are produced and characterized via electron microscopy, conductivity measurements and quartz crystal microbalance gravimetry. Electrochemically, even very thin films of hydrophilized carbon nanofibers are readily detected due to their (pseudo-)capacitive current response.

Electrochemical reactions, such as the oxidation of hydroquinone in an aqueous phosphate buffer, are shown to be catalyzed by the presence of hydrophilized carbon nanofiber fragments (with respect to the same process at a bare ITO electrode). Based on voltammetric responses, the electron transfer across a barrier between carbon nanofibers and the ITO substrate is identified as rate limiting in the presence of chitosan. Li and Liu (2009) reported the preparation of multilayer films from the layer-by-layer deposition of block copolymer nanofibers bearing carboxyl and amine groups in their coronas. The film formation process was followed by AFM and UV absorbance measurements.

6.4.4 Molecular imprinting

Molecular imprinting (Andersson *et al.*, 2006; Alexander *et al.*, 2003, Komiyama *et al.*, 2003) is perceived as a simple way to obtain molecular recognition materials, which can be applicable to chromatography, membranes, sensors, catalysts and so forth. Specialty membranes with various polymeric materials, such as synthetic polymers, derivatives of natural polymers, oligopeptide derivatives and natural polymers, can be converted into molecular recognition materials or membranes by applying an alternative molecular imprinting.

Yoshikawa *et al.* (2007) demonstrated that polymeric materials can be directly converted into molecular (chiral) recognition nanofiber membranes by simultaneously applying an electrospray deposition and an alternative molecular imprinting during the membrane preparation process. Polysulfone with a degree of substitution of 0.88 was adopted as the candidate polymeric material for molecularly imprinted nanofiber membranes. Molecularly imprinted nanofiber membranes imprinted by N-benzyloxycarbonyl-D-glutamic acid (Z-D-Glu) recognize the D-isomer in preference to the corresponding L-isomer and vice versa. The amino acid preferentially incorporated into the membrane is selectively permeated through the membrane by using a concentration gradient as a driving force for membrane transport.

Chronakis *et al.* (2006) demonstrate that electrospun template-directed molecular imprinting is a viable method for creating robust, molecularly imprinted nanofibers that can selectively rebind the target molecule. Molecular recognition sites in polymer nanofibers were prepared with a simple electrospinning method using a combination of structural and functional polymers as the starting material with PET and polyallylamine. The two 4-D imprinted nanofibers displayed favorable binding characteristics in aqueous solution. The imprinted nanofibers also had a well-defined morphology and physical stability. Overall, the electrospun nanofibers provided a porous matrix with very high surface area-to-volume ratio, which is desirable for the selective rebinding of the target molecule. Considering nanofiber processing, the present electrospinning-imprinting method is promising for its simple operation and high efficiency. It is expected that molecularly imprinted nanofibers will find interesting applications in affinity-related separation areas. It should be pointed out that the choice of the functional polymers used for imprinting electrospinning must be made according to the targeted compounds. It is also possible to use postspinning cross-linking to further stabilize the imprinted sites by different chemical means. Furthermore, the electrospun nanofiber itself may be modified with an *in situ* generated molecularly imprinted polymer (MIP) layer, such as via the grafting-from approach using a surface immobilized initiator system.

Wan *et al.* (2006) prepared copolymer of acrylonitrile with vinyl porphyrin through solution polymerization with a high yield and porphyrin conversion ratio. Electrospinning the copolymer solutions at low concentration only gives microspheres, but nanofibers with a diameter of around 330 nm can be obtained by increasing the solution concentration. It is speculated that this type of nanofiber may be a latent support of porphyrins for various purposes such as catalysis, molecular imprinting, sensors and light/energy conversion.

Molecularly imprinted nanoparticles were encapsulated into polymer nanofibers with a simple electrospinning method (Yoshimatsu *et al.*, 2008). The composite nanofibers form nonwoven mats that can be used as an affinity membrane to greatly simplify solid phase extraction of drug residues in analytical samples. Almost 100% of propranolol-imprinted nanoparticles can be easily encapsulated into poly(ethylene terephthalate) nanofibers, ensuring the composite materials to have a high specific binding capacity. As confirmed by radioligand binding analysis, the specific binding sites in the composite materials remain easily accessible and are chiral-selective. Using the new composite nanofiber mats as solid phase extraction materials, trace amounts of propranolol (1 $ngmL^{-1}$) in tap water can be easily detected after a simple sample preparation. As validated in this study, there is no problem of template leakage from the composite nanofibers. Without the solid phase extraction, the existence of propranolol residues in water cannot be confirmed with even tandem high-performance liquid chromatography–mass spectrometry (HPLC–MS/MS) analysis.

6.5 Surface functionalization using biotechnology

A great deal of attention has been given to immobilizing bioactive molecules on the surface of electrospun nanofibers in recent years in order to achieve biologically functionalized materials for drug delivery, tissue engineering and biosensors. Various biomolecules can be incorporated within tissue-engineered scaffolds to enhance their functional properties for biomedical applications. In order to make polycaprolactone (PCL) more suitable for tissue engineering, Mattanavee *et al.* (2009) tried to first modify the PCL electrospun fibrous scaffolds with 1,6-hexamethylenediamine to introduce amino groups on their surface. Various biomolecules – that is, collagen, chitosan and Gly-Arg-Gly-Asp-Ser (GRGDS) peptide – were then immobilized on their surface, using N,N′-disuccinimidylcarbonate as the coupling agent. This study also investigated the appropriateness of these PCL fibrous scaffolds for tissue/cell culture, which was evaluated in vitro with three different cell lines – for example, mouse fibroblasts (L929), human epidermal keratinocytes (HEK001) and mouse calvaria-derived preosteoblastic cells (MC3T3-E1). It was found that, among the various biomolecule-immobilized PCL fibrous scaffolds, the materials immobilized with type I collagen and an Arg-Gly-Asp-containing protein showed the greatest ability to support both the attachment and the proliferation of all of the investigated cell types, followed by those that had been immobilized with GRGDS peptide.

Enzymes are well-known green biocatalysts that possess a high degree of specificity. The specificity involves discrimination between substrates, similar parts of molecules and optical isomers. The mildness and specificity of enzymes endow them with a high efficiency for applications in fine-chemical/pharmaceutical synthesis, food processing, biosensors fabrication, bioremediation and protein digestion in proteomic analysis. The applications of enzymes often suffer from various problems – for example, instability, non-reusability and high cost. Enzyme immobilization has become an effective way to overcome these limitations to some extent. Insoluble supports can be recycled much more easily than soluble enzymes. Improvement of catalytic efficiency of immobilized enzymes via materials engineering was demonstrated through the preparation of bioactive nanofibers (Jia *et al.*, 2002). Bioactive PS nanofibers with a typical diameter of 120 nm were prepared and examined for catalytic efficiency for biotransformations. The electrospun PS nanofibers were functionalized by the chemical attachment of a model enzyme, R-chymotrypsin. It was revealed that the apparent hydrolytic activity of the nanofibrous enzyme in aqueous solutions was over 65% of that of the native enzyme, indicating a high catalytic efficiency as compared to other forms of immobilized enzymes. It was also found from the experimental results that the nanofibrous R-chymotrypsin exhibited a significantly improved nonaqueous activity that was over three orders of magnitude higher than that of its native counterpart suspended in organic solvents, including hexane and isooctane. It appeared that the covalent binding also improved the enzyme's stability.

The group at Jiangnan University has explored the immobilization of enzymes onto the electrospun PAN nanofibrous membranes. AOPAN nanofibrous membranes were prepared through treatment of electrospun PAN nanofibrous membranes in hydroxylamine hydrochloride aqueous solution. Then, Fe(III) ions were incorporated into the amidoxime groups of the AOPAN nanofibrous membranes via metal chelation, and subsequently, catalases were immobilized onto the metal chelated nanofibrous membranes. A schematic of catalases immobilized onto the metal chelated AOPAN nanofibrous membrane is presented in Fig. 6.12. Fe(III) can chelate with amidoxime groups of the AOPAN nanofibrous membranes and the imidazole groups, amino groups on catalases by coordination bond.

The fibrous structures and surface morphologies of the fibers were observed by SEM, as presented in Fig. 6.13. The electrospun PAN nanofibers formed a fibrous membrane with random orientation. The nanofibers looked very even and the average diameter ranged between 220 and 320 nm, as presented in Fig. 6.13a. After reacting with hydroxylamine hydrochloride

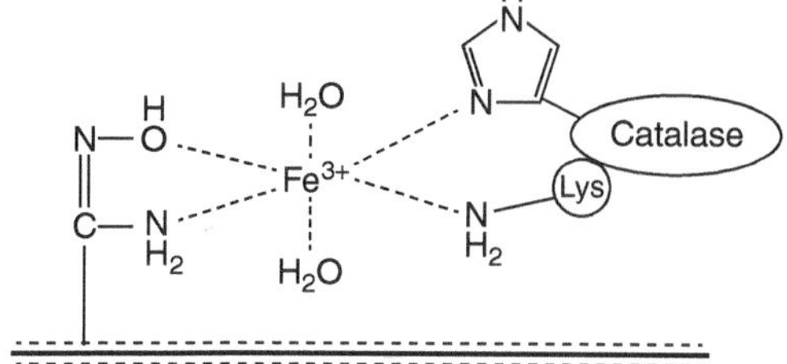

6.12 Schematic of catalases immobilized onto the metal chelated AOPAN nanofibrous membrane.

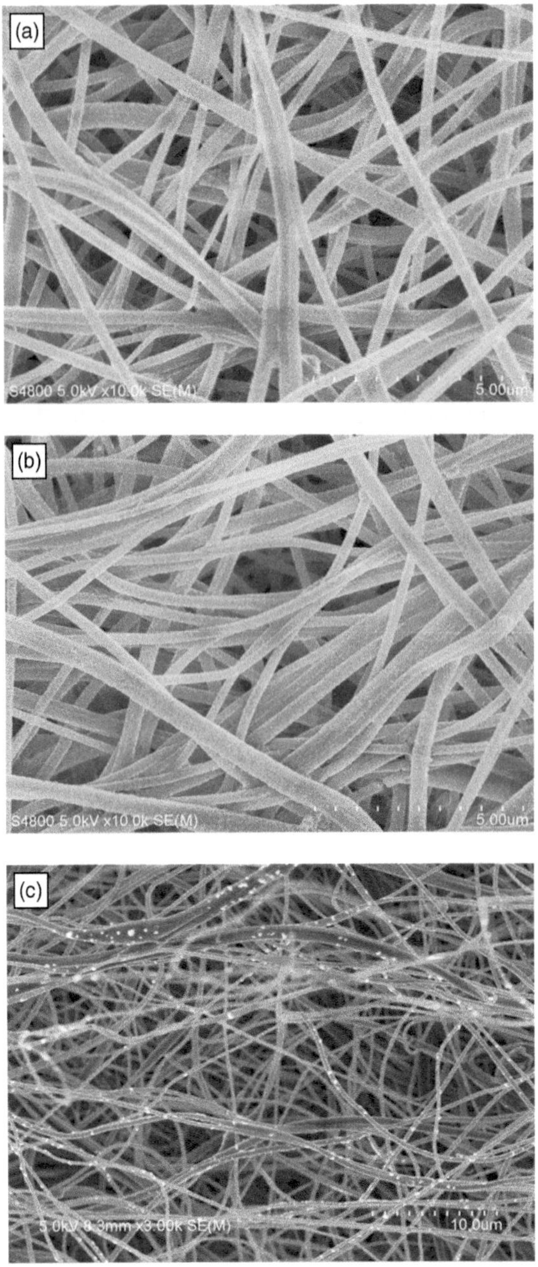

6.13 SEM images of (a) PAN nanofibrous membrane, (b) AOPAN nanofibrous membrane and (c) metal chelated AOPAN nanofibrous membrane with immobilized catalase.

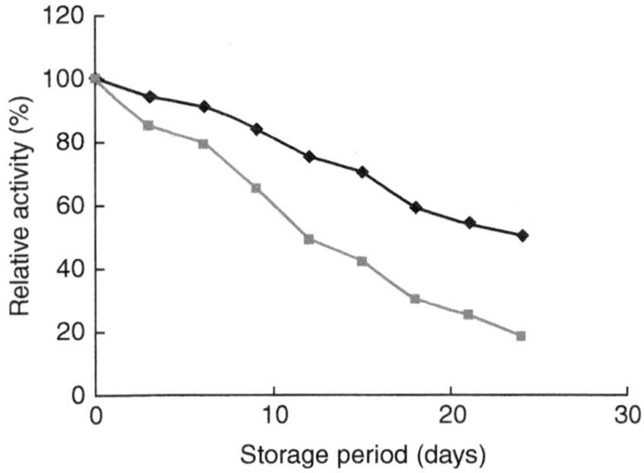

6.14 Storage stability of catalases: (■) immobilized catalases: (◆) free catalases.

for 2 h, the diameter of AOPAN nanofibers did not change substantially and the fibrous structures had not been distorted obviously, as revealed in Fig. 6.13b. The catalases immobilized on the electrospun nanofiber membrane could be clearly recognized, as revealed in Fig. 6.13c.

Storage stability and reusability are important advantages of immobilized enzymes. The storage stability of immobilized and free enzymes is presented in Fig. 6.14. The residual activity of immobilized and free catalase was 70% and 42% after 15 days. Immobilized catalases lost 50% of its activity within 24 days, and the free catalase lost 82% of its activity during the same period. The result indicated that the storage stability of immobilized catalases was improved compared to the free catalases, which might be attributed to the fact that immobilization of an enzyme to a matrix limited its freedom to undergo conformational changes resulting in increased stability.

6.6 Future trends

A considerable amount of research is taking place in the field of surface functionalization of polymer nanofibers by various technologies. In the case of other modified processes, the trend of surface modification will be focused on the combination of several modified technologies and the expansion of final products into technical applications by different surface modification. The following are potential research topics for future work.

- Use of naturally abundant substances, such as chitin and chitosan, as modification materials will be researched further. The modified nanofibers can be applied in cell scaffold, tissue engineering and so on.

- New technologies, such as self-assembled technology, will be applied in surface modification of nanofibers. These technologies will introduce a monolayer of a desired functional group without causing irregular etching or producing significant hazards.
- Nanotechnology, especially atomic deposition, will attract more attention in the research of surface functional nonofibers. Controllability for desired purposes will be the hot topic in these research areas.
- Improving the commercial viability of equipment by manufacturers and researchers will be a potential topic for the future.

6.7 References

Alexander, C., Andersson, H. S., Andersson, L. I., Ansell, R. J., Kirsch, N., Nicholls, I. A. and O'Mahony, J. (2006), Molecular imprinting science and technology: A survey of the literature for the years up to and including 2003, *Journal of Molecular Recognition*, **19**, 106–80.

Alexander, C., Davidson, L., and Hayes, W. (2003), Imprinted polymers: artificial molecular recognition materials with applications in synthesis and catalysis, *Tetrahedron*, **59**, 2025–57.

Baumgarten, P. K. (1971), Electrostatic spinning of acrylic microfibres, *Journal of Colloid and Interface Science*, **36**, 71–9.

Chronakis, I. S., Milosevic, B., Frenot, A. and Ye, L. (2006), Generation of molecular recognition sites in electrospun polymer nanofibres via molecular imprinting, *Macromolecules*, **39**, 357–61.

Demir, M. M., Gulgun, M. A. and Menceloglu, Y. Z. (2004), Palladium nanoparticles by electrospinning from Poly(acrylonitrile-co-acrylic acid)-PdC12 solutions relations between preparation conditions, particle size, and catalytic activity, *Macromolecules*, **37**(5), 1787–92.

Feng, Q., Xia, X., Wei, A. F., Wang, X. Q., Wei, Q. F. and Huo, D. Y. (2011), Preparation of Cu(II)-chelated poly(vinyl alcohol) nanofibrous membranes for catalase immobilization, *Journal of Applied Polymer Science*, **120**, 3291–6.

Ge, L. Q., Wang, X., Tu, Z. C., Pan, C. and Wang, C. (2007), Fabrication of multilayered hollow nanofibres and estimation of its Young's modulus, *Japanese Journal of Applied Physics Part 1-Regular Papers Brief Communications & Review Papers*, **46**, 6790–5.

Hocker, H. (2002), Plasma treatment of textile fibres, *Pure and Applied Chemistry*, 74(3), 423–7.

Huang, D., Wang, Y. J., Yang, L. M. and Luo, G. S. (2006), Direct synthesis of mesoporous TiO2 modified with phosphotungstic acid under template-free condition, *Microporous and Mesoporous Materials*, **96**, 301–6.

Huang, F. L., Wang, Q. Q., Wei, Q. F., Gao, W. D., Shou, H. Y. and Jiang, S. D. (2010), Dynamic wettability and contact angles of poly(vinylidene fluoride) nanofibre membranes grafted with acrylic acid, express, *Polymer Letters*, **4**(9), 551–8.

Hyun, T. S., Kang, J. E., Kim, H. G. and Hong, J. M. (2009), Electrochemical properties of MnOx-RuO2 nanofibre mats synthesized by co-electrospinning, *Electrochemical and Solid State Letters*, **12**, A225–8.

Jia, H. F., Zhu, G. Y., Vugrinovich, B., Kataphinan, W., Reneker, D. H. and Wang, P. (2002), Enzyme-carrying polymeric nanofibres prepared via electrospinning for use as unique biocatalysts, *Biotechnology Progress*, **18**, 1027–32.

Komiyama, M., Takeuchi, T., Mukawa, T. and Asanuma, H. (2003), *Molecular imprinting*, Wiley-VCH.

Li, X. Y. and Liu, G. J. (2009), Layer-by-layer deposition of block copolymer nanofibres and porous nanofibre multilayer films, *Langmuir*, **25**(18), 10811–9.

Liu, Z. F., Jin, Z. G., Li, W., and Liu, X. X. (2006), Ordered porous ZnO thin films formed by dip- coating method using PS templates, *Journal of Sol-Gel Science and Technology*, **40**, 25–30.

Ma, Z. W., He, W. and Yong, T. (2005), Grafting of gelatin on electrospun poly(caprolactone) nanofibres to improve endothelial cell spreading and proliferation and to control cell orientation, *Tissue Engineering*, **11**(7–8), 1149–58.

Ma, Z. W., Kotaki, M. and Ramarkrishna, S. (2006), Surface modified nonwoven polysulphone (PSU) fibre mesh by electrospinning: A novel affinity membrane, *Journal of Membrane Science*, **272**(1–2), 179–87.

Marx, S., Jose, M. V., Andersen, J. D. and Russell, A. J. (2011), Electrospun gold nanofibre electrodes for biosensor, *Biosens Bioelectron*, **26**(6), 2981–6.

Mattanavee, W., Suwantong, O., Puthong, S., Bunaprasert, T., Hoven, V. P., Supaphol, P. (2009), Immobilization of biomolecules on the surface of electrospun polycaprolactone fibrous scaffolds for tissue engineering, *ACS Applied Materials and Interfaces*, **1**(5), 1076–85.

Murphy, M. A,, Wilcox, G. D., Dahm, R. H. and Marken, F. (2005), Electrochemical characterisation of ultrathin carbon nanofibre-chitosan multi-layer films, *Indian Journal of Chemistry Section A-Inorganic Bio-Inorganic Physical Theoretical & Analytical Chemistry*, **44**, 924–31.

Ochanda, F. and Jones, W. E. (2005), Sub-micrometer-sized metal tubes from electrospun fibre templates, *Langmuir*, **21**(23), 10791–6.

Pan, C., Dong, L., Ge, L. Q., Wang, J. and Gu, Z. Z. (2009), Highly active TiO_2/polyelectrolytes hybrid multilayered hollow nanofibrous photocatalyst prepared from electrospun fibres using electrostatic layer-by-layer technique, *Journal of Macromolecular Science Part B-Physics*, **48**, 92–105.

Park, H., Lee, K. Y. and Lee, S. J. (2007), Plasma-treated poly(lactic-co-glycolic acid) nanofibres for tissue engineering, *Macromolecular Research*, **15**(3), 238–43.

Park, J. H., Kim, B. S., Tae, H. J., Kim, I. S. and Kim, H. Y. (2009), Polyelectrolyte multilayer coated nanofibrous mats: Controlled surface morphology and cell culture, *Fibres and Polymers*, **10**, 419–24.

Park, J. H., Kim, B. S., Yoo, Y. C. and Khil, M. S. (2008), Enhanced mechanical properties of multilayer nano-coated electrospun nylon 6 fibres via a layer-by-layer self-assembly, *Journal of Applied Polymer Science*, **107**, 2211–6.

Pinto, N. J., Carrión, P. and Quiñones, J. X. (2004), Electroless deposition of nickel on electrospun fibres of 2-acrylamido-2-methyl-1-propanesulfonic acid doped polyaniline, *Materials Science and Engineering*, **366**(1), 1–5.

Reneker, D. H. and Chun, I. (1996), Nanometre diameter fibres of polymer, produced by electrospinning, *Nanotechnology*, **7**, 216.

Renuga, G., Ma, Z. W., Satinderpal, K. and Seeram, R. (2007), Surface modification and application of functionalized polymer nanofibres, *Topics in Applied Physics*, **109**, 72–91.

Shao, D. F., Wei, Q. F., Zhang, L. W., Cai, Y. B. and Jiang, S. D. (2008), Surface functionalization of carbon nanofibres by sol–gel coating of zinc oxide, *Applied Surface Science*, **254**, 6543–6.

Tao, D., Wei, Q. F., Cai, Y. B., Xu, Q. X. and Sun, L. Y. (2008), Functionalization of polyamide 6 nanofibres by electroless deposition of copper, *Journal of Coatings Technology and Research*, **5**(3), 399–403.

Wan, L. S., Wu, J. and Xu, Z. K. (2006), Porphyrinated nanofibres via copolymerization and electrospinning, *Macromolecular Rapid Communications*, **27**, 1533–8.

Wang, X., Chen, X., Yoon, K., Fang, D., Hsiao, B. S. and Chu, B. (2005a), High flux filtration medium based on nanofibrous substrate with hydrophilic nanocomposite coating, *Environmental Science & Technology*, **39**, 7684–91.

Wang, Y. Z., Yang, Q. B., Shan, G. Y., Wang, C., Du, J. S., Wang, S. G., Li, Y. X., Chen, X. S., Jing, X. B. and Wei, Y. (2005b), Preparation of silver nanoparticles dispersed in polyacrylonitrile nanofibre film spun by electrospinning, *Materials Letters*, **59**, 3046–9.

Wei, Q. F., Gao, W, D. and Hou, D. Y. (2005), Surface modification of polymer nanofibres by plasma treatment, *Applied Surface Science*, **245**(1–4), 16–20.

Wei, Q. F., Tao, D., Gao, W. D. and Huang, Y. B. (2008), Scanning electron microscopy and atomic force microscopy of composite nanofibres, *Microscopy and Analysis*, **22**(2), 11–2.

Wei, Q. F., Wang, H. F., Xu, Y. and Den, B. Y. (2010), Surface functionalization of polymer nanofibres by ITO sputter coating, *Journal of Coatings Technology and Research*, **7**(4), 511–4.

Wong, K. H., Hutter, J. L., Zinke-Allmang, M. and Wan, W. K. (2009), Physical properties of ion beam treated electrospun poly(vinyl alcohol) nanofibres, *European Polymer Journal*, **45**(5), 1349–58.

Wu, N., Shao, D. F., Wei, Q. F., Cai, Y. B. and Gao, W. D. (2009), Characterization of PVAc/TiO2 hybrid nanofibres: From fibrous morphologies to molecular structures, *Journal of Applied Polymer Science*, **112**, 1481–5.

Xiao, X. L., Chen, F., Wei, Q. F. and Wu, N. (2009), Surface modification of polyester nonwoven fabrics by Al2O3 sol–gel coating, *Journal of Coatings Technology and Research*, **6**, 537–41.

Yoon, Y. I., Moon, H. S., Lyoo, W. S., Lee, T. S., and Park, W. H. (2009), Superhydrophobicity of cellulose triacetate fibrous mats produced by electrospinning and plasma treatment, *Carbohydrate Polymers*, **75**, (2), 246–50.

Yoshikawa, M., Nakai, K., Matsumoto, H., Tanioka, A., Guiver, M. D. and Robertson, G. P. (2007), Molecularly imprinted nanofibre membranes from carboxylated polysulfone by electrospray deposition, *Macromolecular Rapid Communications*, **28**, 2100–5.

Yoshimatsu, K., Ye, L., Lindberg, J. and Ioannis, S. C. (2008), Selective molecular adsorption using electrospun nanofibre affinity membranes, *Biosensors and Bioelectronics*, **23**, 1208–15.

Part II
Applications

7
Functional nanofibers for filtration applications

G. G. CHASE, The University of Akron, USA,
S. SWAMINATHAN, Hollingsworth and Vose Company,
USA and B. RAGHAVAN, Parker Hannifin Corporation, USA

Abstract: In this chapter the advantages of incorporating nanofibers into filter media are discussed, looking at the historical uses of rudimentary filters and their subsequent development. Key properties of nanofibers that are useful for enhancing filter performance are described. A filter model is used to demonstrate the effects of nanofibers on capture efficiency and pressure drop for a filter. Filter media constructed with catalysts supported on ceramic nanofibers are described and experimental results for conversion of NO and CO to NO_2 and CO_2 are discussed.

Key words: filtration, filter media, nanofibers, catalyst.

7.1 Introduction

For many years filter manufacturers have used nanofibers in filter media to enhance performance. Only in the last few decades have methods been developed to produce nanofibers on a large scale. With greater access to nanofibers, the filter industry is actively looking for ways to further enhance filter performances through functionalization. In this chapter the advantages of nanofibers in filter media are discussed and a few examples of functionalized fibers are described.

Air filtration has a history that is at least 2000 years old, with the earliest-known records from the time of the Romans (Secundus, AD 77) in which Pliny the Elder describes problems with dust in mines in Egypt. Plinius describes the use of loose bladders to prevent dust inhalation. The literature of the 1500s suggests that fiber fabrics were used for protection from fumes and dust (Agricola, 1556). Respirators using cotton wool were available at the beginning of the twentieth century (Feldhaus, 1929).

In the twentieth century significant advances were achieved in fibrous filter media, driven in part by the need for respirators for personal protection of soldiers from chemical warfare agents during World War I. Harmful gases were found to be separated relatively well using fibrous pads to remove particles from the air and charcoal granules for adsorption of chemical vapors (Albrecht, 1931). During World War II it was found empirically that asbestos fibers formed very effective aerosol filters due to their small fiber diameters (Spurny, 1998). The use of asbestos fibers in filters ended in the 1970s when it was determined that the carcinogenic asbestos fibers were released from the filters. Other submicron-sized fiber materials subsequently replaced the asbestos fibers, including glass, carbon and ceramic.

Filter manufacturers have used polymeric fibers in filter media since about the 1950s, particularly with the development of the melt blowing process to fabricate micron-sized nonwoven fiber mats. Patents in the early 1900s describe electrostatic spinning to make submicron-sized fibers, but production rates were too low to be of practical interest for filter manufacturing. Some manufacturers make filter media of microfibers. To apply nanofibers they add a layer of nanofibers onto the surface of the microfiber media. The manufacturers do not divulge the process by which they do this (H. Y. Chung, Donaldson Company, personal communication, 2000). Since the mid-1990s interest has grown significantly in polymer nanofibers produced by electrospinning. Reneker *et al.* published a seminal paper that describes and models the electrospinning process (Reneker *et al.*, 2000).

Filter efficiency is known to improve as the size of the fibers decreases. Hinds (1999) calculated the filter capture efficiency as a function of fiber size as shown in Fig. 7.1. The efficiency curves have a 'V' shape with the apex of the 'V' indicating the most penetrating particle size. As the fiber diameter decreases, the most penetrating particle size decreases and the capture efficiency of the most penetrating particle size increases.

Nanofibers have a very high surface area per unit mass that enhances capture efficiency and other surface area-dependent phenomena that may be engineered into the fiber surfaces (such as catalysis or ion exchange).

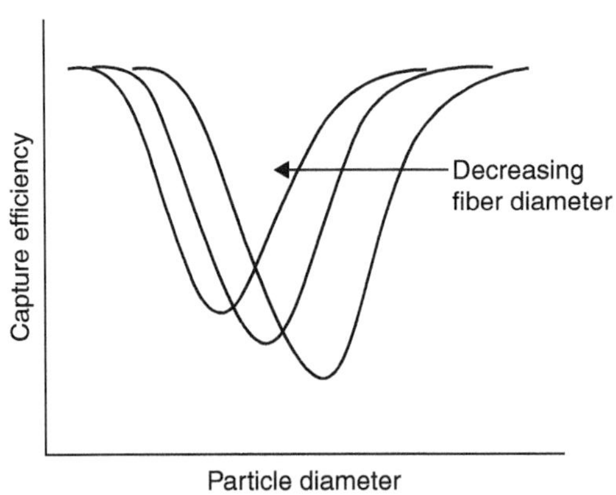

7.1 Filter capture efficiency as a function of particle diameter and fiber size (Hinds, 1999). The most penetrating particle size typically occurs between 100 and 500 nm.

Society needs enhanced filters to remove contaminants and to protect people and the environment. Recent developments in nanoscale materials have elevated general public awareness and concerns about potential harmful effects of man-made nanoparticles. Nanofibers enhance filter performance for capture of naturally occurring nanoparticles such as viruses, as well as micron-sized particles such as bacteria or man-made particles such as soot from diesel exhaust.

Society needs efficient filters for many applications; three in particular are mentioned here. First is air filtration for personal protection via respirators and face masks. In 1918–19, estimates show up to about 40 million people died from the 'Spanish Flu' H1N1 pandemic. The 'Asian Flu' H2N2 influenza in 1957–58 killed 1 million people worldwide and the 'Hong Kong Flu' H3N2 virus in 1968–69 killed more than 1 million people worldwide (Institute of Medicine of the National Academies, 2006). There is significant concern about future potential human pandemics from the 'Bird Flu' H5N1 virus or other agents. According to the Centers for Disease Control (CDC), flu seasons are unpredictable and can be severe. Over a period of 30 years, between 1976 and 2006, estimates of flu-associated deaths range from a low of about 3000 to a high of about 49 000 people.

It is estimated that the 2009 H1N1 pandemic resulted in more than 12 000 flu-related deaths in the United States. In contrast to seasonal flu, nearly 90% of the deaths occurred among people younger than 65 years of age (Centers for Disease Control and Prevention, 2010). There are also significant potential hazards from bioterrorism threats involving dangerous airborne viruses, toxins, chemicals and pathogens.

In a bioterrorism attack, facemasks and air filters that capture nanoparticles, pathogens and toxins will play a key role in ensuring the safety of the population. In a pandemic situation, vaccines and antiviral medications may not exist or may be available in limited quantities; thus increased reliance must be placed on face masks and air filters to minimize disease transmission. In addition to face masks, point-of-use and point-of-entry air filters in homes and buildings can provide an extra measure of safety. Large-scale production of nanomaterials (e.g., carbon black, titanium dioxide) has been ongoing for decades; the industry is now poised for rapid growth as new potential applications are discovered in electronics, medicine, energy and the environment. Therefore, transformative research on advanced filtration systems is required to provide necessary protection from airborne threats.

The second application is safe potable water. According to the World Health Organization (WHO, 2008), approximately 1 billion of the world's total population of 6.1 billion lacks access to potable water, and about 2.2 million, mostly children in developing countries, die each year from waterborne diseases (WHO, 2008). Among the most important diseases transmitted by water are cholera, typhoid, malaria and hepatitis A. Access to potable water would transform the lives of people in developing countries (Hillie & Hlophe, 2007). The US Geological Survey (USGS, 2010) points out that more safe water would be available for consumption if technologies can be developed to efficiently detect and remove contaminants from water. Environmental regulations require minimum contaminant levels of a wide range of microorganisms, organic molecules (pesticides, drugs, industrial by-products) and inorganic molecules (metals) (USEPA, 2010).

The third application is in biopharmaceuticals. Biopharmaceuticals are therapeutic molecules of biological origin (antibodies, antibody fragments, peptides, viral vaccines) and are the fastest-growing segment of the pharmaceutical industry (with ~ $75 billion in worldwide sales in 2007) (Levine, 2009). Approximately $9.7 billion of biopharmaceutical sales in 2007 were from therapeutics made directly from pooled human plasma from donors (Belay & Schonberger, 2005). These molecules include albumin, polyclonal antibodies, alpha-1 proteinase inhibitor, fibrinogen and blood clotting factors. The concern with application of biopharmaceuticals is the potential for diseases transmittable via infected blood and plasma, for example variant Creuzfeldt Jakob Disease (vCJD, the human form of mad cow disease transmitted by prion proteins), Dengue viruses and the Babesia species of protozoa. As of 2007, there were 203 reported cases of vCJD in the United Kingdom France, US, Ireland, the Netherlands, Italy, Japan, Canada, Portugal, Spain and Saudi Arabia. It is estimated that one in 4000–10 000 blood donors in the United Kingdom are infected with vCJD (Williamson, 2009). In addition, several bioterrorism agents were classified by the CDC as risks to the blood supply – for example, anthrax, botulism, plague, small pox, tularemia and viral hemorrhagic fevers. Efficient protein capture and purification filtration devices are needed that will reduce the risk of disease transmission through biopharmaceutical therapies. Filter media, constructed with selectively functional nanofibers, have potential for removal of proteins and other contaminants to prevent transmission of diseases.

In this chapter the key principles of filtration are explained and the advantages of nanofibers in filter media are discussed. A simplistic model is described that shows the improved performance due to the presence of nanofibers in microfiber media.

Filter media with functionalized nanofibers are discussed by way of the example of catalysts supported on ceramic nanofibers. A ceramic filter medium for use in filtering and reacting with gases in engine exhaust is described, modeled and experimental data are presented.

Some future trends in filter media development are discussed in the last section of the chapter.

7.2 Key principles of nanofibers for filtration

7.2.1 Specific surface area

The small diameters of nanofibers are useful in filter media primarily due to their increase in surface area per unit mass of fiber. Figure 7.2 shows the external surface area per gram of fiber as a function of fiber diameter for typical polymer fibers having a mass density of 900 kg/m^3 and typical ceramic fibers having a mass density of 2500 kg/m^3. The plot shows that 100 nm diameter fibers have 100 times as much surface area as 10 micron (10 000 nm) diameter fibers. This significant increase in surface area has a large effect on momentum, energy and mass transfer in filter media. Figure 7.3 shows the significant size contrast between the electrospun fibers and typical microfibers. Figure 7.4 shows nylon nanofibers mixed among glass microfibers in a filter medium; the nanofibers cross over the pore openings between the microfiber in the medium.

7.2.2 Continuum, slip and molecular flows

The size of the fibers affects filter performance in a number of ways. In fluid flow, momentum is transferred from the moving fluid to stationary solid surfaces at the fluid boundaries. For flow through a filter, the fluid transfers significant force to the fibers as it flows around the stationary fiber surfaces. The greater the surface area, the more momentum is transferred. Hence, holding the mass of fibers in a filter medium constant, as the fiber diameter decreases the area increases and the momentum transfer increases. This

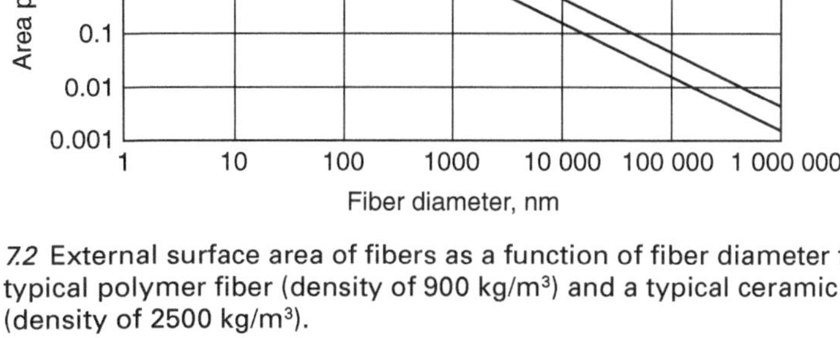

7.2 External surface area of fibers as a function of fiber diameter for a typical polymer fiber (density of 900 kg/m^3) and a typical ceramic fiber (density of 2500 kg/m^3).

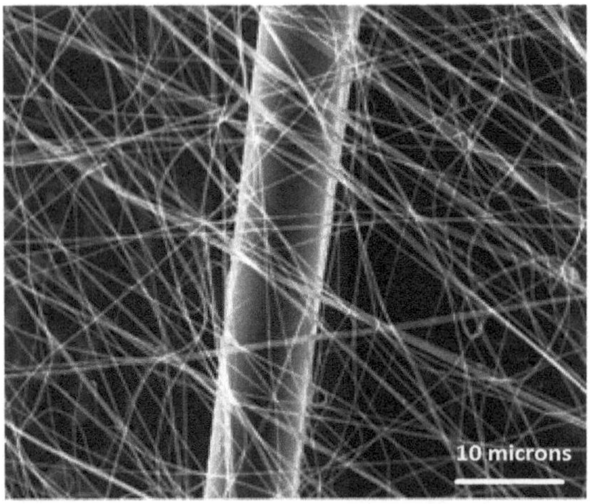

7.3 Electrospun Nomex™ nanofibers layered across an 8 micron glass fiber.

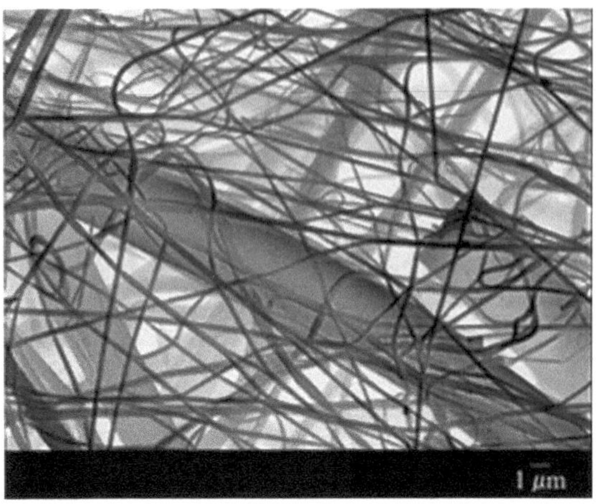

7.4 Nylon nanofibers are mixed among glass microfibers in a filter medium. The nanofibers span the pore openings between the microfibers.

results in an increase in the pressure drop across the filter which is directly related to the operating cost of the filter.

Gases can be modeled as continuous fluids provided that objects in the flow path are large compared to the distance between the gas molecules (i.e., mean free path). This is referred to as 'continuum flow'. The Navier–Stokes equations (Stokes, 1845; Bird *et al.*, 2002) are used to model continuum flows, accounting for mass continuity, linear and angular momentum and mechanical energy. Commercial computational fluid dynamics (CFD) software may be used to rigorously model flow and pressure profiles in fibrous media. The most realistic predictive models are those that apply the CFD to fiber media geometries determined from measurements of actual media taking into account small-scale inhomogeneities (Jaganathan *et al.*, 2008).

Historically, models based on creeping flow assumptions have been applied to simplified fiber media geometries to deduce pressure drop correlations. The correlations were adjusted by fitting to experimental data to better predict the relationship between pressure drop and flow rate. One of the more popular approaches is the cell model approach by Happel (1959) and Kuwabara (1952) in which the flow around the fiber is assumed to be symmetric (i.e., perpendicular to the fiber axis and hence is two-dimensional in angle and radius). The creeping flow and symmetry assumptions allow the flow equations to be simplified to a simple biharmonic equation

$$\nabla^4 \psi = 0 \tag{7.1}$$

where ψ is the stream function related to the velocities (Bird *et al.*, 2002). Differential equations of this type have an infinite number of solutions. A functional form fitting the solution is assumed and the parameters in the function are fitted. The stream function solution is used directly to determine the pressure drop (Imai, 1951). Further discussion on this topic is given in Brown (1993).

The continuum flow theory described above assumes that the gas velocity in the vicinity of the surface of the stationary fibers has zero velocity. Given the molecular nature of the gas, this assumes that the fiber diameter is much larger than the mean distance between the gas molecules. Many gas molecules are in contact with the fiber surface and hence the mass-averaged velocity (continuum velocity) of the gas is the same velocity as the fiber surface (zero velocity). As the fiber diameter decreases this is not necessarily true. The mean free path between air molecules at 1 atm absolute pressure and 21°C is about 63 nm. Typical electrospun fibers can have diameters less than this dimension. The small dimension of the fibers allows many of the gas molecules to 'slip' past the fiber without colliding with it. The net result is that some of the gas molecules near the fiber surface do not take on the velocity of the fiber surface. When all of the molecular velocities are averaged in a small volume element at the fiber surface to determine the continuum velocity, the velocity of the gas at the fiber surface is not zero.

As indicated in Fig. 7.5, the gradient in the gas velocity at the fiber surface is significantly smaller for slip flow than for continuum flow conditions. The velocity gradient is proportional to the drag force of the moving fluid on the fiber. Hence, less drag force per unit length of fiber is required for the gas to flow around a fiber in slip flow than in continuum flow.

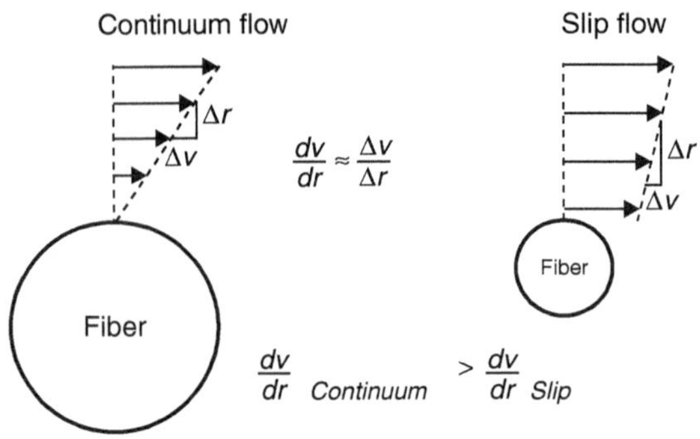

7.5 Gas velocities for flow past fibers in continuum flow and slip flow. The velocity gradient in slip flow is less than the gradient in continuum flow, hence the force per unit area on the fiber in slip flow is less than that in continuum flow.

Continuum and slip flows are characterized by the Knudsen number (Brown, 1993)

$$Kn = \frac{2\lambda}{d_f} \quad [7.2]$$

where λ is the mean free path between the gas molecules and d_f is the fiber diameter. The mean free path may be calculated by the expression (Bird *et al.*, 2002)

$$\lambda = \frac{1}{\sqrt{2\pi d^2 n}} \quad [7.3]$$

where d is the molecule diameter and n is the number of molecules per unit volume. n is calculated as the product of the molar concentration obtained from the ideal gas law and Avogadro's number.

The flow around a fiber is considered to be in the continuum flow regime when $Kn < 0.01$ and to be in the slip flow regime when $0.01 < Kn < 0.25$. When $10 < Kn$ the flow is in the molecular flow regime in which free molecular theory applies. Between the slip and molecular flow regimes is a transitional flow regime. Correlations relating pressure drop to flow rates for gas flows through fiber media are available in literature for the continuum, slip and molecular flow regimes (Brown, 1993). Few correlations are available for the transition flow regime (Zhao & Povitsky, 2009).

7.2.3 Fiber and mat strength

When you reduce the size of a material object from meters to microns in size, the material properties such as density, elasticity and thermal conductivity remain the same as the bulk material property. However, when the material size is reduced to a few tens of nanomaterials, the molecules can orient, which changes the properties from the bulk. Furthermore, if molecules on the surface have different properties than the interior, at the nanoscale the ratio of surface molecules to interior molecules significantly increases and hence the observed material properties of the nanoscale materials can differ significantly from bulk properties.

Garcia *et al.* (2010) describe how aligned molecules in small-diameter polyamides increase the fiber tensile strength over that of bulk material strength. This means that nanofibers can bridge across distances greater than could be possible if they had the same strength as the bulk materials. In filter media, nanofibers are commonly applied bridging gaps of tens or hundreds of microns.

7.2.4 Mass and heat transfer

Mass transfer and heat transfer can be modeled through transfer coefficients; heat and mass transfer depend on total surface area (Bird *et al.*, 2002). In heat transfer the total thermal energy crossing a fluid–solid interface is given by

$$Q = hA\Delta T \quad [7.4]$$

where h is the heat transfer coefficient, A is the surface area of the interface and ΔT is the temperature difference driving the heat transfer. Similarly, for mass transfer in a dilute system, the total rate of mass transfer across a fluid–solid interface is given by

$$W_A = kA\Delta x_A \quad [7.5]$$

where k is the mass transfer coefficient and Δx_A is the difference in molar fraction of chemical species A that drives the mass transfer. The transfer coefficients, h and k, are correlated to dimensionless groups such as Reynolds, Prandtl and Schmidt numbers. The mechanism of slip flow affects the rates of heat and mass transfer due to the different motions of the molecules in slip and continuum flows at the solid surface. Correlation dependence of the transfer coefficients on the Knudsen number for parallel plates (Inman, 1965), spherical particles (Chen & Pfender, 1983) and microchannels (Castelloes *et al.*, 2007) have been reported. Correlations of the effect of slip flow on the heat and mass transfer coefficients for flows around nanofibers are not available.

For filters with nanofibers the increase in surface area provides more surface for capture of nanoparticles. Small fiber diameters mean that chemical reactant and product species have much smaller distances to diffuse to reach catalyst reaction sites.

Typical catalyst pellets have small pores through which the chemical reactants and products diffuse; the pellets have characteristic dimensions of several millimeters. For a first order reaction on the internal surfaces of a spherical catalyst pellet having an outer radius R, the catalyst effectiveness factor, η, is given by (Thiele, 1939; Bird *et al.*, 2002; Fogler, 2006)

$$\eta = \frac{3}{K^2}(K \coth K - 1) \qquad [7.6]$$

where K is a dimensionless parameter given by

$$K = R\sqrt{\frac{k_1 a}{D}} \qquad [7.7]$$

Here, k_1 is the reaction rate constant, a is the pellet internal surface area per pellet volume and D is the effective diffusivity of the reactant chemical species through the pellet. The effectiveness factor is a multiplicative factor of how much the reaction rate decreases due to the resistance of diffusion in the pellet.

The usefulness of Equation [7.6] becomes apparent when the equation is plotted as shown in Fig. 7.6. When K is small, diffusion has negligible effect and η is approximately unity. When K is large the effectiveness factor decreases significantly, meaning that much of the internal pore surfaces do not contribute to the reaction. Effectively, this means that only the catalyst sites near the external surface of the pellet contribute to the reaction and the catalyst material in the interior of the pellet does not contribute to the reaction. When catalyst metals cost thousands of dollars per ounce this can be a significant ineffective use of resources.

For nanofibers of diameter d_f, the value of K becomes

$$K = \frac{3}{4} d_f \sqrt{\frac{k_1 a}{D}} \qquad [7.8]$$

and Equation [7.6] approximates the relation between K and η in cylindrical geometry. Hence, the K value for a 100 nm nanofiber is 10 000 times smaller than the K value for a 0.75 mm pellet. This means that a pellet of a size and material that has a low effectiveness factor can be much more effective if made into nanofibers. This can result in a significant reduction of the amount of catalyst needed for a particular application and a significant reduction in cost.

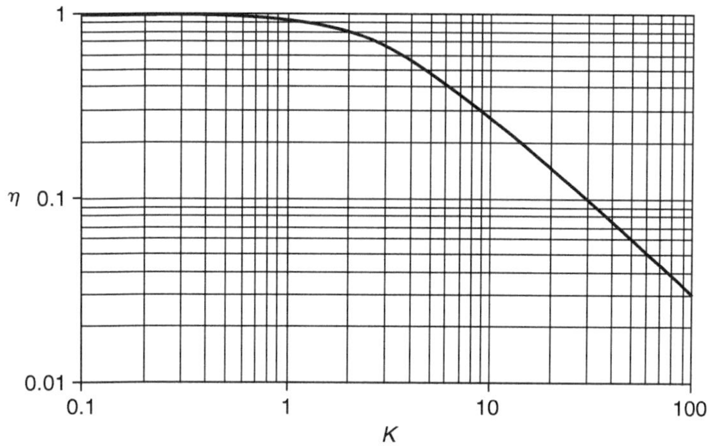

7.6 Effectiveness factor for a spherical catalyst pellet as a function of the dimensionless *K*.

7.3 Filtration of nanoparticles from gas streams

Nanoparticles can be man-made or naturally occurring. Examples of materials and approximate size ranges include tobacco smoke (20–1000 nm), carbon black (20–100 nm), colloidal silica (5–800 nm), viruses (5–200 nm) and albumin protein (9–100 nm). Fibrous filter media can efficiently and economically remove these particles from gas streams.

7.3.1 Filtration mechanisms

Filter performance is characterized by effectiveness in separating particles from the fluid stream and the pressure difference between the inlet stream and the effluent exiting the filter. The pressure difference is directly related to the energy required to operate the filter and thus is a metric of the operating cost.

Several mechanisms contribute to the overall separation performance of a filter. The most obvious mechanism is straining, which is when a particle is larger than a pore opening. Straining may occur at the inlet surface of the filter (and results in the formation of a cake on the filter surface) or it may occur in the interior of the filter where a pore throat is smaller than the particle. Straining may also occur when several particles simultaneously try to pass through an opening that is large enough for one particle at a time but not multiple particles simultaneously. In this latter occurrence the particles bridge over the pore opening and may be held in place at the pore opening due to pressure and fluid-dynamic forces. In fibrous media straining mostly occurs with particles larger than about 10 microns.

Fibrous media are normally designed for low pressure drop which requires pore openings of tens or hundreds of microns. These pore openings are much larger than the typical particles that are separated from gases. For separation of nanoparticles the pore openings are many orders of magnitude larger. The larger pore openings result in a lower pressure drop across the filter and hence reduced operating cost. Yet, fibrous filters can be very effective at separating nano- and micron-sized particles. The straining mechanism does not contribute significantly to the separation of these small particles, but there are several single fiber mechanisms that contribute to the separation effectiveness of these filters. In general, very small particles can be strongly influenced by weak surface forces such as van der Waal's forces and hydrogen bonding. If a small particle, less than about 10 microns, comes in contact with a fiber surface then these weak forces can cause the particle to attach to the fiber surface. The single fiber mechanisms of particle capture cause the particles to come into contact with the fiber surface to be captured.

The single fiber mechanisms comprise of direct interception, inertial impaction, diffusional deposition, gravitational settling and electrostatic charge. These mechanisms are well described in references (see Brown, 1993) and only a brief description is given here. As shown in Fig. 7.7, direct interception involves a particle following a fluid stream line that results in the particle coming into contact with the fiber surface. Inertial impaction occurs when the fluid stream line bends around a fiber but the particle has enough mass that its inertia causes it to deviate from the stream line and to contact the fiber surface. Diffusional deposition is a combined action of fluid flow and Brownian motion that brings a particle into contact with a fiber surface. In gravitational settling the particle's movement deviates from the flow stream line due to gravitational force acting on the particle and results in the particle contacting the fiber surface. The last mechanism of electrostatic charge is the effect of the electrostatic attraction force between the particle and the fiber when the particle and fiber have opposite charges.

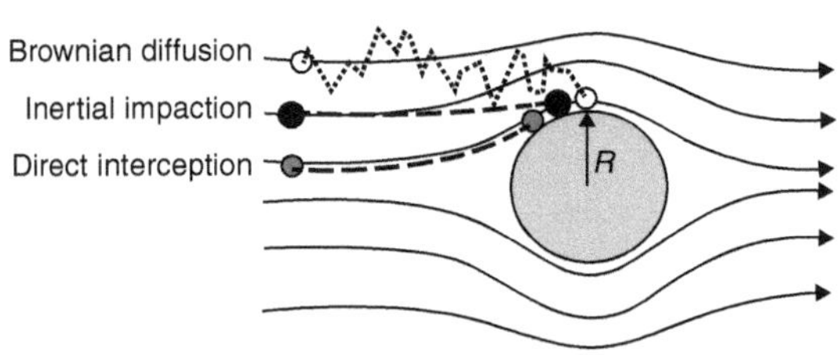

7.7 Particle capture mechanisms for particles carried by gas flow past a fiber of radius R. The particles follow or deviate from the gas streamlines (lines with arrows) based upon the mechanisms of direct interception, inertial impaction, or Brownian diffusion.

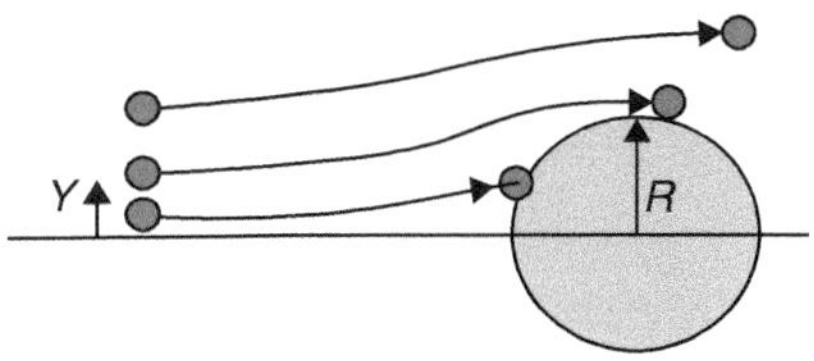

7.8 Concept of fiber capture efficiency. Particles initially carried by streamlines within distance *Y* of the centerline are captured on the fiber surface.

Correlations for calculating single fiber capture efficiency are reported in literature (Brown, 1993, 1998). The single fiber efficiency is defined based upon the fiber's projected area. If a fiber is oriented with its axis perpendicular to the flow, the projected area is the product of length multiplied by diameter. A fiber that has an efficiency of unity removes from the air all of the particles that lie within the volume swept out by its area and the velocity vector of the air, the latter assumed to be flowing uniformly. In general, a fiber does not remove all of these particles, and the single fiber efficiency, E, is defined as the quotient of the number of particles actually removed and the number that would be removed by a 100% efficient fiber. This value is proportional to the area through which the stream lines flow, YL, divided by the projected area RL where R and Y are indicated in Fig. 7.8 and L is the length of the fiber. The efficiency is given by

$$E = \frac{Y}{R} \tag{7.9}$$

Correlations are available that can be used to calculate the single fiber efficiency, E, as a function of the fluid properties, particle size, fiber size, volume fraction of fibers in a filter and the fluid approach velocity (Brown, 1993). To model filters with nanofibers we need to relate the single fiber efficiency to the overall filter performance.

7.3.2 Layer efficiency

The penetration of particles of a particular size through a depth filter of homogeneous properties is a function of the distance into the filter. A filter layer of thickness, dz, challenged by an incident number of particles, N, will capture dN number of particles. The number of captured particles is proportional to the thickness and the incident number of particles. The proportionality constant, α, is called the filter coefficient, defined by the expression

$$\alpha = \frac{-1}{N}\frac{dN}{dz} \tag{7.10}$$

Let α_i be defined as the projected area of fibers of diameter d_i per volume of filter of layer thickness Δz. By definition of E, then, the filter coefficient is related to α_i by the expression

$$\alpha = E_i \alpha_i \qquad [7.11]$$

where E_i is the single fiber efficiency for the fibers of size d_i. If the filter has more than one fiber size then

$$\alpha = \sum E_i \alpha_i \qquad [7.12]$$

To determine α_i, let ε_i represent the volume fraction of fibers of size d_i in the filter layer and let L_i be the sum of lengths of all of the fibers of size d_i in the layer. This analysis assumes that all of the fiber surfaces are available for particle capture and neglects fiber–fiber contacts that may cause shadow effects that cover or block surfaces of fibers.

By definition of α_i

$$\alpha_i = \frac{\text{Projected area of fibers of size } d_t}{\text{Volume of the filter layer}} = \frac{d_i L_i}{\text{Vol}} \qquad [7.13]$$

and by definition of ε_i

$$\varepsilon_i = \frac{\text{Volume of fibers of size } d_i}{\text{Volume of the filter layer}} = \frac{\left(\pi d_i^2/4\right) L_i}{\text{Vol}} \qquad [7.14]$$

Equations [7.13] and [7.14] combine to yield an expression for α_i as

$$\alpha_i = \frac{4\varepsilon_i}{\pi d_i} \qquad [7.15]$$

Hence, combining Equations [7.15] and [7.12] the filter coefficient for a filter layer of multiple fiber sizes can be estimated by

$$\alpha = \sum \frac{4E_i \varepsilon_i}{\pi d_i} \qquad [7.16]$$

7.3.3 Species balance and prediction of particle concentration in the gas stream

A volume-averaged continuum theory is used to model the process (Chase, 2002). To simplify the model equations, the following assumptions are made:

the medium fibers are rigid and stationary, the medium is incompressible, the process is isothermal, the flow is constant and unidirectional in the z direction, the particles are captured by the fibers, the fibers are rigid and stationary, the fibers are randomly oriented, the fibers are of one diameter, the particles are of one diameter, gravity is neglected, the permeability of medium is uniform and isotropic and the porosity of the medium is uniform and constant. The effect of particle loading is neglected in this model. The filter medium is assumed to be constructed of micro- and nanosized fibers. The nanofibers are assumed to be uniformly mixed with the microfibers.

The species conservation balance of gas phase and the momentum balance of the gas phase are the starting equations for the model. The species balance determines the aerosol outlet concentration. The gas phase momentum balance determines the pressure drop across the media.

The full set of conservation equations are too complex to model directly. The above assumptions are applied to simplify the equations. The simplified gas particle species balance and gas phase momentum balances are

$$\frac{\partial(\varepsilon\rho_A v_z)}{\partial z} + I = 0 \qquad [7.17]$$

$$\varepsilon\frac{\partial P}{\partial z} + F_z = 0 \qquad [7.18]$$

The momentum balance is used later to determine the pressure drop. The mass fraction of particles in the gas phase is the ratio of particle mass concentration to the gas phase density,

$$w = \frac{\rho_A}{\rho} \qquad [7.19]$$

Rewriting the conservation Equation [7.17] gives

$$\varepsilon\rho v_z \frac{\partial w}{\partial z} + I = 0 \qquad [7.20]$$

The species balance is solved by introducing a constitutive equation for the mass transfer term, I, as

$$I = \alpha w \rho \varepsilon v_z \qquad [7.21]$$

Equations [7.20] and [7.21] are combined to obtain

$$\frac{dw}{dz} = -\alpha w \qquad [7.22]$$

Upon integrating Equation [7.22] from the inlet at $z = 0$ to an arbitrary position, z, we get

$$w(z) = w(0)\exp(-\alpha z) \quad [7.23]$$

and the outlet concentration is obtained at $z = L$

$$w_{OUT} = w_{IN}\exp(-\alpha L) \quad [7.24]$$

Correlations for the single fiber capture efficiencies, E_i, are available from literature (Brown, 1993). For submicron-sized particles the dominant mechanisms are direct interception and Brownian diffusion. Other mechanisms such as inertial impaction are present but are negligible.

A formula for calculating the single fiber capture efficiency by diffusion and direct interception and accounting for aerodynamic slip is

$$E_{DR} + E_D + E_R + E'_{DR} \quad [7.25]$$

where the Brownian diffusion efficiency is calculated from

$$E = 2.7\, Pe^{-2/3} \lfloor 1 + 0.39\, \zeta^{-1/3}\, Pe^{-1/3}\, \mathrm{Kn} \rfloor \quad [7.26]$$

$$\zeta = -0.5\ln(\varepsilon_i) - 0.52 + 0.64\varepsilon_I + 1.43(1 - \varepsilon_i)Kn \quad [7.27]$$

and the direct interception efficiency is calculated from

$$E_R = \frac{1}{2\zeta\left[(1+N_R)^{-1} - (1+N_R) + 2(1+N_R)\ln(1+N_R) + 2.86Kn(2+N_R)N_R(1+N_R)^{-1}\right]} \quad [7.28]$$

The combination term is

$$E'_{DR} = 1.24\zeta'^{-1/2} Pe^{-1/2} N_R^{2/3} \quad [7.29]$$

where N_R is the ratio of the diameters of the particles to the fibers,

$$N_R = \frac{d_p}{d_i} \quad [7.30]$$

and the Knudsen number, Kn, is defined in Equation [7.2].

The Peclet number is

$$\mathrm{Pe} = \frac{d_i v_i}{D} \quad [7.31]$$

where the diffusion coefficient for the particles is given by

$$D = \eta k_B T \quad [7.32]$$

Here, η is the mobility given by

$$\eta = \frac{\mathrm{Cn}}{3\pi\eta d_p} \quad [7.33]$$

and Cn is the Cunningham factor

$$\mathrm{Cn} = 1 + \frac{2A\lambda}{d_p} + \frac{2Q\lambda}{d_p}\exp\left(\frac{Bd_p}{2\lambda}\right) \quad [7.34]$$

The constants are $A = 1.246$, $Q = 0.42$ and $B = 0.87$ and $\lambda = 0.069$ μm is the mean free path of air molecules at 25°C and 1 atm pressure. Hence, with these equations we can estimate the filter coefficient, α. With the filter coefficient we can estimate the outlet concentration of the filter using Equation [7.24].

7.3.4 Calculation of pressure drop

The total force acting on all fibers, F_{tot}, is related to the pressure drop by

$$F_{\mathrm{tot}} = A\Delta P \quad [7.35]$$

The total force is related to the drag force per volume of filter, F_z in Equation [7.18], assuming the filter has uniform properties, by the expression

$$F_{\mathrm{tot}} = ALF_z \quad [7.36]$$

We define the force per unit length of fiber as f_i. For a medium with multiple fiber diameters,

$$F_{\mathrm{tot}} = \sum L_i f_i \quad [7.37]$$

In integration of Equation [7.18], assuming uniform properties over the filter thickness, the pressure drop is related to the drag force by

$$\Delta P = \frac{F_z L}{\varepsilon} \qquad [7.38]$$

Combining Equations [7.35]–[7.38] we get the expression for the drag force to be

$$F_z = \frac{\varepsilon \sum L_i f_i}{AL} \qquad [7.39]$$

With help of Equation [7.14], the drag force becomes

$$F_z = \frac{4\varepsilon}{\pi} \sum \left(\frac{\varepsilon_i}{d_i^2} f_i \right) \qquad [7.40]$$

For a medium of just one fiber size, combining Equations [7.38] and [7.40] gives an expression relating the force per length of fiber to the pressure drop

$$f_i = \frac{\Delta P}{L} \frac{\pi d_i}{4\varepsilon_i} \qquad [7.41]$$

Using correlations for the pressure drop for a filter medium of fibers of one size, then by inspection with Equation [7.41] we can deduce an expression for f_i. The correlations for f_i account for continuum, slip or molecular flow phenomena, depending on the diameters of the fibers.

For $Kn < 0.25$ (for continuum and slip flow of the microfibers), we get a correlation for the pressure drop to be

$$\Delta P = \frac{16\mu L v_Z^G (1 + 1.996Kn)\varepsilon_i}{d_i^2\left[\left(-0.5\ln(\varepsilon_i)\right) - 0.75 + \varepsilon_i - \left(\varepsilon_i^2/4\right) + 1.996Kn\left(-0.5\ln(\varepsilon_i) - 0.25 + \left(\varepsilon_i^2/4\right)\right)\right]} \qquad [7.42]$$

By inspection, the expression for f_i is

$$f_i = \frac{4\mu\pi v_V^G (1 + 1.996Kn)}{\left[-0.5\ln(\varepsilon_i) - 0.75 + \varepsilon_i - \left(\varepsilon_i^2/4\right) + 1.996Kn\left(-0.5\ln(\varepsilon_i) - 0.25 + \left(\varepsilon_i^2/4\right)\right)\right]} \qquad [7.43]$$

For $Kn > 10$ (for molecular flow with very small nanofibers) we have

$$\Delta P = \frac{4.58\mu\varepsilon_i L v_z}{d_i \lambda} \quad [7.44]$$

Hence, the force per unit length of fiber becomes

$$f_i = \frac{2.29\pi\mu v_z}{Kn} \quad [7.45]$$

Correlations for the transition flow range between the slip flow regime $0.25 < Kn < 10$ are not readily available in literature. For fibers in the intermediate range in this model, we assume a linear interpolation between the slip flow value at $Kn = 0.25$ and the molecular flow value at $Kn = 10$ to estimate the intermediate value of f_i. These equations allow us to calculate the pressure drop for a medium with multiple fiber diameters, by the equation

$$\Delta P = \frac{4L}{\pi} \sum \frac{\varepsilon_i}{d_1^2} f_i \quad [7.46]$$

7.3.5 Calculation of quality factor

The quality factor (QF, also known as figure of merit) is given by Brown (1993) as

$$\mathrm{QF} = \frac{-\ln\left(w_{\mathrm{out}}/w_{\mathrm{in}}\right)}{\Delta P} \quad [7.47]$$

and represents the ratio of the capture performance to the power required to achieve the capture performance. In general, media with higher QFs perform better than media with lower QFs. A filter with a larger QF has a greater capture efficiency than a filter with a lower QF when operated at the same pressure drop. Substitution of Equations [7.24] and [7.46] into [7.47] gives the working equation,

$$\mathrm{QF} = \frac{\alpha}{(4/\pi)\sum\left(\left(\varepsilon_i/d_i^2\right)f_i\right)} \quad [7.48]$$

for calculating the QF. Of most interest is how much change occurs in the QF due to the presence of the nanofibers. We define the relative quality factor, RQF, as

$$RQF = \frac{QF}{QF_{Basis}} \tag{7.49}$$

where the basis QF is the QF for the filter medium without nanofibers.

7.3.6 Model results

The model is easily programmed into a spreadsheet for calculations. As an example, the filter is modeled at room temperature, atmospheric pressure, with air face velocity of 2.1 m/s and constant, uniform 10% saturation. All filters are 1 cm thick. Porosity is determined from the volumes of the fibers present in the filter.

The amount of nanofibers in the filter is calculated as the ratio of the external surface area of the nanofibers divided by the external surface area of the microfibers in the filter medium. An area ratio of zero means no nanofibers are present in the filter.

Figure 7.9 shows the RQF curves for three particle sizes for 3 micron diameter microfibers and 300 nm diameter nanofibers. The RQF increases rapidly with small additions of nanofiber, reaches a maximum at an area ratio of about 1–3 (i.e., 1:1 to 3:1), then slowly declines. The rapid increase in RQF is due to the significant increase in surface area available for capture of particles. However, as more nanofibers are added to the filter, the pressure drop due to the added fibers increases faster than the corresponding increase in particle capture and the RQF passes through a maximum and then declines. Figure 7.9 also shows that the RQF is sensitive to the size of the particles being captured.

Figure 7.10 is similar to Fig. 7.9, but three sizes of microfibers are compared. The 5 micron fiber curve shows the largest RQF because these fibers have the largest difference in size between the micro and nanofibers. If the magnitudes of the QFs were compared instead of the RQFs, the 1 micron fibers have the highest QFs. This plot shows that the filters made of larger microfibers will have the greatest increase in RQF with the addition of nanofibers. These plots show the same trend of a rapid rise in RQF, with a maximum occurring area ratios of about 1–3.

Figure 7.11 shows a similar plot for different sizes of nanofibers. For the 150 nm particle size, the 100 nm fibers have the highest RQF of those compared. Maximums in the RQF curves occur in the range of about 1–3 area ratios.

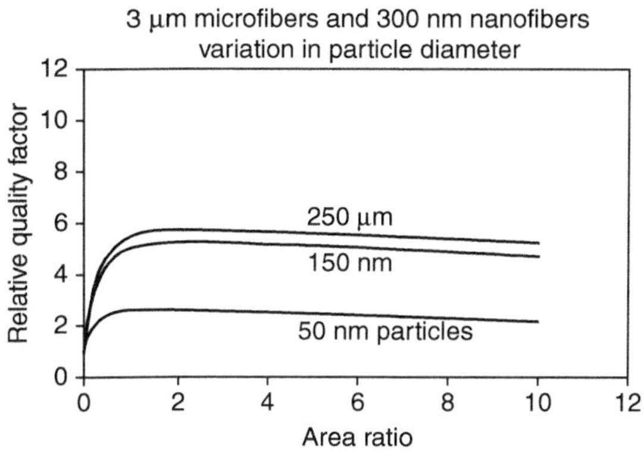

7.9 RQFs for different sizes of particles.

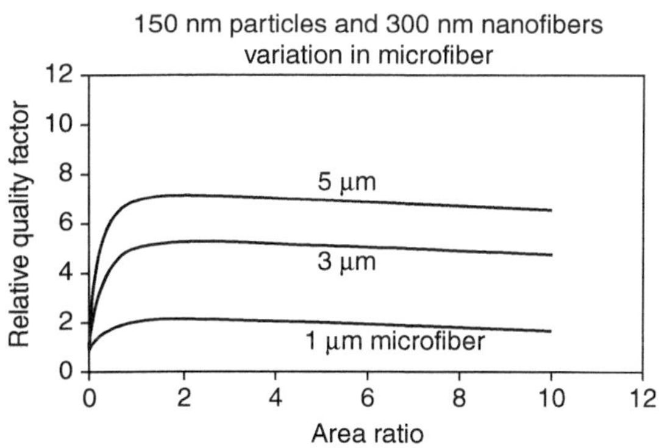

7.10 RQFs for different size microfibers.

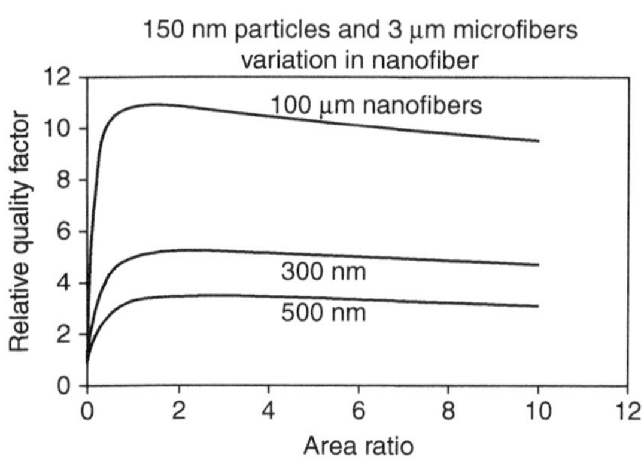

7.11 RQFs for different size nanofibers.

The plots in Figs 7.9 through 7.11 show that the RQF has a maximum relative to the amount of nanofibers added to the filter medium. The RQF also varies with fiber sizes and particle sizes. These variations show it is possible to optimize the filter design for the capture of particular particle sizes.

7.3.7 Summary

A model was developed to calculate the RQF for filter media mixed with micro- and nanofibers. The model was based on assumptions such as uniform porosity and neglects particle loading. The single fiber mechanisms of drag force and particle capture are linearly summed to estimate the performance of a bicomponent fiber filter medium. The mechanisms account for the effects of slip and molecular flow phenomena.

The model results show a significant increase in the RQF with the addition of small amounts of nanofibers. The RQF plots pass through a maximum and gradually decline as greater amounts of nanofibers are added to the filters. The dependence of the RQF on the fiber amounts, fiber sizes and particle sizes suggest that it may be possible to optimize the filter design.

7.4 Nanofiber catalyst support structures

The abatement of nitrogen oxides (NO, NO_2) and carbon monoxide caused by automotive and industrial exhausts is a topic of scientific and academic research. These gases are potential health hazards and greenhouse gases. Noble metal oxides can be supported by nanofibers and are used as catalysts for such heterogeneous catalytic reactions. These nanofibers have special electronic and morphological properties and high surface energy, which results in their unique catalytic selectivity. The noble metals such as palladium, platinum and rhodium have been incorporated into alumina nanofiber support by sol-gel processing and electrospinning techniques. Only palladium and rhodium catalysts are discussed in this work. The diameters of the nanofibers and nanoparticles were in the range of 60–160 nm and 2–10 nm, respectively. These nanofibers have been incorporated into alumina microfibrous filter media by a vacuum molding technique.

This filter media is successfully able to convert NO to N_2 and CO to CO_2. These filters can simultaneously remove particles from gas streams on the basis of size, shape and charge of the particles with respect to the properties of the filter medium. In this section the catalyst performance of the filter is discussed.

Palladium and rhodium are increasingly used in catalytic converters to reduce emissions due to nitric oxides (NO_x), carbon monoxide and hydrocarbons. Both NO_x and CO are harmful greenhouse gases and are

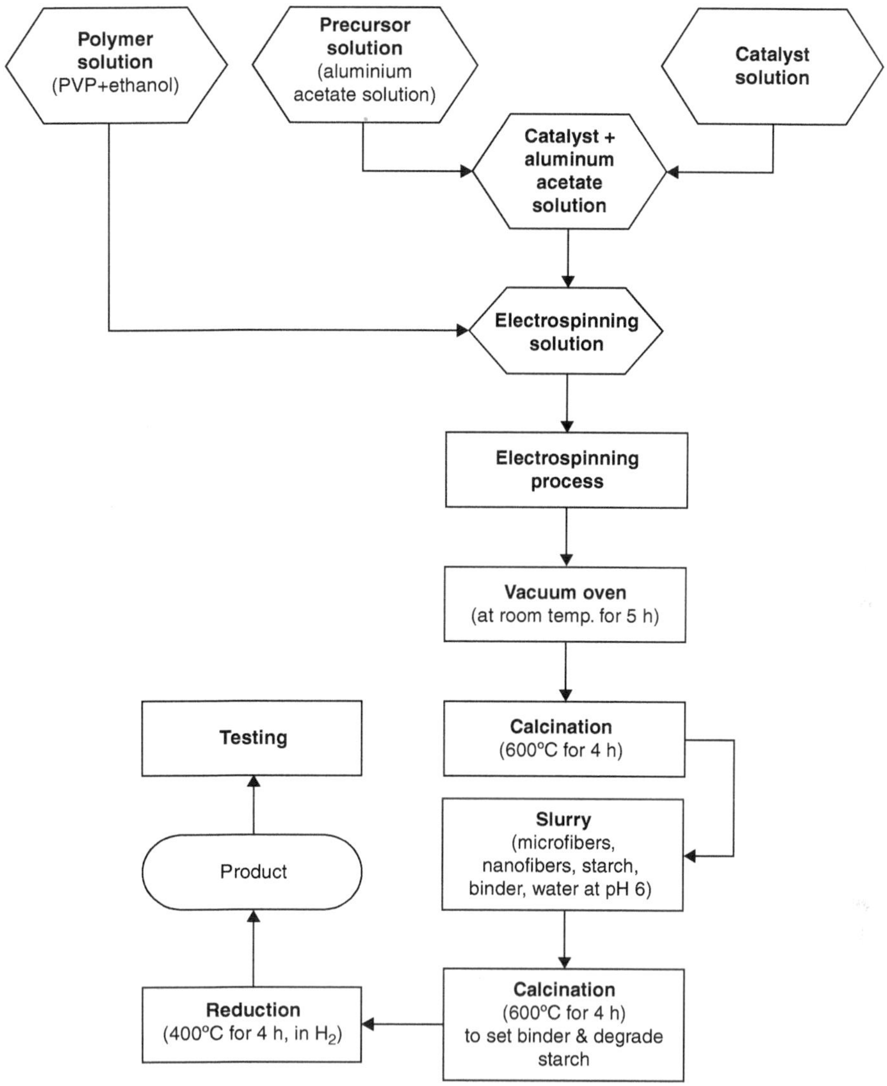

7.12 Process diagram for the manufacture of catalytic filter.

environmental hazards. Of the metals, Rh is primarily responsible for the reduction of NO_x. This is because of the greater ability of metallic Rh to dissociate NO as compared to that of Pd (Alkemade & Schumann, 2006; Belton & Taylor, 1999). Within the family of precious metals, Pd is usually less expensive than Rh. Hence the optimization of concentration of these metals is important. Pd is also less resistant to deactivation by poisons such as Pb and S compared to Rh (Farrauto & Heck, 1999).

Noble metals can be used as nanocatalysts to convert these toxic substances into less harmful forms from the various exhausts. Catalytic activity of supported metal particle catalysts is strongly dependent on the size and shape of the particles. Hence, nanocatalysts are very active since most of the particle surface is available for catalysis (Gai *et al.*, 2002). Metal oxide nanofibers are potential candidates as catalyst supports for heterogeneous catalytic reactions (Park & Baker, 1998; Patel *et al.*, 2007; Park *et al.* 2008). Nanofibers can be produced by electrospinning (Reneker *et al.*, 2000); as a catalyst support structure, they can also provide a means of constructing catalytic systems that are lightweight, thermally and chemically stable and with large surface areas. The small fiber diameters provide the structure with a high effectiveness factor, as discussed in Section 7.2.

7.4.1 Experimental description

The process of fabrication of a catalytic filter is described in Fig. 7.12. The nanofibers are first formed as polymer nanofibers with the ceramic precursors contained in the polymer. The polymer nanofibers are calcined to remove the organics and to convert the metals to metal oxides. The calcined nanofibers are characterized using scanning electron microscopy (SEM) and transmission electron microscopy (TEM). SEM images are used to obtain the diameter of the nanofiber mat and TEM is used to determine the catalytic particle size distribution.

The calcined nanofibers are immobilized into a microfiber filter media by a vacuum molding technique. The filter is characterized for porosity and permeability. The porosity is measured using a specially made pycnometer. The permeability is measured using a Frazier test. The noble metal oxides are reduced to noble metal by reaction with hydrogen gas to enhance the reactivity and selectivity of the catalyst particles on the alumina nanofibers. The reduced filter media are tested for activity in a chemical reactor.

Electrospinning of palladium doped alumina nanofibers

To synthesize the nanofibers with palladium catalyst, a 9 wt% solution of polyvinylpyrrolidone (PVP, Aldrich, molecular weight (MW) 1 300 000) in ethanol (Pharmco-AAPER alcohol) was prepared. A separate aluminum acetate precursor solution (AA solution) was prepared by mixing aluminum acetate (Alfa Aesar, Basic hydrate), water and formic acid in the weight ratio of 1:2.5:1, respectively. The PVP solution and AA solution were mixed in the weight ratio of 1:2, respectively. Palladium chloride (Sigma-Aldrich, 60% Pd) was combined in the resulting mixture in the ratio of 5 wt% $PdCl_2$ with respect to aluminum acetate and the mixture was stirred overnight at 40°C using a magnetic stirrer. The solution was electrospun with a flow rate

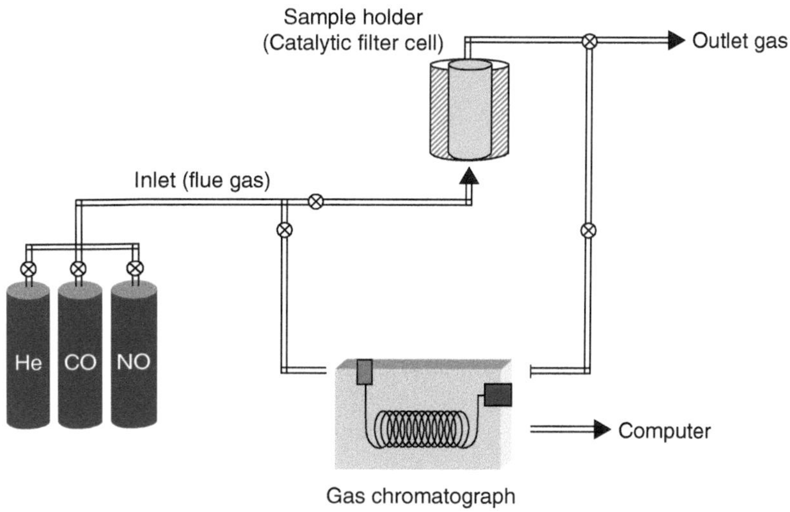

7.13 Schematic of the testing setup used to determine the catalytic activity of filter media.

of 2 μL/min and an electric field of 1 kV/cm. The electrospun nanofibers were calcined to 600°C to degrade the polymer and convert the metal precursors to metal oxides. The PdO particles were reduced to Pd by heating to 400°C in a hydrogen atmosphere.

Electrospinning of rhodium doped alumina nanofibers

The fibers with Rh catalyst started with similar PVP and AA solutions. The PVP and AA solutions were mixed in the ratio of 1:4 by weight, respectively. Rhodium chloride ($RhCl_3$, 99.9%, Sigma-Aldrich) was dissolved in ethanol at a solution concentration of 3.7 wt of rhodium chloride with respect to the aluminum acetate. This solution was mixed for 12 h at 50°C. The solution was electrospun and calcined as above to form Rh_2O_3/Al_2O_3 nanofibers. The Rh_2O_3 particles were reduced to Rh by heating in a hydrogen atmosphere.

Fabrication of nanofiber-based filter media

Filter media were prepared by vacuum molding an aqueous slurry of fibers and binder. The vacuum molding apparatus consists of a mixing tank, collecting tank and vacuum pump. The slurry was prepared using a mixture of microfibers, ceramic nanofibers, corn starch and alumina binder in water. Dilute acid was added to maintain the slurry at a pH of about 6.0 and starch was added to enhance the binding capability of the binder and fibers. The slurry was mixed overnight using a stirrer to obtain a uniform dispersion of materials. The mixed slurry was vacuum pulled through a hollow Plexiglas mold with a steel wire mesh to capture the fibers to form 2.3 cm diameter and 2.3 cm thick disk-shaped wet fibrous filter cakes. The wet cake was dried at 120°C to remove moisture and heated to 600°C to thermally set the fibers.

Testing

Continuous flow was established by passing three gases (He, CO and NO) at a controlled flow rate through a catalytic filter reactor. Helium was used as the carrier gas and to increase the velocity of the NO and CO gases. The catalytic filter was heated in the reactor to the desired operating temperature. The outlet gases from the reactor were analyzed using a gas chromatograph equipped with a thermal conductivity reactor. Figure 7.13 shows a schematic of the experimental setup.

7.4.2 Results from testing nanofiber catalyst support structures

The TEM images of the PdO/Al_2O_3 and Rh_2O_3/Al_2O_3 fibers are shown in Fig. 7.14. The diameters of the nanofibers and the particle size are summarized in Table 7.1. The permeability of the nanofibrous media was 1.2 $\times 10^{-11}$ m^2. The porosity was calculated to be 0.98.

The catalytic activities of the Pd/Al_2O_3 and Rh/Al_2O_3 filters are shown in Figs 7.15 and 7.16, respectively. For the Pd/Al_2O_3 catalyst the degradation temperature of NO was 350°C and for the Rh/Al_2O_3 catalyst the degradation of NO was at 250°C. The lower degradation temperature by the Rh/Al_2O_3

7.14 TEM micrographs of (a) PdO/Al and (b) Rh_2O_3/Al_2O_3.

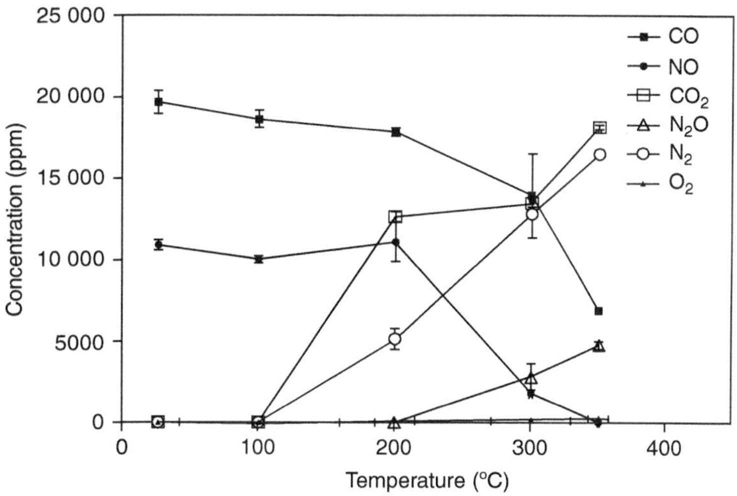

7.15 Temperature versus concentration for Pd/Al_2O_3.

Table 7.1 Diameter and particle size of metal doped alumina nanofibers

Material	PdO/Al_2O_3	Rh_2O_3/Al_2O_3
Fiber diameter (nm)	100–120	80–100
Catalyst particle size (nm)	1.5–4.5	7–10

catalyst is due to the greater ability of Rh to dissociate NO in comparison to Pd (Belton & Taylor, 1999; Alkemade & Schumann, 2006).

7.4.3 Summary

Alumina nanofibers with noble metal oxide (PdO and Rh_2O_3) nanoparticles were synthesized successfully by the electrospinning technique. The diameters of alumina nanofibers were in the range of about 50~150 nm and the 1.5~10 nm-sized catalyst particles were well dispersed in the nanofibers. The filter media were constructed using vacuum mold apparatus. The catalyst precursors were reduced to pure metals by heating in a hydrogen atmosphere and tested for their activity. These catalysts showed activity when exposed to a mixture of CO and NO. They successfully converted the NO to N_2 and CO to CO_2.

7.5 Future trends and conclusions

Nanofibers are useful for enhancing filter performance. The nanofibers increase the capture efficiency, as demonstrated in the model calculations,

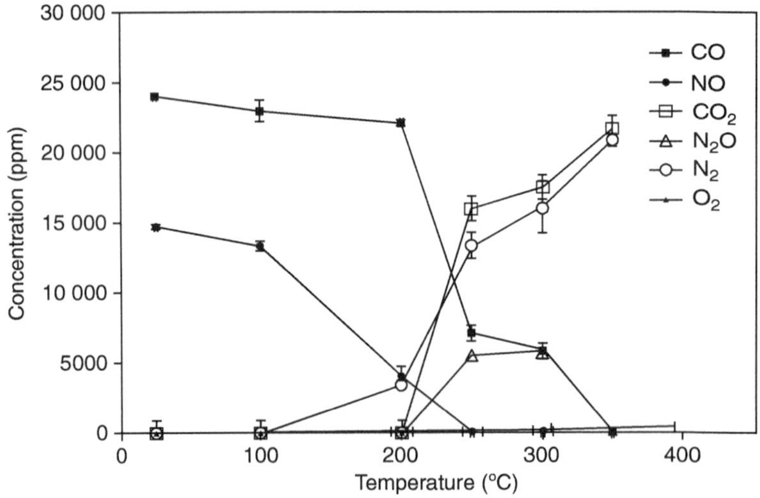

7.16 Temperature versus concentration for Rh/Al_2O_3 catalytic filter.

without a proportional increase in pressure drop. The QF calculations show an optimum filter design occurs with a nanofiber/microfiber area ratio of about unity.

The high surface area to mass ratio of the nanofibers makes them useful as catalyst support structures. The catalyst filter media were effective in converting NO to NO_2 and CO to CO_2 as described in Section 7.4.

Future trends of nanofibers in filter media are expected to apply the advantages of the small diameters and the large surface areas. The following are potential research topics for future work.

- New or improved correlations are needed for predicting particle capture on nanofibers. In particular, correlations are needed for the transition flow range for $0.25 < Kn < 10$.
- The scale of roughness of nanofiber mats is much smaller than the size of typical aerosol droplets. Such small-scale roughness can be used to create surfaces that are superhydrophobic (contact angles exceeding 150°) (Rutledge, 2008). Such structures may be useful for separation of mists from gas streams.
- Alignment of microfibers in filter media is known to affect filter performance (Chase *et al.*, 2000). Nanofibers can also be aligned (Katta *et al.*, 2004), but have not yet been explored for their alignment effect on filter performance.
- Besides catalysis, nanofiber surfaces can be functionalized to perform specific reactions, such as ion exchange (An *et al.*, 2006). This presents opportunities to fabricate filters that can react with and chemically

remove particles from liquid streams and regenerate the filter. Filters may be constructed to detect, capture and neutralize viruses and microorganisms.
- Ceramic nanofibers may be used in filtration of high-temperature gases in the range of 1000–2000°C for power production and filtration of exhaust gases from large furnaces (Raghavan & Chase, 2009).

7.6 Acknowledgments

This work was supported by the Coalescence Filtration Nanomaterials Consortium: Ahlstrom Paper Group, Donaldson Company, Cummins Filtration, Elmarco, Parker Hannifin and SNS Nanofiber Technology. This work was also supported by the National Science Foundation, grant number STTR 0822914, and the MemPro Ceramics Corporation.

7.7 References

Agricola, G. (1556), *De re Metallica*, libri xii, available from: http://echo.mpiwg-berlin.mpg.de/ECHOdocuView?mode=texttool&url=/mpiwg/online/permanent/library/5CTEBAHQ&pn=5 (accessed 4 January 2011).

Albrecht, F. (1931), Theoretische untersuchungen übfer die ablagerung von staub aus der luft und ihre anwendung auf die theorie der staubfilter [Theoretical research concerning the settling of dust from streaming air and its application to the theory of the dust filter], *Physikalische Zeitschrift*, **32**, 48–56.

Alkemade, U. G. and Schumann, B. (2006), Engines and exhaust after treatment systems for future automotive applications. *Solid State Ionics*, **177**, 2291–6.

An, H., Shin, C. and Chase, G. G. (2006), Ion exchanger using electrospun polystyrene nanofibers, *Journal of Membrane Science*, **283**, 84–7.

Belay, E. D. and Schonberger, L. B. (2005), The public health impact of prion diseases, *Annual Review of Public Health*, **26**, 191.

Belton, D. N. and Taylor, K. C. (1999), Automobile exhaust emission control by catalyst, *Current Opinion in Solid State and Materials Science*, **4**, 97–102.

Bird, R. B., Stewart, W. E. and Lightfoot, E. N. (2002), *Transport Phenomena* (2nd edn), New York, Wiley.

Brown, R. C. (1993), *Air Filtration*, Oxford, Pergamon Press.

Brown R. C. (1998), Airflow through Filters: Beyond Single Fiber Theory, in K. R. Spurny (ed.), *Advances in Aerosol Filtration*, Boca Raton, CRC Press LLC.

Castelloes, F. V., Cardoso, C. R., Couto P. and Cotta, R. M. (2007), Transient analysis of slip flow and heat transfer in microchannels, *Heat Transfer Engineering*, **28**(6), 549–58.

Centers for Disease Control and Prevention (2010), Key facts about influenza and the influenza vaccine, available from: http://www.cdc.gov/flu/keyfacts.htm (accessed 4 January 2011).

Chase, G. G. (2002), Transport phenomena in porous media, in J. Saleh (ed.), *Fluid Flow Handbook*, New York, McGraw Hill.

Chase, G. G., Beniwal, V. and Venkataraman, C. (2000), Measurement of uni-axial fiber angle in non-woven fibrous media, *Chemical Engineering Science*, **55**(12), 2151–60.

Chen, X. and Pfender, E. (1983), Effect of the Knudsen number on heat transfer to a particle immersed into a thermal plasma, *Plasma Chemistry and Plasma Processing*, **3**(1), 97–113.

Farrauto, R. J. and Heck, R. M. (1999), Catalytic converters: State of the art and perspectives, *Catalysis Today*, **51**, 351–60.

Feldhaus, G. M. (1929), Schutzmasken in vergangenen Jahrhunderten, *Gasmaske*, **1**, 104.

Fogler, H. S. (2006), *Elements of chemical reaction engineering* (4th edn), New York, Prentice Hall.

Gai, P. L., Roper, R. and White, M. G. (2002), Recent advances in nanocatalysis research, *Current Opinion in Solid State and Materials Science*, **6**, 401–6.

Garcia, J. M., Garcia, F. C., Serna, F. and de la Peña, J. L. (2010), High-performance aromatic polyamides, *Progress in Polymer Science*, **35**, 623–86.

Happel, J. (1959), Viscous flow relative to arrays of cylinders, *AIChE Journal*, **5**(2), 174–7.

Hillie, T. and Hlophe, M. (2007), Nanotechnology and the challenge of clean water, *Nature Nanotechnology*, **2**, 663–4.

Hinds, W. C. (1999), *Aerosol Technology*, New York, John Wiley & Sons, Inc.

Imai, I. (1951), On the asymptotic behavior of viscous fluid flow at a great distance from a cylindrical body, with special reference to Filon's paradox, *Proceedings of the Royal Society A*, **208**, 487–516.

Inman, R. M. (1965), Heat transfer in thermal entrance region with laminar slip flow between parallel plates at unequal temperatures, NASA-TN-D-2980.

Institute of Medicine of the National Academies (2006), *Reusability of facemasks during an influenza pandemic*, Board on Health Sciences Policy, Washington, D.C., National Academies Press.

Jaganathan, S., Tafreshi, H. V. and Pourdeyhimi, B. (2008), A case study of realistic two-scale modeling of water permeability in fibrous media, *Separation Science and Technology*, **43**(8), 1901–16.

Katta, P., Alessandro, M., Ramsier, R. D. and Chase, G. G. (2004), Continuous electrospinning of aligned polymer nanofibers onto a wire drum collector, *Nano Letters*, **4**(11), 2215–8.

Kuwabara, S. (1952), The forces experienced by randomly distributed parallel circular cylinders or spheres in a viscous flow at small reynolds numbers, *Journal of the Physical Society of Japan*, **14**(4), 527–32.

Levine, H. L. (2009), Challenges and solutions for biopharmaceutical manufacturing, Cambridge Health Tec Pep Talk Conference, San Diego, CA, January 15.

Park, C. and Baker, R. T. K. (1998), Catalytic behavior of graphite nanofiber supported nickel particles, the influence of the nanofiber structure, *Journal of Physical Chemistry B*, **102**, 5168–77.

Park, S. J., Bhargava, S., Bender, E. T., Chase, G. G. and Ramsier, R. D. (2008), Palladium nanoparticles supported by alumina nanofibers synthesized by electrospinning, *Journal of Materials Research*, **23**, 1193–6.

Patel, C., Li, S., Wang, C., Zhang, W. and Wei, Y. (2007), Electrospinning of porous silica nanofibers containing silver nanoparticles for catalytic applications, *Chemistry of Materials*, **19**, 1231–8.

Raghavan, B. and Chase, G. G. (2009), *Ceramic nanofiber media for hot gas filtration*, Ann Arbor, American Filtration and Separations Society.
Reneker, D. H., Yarin, A. L., Fong, H. and Koombhongse, S. (2000), Bending instability of electrically charged liquid jets of polymer solutions in electrospinning, *Journal of Applied Physics*, **87**(9), 4531–47.
Rutledge, G. C. (2008), A review of recent results on superhydrophobic materials based on micro- and nanofibers, *Journal of Adhesion Science and Technology*, **22**(15), 1799–817.
Secundus, G. P. (ca AD 77), *Naturalis Historia*, available from: http://www.istrianet.org/istria/illustri/non-istrian/pliny/index.htm (accessed 4 January 2011).
Spurny, K. R. (1998), The history of dust and aerosol filtration, in K. R. Spurny (ed.), *Advances in Aerosol Filtration*, Boca Raton, CRC Press LLC.
Stokes, G. G. (1845), *Proceedings of the Cambridge Philosophical Society*, **8**, 287–319.
Thiele, W. E. (1939), Relation between catalytic activity and size of particle, *Industrial & Engineering Chemistry Research*, **31**, 916–20.
USEPA (2010), Drinking water contaminants, US Environmental Protection Agency, available from: http://water.epa.gov/drink/contaminants/index.cfm (accessed 4 January 2011).
USGS (2010), Earth's water distribution. United States Geological Survey., available from: http://ga.water.usgs.gov/edu/waterdistribution.html (accessed 4 January 2011).
WHO (2008), *Guidelines for drinking water quality* (3rd edn), Geneva, World Health Organization.
Williamson, L. (2009), Impact of vCJD on UK blood services: would you be tested?, UK NHS, *Blood and Transplant*, available from: http://www.transfusionguidelines.org.uk/docs/pdfs/rtc-ne_edu_williamson.pdf (accessed 4 January 2011).
Zhao, S. and Povitsky, A. (2009), A hybrid molecular and continuum method for low-Reynolds-number flows, *Nonlinear Analysis: Theory, Methods & Applications*, **71**(12), 2551–64.

7.8 Appendix: notation

A = Cross-sectional area of filter, heat or mass transfer area
a = Internal catalyst pellet surface area per unit volume
Cn = Cunningham factor
D = Effective coefficient of diffusion
d = Molecule diameter
d_i, d_f = Fiber diameter
d_p = Particle diameter
E, E_i = Single fiber capture efficiency, capture efficiency of fiber of size d_i
E_R, E_D, E_{DR} = Single fiber efficiencies by direct interception, diffusion and combined diffusion and interception mechanisms
E'_{DR} = Combination term in Equation [7.25]
F_z = Drag force
F_{tot} = Total force acting on the filter
f_i = Drag force per unit length of fiber of diameter
h = Heat transfer coefficient
I = Rate of mass transfer of aerosol particles captured from gas phase onto fibers per unit volume of filter

K = Dimensionless parameter given by Equation [7.7]
Kn = Knudsen number
k = Mass transfer coefficient
k_1 = First-order reaction rate constant
k_B = Boltzmann's constant, 1.38066×10^{-23} J/K
L = Filter thickness or length of fiber
L_i = Total length of fibers of diameter within elemental volume of the filter medium
N_R = Ratio of diameters, d_p/d_i
n = Number of molecules per unit volume
P, ΔP = Gas phase pressure, pressure drop across filter
Pe = Peclet number
Q = Amount of thermal energy transfer due to conduction
QF = Quality factor
RQF = Relative quality factor (QF of the medium/QF of medium of only microfibers)
R = Outer radius of catalyst pellet, fiber radius
T, ΔT = Temperature, temperature difference
v_z = Intrinsic gas phase velocity in the z direction
w = Mass fraction of particles in the gas phase
w_{in}, w_{out} = Inlet and outlet mass fractions of particles in gas phase
W_A = Amount of mass transfer per time of species A
Δx_A = Difference in molar fraction concentrations of chemical species A
Y = Distance through which stream lines flow and particles are captured
α = Overall filter coefficient for the filter medium
α_i = Projected area of fibers of diameter d_i in a filter layer
ε = Porosity of the media
ε_i = Volume fraction of fibers of size d_i in the filter
ζ = Hydrodynamic factor
ψ = The stream function of the flow velocities
λ = Mean free path
η = Catalyst effectiveness factor
η = Particle mobility
μ = Viscosity of gas phase
ρ = Gas phase density
ρ_A = Mass concentration of aerosol particles in the gas phase, kg/m^3

8
Functional nanofibers for drug delivery applications

Q. WEI, Jiangnan University, China
and A. WEI, Anhui Polytechnic University, China

Abstract: In this chapter the physical and chemical mechanisms of drug delivery for nanofibers are introduced; the varying characteristics of these mechanisms and their subsequent strengths and weaknesses for drug delivery are studied. Poly (L-lactide) (PLLA)/Clarithromycin electrospun composite nanofibers are used as an example in discussing the feasibility of electrospun nanofiber application in drug delivery systems. The final conclusion shows that electrospun nanofibers could be very suitable for sustained drug delivery systems.

Key words: applications, electrospun nanofibers, drug delivery.

8.1 Introduction

Since almost all drugs are toxic at the wrong dose, conventional pills and injections may not be the best modes of the drug administration. As Fig. 8.1 shows, drug concentration in the blood will change significantly after taking the required dose. This means patients must often take a higher dose in order to sustain the therapy of the drug. The toxicity of the drug will increase, especially for very toxic drugs, meaning more harm to patients. If the amount of the drug could be reduced without compromising therapeutic effectiveness, the possible toxic side effects of the drug will decrease to patients' benefit. This is called a sustained drug delivery system. In such a system, the drug is designed to be released in a controlled fashion to maintain an appropriate concentration for a long period of time, which eliminates the potential for both under- and overdosing, achieving more effective therapeutic outcomes. It permits the maintenance of drug levels within a desired range, thus leading to a better use of the drug and to a reduction of the required drug administration.

Sustained drug delivery systems have gained much attention in the last few decades. Pharmacologists, polymer scientists, chemicals and medical researchers have done much work on these systems and invented different devices that encapsulate a drug and release it at controlled rates for relatively long periods of time. Delivery devices include micro- or nanoparticles, polymer-based disks, rods, pellets, plasters and osmotic pumps, with different

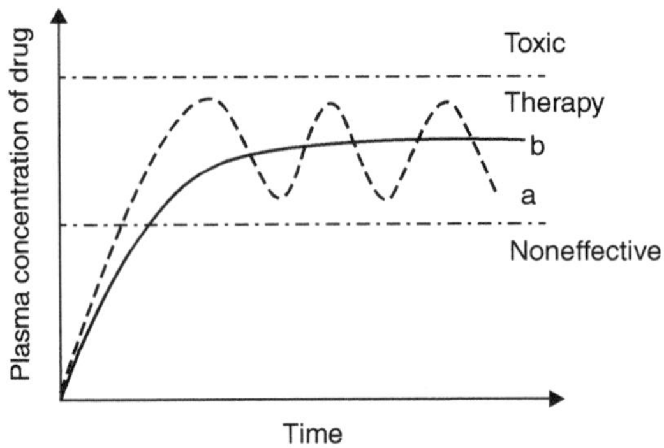

8.1 Blood concentration changes after taking (a) the traditional doses and (b) sustained-release preparation.

types of release (e.g., continuous, pulsatile), according to the medicine that has to be taken (Petitti *et al.*, 2008).

Most of these delivery systems aim at optimizing the preparation method in order to increase the encapsulation efficiency and to obtain a desired release profile. In order to do this, experimental investigations on different preparation methods, process conditions and materials are carried out to identify the influence of the different parameters on drug loading, on microparticles morphology, on polymer structure and consequently on the resulting release curve. Birnbaum *et al.* (2000) studied the release of β-estradiol from poly(DL-lactic acid-co-glycolic acid) (PLAGA) microparticles and the effect of organic phase solvent on encapsulation and release, Jain *et al.* (2000) analyzed the effect of various formulation variables on controlled release of drugs from injectable *in situ* formed biodegradable polylactide-co-glycolide (PLGA) microspheres and Jeong *et al.* (2003) discussed the effects of crystalline microstructure on drug release behavior of poly(ε-caprolactone) microspheres.

It is well known that a drug's delivery vehicle can have a significant impact on its efficacy. Nanoscale manipulation of drug delivery vehicles can substantially improve pharmacokinetics, pharmacodynamics, nonspecific toxicity, immunogenicity and biorecognition properties (Couvreur and Vauthier, 2006; Kopecek, 2003). As a result, applying these technologies to pharmaceutical development has the potential to revolutionize the delivery of biologically active compounds. Electrospinning is a unique processing technique that can be used to fabricate ultrafine fibers, diameters of which are in the submicrometer down to nanometer range. In this process, a continuous strand of a polymer liquid (i.e., solution or melt) is ejected through a nozzle by a high electrostatic force to deposit randomly on a grounded collector as a nonwoven fiber mat. These fibers exhibit several interesting characteristics – for example, a high surface area to mass or volume ratio, small inter-fibrous pore size with high porosity, vast possibilities for surface functionalization (Reneker and Yarin, 2008). These advantages render electrospun polymeric fibers as good candidates for a wide variety of applications, including filters (Nakano and Kameoka, 2007), composite reinforcements (Kim and Eneker, 1999), carriers for topical or transdermal delivery of drugs (Kenawy *et al.*, 2002) and scaffolds for cell and tissue culture (Ji *et al.*, 2006).

As a potential drug delivery carrier, electrospun nanofibers have exhibited many advantages. The drug loading is very easy to implement via an electrospinning process, and the high applied voltage used in the electrospinning process has little influence on the drug activity. The high specific surface area and short diffusion passage length give the nanofiber drug system a higher overall release rate than the bulk material (e.g., film). More than one drug can be loaded into the electrospun nanofibers and the release profile can be finely controlled by modulation of nanofiber morphology, porosity and composition. Since Kenawy *et al.* (2002) first studied drug delivery from poly(ethylene-co-vinyl-acetate), poly(lactic-acid) and a 1:1 blend of the two polymers electrospun from chloroform solution with tetracycline hydrochloride as the model drug, many researchers have focused on this field. Santi Tungprapa studied the release characteristics of four model drugs from drug-loaded electrospun cellulose acetate fiber mats (Tungprapa *et al.*, 2007). Xiuling Xu *et al.* (2009) prepared the ultrafine PEG–PLA fibers loaded with both paclitaxel and doxorubicin hydrochloride and discussed their in vitro cytotoxicity. Verreck *et al.* (2003b) used polymer-based electrospun nanofibers containing dispersions of poorly water-soluble pharmaceuticals as deliveries systems. However, their researches were was most focused on the release regularities of loaded drugs; few studied the surface properties of the nanofiber membranes and analyzed the interactions between the drugs and the macromolecule carriers.

8.2 Drug delivery mechanisms

Nanofibers afford great flexibility in selecting materials for drug delivery applications. Either biodegradable or nondegradable materials can be used to control whether drug release occurs via diffusion alone or diffusion and scaffold degradation. Additionally, due to the flexibility in material selection, a number of drugs can be delivered including: antibiotics, anticancer drugs, proteins and DNA. Using the various electrospinning techniques, a number of different drug-loading methods can also be utilized: coatings, embedding and encapsulating (coaxial and emulsion electrospinning). These techniques can be used to give finer control over drug release kinetics.

For drug delivery applications, several polymers (in terms of degradability and crystallinity) as well as drug/growth factor loading in crystalline or amorphous form in order to fulfill specific requirements of drug-eluting fiber mats have been studied. Usually, good mechanical properties and biocompatibility are required, together with control of drug release and burst effect in order to ensure physical integrity of the construct, long-term delivery or immediate action at the targeted location. There are several factors that can affect the drug release from electrospun fibers:

- fiber construct geometry and thickness (Okuda *et al.*, 2010)
- fiber diameter and porosity (Cui *et al.*, 2006)
- fiber composition
- drug molecular weight (Diller *et al.*, 2007)
- fiber crystallinity (Tungprapa *et al.*, 2007)
- fiber swelling
- drug loading and drug solubility in the release medium (Xie and Diller, 2010)
- drug state (Zamani *et al.*, 2010)
- drug–polymer–electrospinning solvent interactions (Xie and Diller, 2010).

The release characteristics of the fiber mat are highly influenced by the state of the drug and the structure of the polymer that forms the fiber. For example, the crystallinity of the polymer controls the rate of drug release. This is because semicrystalline polymers show, in general, a higher extent of burst for two reasons:

- the instantaneous release of the drug deposited at the fiber surface
- the hindered release of the drug from the fiber bulk due to limited water uptake in the semicrystalline regions.

The drug state in the fibers is also an important factor, since it was shown that a drug that is incorporated in crystalline form will mainly be deposited outside the fibers and trigger burst release, while a drug in an amorphous state will be loaded inside the fibers and be released in a sustained manner. Drug loading is another factor that can affect drug release: higher loadings will produce faster release. On the one hand, at high loadings, there is more surface segregated drug that dissolves fast; on the other hand, there is an increase in porosity during drug elution proportional to the initial amount of drug.

Drug compatibility with polymer solution was also shown to be an important factor in controlling release, as lipophilic drugs should be incorporated in lipophilic polymers and hydrophilic drugs in hydrophilic polymers in order to avoid drug deposition outside fibers and subsequent burst. Moreover, the interaction between drug and polymer can block the crystallization of the drug in the fibers, if so desired, and can even determine sustained release of drugs in a crystalline state because of chemical interaction with the polymer.

The drug release mechanism can be physical or chemical in nature. Diffusion is always involved. For nonbiodegradable matrix and membrane devices, release is by diffusion and is driven by the concentration gradient. It can also be driven by osmotic pressure or matrix swelling. For matrices which are biodegradable or contain drug conjugates, release is controlled by the hydrolytic or enzymatic cleavage of the relevant chemical bonds, although diffusion of the reactants and the liberated drug molecules may still be rate limiting. Only a qualitative discussion is pursued here to facilitate understanding of the release patterns ensuing from the more common delivery systems.

8.2.1 Physically controlled drug release mechanisms

Release may be controlled by drug diffusion or solvent penetration. The diffusion-controlled systems can be designated as either reservoir or matrix devices. In the former case, the drug is encapsulated by a polymeric membrane. The drug can be in a solid or a liquid state, and the membrane can be microporous or nonporous. If the contained drug is maintained in a saturated state, the transport of drug molecules across the membrane will be kept constant, as the driving force is unchanged. Such a constant or zero-order release would require the drug to remain in a solid or suspension state. The saturated state would be difficult to maintain if the drug has high water solubility.

Even if all the requirements for constant release are met, the release is generally not constant in the initial and end periods. When the device is placed in a releasing medium, it takes a certain time for the system to reach a steady state. Depending on the history of the device, a lag time or a burst effect may be observed. If the membrane is devoid of drug molecules at the time of release, an induction period is needed to saturate the membrane. If, on the other hand, the drug molecules have been accumulating in the membrane as a result of fabrication or storage, the initial release rate will be higher than the steady-state value. Toward the end of the release period, the drug concentration in the core will drop below the saturation value and the release rate will decrease. If a nonperfect sink situation exists, which may occur in many *in vivo* situations, the increased mass-transfer resistance across the membrane will slow down the release.

The simplest and most widely used devices are matrix systems in which the drug is dissolved or dispersed in the polymer. One characteristic of these systems is their release rate, which decreases with time as a result of increasing diffusion distance for the drug solutes to travel from the core to the surface. The amount of drug released is often proportional to the square root of elapsed time. In addition to permeation through the polymer phase, the drug molecules can also diffuse through channels created by dissolution of the drug phase. Release of macromolecules which have low permeability generally occurs via these pores (Chew *et al.*, 2005; Siegel and Langer, 1984).

Since the extent and size of the pores and channels created are determined by the drug incorporated, the loading level and particle size of the drug solutes have a profound influence on release kinetics. Below a critical loading level, some of the isolated drug particles may be trapped inside the matrix. The release behavior of these diffusion-controlled systems is highly dependent on the physical properties of the drug. In addition to loading level and particle size of the solutes, the drug solubility in the polymer and drug diffusivity in the polymer phase are both important parameters. The shape of the device, which determines the surface area available and the path length for diffusion, is also critical. In solvent-controlled systems, the release of the drug is regulated by the permeation of water through the polymer.

In order to study the drug release mechanism, different equations (Equations [8.1] and [8.2]) were used to model the release data. The equations were fitted to the data using nonlinear regression and the results were compared in terms of goodness of the fit. The power law equation (Equation [8.1]) is one of them and was chosen because it is the most widely used equation in works concerning drug release (Peppas and Brannon-Peppas, 1994).

$$\frac{m_t}{m_{\text{tot}}} = \alpha_0 + kt^n \qquad [8.1]$$

In Equation [8.1], m_t means the mass of the drug released at time t, m_{tot} is the total drug amount and m_t/m_{tot} is the fractional release of the drug at time t, α_0 is a constant, representing the percentage of burst release, k is the kinetic constant and n is the release exponent, indicating the mechanism of drug release. In most models, the release mechanism has been attributed to diffusion of the drug from the polymers and, under this assumption, a 100% release of the drug is expected in a certain time. In the desorption model, the authors suggest that release is not controlled by solid-state diffusion, but by the desorption of the drug from pores in the fibers or from the outer surface of the fibers. Thus, only the drug on the fiber and pore surfaces can be released, whereas the drug from the bulk cannot be released within the time scales characteristic of the release experiments. Equation [8.2] is based on a pore model, in which the effective drug diffusion coefficient, D_{eff}, is considered and not the actual diffusion coefficient in water, D (with D_{eff}/D <<1), because desorption from the pore is the rate-limiting step and not drug diffusion in water, which is relatively fast.

$$\frac{m_t}{m_{tot}} = \left[1 - \exp\left(-\frac{\pi^2}{8}\frac{t}{t_r}\right)\right] \qquad [8.2]$$

where the porosity factor $\alpha = m_{s0}/(m_{s0} + m_{b0}) < 1$, with m_{s0} and m_{b0} being the initial amount of drug at the fiber surface and the initial amount of drug in the fiber bulk, respectively; m_t is the drug amount released at time t, while the total initial amount of drug in the fiber is $m_{tot} = m_{s0} + m_{b0}$ and τ_r is the characteristic time of the release process (Srikar *et al.*, 2008).

8.2.2 Chemically controlled drug release mechanisms

The two common chemically sustained delivery systems are a biodegradable matrix in which the drug is dispersed and a polymer–drug conjugate in which the drug molecules are linked to the side chains of the polymer (Heller, 1980). The dissolution of the matrix is affected by hydrolytic or enzymatic cleavage of the backbone of the polymer. The cleavage can also occur in the cross-linking bonds, rendering soluble an initially cross-linked polymer. Alternatively, the dissolution originates from hydrolysis, ionization or protonation of the side chains of the polymer. If drug release for these systems is controlled solely by the biodegradation of the matrix, a constant release rate will be obtained, provided the surface area of the device is maintained constant. However, that represents an idealized case. In practice the drug molecules can also diffuse through the matrix.

The release is commonly intermediate between zero-order and first-order kinetics. This system enjoys the advantage of having release kinetics that are less dependent on the properties of the drugs than are the nonbiodegradable systems. The release rate of the polymer–drug conjugate, which has been called a pendant system, is dependent on the cleavage of the polymer bond. If the drug is attached to the polymer via a spacer, the hydrolysis of the polymer spacer and the spacer-drug bonds are both relevant. The spacer approach provides an effective means of controlling the release rate. Although penetration of water into the matrix and outward diffusion of the cleaved drug molecules constitute part of the rate barrier, the cleavage of the polymer–drug bond as the rate-determining step is preferred for better control. Generally, the release rate drops with time as the drug concentration decreases.

8.3 Applications of functional nanofibers in drug delivery: a case study

Clarithromycin (CLM) is a new kind of semisynthetic macrolide antibiotic which is widely used clinically to treat the infection of respiratory tract, such as pneumonia, bronchitis and tonsillitis (Chu *et al.*, 1993). It also has the advantage of treating gastric ulcer in combination with other drugs (Guay *et al.*, 2001). CLM has low bioavailability *in vivo* due to its poor water solubility and short half life. What's more, it also can lead to some severe side effects such as nausea, diarrhea and abdominal pain (Gaynor and Mankin, 2003; Sibley and Pedersen, 2008). So, it is necessary to develop a new form of CLM in order to avoid these defects. One promising approach is to incorporate CLM into the polymer carrier by electrospinning. The advantages of this method are not only improvement of the drug's bioavailability but also decrease of its side effects (Ngawhirunpat *et al.*, 2009; Verreck *et al.*, 2003a).

In this study PLLA was used to prepare the composite nanofiber scaffolds loaded with CLM by electrospinning. Low toxic solvents of acetone and dichloromethane were selected; their ratios and the effects on the morphology of electrospun fibers were investigated. The structures and morphology of the scaffolds were studied by scanning electron microscopy (SEM), Fourier transform infrared spectrometer (FT-IR) and X-ray diffraction (XRD).

8.3.1 Materials and methods

Materials

CLM (purity more than 99%) was provided by Xinyu Chemical Engineering Co. Ltd (Zhenjiang, China) and stored in the refrigerator at –20°. Its chemical structure is shown in Fig. 8.2. PLLA (M_η = 100 000; 83 000) were purchased from Bright China Industrial Co. Ltd (Shenzhen, China). Acetone and dichloromethane were obtained from Sinopharm Chemical Reagent Co. Ltd (China) and were used without further purification.

Solvent selection

Acetone and dichloromethane, which are both low toxic solvents, were chosen. Their weight ratios varied from 1/1, 2/1, 1/2, 1/0, 0/1, 1 to 0. Chips of PLLA (M_η = 83 000) were dissolved to prepare particular concentrations of the solutions. By observation of the morphology of electrospun scaffolds using SEM, the ratio of acetone and dichloromethane was determined. The larger molecular weight of PLLA (M_η = 100 000) was selected to prepare

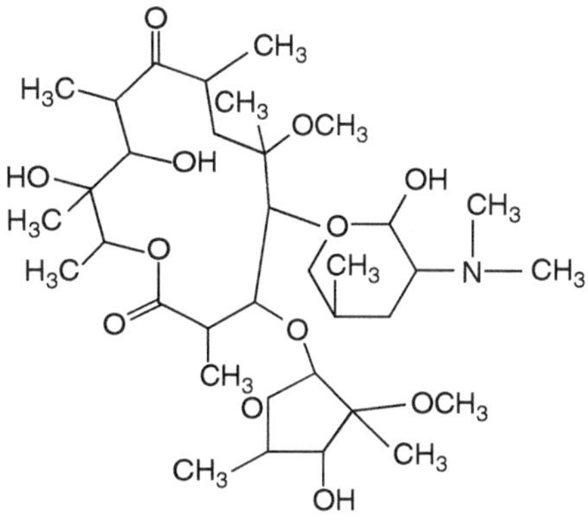

8.2 Chemical structure of Clarithromycin.

the composite nanofiber scaffolds. Weighed amounts of CLM power were then added into the above solvents system, the proportions of which (with respect to the polymer used) varied from 5% to 10%, 15% and 30% ; this was then stirred for 3–4 h before electrospinning.

Electrospinning

The polymer solution was transferred to a 20 ml syringe with a flat-end metal needle cut, and then this syringe was set in a pump for controlled feeding rates. A high-voltage DC power supply was used to initiate the jet and the electrospun scaffolds were collected onto a grounded cylindrical stainless steel mandrel. The electric field strength was 13–15 kV; the distance between the needle tip and the collector was 18–22 cm. All electrospinning experiments were carried out at about 20°C in air. In order to remove the residual solvent, the fiber mats collected were dried in a vacuum at room temperature for 24 h. The control nanofibers were also fabricated by the same method without CLM incorporation.

Characterization

The surface topography of the composite nanofiber scaffolds was observed by a scanning electron microscope (Hitachi S-4800, Japan). The scaffolds were placed on a metal stud, fixed with double-sided conductive tape. A very thin layer of gold was applied to the fibers by a sputtering unit for 40 s. The gold-coated scaffolds were then placed in the microscope and scanned in 5 kV mode. The Image J software was used to analyze the distributions of the electrospun fibers' diameters. CLM, the control and the composite scaffolds were prepared in KBr pellet and scanned by FT-IR spectrophotometer (Nicolet NEXUS and America). XRD patterns were obtained with a BRUKERS-AXS D 8 FOCUS diffractometer. Scans were obtained in the 2θ range of 5–60° with a step size of 0.02° every 1 s.

8.3.2 Results and discussion of the applications of functional nanofibers in drug delivery

Morphology of the nanofibers

As is known, the solvents are very important factors for electrospinning. Based on laboratory trials, the acetone and dichloromethane were chosen as the solvents and the different ratios of the two solvents were used.

As illustrated in Fig. 8.3, the collected material contained few nanofibers formed among the beads when only the solvent of acetone was used. The shape of beads changed from sphere to spindle and the numbers of fibers between the beads seemed to increase, but there were still a few when the PLLA concentration was increased. Low viscosity and high surface tension were the main factors affecting the formation of nanofibers (Fong *et al.*, 1999).

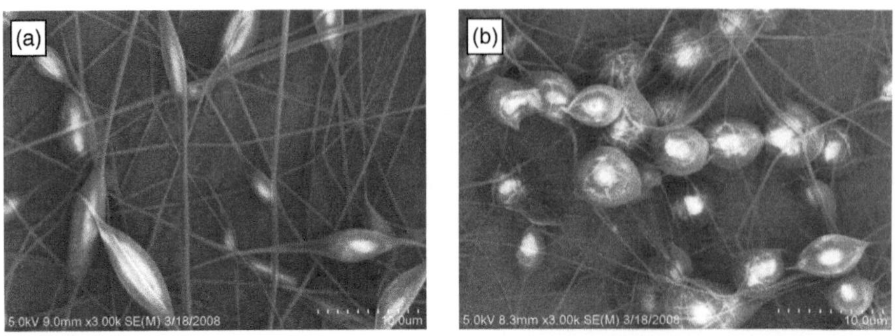

8.3 SEM photographs of PLLA (M_η = 83 000, 30% w/w) electrospun nanofibers in different solvents: (a) acetone/dichloromethane (1/2), (b) acetone/dichloromethane (1/1), (c) acetone/dichloromethane (2/1) and (d) dichloromethane.

PLLA and acetone were both polar and their interactions increased the surface tension of the solution. In addition, the viscosity-average molecular weight of PLLA (M_η = 83 000) was lower, which also decreased the viscosity of the whole solution.

As Table 8.1 shows, when an amount of dichloromethane was added into the solvent, the surface tension of the whole solution system decreased, which facilitated the formation of fibers in electrospinning. However, the electrical conductivity of dichloromethane was lower than that of acetone; the addition of dichloromethane also reduced the electrical conductivity of the whole solution, which would have an adverse effect on splitting of PLLA in electrospinning. The images in Fig. 8.3 showed the effect of solvent on electrospinning. It was found that the nanofibers with a smooth surface and few beads were obtained when the ratio of acetone to dichloromethane was 1:2, as shown in Fig. 8.3.

The nanofibers prepared by using the ratio of acetone to dichloromethane at 1:2 are shown in Fig. 8.4; no presence of drug aggregates can be observed on the surfaces of the electrospun fibers. The diameters and the diameter distributions of the electrospun fibers analyzed by Image J are listed in Table 8.2. The average diameter of PLLA control electrospun fibers was 1091.8 nm, while the average diameter of the composite fibers loaded with 5 wt% CLM decreased to 408.4 nm. The average diameter of the electrospun fibers dropped further to about 261.4 and 210.7 nm as the drug-loading amount was increased to 10% and 15%, respectively. It was also observed that the standard deviations of fiber diameter decreased, indicating evener distribution of the fiber diameters as the drug content was increased.

As it was known, the concentration of PLLA in the solution was relatively reduced when PLLA was loaded with some drugs, and its interaction with acetone was weakened so that the evaporation of acetone became easier. The addition of CLM not only decreased the concentration of PLLA in the

Table 8.1 Measurement results of basic parameters for 30 wt.% PLLA (M_η = 83 000) in the different solutions

Different solutions ($w.w^{-1}$)	Surface tension ($mN.m^{-1}$)	Viscosity (mPa.S)	Electrical conductivity ($\times 10^{-6}$ $S.m^{-1}$)
Acetone	30.60	11.0	13.6
Dichloromethane	26.61	45.0	0.8
Acetone/ dichloromethane(1/1)	27.48	20.0	14.5
Acetone/ dichloromethane(1/2)	26.35	22.5	5.9
Acetone/ dichloromethane(2/1)	27.51	16.5	9.4

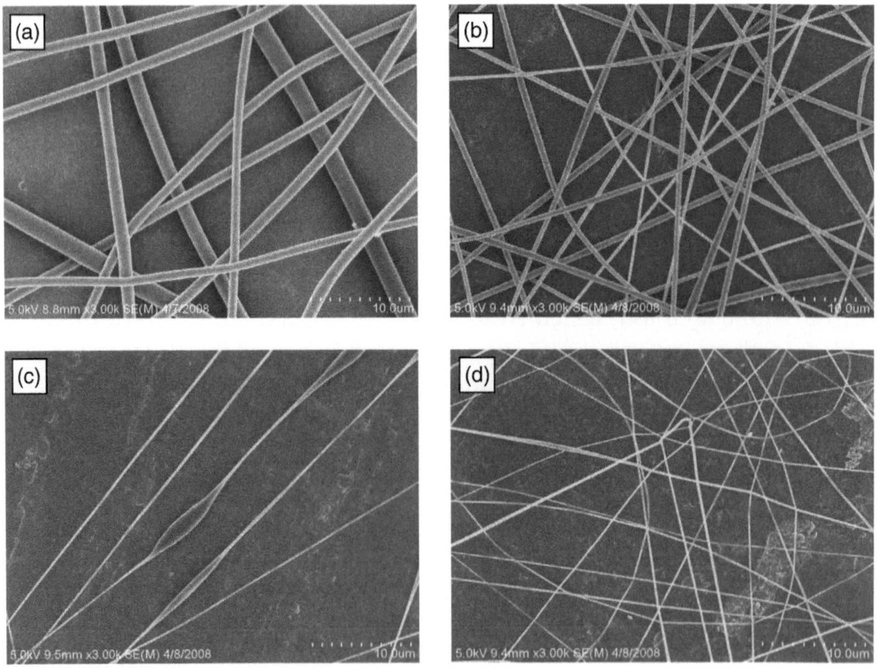

8.4 SEM photographs of PLLA (M_η = 100 000, 10%w/w) nanofibers containing (a) 0%, (b) 5%, (c) 10% and (d) 15% w/w of Clarithromycin.

Table 8.2 Distributions of electrospun fibers' diameters with different drug-loading amount

Number	Drug-loading amount (wt.%)	Mean values of fibers' diameters (nm)	Standard deviations of fibers' diameters
1	0	1091.8	223.0
2	5	408.4	112.8
3	10	261.4	79.6
4	15	210.7	50.4

solution, but also reduced the surface tension of the solution, which resulted in instability of the jet and uneven distribution of the fibers' diameters. On the other hand, CLM was also polar, and it could form the interactions between the solvents and PLLA. When the content of CLM was low, the interactions were not very strong, but the interactions became stronger and increased the surface tension of the solution with an increase of the CLM content. Because the viscosity of the solution decreased and the surface tension increased, the fiber diameters decreased, the fiber orientation tangled and more beads were easy to form, as revealed in Fig. 8.4.

Fourier transform infrared spectrometer (FT-IR) and X-ray diffraction (XRD) analyses

CLM was incorporated into the composite nanofiber scaffolds by electrospinning. In order to investigate the physical state of the drug in the polymer matrix and whether the chemical structure of this drug changed, FT-IR spectrophotometer and XRD were used.

As shown in Fig. 8.5, both characteristic peaks of CLM and PLLA could be found in the FT-IR spectra of the drug-loaded fiber and some characteristic peaks migrated, but no other new peaks appeared, which indicated that the drug had no chemical reaction with PLLA. The hydroxyl stretching peak of CLM was migrated from 3506.60 to 3462.56 cm^{-1} and the carbonyl stretching peak of PLLA was also shifted from 1756.43 to 1752.39 cm^{-1} in the composite nanofiber scaffolds, which was caused by the formation of the hydrogen bond between the hydroxyl of CLM and the carbonyl of PLLA (Deitzel *et al.*, 2001).

This strong action was helpful in encapsulating the small molecules of CLM into the large molecules of PLLA. In fact, it was found that the spectrum of the composite nanofiber scaffolds was almost the same as that of PLLA when the FT-IR-ATR (attenuated total reflection) was used to examine the chemical structure of the composite nanofiber scaffolds. However, the spectrum revealed the composite characteristic peaks of PLLA and CLM when the KBr pellet method was adopted. This interesting phenomenon also indicated that the small molecules of CLM could be encapsulated into the large molecules of PLLA, which facilitated the controlled release of the drug.

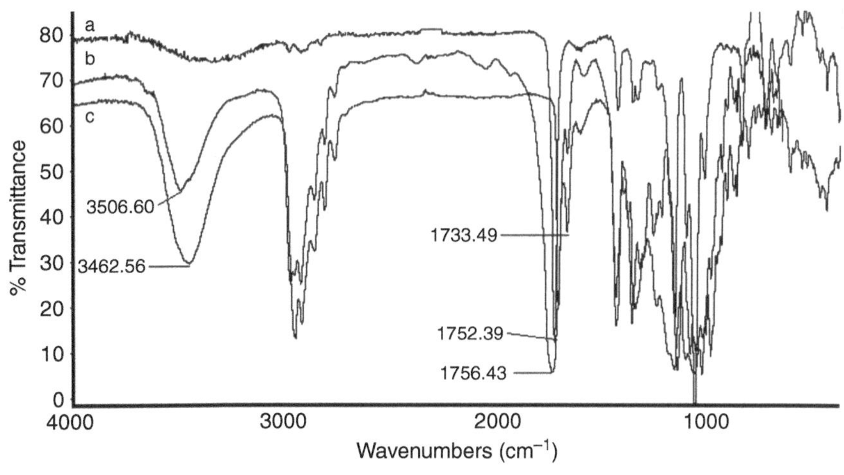

8.5 IR spectrums of (a) blank PLLA fiber, (b) Clarithromycin-loaded fiber and (c) Clarithromycin.

The XRD patterns of CLM, cast film of PLLA and electrospun PLLA nanofibers are shown in Fig. 8.6. The characteristic peak of cast film of PLLA was 16.6°, but, while PLLA was electrospun into nanofiber scaffolds, its characteristic peak disappeared, which was consistent with the reports in some literature (Deitzel *et al*., 2001; Sheetal and Tejraj, 2006).

Electrospinning was such a fast process that molecular chains of PLLA could not crystallize in a very short time. The XRD patterns of different content of CLM in PLLA nanofiber scaffolds are shown in Fig. 8.7. There were many characteristic peaks on the XRD pattern of CLM, which indicated

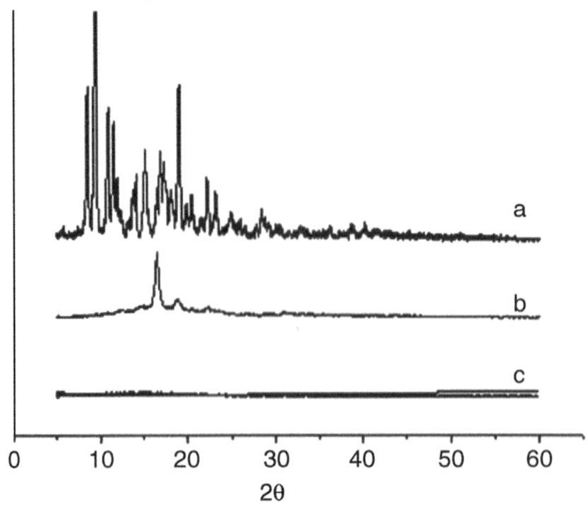

8.6 XRD patterns of (a) CLM, (b) cast films and (c) PLLA nanofiber scaffolds.

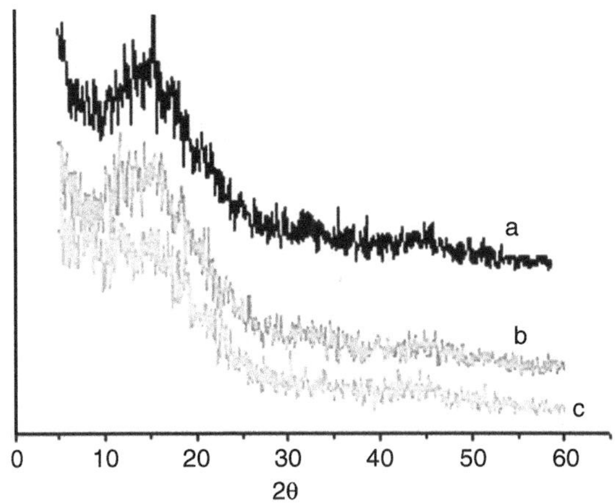

8.7 XRD patterns of PLLA nanofiber scaffolds containing (a) 0, (b) 5 and (c) 10 wt.% CLM.

this drug was polycrystal, but the characteristic peaks almost disappeared when it was loaded in PLLA electrospun fibers. The variation of drug-loading amount did not change the pattern, which implied the dispersions of the drug were amorphous in PLLA, probably as a solid solution or amorphous molecular aggregates in the fibers. So incorporation of CLM into biodegradable PLLA nanofibers was feasible and the distribution state of the drug was amorphous, which developed the bioavailability of the drug and could be helpful for the drug release in the polymer carrier of PLLA nanofibers.

8.3.3 Summary

CLM/poly (L-lactic acid) (PLLA) composite nanofiber scaffolds were prepared by electrospinning. When the loading amount was lower, such as 5 wt%, the diameters of electrospun fibers were smaller than the diameters of the control fibers. However, the distribution of the fibers tended to become uneven and more beads appeared with the increase in CLM loading amount. The FT-IR spectra indicated the formation of a hydrogen bond between the drug and PLLA in the composite nanofiber scaffolds.

The XRD patterns of CLM/PLLA composite nanofiber scaffolds revealed that CLM was incorporated in an amorphous state into the carrier of PLLA nanofibers. The study revealed that CLM/PLLA composite nanofiber scaffolds could develop the bioavailability of CLM and control the release of CLM, which could be encapsulated into the PLLA nanofibers.

8.4 Future trends

Sustained drug delivery systems offer several potential advantages over traditional methods of administration. First, drug release rates can be tailored to the needs of a specific application – for example, providing a constant rate of delivery or pulsatile release. Second, these systems provide protection for drugs, especially those involving proteins that are otherwise rapidly destroyed by the body. Finally, these systems can increase patient comfort and compliance by replacing frequent (e.g., daily) doses with infrequent (once per month or less) injection.

There is much room for growth and improvement in sustained delivery systems using nanofibers. The search for ideal carriers, especially for biodegradable carriers, should go on. This should be aided by an understanding of the transport characteristics in polymers and the relationship between structure and biocompatibility. To achieve the desirable release patterns, more attention should be paid to the design of dosage forms. For instance, laminated, macroporous nanofibers can theoretically provide distinct release profiles, and techniques for their manufacture should be perfected. One area warranting extensive study is the theoretical development of microencapsulation. Microencapsulation promises ease of administration as well as potential advantages derived from passive targeting by intravenous or intra-arterial injection. However, the process in general is too empirical; a more unifying framework to account for the process parameters is needed. To complement these advances, further development of mathematical models is needed to assist in understanding the release mechanisms of different formulations and aid in the design of better and more predictable systems.

8.5 References

Birnbaum, D.T., Kosmala, J. D., Henthorn, D. B. and Brannon-Peppas, L. (2000), Controlled release of estradiol from PLAGA microparticles: The effect of organic phase solvent on encapsulation and release, *Journal of Controlled Release*, **65**, 375–87.

Chew, S. Y., Wen, J., Yim, E. K. F. and Leong, K. W. (2005), Sustained release of proteins from electrospun biodegradable fibers, *Biomacromolecules*, **6**, 2017–24.

Chu, S., Wilson, D. S., Deaton, R. L., Mackenthun, A. V., Eason, C. N. and Cavanaugh, J. H. (1993), Single and multiple-dose pharmacokinetics of CLM, a new macrolide antimicrobial, *Journal of Clinical Pharmacology*, **33**, 719–26.

Couvreur, P. and Vauthier, C. (2006), Nanotechnology: Intelligent design to treat complex disease, *Pharmaceutical Research*, **23**, 1417–50.

Cui, W., Li, X., Zhu, X., Yu, G., Zhou, S. and Weng, J. (2006), Investigation of drug release and matrix degradation of electrospun poly(DL-lactide) fibers with paracetanol inoculation, *Biomacromolecules*, **7**, 1623–9.

Deitzel, J. M., Kleinmeyer, J., Harris, D. and Beck Tan, N. C. (2001), The effect of processing variables on the morphology of electrospun nanofibers and textiles, *Polymer*, **42**, 261–72.

Diller, G. B., Cooper, J., Xie, Z., Wu, Y., Waldrup J. and Ren, X. (2007), Release of antibiotics from electrospun bicomponent fibers, *Cellulose*, **14**, 553–62.

Fong, H., Chun, I. and Reneker, D. H. (1999), Beaded nanofibers formed during electrospinning, *Polymer*, **40**, 4585–92.

Gaynor M. and Mankin A. S. (2003), Macrolide antibiotics: Binding site, mechanism of action, resistance, *Current Topics in Medicinal Chemistry*, **3**, 949–61.

Guay, D. R., Gustavson, L. E., Devcich, K. J., Zhang, J., Cao, G. and Olson, C. A. (2001), Pharmacokinetics and tolerability of extended-release CLM, *Clinical Therapeutics*, **23**, 566–77.

Heller, J. (1980), Controlled release of biologically active compounds from bioerodible polymers, *Biomaterials*, **1**, 51–7.

Jain, R. A., Rhodes, C. T., Railkar, A. M., Malick, A. W. and Shah, N. H. (2000), Controlled release of drugs from injectable in situ formed biodegradable PLGA microspheres: Effect of various formulation variables, *European Journal of Pharmaceutics and Biopharmaceutics*, **50**, 257–62.

Jeong, J. C., Lee, J. and Cho, K. (2003), Effects of crystalline microstructure on drug release behaviour of poly(ε-caprolactone) microspheres, *Journal of Controlled Release*, **92**, 249–58.

Ji, Y., Ghosh, K., Shu, X. Z., Li, B., Sokolov, J. C., Prestwich, G. D., Clark, R. A. F. and Rafailovich, M. H. (2006), Electrospun three-dimensional hyaluronic acid nanofibrous scaffolds, *Biomaterials*, **27**, 3782–92.

Kenawy, El-R., Bowlin, G. L. and Mansfield, K. (2002), Release of tetracycline hydrochloride from electrospun poly(ethylene-co-vinylacetate), poly(lactic acid), and a blend, *Journal of Control Release*, **81**, 57–64.

Kim, J. S. and Eneker, D. H. (1999), Mechanical properties of composites using ultrafine electrospun fibres, *Polymer Composites*, **20**, 124–31.

Kopecek, J. (2003), Smart and genetically engineered biomaterials and drug delivery systems, *European Journal of Pharmaceutical Sciences*, **20**, 1–16.

Nakano, K. and Kameoka, J. (2007), Self-cleaning filtration nanofibre membrane, US Patent, 0151921.

Ngawhirunpat, T., Opanasopit, P., Rojanarata, T., Akkaramongkolporn, P., Ruktanonchai, U. and Supaphol, P. (2009), Development of meloxicam-loaded electrospun polyvinyl alcohol mats as a transdermal therapeutic agent, *Pharmaceutical Development and Technology*, **14**, 70–9.

Okuda, T., Tominaga, K. and Kidoaki, S. (2010), Time-programmed dual release formulation by multilayered drug-loaded nanofiber meshes, *Journal of Controlled Release*, **143**, 258–64

Peppas, N. A. and Brannon-Peppas, L. (1994), Water diffusion sorption in amorphous macromolecular systems and foods, *Journal of Food Engineering*, **22**, 189–210.

Petitti, M., Vanni, M. and Barresi, A. A. (2008), Controlled release of drug encapsulated as a solid core: Theoretical model and sensitivity analysis, *Chemical Engineering Research and Design*, **86**, 1294–300.

Reneker, D. H. and Yarin, A. L. (2008), Electrospinning jets and polymer nanofibers, *Polymer*, **49**, 2387–425.

Sheetal, S. J. and Tejraj M. A. (2006), Molecular modeling simulations and thermodynamic approaches to investigate compatibility/incompatibility of poly(L-lactide) and poly(vinyl alcohol) blends, *Polymer*, **47**, 8061–71.

Sibley, S. D. and Pedersen, J. A. (2008), Interaction of the macrolide antimicrobial CLM with dissolved humic acid, *Environmental Science & Technology*, **42**, 422–8.

Siegel, R. and Langer, R. (1984), Controlled release of polypeptides and other macromolecules, *Journal of Pharmacy Research*, **1**, 2–10.

Srikar, R., Yarin, A. L., Megaridis, C. M., Bazilevsky, A. V. and Kelley, E. (2008), Desorption limited mechanism of release from polymer nanofibers, *Langmuir*, **24**, 965–74.

Tungprapa, S., Jangchud, I. and Supaphol, P. (2007), Release characteristics of four model drugs from drug-loaded electrospun cellulose acetate fiber mats, *Polymer*, **48**, 5030–41.

Verreck, G., Chun, I., Peeters, J., Rosenblatt, J. and Brewster, M. E. (2003a), Preparation and characterization of nanofibers containing amorphous drug dispersions generated by electrostatic spinning, *Journal of Pharmacy Research*, **20**, 810–7.

Verreck, G., Chun, I., Rosenblatt, J., Peeters, J., Dijck, A. V., Mensch, J., Noppe, M. and Brewster, M. E. (2003b), Incorporation of drugs in an amorphous state into electrospun nanofibers composed of a water-insoluble, nonbiodegradable polymer, *Journal of Controlled Release*, **92**, 349–60.

Xie, Z. and Diller, G. B. (2010), Electrospun poly(d,l-lactide) fibers for drug delivery: The influence of cosolvent and the mechanism of drug release, *Journal of Applied Polymer Science*, **115**, 1–8.

Xu, X., Chen, X., Xu, X., Lu, T., Wang, X., Yang, L. and Jing, X. (2006), BCNU-loaded PEG–PLLA ultrafine fibers and their in vitro antitumor activity against Glioma C6 cells, *Journal of Controlled Release*, **114**, 307–16.

Zamani, M., Morshed, M., Varshosaz, J. and Jannesari, M. (2010), Controlled release of metronidazole benzoate from poly(-caprolactone) electrospun nanofibers for periodontal diseases, *European Journal of Pharmaceutics and Biopharmaceutics*, **75**, 179–85.

9
Functional nanofibers for tissue engineering applications

I. A. RODRIGUEZ, J. M. MCCOOL and G. L. BOWLIN,
Virginia Commonwealth University, USA

Abstract: Tissue engineering aims to restore function or replace damaged tissue through biological principles and engineering. Electrospinning is a scaffold fabrication technique that produces continuous nanofiber meshes that mimic the extracellular matrix (ECM) of native tissue. This chapter briefly introduces various tissues and their respective ECM structure and composition. It then discusses the use of electrospun scaffolds intended for tissue engineering of each tissue, specifically focusing on cell response/growth to various electrospun nanofibers (natural/synthetic polymers, composites, incorporation of biological factors, effects of surface modifications, etc.).

Key words: tissue engineering, extracellular matrix, electrospinning, cell–matrix interaction, biodegradable.

9.1 Introduction

The field of tissue engineering aims to develop viable substitutes with the ability to repair and regenerate the functions of damaged tissue (Langer and Vacanti, 1993). Cells respond to environmental cues such as composition, topographical structure, mechanical properties and presence of biological factors. Therefore, within tissue engineering, cellular response and subsequent tissue growth are dependent on the characteristics of the scaffolding system. Tissue scaffolds are designed to serve as a temporary structure onto which new tissues grow and eventually replace the scaffold. It is important to fabricate scaffolds that mimic the structure and composition of the body's native ECM (Liao *et al.*, 2008). The ECM is the framework of each tissue that provides cues to promote cellular adhesion, migration, differentiation and proliferation. Cells continually produce, secrete and modify ECM components, which attribute to a unique ECM structure and composition for each tissue.

Several scaffold fabrication techniques for tissue engineering applications have been developed using synthetic and/or natural polymers. Some of these methods include freeze-drying, plates/meshes, solvent casting/particulate leaching, gas foaming/particulate leaching, solvent casting/salt leaching, salt leaching/solid–liquid phase separation and electrospinning. This chapter will specifically focus on the cellular responses to electrospun scaffolds for skin, vascular, bone, peripheral nerve, cartilage, ligament and tendon tissue engineering. Each section will discuss the ECM, applications of electrospun scaffolds and future trends for each tissue.

9.2 Electrospinning of nanofibers

To achieve cell attachment, growth and resulting tissue regeneration in three dimensions, scaffolds should be engineered to be porous and have a large surface area (Zhang and Ma, 1999; Murphy *et al.*, 2000; Liao *et al.*, 2008; Zhang *et al.*, 2008a). Electrospinning is a reliable, simple and economic method used to create nanofibrous ECM mimicking scaffolds with high surface area, high porosity and an interconnected pore network. Components of the electrospinning process can be altered to control and modify the microstructure of the porous nanofibrous scaffolds (Chen *et al.*, 2006; Zhang *et al.*, 2008a). For example, fiber diameters can be controlled by varying the concentration and composition of the polymer solution, fibers can be randomly oriented or aligned depending on the rotation speed of the mandrel, and scaffold shape can be tailored, depending on the shape of the mandrel (rectangular, cylindrical, etc.) (Sell *et al.*, 2008).

Electrospinning is a versatile method of fabricating scaffolds for tissue engineering as these nonwoven mats can be tailored for the regeneration of various tissues. Synthetic and/or natural polymers have been electrospun independently and in combination, which provides advantages for mechanically stable bioactive scaffolds. In addition, other components such as growth factors, proteins and minerals have been incorporated within the electrospun nanofibers to enhance cellular response. The nanofibers of electrospun scaffolds can also be easily modified via cross-linking and/or surface modifications to enhance mechanical properties and cell–ECM interactions. The properties and characteristics of electrospun scaffolds allow researchers to reproduce scaffolds specifically tailored for regeneration of a targeted tissue.

9.3 Applications of nanofibers in tissue engineering

9.3.1 Skin and wound healing

Skin is the largest organ of the body and functions mainly as a protective barrier against external threats entering the body. Other functions of skin include sensory detection, self-healing and fluid regulation. Skin is composed of two layers, the epidermis and dermis, which, combined, are responsible for the biomechanical and functional properties. The epidermis is composed of layers of keratinocytes while the dermis consists of mainly fibroblasts which secrete ECM proteins such as collagen and elastin (Pareanteau *et al.*, 2000; Falanga *et al.*, 2007). Skin wounds can be caused by a variety of situations or conditions, including burns, mechanical trauma, surgical operations, poor blood circulation or aging (Allgower *et al.*, 1995). Most skin wounds can heal naturally; however, many natural skin substitutes (allografts, autografts and xenografts) have been used for wound healing. These grafts currently fall short in efficiently healing the wound and present disadvantages such as limited donor sites, slow healing and possible disease transfer. The porosity and high surface to volume ratio of biodegradable electrospun scaffolds are desirable characteristics for wound healing, specifically allowing oxygen permeability, barrier properties to prevent bacterial infiltration and infiltration of cells to migrate and repopulate (Khil *et al.*, 2003; Chong *et al.*, 2007).

Commonly electrospun synthetic polymers such as polyurethane (PU), poly(methyl methacrylate), poly(D,L)-lactide-co-glycolide (PLGA), poly (epsilon-caprolactone) (PCL) and copolymers have been proposed for wound dressing and skin tissue engineering applications (Khil *et al.*, 2003; Suwantong *et al.*, 2007; Blackwood *et al.*, 2008; Kumbar *et al.*, 2008; Zhu *et al.*, 2008; Liu *et al.*, 2009; Lowery *et al.*, 2010). One such study successfully utilized a controlled vacuum seeding technique to seed National Institutes of Health (NIH) 3T3 fibroblasts in electrospun PCL scaffolds at various depths. Under vacuum, fibroblast proliferation and collagen secretion was not negatively affected. These results suggest that this technique can be used to create an electrospun bilayer skin model with fibroblasts seeded in an intermediate layer and keratinocytes on the top (Chen *et al.*, 2009b).

Synthetic polymers are often combined with other materials and biological factors to enhance cellular response to scaffolds. Electrospun Dextran/PLGA scaffolds proved to provide a hospitable environment for dermal fibroblasts. The cells migrated into the scaffolds and formed a dense multilayered organization that resembled native dermal architecture (Pan *et al.*, 2006). More recently, modifications such as heparin-coated and growth factor-loaded electrospun synthetic scaffolds have been found to enhance endothelial cell infiltration and fibroblast collagen deposition, respectively (Kurpinski *et al.*, 2010; Yang *et al.*, 2011).

Electrospun natural polymers such as collagen, gelatin, silk and chitin have been determined to be viable candidates for fabricating scaffolds intended for skin tissue engineering (Min *et al.*, 2004; Noh *et al.*, 2006; Powell & Boyce, 2008). Specifically, collagen is an attractive natural polymer for skin tissue engineering as it is found in skin ECM. Powell *et al.* compared freeze-dried and electrospun collagen skin substitutes both *in vitro* and *in vivo*. *In vitro*, both scaffolds showed excellent cellular organization, yet interestingly there were no significant differences between cell proliferation. However, when implanted in mice, the electrospun collagen scaffolds exhibited superior cellular organization and the potential to reduce wound contraction when compared to the freeze-dried scaffolds (Powell *et al.*, 2008).

Surface modification of nanofibers is a popular technique to increase functional activity of electrospun scaffolds. Rho *et al.* coated electrospun collagen nanofibers with ECM proteins (specifically a type I collagen coating) and found that this combination accelerated wound healing in early stages (Rho *et al.*, 2006). Jeong *et al.* concluded that electrospun silk fibers treated with plasma in the presence of oxygen had enhanced cellular activities for normal human epidermal keratinocytes (Fig. 9.1) and fibroblasts (Fig. 9.2) when compared to untreated controls (Jeong *et al.*, 2009). In addition to surface modifications, others have explored the possibilities of combining collagen with other natural polymers to tailor the morphology, mechanical and bioactive properties of scaffolds for skin tissue engineering (Yeo *et al.*, 2008; Dhandayuthapani *et al.*, 2010).

Several natural and synthetic composites, including collagen/PCL and copolymers, gelatin/PCL and copolymers, chitsan/PLGA, and chitosan/PCL have been electrospun and characterized for their potential in wound healing and skin tissue engineering. All blends mentioned above showed enhanced

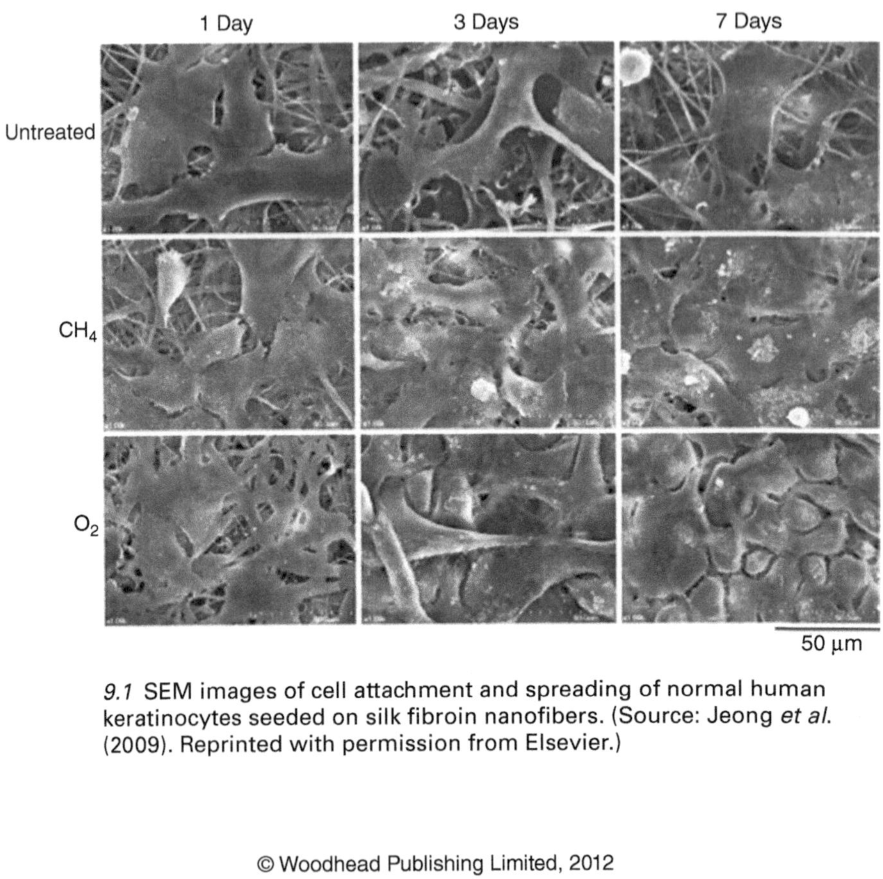

9.1 SEM images of cell attachment and spreading of normal human keratinocytes seeded on silk fibroin nanofibers. (Source: Jeong *et al.* (2009). Reprinted with permission from Elsevier.)

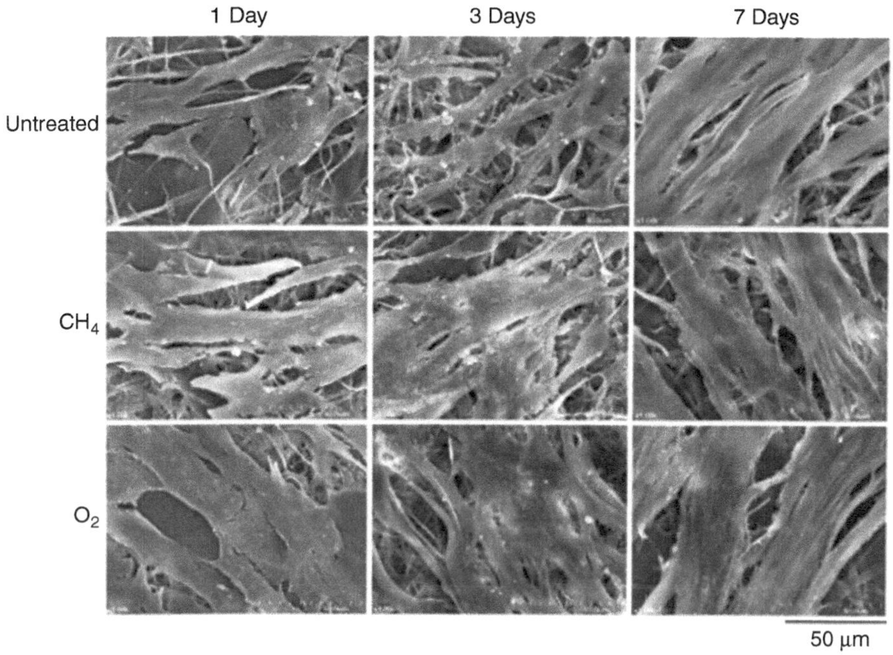

9.2 SEM images of cell attachment and spreading of normal human fibroblasts seeded on silk fibroin nanofibers. (Source: Jeong *et al.* (2009). Reprinted with permission from Elsevier.)

cellular responses when compared to controls (Venugopal *et al.*, 2006; Jeong *et al.*, 2008; Wu *et al.*, 2010b; Chandrasekaran *et al.*, 2011; Jin *et al.*, 2011; Shalumon *et al.*, 2011). Taking it a step further, Tigli *et al.* immobilized epidermal growth factor (EGF) on electrospun gelatin/PCL nanofibrous scaffolds. This study determined that mouse fibroblasts on gelatin/PCL/EGF scaffolds showed early cell spreading and rapid proliferation when compared to PCL/EGF, gelatin/PCL and PCL scaffolds (Tigli *et al.*, 2011).

Surface modifications and the incorporation of natural polymers and growth factors in synthetic electrospun scaffolds have proven to enhance the activity of various skin cells. A combination of the previously mentioned factors could result in an ideal scaffold for wound healing and skin tissue engineering considering biomechanical, structural and bioactive properties.

9.3.2 Vascular tissue

Engineering vascular grafts requires an understanding of tri-layered vessel architecture. When designing vascular grafts, the pulsating pressure of blood pumping through the grafts must also be taken into consideration. The challenging environment surrounding vascular engineering calls for materials that can not only withstand the demanding mechanical needs, but also must encourage quick and efficient endothelialization.

Natural polymers, such as silk, gelatin and collagen, promote cellular attachment and proliferation while sustaining the mechanical properties required of a vascular graft. Electrospun silk fibroin grafts may promote adhesion and elongation of human coronary artery smooth muscle cells (hCASMC) within five days (Zhang *et al.*, 2008b). Culturing vascular cells under dynamic flow conditions may improve cellular alignment and proliferation due to more effective nutrient and oxygen distribution to cells (Zhang *et al.*, 2009b). Besides the biocompatibility of electrospun silk fibroin, the burst strength and tensile properties are comparable to native vessels (Soffer *et al.*, 2008). Gelatin is another potential vascular biomaterial that, like silk, can promote cellular proliferation and adhesion in preliminary studies. Also like silk, electrospun gelatin must be cross-linked post-fabrication. Cross-linking agents like glutaraldehyde have cytotoxic effects that can decrease the effectiveness of an electrospun gelatin vascular graft. However, the naturally occurring cross-linking agent genipin may actually improve the mechanical properties while preserving the fiber morphology and cellular compatibility of the scaffold. Studies show that cross-linking with genipin may improve the mechanical properties of natural electrospun gelatin grafts (Panzavolta *et al.*, 2011).

Synthetic polymers, such as poly(L-lactide-co-epsilon-captrolactone) (P(LLA-CL)) are mechanically suitable for vascular grafts and promote cellular attachment and proliferation (Xu *et al.*, 2004b). When P(LLA-CL) is electrospun into aligned nanofibers, human coronary artery smooth muscle cells attach and migrate along the fiber axis which mimics the circumferential orientation of the arterial medial layer (Xu *et al.*, 2004a). PCL is another synthetic polymer that can be fabricated in a nanofibrous tubular scaffold with a controllable orientation. Nanofibers can be aligned in circumferential and/or axial directions to enhance endothelial cell attachment. Studies suggest that multiple layers of aligned nanofibers in different orientations may improve cellular attachment and proliferation for vascular engineering (Wu *et al.*, 2010a).

An electrospun graft with an outer polylactide shell and a silk fibroin/gelatin core exhibits acceptable mechanical properties while promoting adherence and spread throughout the graft of endothelial cells and fibroblasts (Wang *et al.*, 2009b). A novel approach to electrospinning vascular grafts was reported by Rainer *et al.*; this consists of a three-layered graft of synthetic and natural materials. The middle layer of collagen is between two functionalized layers of poly-L-lactide (PLLA). The inner PLLA layer acts like a drug delivery system to promote angiogenesis and prevent thrombosis,

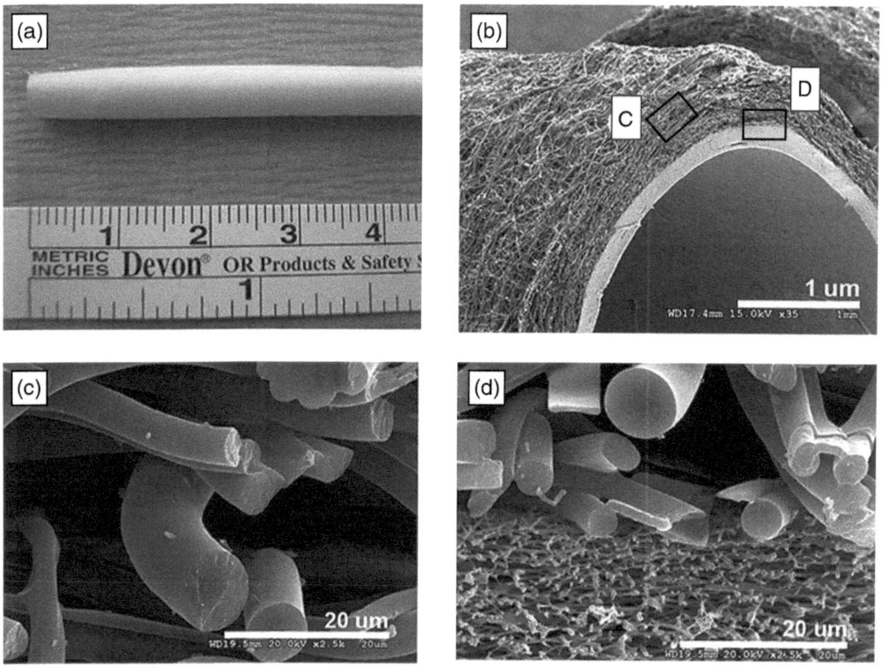

9.3 Bilayered electrospun PCL/collagen vascular scaffolds: (a) gross appearance. SEM images of cross-sectional for (b) entire, (c) outer layer, and (d) interface between outer and inner layers. (Source: Ju *et al.* (2010). Reprinted with permission from Elsevier.)

while the outer layer can induce media and adventitia generation (Rainer *et al.*, 2010). Other combinations such as chitosan/PCL, PU/gelatin, PCL/collagen (Fig. 9.3) and polydioxanone/elastin have all been evaluated as possible electrospun vascular grafts (Smith *et al.*, 2008; Garg *et al.*, 2009; Detta *et al.*, 2010; Ju *et al.*, 2010; Yang *et al.*, 2010).

Utilizing the electrospinning process allows the opportunity to introduce specific modifications to improve vascular grafts. Growth factors, such as fibroblast growth factor-2, can be incorporated into nanofibrous scaffolds by electrospinning from an aqueous solution (Ye *et al.*, 2011). Vascular endothelial growth factor, platelet-derived growth factor-BB and heparin can be incorporated into the vascular grafts to improve endothelialization of the implanted grafts while preventing occlusion and failure (Luong-Van *et al.*, 2006; 2010; Li *et al.*, 2010 Jia *et al.*, 2011).

9.3.3 Bone

Bone is a natural composite of collagenous organic matrix reinforced by an inorganic mineral phase of hydroxyapatite (HA). Bone also contains other components including calcium phosphates, water, proteins, etc. (Ciardelli *et al.*, 2010). Natural bone is a highly organized structure of parallel (mainly type I) collagen nanofibrils, and carbonated apatite (nano-HA crystals, nHA) located within the fibrils and precipitated on the surface (Weiner and Traub, 1992; Ciardelli *et al.*, 2010). These nHA crystals are platelet- or spindle-shaped and up to 200 nm in length, which provides a large surface area for effective mineral exchange (Salo, 2002). The ECM is the foundation upon which minerals are deposited during the process of forming new bone. This organic–inorganic composition is ultimately responsible for the biological function and biomechanical properties of bone (Jang *et al.*, 2009). For the application of electrospun scaffolds for bone engineering, one of the main challenges is that the consistency of the fibers does not match the ECM of the bone microenvironment. The scaffold must have organic mineral components (such as HA) and/or be treated postfabrication (commonly with a carbonated apatite coating) to enhance bioactivity. Moreover, the mechanical demands of a bone implant can be an obstacle for an electrospun scaffold. Usually, metals, glass and ceramics are used for bone applications; therefore, electrospun polymers must offer mechanical stability as well as advanced cellular interactions.

Early electrospinning bone tissue engineering experiments focused on the use of biodegradable synthetic polymers, specifically PCL, as nanofibrous scaffolds. In 2003, Yoshimoto *et al.* seeded rat bone marrow-derived mesenchymal stem cells (MSCs) in osteogenic differentiation medium on electrospun PCL scaffolds. After four weeks in culture, cell multilayers, a calcified ECM and the presence of type I collagen were detected (Fig. 9.4) (Yoshimoto *et al.*, 2003). In a follow-up study, the same lab implanted the four-week MSC preseeded PCL scaffolds in the omenta of rats. After four weeks, *in vivo* scaffolds maintained their original size and shape, cells were observed throughout and the ECM produced resembled that of woven bone (Shin *et al.*, 2004). Electrospun PCL nanofibers have been compared to smooth PCL films by seeding rat-derived bone marrow stromal cells on surfaces and quantifying alkaline phosphatase (ALP) activity, mineral deposition and ECM protein production. After three weeks in culture, the PCL nanofiber scaffolds supported greater adhesion and proliferation, and demonstrated significantly higher values of ALP activity, calcium phosphate mineralization and ECM protein (osteocalcin and osteopontin) production (Ruckh *et al.*, 2010). In addition to rat-derived cells, human MSCs (hMSCs) and human amniotic fluid stem (hAFS) cells on electrospun PCL scaffolds (591 nm average fiber diameter) have been studied. Both cell types demonstrated strong osteogenic potential; however, compared to hMSCs, hAFS cells showed delayed ALP activity, but elevated mineralization. When compared to tissue culture plastic, both cell types showed delayed attachment and proliferation on meshes, but more importantly enhanced

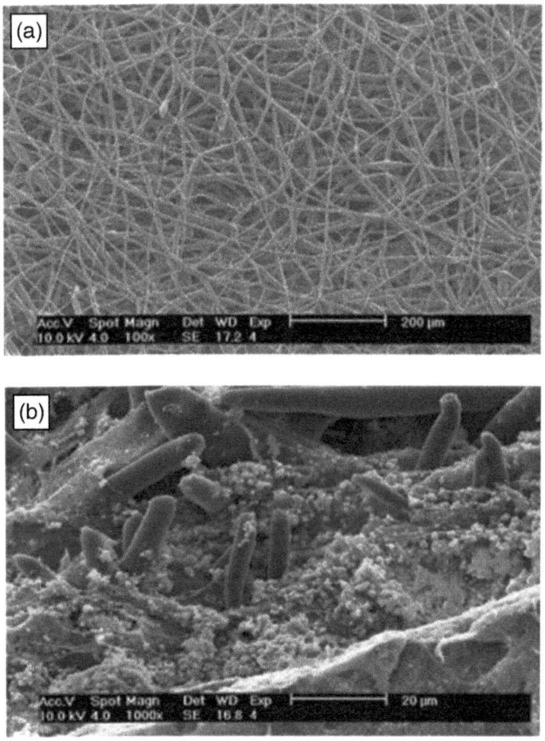

9.4 SEM images of MSCs seeded on electrospun PCL scaffolds. (a) Scaffold with randomly oriented nanofibers prior to seeding. (b) High magnification after 4 weeks of culture. Abundant mineralization and collagen bundles are visible. (Source: Yoshimoto *et al.* (2003). Reprinted with permission from Elsevier.)

mineralization over a longer period of time (Kolambkar *et al.*, 2010). A separate study was conducted on electrospun poly(lactic acid) (PLA), poly(ethylene glycol) and copolymers specifically focusing on mouse-calvarian (MC3T3-E1) osteoprogenitor cell responses to different fiber diameters (140 nm–2.1 μm). It was determined that topographical factors can be tailored to regulate morphology and proliferation of osteoblast-like cells (Badami *et al.*, 2006).

Although purely synthetic scaffolds show promise, they do not truly mimic bone ECM. Recent studies have focused on blended electrospun scaffolds comprised of synthetic polymers (PCL, PLA, PLGA, etc.), natural polymers (collagen, gelatin, chitosan, etc.) and/or minerals (nHA, calcium phosphates, etc.). Natural polymers are bioactive, possess additional mineral nucleation and cell-binding cites, but lack mechanical integrity when electrospun alone. Collagen is an attractive material to incorporate within scaffolds as it is the main organic component of bone ECM. Liao *et al.* explored the bioactivity of collagen versus PCL by subjecting scaffolds to a mineralizing solution. Electrospun collagen was determined to be more suitable for biomimetic bone tissue engineering due to the presence of active surface functional groups (such as carboxyl and carbonyl) which favor apatite nucleation and subsequent crystal growth (Liao *et al.*, 2008). Ekaputra *et al.* studied the cellular effects of adding collagen to PCL solutions and electrospinning PCL (564 nm fiber diameter) and PCL/collagen (513 nm fiber diameter) scaffolds. Results indicated that pig bone marrow mesenchymal cells on PCL/collagen scaffolds showed higher proliferation, produced osteocalcin and formed denser cell layers, collagen deposition and mineralization (Ekaputra *et al.*, 2009). Taking a different approach, Haslauer *et al.* added a uniform collagen (type I) coating to the surface of electrospun PCL nanofibers to determine effects on viability, proliferation and osteogenesis of human adipose-derived stem cells (hASCs). After two weeks in osteogenic culture, PCL/collagen scaffolds showed greater cell spreading at earlier time points and five-times higher calcium accretion (Haslauer *et al.*, 2011). Overall the incorporation of collagen in electrospun PCL scaffolds proves to be a valuable addition.

In order to mimic the inorganic component of bone ECM, the incorporation of HA within electrospun scaffolds is of interest. In general, the addition of HA decreases fiber diameter, creates a coarser fiber surface, supports long-term culture and enhances cell proliferation, protein adsorption, mineralization and ALP activity (Sui *et al.*, 2007; Lee *et al.*, 2010; Bakhshandeh *et al.*, 2011; Fang *et al.*, 2011; Peng *et al.*, 2011). Chen *et al.* determined that the enhancement of MSC osteogenic differentiation is positively correlated with nHA concentration in electrospun PCL scaffolds with 50 wt% nHA exhibiting the highest mineralization (Chen and Chang, 2011). HA (30 wt%) has also been electrospun with natural chitosan in order to mimic the mineralized nanofibril structure of bone ECM. After 10 and 15 days in culture, human fetal osteoblasts showed increased proliferation and mineralization in chitosan/HA scaffolds when compared to pure chitosan scaffolds (Zhang *et al.*, 2008c).

Electrospun organic–inorganic nanocomposite scaffolds intended to mimic the structure, composition and biological function of bone ECM have become more popular throughout the past few years. In 2008, Li *et al.* seeded preosteoblastic MC3T3-E1 cells on three different electrospun scaffold types – PCL, PCL coated with gelatin, PCL coated with gelatin then incubated in 10× simulated body fluid to further coat the nanofibers with a thin calcium phosphate apatite layer via mineralization. After seven days, the cells on the surface-modified PCL-gelatin-calcium phosphate scaffolds had higher proliferation rates (1.9 times) than PCL scaffolds (Li *et al.*, 2008). MC3T3-E1 cells have also been seeded on electrospun gelatin-apatite-poly(lactide-co-caprolactone) (PLCL) blended scaffolds under osteogenic conditions and it was found that osteogenic differentiation of cells was significantly enhanced when compared to the pure PLCL scaffold. Microcomputed tomography revealed that gelatin-apatite-PLCL scaffolds also supported defect closure and regenerated new bone better than PLCL scaffolds when implanted in a rat calvarium for six weeks (Jegal *et al.*, 2011). PLLA-collagen-HA and PLGA-collagen-calcium phosphate scaffolds have been seeded with hFOBs and hMSCs, respectively, and were both determined to support osteogenic differentiation and mineralization of cells when compared to pure synthetic polymer scaffold controls (Prabhakaran *et al.*, 2009b; Hild *et al.*, 2011).

Combining natural polymers, synthetic polymers and inorganic minerals best mimics the structural, mechanical and biological combination of bone ECM. These organic–inorganic scaffolds prove to be great candidates for bone tissue engineering substrates. Other areas of interest that show promise include biomimetic apatite coated nanofibers, demineralized bone powder blended with electrospun synthetic polymers, and electrospun PCL tubes with an internal alginate hydrogel structure capable of incorporating growth factors (Ko *et al.*, 2008; Kolambkar *et al.*, 2011; Rodriguez *et al.*, 2011). The latter experiment combined the guiding regeneration capabilities of electrospun nanofibers with the sustained release advantages of an alginate hydrogel to create a novel scaffold for repair of large bone defects. *In vivo* functional regeneration of bone was most prominent in a perforated PCL mesh with a recombinant bone morphogenetic protein-2 loaded alginate hydrogel core (Kolambkar *et al.*, 2011).

9.3.4 Peripheral nerve

The use of autografts and allografts for functional recovery of peripheral nerves presents the standard challenges of donor site morbidity, limited availability and possible disease transfer. As a result, the field of scaffold-based nerve tissue engineering has focused on developing analogues to facilitate axonal guidance ultimately achieving nerve regeneration. Peripheral nerve tissue is composed of myelinated and unmyelinated axons, Schwann cells, capillary wall endothelial cells, connective tissue cells, collagen fibers, fibroblasts and other specific cell types (Fazan *et al.*, 1997).

It has been found that randomly oriented electrospun fiber scaffolds result in nondirectional growth of neurites (Xie *et al.*, 2009) and even suggested that such crossing of fibers hinder directed axonal growth (Wang *et al.*, 2009a). Randomly oriented fibrous scaffolds have been modified via surface functionalization, encapsulation of growth factors and blends of natural polymers to construct scaffolds more suitable for nerve regeneration compared to their respective synthetic polymer controls (Koh *et al.*, 2008;

Nisbet *et al.*, 2008; Prabhakaran *et al.*, 2008, 2009a; Alvarez-Perez *et al.*, 2010; Liu *et al.*, 2011; Wang *et al.*, 2011b). However, since axonal guidance is a key factor in nerve regeneration, the majority of electrospun scaffolds intended for nerve tissue engineering consist of aligned nanofibers to direct axonal growth. Aligned electrospun nanofibers can be fabricated by increasing the rotational speed of the grounded target (Corey *et al.*, 2007; Chew *et al.*, 2008; Subramanian *et al.*, 2011) or by electrospinning techniques such as air gap electrospinning (Jha *et al.*, 2011). Wang *et al.* studied the effects of varying the diameter of aligned PLLA fibers on neurite outgrowth and Schwann cell migration (Fig. 9.5). It was determined that neurite length was shorter on small diameter fibers and Schwann cell migration was superior in small (293 ± 65 nm) and large (1325 ± 383 nm) diameters, suggesting fiber diameter is an important parameter to consider when designing aligned scaffolds (Wang *et al.*, 2010).

As previously mentioned, surface modification, incorporation of growth factors and blending of natural polymers are all methods which have been used to further improve the effectiveness and bioactivity of aligned electrospun synthetic polymer scaffolds. Separate *in vivo* rat studies showed that the encapsulation of glial cell-derived neurotrophic factors and nerve growth factors promoted more nerve regeneration when compared with unmodified synthetic scaffold controls (Chew *et al.*, 2007; Wang *et al.*, 2012). Zander *et al.* demonstrated that PCL scaffolds surface functionalized with ECM proteins (collagen and laminin) enhanced neuron-like PC12 cellular adhesion and neurite outgrowth compared to pure PCL scaffolds (Zander *et al.*, 2010). Aligned synthetic scaffolds have also been blended with collagen, gelatin and silk to enhance axonal guidance, nerve differentiation and proliferation, and peripheral nerve regeneration (Schnell *et al.*, 2007; Ghasemi-Mobarakeh *et al.*, 2008; Wang *et al.*, 2011a).

The majority of scaffolds intended for nerve tissue engineering focus on using variations of aligned synthetic or natural/synthetic fibers to regenerate nerve and direct axonal growth. Few studies have explored the use of natural polymer scaffolds as nerve-guiding conduits (Wang *et al.*, 2009c; Timnak *et al.*, 2011). An interesting approach has been to electrospin aligned electroactive nanofibers (such as polypyrrole/poly(styrene-beta-isobutylene-beta-styrene) and polyaniline) which have been shown to support and guide nerve regeneration *in vitro*. Electrical stimulation can be directly applied to these aligned electrospun conductive nanofibrous scaffolds to enhance cell proliferation and neurite growth of nerve stem cells (Ghasemi-Mobarakeh *et al.*, 2009; Liu *et al.*, 2010). Several materials have been suggested as viable candidates for scaffolds intended for nerve tissue engineering; however, the literature suggests that constructing aligned electrospun scaffolds will enhance nerve regeneration and direct axonal growth when compared to randomly oriented fibers.

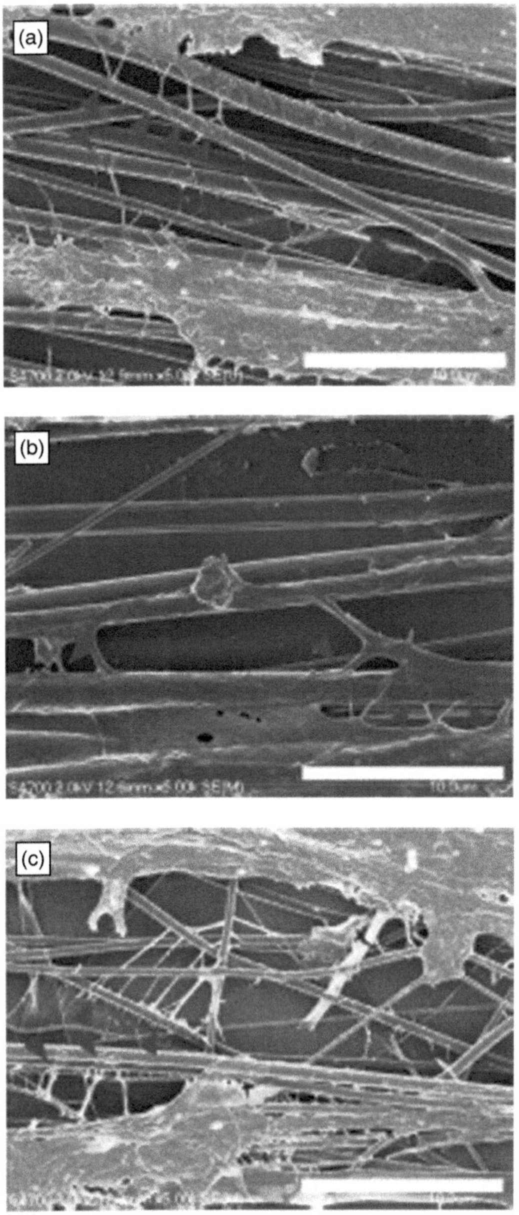

9.5 SEM images of dorsal root ganglia cultured on electrospun PLLA aligned nanofibrous scaffolds with different diameters: (a) neurites on large, (b) intermediate and (c) small diameter fibers. After 5 days in culture, neurites covered the surface of the fiber instead of existing between fibers. However, small filopodial extensions attached to nearby fibers for each condition. Scale bar = 10 μm. (Source: Wang *et al.* (2010). Reprinted with permission from Elsevier.)

9.3.5 Cartilage

The cartilage of joints withstands compressive, tensile and shear forces while providing a nearly frictionless surface between the bones in the joints (Riesle *et al.*, 1998; Shields *et al.*, 2004; Woodfield *et al.*, 2004). When designing cartilage implants, the mechanical requirements differ from other types of tissue grafts with compressive forces. In terms of electrospinning, a polymer that fabricates a porous, mechanically stable graft while promoting cellular attachment and proliferation would be ideal.

PCL is a biodegradable synthetic polymer that may be used to fabricate nanofibrous scaffolds for cartilage repair. Studies show that it promotes proliferation and maintenance of seeded chondrocytes (Li *et al.*, 2003). It has also been shown that electrospun PCL grafts support chondrogenesis of MSCs when cultured in the presence of transforming growth factor-beta (TGF-beta, Fig. 9.6) (Li *et al.*, 2005). PCL may also be compounded with starch prior to electrospinning in an attempt to enhance ECM formation *in vitro*. When seeded with bovine articular chondrocytes, ECM formation and

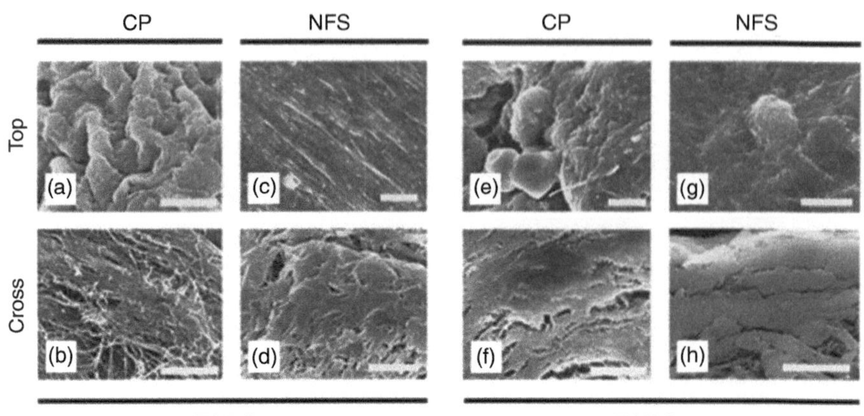

9.6 Morphology of day 21 cell pellet (CP) and PCL nanofibrous scaffold (NFS) MSC cultures examined by SEM. (a–d) Cultures maintained without TGF-β1; (e–h) cultures treated with TGF-β1. (a) Top view of CP, revealing a roughened surface, (b) cross-sectional view of CP, showing the presence of native collagen-like fibers, (c) top view of NFS, showing fibroblast-like cells covering the surface, (d) cross-sectional view of NFS, showing cells covered with ECM and integrated with PCL nanofibers, (e) top view of CP, with round chondrocyte-like cells on the surface, (f) cross-sectional view of CP, showing thick, ECM, (g) top view of NFS, showing the presence of round, ECM-embedded chondrocyte-like cells, and (h) cross-sectional view of NFS, showing a thick and dense ECM rich layer. Scale bar = 10 μm. (Source: Li *et al.* (2005). Reprinted with permission from Elsevier.)

glycosaminoglycan were detected throughout both PCL and PCL-starch scaffolds (da Silva *et al.*, 2009). Liao *et al.* designed a decellularized cartilaginous ECM/PCL composite for a cartilage graft. The ECM coating in addition to the microfibrous scaffold further enhanced the chondrogenesis of the MSCs (Liao *et al.*, 2010). These findings further support the adequacy of PCL for cartilage repair.

Electrospun natural polymers, such as collagen and chitosan are also suitable candidates for cartilage repair. Collagen type II can be electrospun into nanofibrous grafts that support chondrocyte attachment, proliferation and infiltration throughout the scaffold. Moreover, the mechanical properties can withstand the tensile stresses associated with the cartilage environment once cross-linked (Shields *et al.*, 2004). Chitosan may also be used to fabricate electrospun nanofibrous grafts for cartilage repair. Like collagen, chitosan exhibits suitable tensile strength for a cartilage implant, and is highly compatible with chondrocytes (Subramanian *et al.*, 2005).

A major hurdle for designing cartilage constructs is the ability of the implant to withstand the mechanical stresses while promoting cellular attachment, proliferation and infiltration. When focusing on electrospun scaffolds for the graft, a major concern is the ability to withstand the compressive forces. While a composite consisting of a hydrogel and electrospun shell may be an option for cartilage repair, another possibility may be to fabricate the electrospun scaffold, seed it with chondrocytes and culture it in a bioreactor to enhance the cellular response prior to implantation (Alves da Silva *et al.*, 2010). A further variation that may improve electrospun cartilage constructs is the fabrication of micron-sized fibers instead of nanofibers. Studies show that, with the increased fiber size and resulting increased pore size, cellular differentiation and chondrogenesis may occur (Liao *et al.*, 2010; Shanmugasundaram *et al.*, 2011). If the physical architecture of the scaffold plays a key role in chondrogenesis, then another possible avenue would be combining micro- and nanofibrous structures that provide an interconnected open pore structure. This modification has exhibited an increased chondrocyte proliferation and glycosaminoglycan production (Shim *et al.*, 2009).

9.3.6 Ligament and tendon

Ligaments and tendons are connective tissues that bridge bone-to-bone and muscle-to-bone, respectively. Ligaments and tendons are composed primarily of parallel collagen type I fiber bundles and contain various cell types (mainly fibroblasts) and other matrix materials (Sharma and Maffiulli, 2005; Doroski *et al.*, 2007; Chen *et al.*, 2009a). Since both of these tissues are similar in composition and structure, the design of a scaffold can be applied to either tissue (connective tissue engineering). One such experiment seeded bone marrow-derived MSCs on microfibrous silk coated with electrospun PLGA (Sahoo *et al.*, 2010). A separate study seeded porcine bone marrow stromal cells on knitted PLGA scaffolds coated with electrospun PLGA nanofibers (Sahoo *et al.*, 2006). Both take advantage of the biomimetic properties of the nanofibrous coating while maintaining a stable internal scaffold structure important for support to the injured ligament/tendon.

Guiding cell alignment and oriented ECM deposition are critical for regeneration of anisotropic tissues such as ligaments and tendons. Earlier studies on ligament tissue engineering using nanofibers focused on creating aligned electrospun synthetic polymer scaffolds (Lee *et al.*, 2005). In 2009, Bashur *et al.* determined that the formation of ligament-like tissue on electrospun poly(ester urethane)urea scaffolds was enhanced in scaffolds consisting of submicron fibers when compared to micron-sized fibers (Bashur *et al.*, 2009). More recently, silk has been explored as a potential material for electrospun scaffolds intended for ligament tissue engineering (Teh *et al.*, 2011). Sell *et al.* used an air gap electrospinning technique to create aligned silk nanofibers and determined that these scaffolds significantly enhanced human dermal fibroblast penetration when compared to traditionally electrospun scaffolds (Sell *et al.*, 2010). Other electrospun natural polymers, such as gelatin, have shown acceptable attachment, proliferation and growth of periodontal ligament cells *in vitro* (Zhang *et al.*, 2009a).

Since tendon is very similar in composition and structure to ligament, many of the scaffold design parameters overlap. Synthetic polymer scaffolds have been electrospun to mimic the structural organization of collagen fibers of tendons and have proven to be viable candidates for tendon regeneration (Fig. 9.7) (Xie *et al.*, 2010; Yin *et al.*, 2010; James *et al.*, 2011). Within the past year, the potential application of synthetic/natural polymer blends such as PCL/collagen, PLLA/collagen and polyvinyl alcohol (PVA)/wild antheraea pernyi silk fibroin has been investigated and determined to have acceptable cell compatibility with myoblasts, fibroblasts and rat tendon cells *in vitro* (Ladd *et al.*, 2011; Wu *et al.*, 2011). The field of ligament and tendon tissue engineering will continue to advance, especially if studies continue focusing on aligned nanofibrous scaffolds which mimic the collagen bundles observed in the native tissue.

9.4 Conclusions

With the wide variety of polymers suitable for electrospinning, it is possible to construct various implants tailored to a specific type of damaged or diseased tissue. It is evident that various types of cells respond differently to the microenvironments of electrospun fibers. Modifications, such as introducing growth factors and mechanical stimuli, can greatly change the cellular response to the scaffolds. A small change in fiber size, orientation

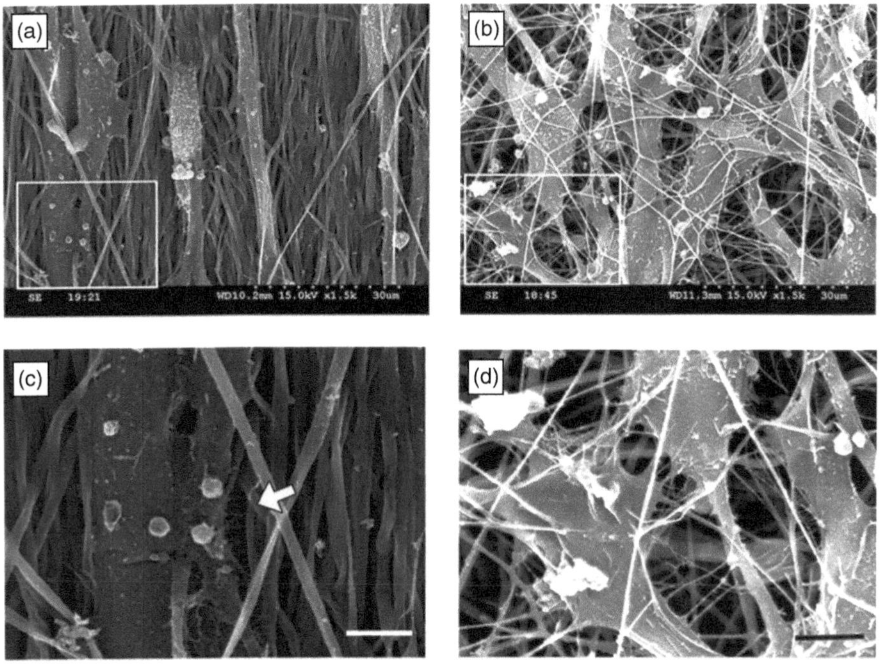

9.7 SEM images of morphological changes of human tendon stem/progenitor cells (hTSPCs) growing on (a) aligned and (b) randomly oriented PLLA scaffolds. (c) and (d) are high magnifications of boxed areas in (a) and (b), respectively, and highlight the cell–matrix adhesion between hTSPCs and PLLA nanofibers. The arrow in (c) indicates filament-like structure on aligned scaffold. Scale bars = 30 μm (a and b) and 5 μm (c and d). (Source: Yin *et al.* (2010). Reprinted with permission from Elsevier.)

or pore size can significantly affect the cellular orientation and migration around and into the scaffold. While the best combination of fiber diameter and pore size remains unknown, many variations have been explored.

The electrospinning process allows for rapid production of various scaffolds, but there are still many improvements that must be made in order to close the gap between the native ECM and the artificial one. Electrospun scaffolds still fall short of the mechanical demands of the natural microenvironment. The compressive forces for cartilage and bone, as well as the pulsating pressure for a vessel remain challenges for electrospun scaffolds to overcome. Overall, electrospun grafts are designed to more closely mimic the native ECM to encourage natural tissue repair.

9.5 References

Allgower, M., Schoenenberger, G. A. and Sparkes, B. G. (1995), Burning the largest immune organ, *Burns*, **21**(Suppl 1), S7–47.

Alvarez-Perez, M. A., Guarino, V., Cirillo, V. and Ambrosio, L. (2010), Influence of gelatin cues in PCL electrospun membranes on nerve outgrowth, *Biomacromolecules*, **11**(9), 2238–46.

Alves da Silva, M. L., Martins, A., Costa-Pinto, A. R., Costa, P., Faria, S., Gomes, M. *et al.* (2010), Cartilage tissue engineering using electrospun PCL nanofiber meshes and MSCs, *Biomacromolecules*, **11**(12), 3228–36.

Badami, A. S., Kreke, M. R., Thompson, M. S. Riffle, J. S. and Goldstein, A. S. (2006), Effect of fiber diameter on spreading, proliferation, and differentiation of osteoblastic cells on electrospun poly(lactic acid) substrates, *Biomaterials*, **27**(4), 596–606.

Bakhshandeh, B., Soleimani, M., Ghaemi, N. and Shabani, I. (2011), Effective combination of aligned nanocomposite nanofibers and human unrestricted somatic stem cells for bone tissue engineering, *Acta Pharmacologica Sinica*, **32**(5), 626–36.

Bashur, C. A., Shaffer, R. D., Dahlgren, L. A., Guelcher, S. A. and Goldstein, A. S. (2009), Effect of fiber diameter and alignment of electrospun polyurethane meshes on mesenchymal progenitor cells, *Tissue Engineering Part A*, **15**(9), 2435–45.

Blackwood, K. A., McKean, R, Canton, I., Freeman, C. O., Franklin, K. L., Cole, D. *et al.* (2008), Development of biodegradable electrospun scaffolds for dermal replacement, *Biomaterials*, **29**(21), 3091–104.

Chandrasekaran, A. R., Venugopal, J., Sundarrajan, S. and Ramakrishna, S. (2011), Fabrication of a nanofibrous scaffold with improved bioactivity for culture of human dermal fibroblasts for skin regeneration, *Biomedical Materials*, **6**(1), 015001.

Chen, J. P. and Chang, Y. S. (2011), Preparation and characterization of composite nanofibers of polycaprolactone and nanohydroxyapatite for osteogenic differentiation of mesenchymal stem cells, *Colloids and Surfaces B: Biointerfaces,* **86**(1), 169–75.

Chen, J., Chu, B. and Hsiao, B. S. (2006), Mineralization of hydroxyapatite in electrospun nanofibrous poly(L-lactic acid) scaffolds, *Journal of Biomedical Materials Research Part A*, **79**(2), 307–17.

Chen, J., Xu, J. , Wang, A. and Zheng, M. (2009a), Scaffolds for tendon and ligament repair: review of the efficacy of commercial products, *Expert Review of Medical Devices*, **6**(1), 61–73.

Chen, M., H. Michaud and S. Bhowmick (2009b), Controlled vacuum seeding as a means of generating uniform cellular distribution in electrospun polycaprolactone (PCL) scaffolds. *J Biomech Eng,* **131**(7), 074521.

Chew, S. Y., Mi, R., Hoke, A. and Leong, K. W. (2007), Aligned protein-polymer composite fibers enhance nerve regeneration: A potential tissue-engineering platform, *Advanced Functional Materials*, **17**(8), 1288–96.

Chew, S. Y., Mi, R., Hoke, A. and Leong, K. W. (2008), The effect of the alignment of electrospun fibrous scaffolds on Schwann cell maturation, *Biomaterials*, **29**(6), 653–61.

Chong, E. J., Phan, T. T., Lim, I. J., Zhang, Y. Z., Bay, B. H., Ramakrishna, S. *et al.* (2007), Evaluation of electrospun PCL/gelatin nanofibrous scaffold for wound healing and layered dermal reconstitution, *Acta Biomaterialia*, **3**(3), 321–30.

Ciardelli, G., Gentile, P., Chiono, V., Mattioli-Belmonte, M., Vozzi, G., Barbani, N. *et al.* (2010), Enzymatically crosslinked porous composite matrices for bone tissue regeneration, *Journal of Biomedical Materials Research Part A*, **92**(1), 137–51.

Corey, J. M., Lin, D. Y., Mycek, K. B., Chen, Q., Samuel, S., Feldman, E. L. *et al.* (2007), Aligned electrospun nanofibers specify the direction of dorsal root ganglia neurite growth, *Journal of Biomedical Materials Research Part A*, **83**(3), 636–45.

da Silva, M. A., Crawford, A., Mundy, J., Martins, A., Araujo, J. V., Hatton, P. V. *et al.* (2009), Evaluation of extracellular matrix formation in polycaprolactone and starch-compounded polycaprolactone nanofiber meshes when seeded with bovine articular chondrocytes, *Tissue Engineering Part A*, **15**(2), 377–85.

Detta, N., Errico, C., Dinucci, D., Puppi, D., Clarke, D. A., Reilly, G. C. *et al.* (2010), Novel electrospun polyurethane/gelatin composite meshes for vascular grafts, *Journal of Materials Science: Materials in Medicine*, **21**(5), 1761–9.

Dhandayuthapani, B., Krishnan, U. M. and Sethuraman, S. (2010), Fabrication and characterization of chitosan-gelatin blend nanofibers for skin tissue engineering, *Journal of Biomedical Materials Research Part B: Applied Biomaterials*, **94**(1), 264–72.

Doroski, D. M., Brink, K. S. and Temenoff , J. S. (2007), Techniques for biological characterization of tissue-engineered tendon and ligament, *Biomaterials*, **28**(2), 187–202.

Ekaputra, A. K., Zhou, Y., Cool, S. M. and Hutmacher, D. W. (2009), Composite electrospun scaffolds for engineering tubular bone grafts, *Tissue Engineering Part A*, **15**(12), 3779–88.

Falanga, V., Faria, K., Robert, L. and Joseph, V. (2007), Bioengineered skin constructs, *Principles of Tissue Engineering* (3rd edn), 1167–85.

Fang, R., Zhang, E., Xu, L. and Wei, S. (2011), Electrospun PCL/PLA/HA based nanofibers as scaffold for osteoblast-like cells, *Journal of Nanoscience and Nanotechnology*, **10**(11), 7747–51.

Fazan, V. P., Salgado, H. C. and Barreira, A. A. (1997), A descriptive and quantitative light and electron microscopy study of the aortic depressor nerve in normotensive rats, *Hypertension*, **30**(3 Pt 2), 693–8.

Garg, K., Sell, S. A., Madurantakam, P. and Bowlin, G. L. (2009), Angiogenic potential of human macrophages on electrospun bioresorbable vascular grafts, *Biomedical Materials*, **4**(3), 031001.

Ghasemi-Mobarakeh, L., Prabhakaran, M. P., Morshed, M., Nasr-Esfahani, M. H. and Ramakrishna, S. (2008), Electrospun poly(epsilon-caprolactone)/gelatin nanofibrous scaffolds for nerve tissue engineering, *Biomaterials*, **29**(34), 4532–9.

Ghasemi-Mobarakeh, L., Prabhakaran, M. P., Morshed, M., Nasr-Esfahani, M. H. and Ramakrishna, S. (2009), Electrical stimulation of nerve cells using conductive nanofibrous scaffolds for nerve tissue engineering, *Tissue Engineering Part A*, **15**(11), 3605–19.

Haslauer, C. M., Moghe, A. K., Osborne, J. A., Gupta, B. S. and Loboa, E. G. (2011), Collagen-PCL sheath-core bicomponent electrospun scaffolds increase osteogenic differentiation and calcium accretion of human adipose-derived stem cells, *Journal of Biomaterials Science Polymer Edition*, **22**(13), 1695–712(18).

Hild, N., Schneider, O. D., Mohn, D., Luechinger, N. A., Koehler, F. M., Hofmann, S. *et al.* (2011), Two-layer membranes of calcium phosphate/collagen/PLGA nanofibres, in vitro biomineralisation and osteogenic differentiation of human mesenchymal stem cells, *Nanoscale*, **3**(2), 401–9.

James, R., Kumbar, S. G., Laurencin, C. T., Balian, G. and Chhabra, A. B. (2011), Tendon tissue engineering: adipose-derived stem cell and GDF-5 mediated regeneration using electrospun matrix systems, *Biomedical Materials*, **6**(2), 025011.

Jang, J. H., Castano, O. and Kim, H. W. (2009), Electrospun materials as potential platforms for bone tissue engineering, *Advanced Drug Delivery Reviews*, **61**(12), 1065–83.

Jegal, S. H., Park, J. H., Kim, J. H., Kim, T. H., Shin, U. S., Kim, T. I. *et al.* (2011), Functional composite nanofibers of poly(lactide-co-caprolactone) containing gelatin-apatite bone mimetic precipitate for bone regeneration, *Acta Biomaterialia*, **7**(4), 1609–17.

Jeong, L., Yeo, I. S., Kim, H. N., Yoon, Y. I., Jang da, H., Jung, S. Y. *et al.* (2009), Plasma-treated silk fibroin nanofibers for skin regeneration, *International Journal of Biological Macromolecules*, **44**(3), 222–8.

Jeong, S. I., Lee, A. Y., Lee, Y. M. and Shin, H. (2008), Electrospun gelatin/poly(L-lactide-co-epsilon-caprolactone) nanofibers for mechanically functional tissue-engineering scaffolds, *Journal of Biomaterials Science Polymer Edition*, **19**(3), 339–57.

Jha, B. S., Colello, R. J., Bowman, J. R., Sell, S. A., Lee, K. D., Bigbee, J. W. *et al.* (2011), Two pole air gap electrospinning: Fabrication of highly aligned, three-dimensional scaffolds for nerve reconstruction, *Acta Biomaterialia*, **7**(1), 203–15.

Jia, X., Zhao, C., Li, P., Zhang, H., Huang, Y., Li, H. *et al.* (2011), Sustained release of VEGF by coaxial electrospun dextran/PLGA fibrous membranes in vascular tissue engineering, *Journal of Biomaterials Science Polymer Edition*, **22**(13), 1811–27.

Jin, G., Prabhakaran, M. P. and Ramakrishna, S. (2011), Stem cell differentiation to epidermal lineages on electrospun nanofibrous substrates for skin tissue engineering, *Acta Biomaterialia*, **7**(8), 3113–22.

Ju, Y. M., Choi, J. S., Atala, A., Yoo, J. J. and Lee, S. J. (2010), Bilayered scaffold for engineering cellularized blood vessels, *Biomaterials*, **31**(15), 4313–21.

Khil, M. S., Cha, D. I., Kim, H. Y., Kim, I. S. and Bhattarai, N. (2003), Electrospun nanofibrous polyurethane membrane as wound dressing, *Journal of Biomedical Materials Research Part B: Applied Biomaterials*, **67**(2), 675–9.

Ko, E. K., Jeong, S. I., Rim, N. G., Lee, Y. M., Shin, H. and Lee, B. K. (2008), In vitro osteogenic differentiation of human mesenchymal stem cells and in vivo bone formation in composite nanofiber meshes, *Tissue Engineering Part A*, **14**(12), 2105–19.

Koh, H. S., Yong, T., Chan, C. K. and Ramakrishna, S. (2008), Enhancement of neurite outgrowth using nano-structured scaffolds coupled with laminin, *Biomaterials*, **29**(26), 3574–82.

Kolambkar, Y. M., Peister, A., Ekaputra, A. K., Hutmacher, D. W. and Guldberg, R. E. (2010), Colonization and osteogenic differentiation of different stem cell sources on electrospun nanofiber meshes, *Tissue Engineering Part A*, **16**(10), 3219–30.

Kolambkar, Y. M., Dupont, K. M., Boerckel, J. D., Huebsch, N., Mooney, D. J., Hutmacher, D. W. *et al.* (2011), An alginate-based hybrid system for growth factor delivery in the functional repair of large bone defects, *Biomaterials*, **32**(1), 65–74.

Kumbar, S. G., Nukavarapu, S. P., James, R., Nair, L. S. and Laurencin, C. T. (2008), Electrospun poly(lactic acid-co-glycolic acid) scaffolds for skin tissue engineering, *Biomaterials*, **29**(30), 4100–7.

Kurpinski, K. T., Stephenson, J. T., Janairo, R. R., Lee, H. and Li, S. (2010), The effect of fiber alignment and heparin coating on cell infiltration into nanofibrous PLLA scaffolds, *Biomaterials*, **31**(13), 3536–42.

Ladd, M. R., Lee, S. J., Stitzel, J. D., Atala, A. and Yoo, J. J. (2011), Co-electrospun dual scaffolding system with potential for muscle-tendon junction tissue engineering, *Biomaterials*, **32**(6), 1549–59.

Langer, R. and Vacanti, J. P. (1993), Tissue engineering, *Science*, **260**(5110), 920–6.

Lee, C. H., Shin, H. J., Cho, I. H., Kang, Y. M., Kim, I. A., Park, K. D. *et al.* (2005), Nanofiber alignment and direction of mechanical strain affect the ECM production of human ACL fibroblast, *Biomaterials*, **26**(11), 1261–70.

Lee, J. H., Rim, N. G., Jung, H. S. and Shin, H. (2010), Control of osteogenic differentiation and mineralization of human mesenchymal stem cells on composite nanofibers containing poly[lactic-co-(glycolic acid)] and hydroxyapatite, *Macromolecular Bioscience*, **10**(2), 173–82.

Li, H., Zhao, C., Wang, Z., Zhang, H., Yuan, X. and Kong, D. (2010), Controlled release of PDGF-bb by coaxial electrospun dextran/poly(L-lactide-co-epsilon-caprolactone) fibers with an ultrafine core/shell structure, *Journal of Biomaterials Science Polymer Edition*, **21**(6), 803–19.

Li, W. J., Danielson, K. G., Alexander, P. G. and Tuan, R. S. (2003), Biological response of chondrocytes cultured in three-dimensional nanofibrous poly(epsilon-caprolactone) scaffolds, *Journal of Biomedical Materials Research Part A*, **67**(4), 1105–14.

Li, W. J., Tuli, R., Okafor, C., Derfoul, A., Danielson, K. G., Hall, D. J. *et al.* (2005), A three-dimensional nanofibrous scaffold for cartilage tissue engineering using human mesenchymal stem cells, *Biomaterials*, **26**(6), 599–609.

Li, X., Xie, J., Yuan, X. and Xia, Y. (2008), Coating electrospun poly(epsilon-caprolactone) fibers with gelatin and calcium phosphate and their use as biomimetic scaffolds for bone tissue engineering, *Langmuir*, **24**(24), 14145–50.

Liao, J., Guo, X., Grande-Allen, K. J., Kasper, F. K. and Mikos, A. G. (2010), Bioactive polymer/extracellular matrix scaffolds fabricated with a flow perfusion bioreactor for cartilage tissue engineering, *Biomaterials*, **31**(34), 8911–20.

Liao, S., Murugan, R., Chan, C. K. and Ramakrishna, S. (2008), Processing nanoengineered scaffolds through electrospinning and mineralization suitable for biomimetic bone tissue engineering, *Journal of the Mechanical Behavior of Biomedical Materials*, **1**(3), 252–60.

Liu, J. J., Wang, C. Y., Wang, J. G., Ruan, H. J. and Fan, C. Y. (2011), Peripheral nerve regeneration using composite poly(lactic acid-caprolactone)/nerve growth factor conduits prepared by coaxial electrospinning, *Journal of Biomedical Materials Research Part A*, **96**(1), 13–20.

Liu, X., Chen, J., Gilmore, K. J., Higgins, M. J., Liu, Y. and Wallace, G. G. (2010), Guidance of neurite outgrowth on aligned electrospun polypyrrole/poly(styrene-beta-isobutylene-beta-styrene) fiber platforms, *Journal of Biomedical Materials Research Part A*, **94**(4), 1004–11.

Liu, Y., Ji, Y., Ghosh, K., Clark, R. A., Huang, L. and Rafailovich, M. H. (2009), Effects of fiber orientation and diameter on the behavior of human dermal fibroblasts on electrospun PMMA scaffolds, *Journal of Biomedical Materials Research Part A*, **90**(4), 1092–106.

Lowery, J. L., Datta, N. and Rutledge, G. C. (2010), Effect of fiber diameter, pore size and seeding method on growth of human dermal fibroblasts in electrospun poly(epsilon-caprolactone) fibrous mats, *Biomaterials*, **31**(3), 491–504.

Luong-Van, E., Grondahl, L., Chua, K. N., Leong, K. W., Nurcombe, V. and Cool, S. M. (2006), Controlled release of heparin from poly(epsilon-caprolactone) electrospun fibers, *Biomaterials*, **27**(9), 2042–50.

Min, B. M., Lee, G., Kim, S. H., Nam, Y. S., Lee, T. S. and Park, W. H. (2004), Electrospinning of silk fibroin nanofibers and its effect on the adhesion and spreading of normal human keratinocytes and fibroblasts in vitro, *Biomaterials*, **25**(7–8), 1289–97.

Murphy, W. L., Kohn, D. H. and Mooney, D. J. (2000), Growth of continuous bone-like mineral within porous poly(lactide-co-glycolide) scaffolds in vitro, *Journal of Biomedical Materials Research*, **50**(1), 50–8.

Nisbet, D. R., Yu, L. M., Zahir, T., Forsythe, J. S. and Shoichet, M. S. (2008), Characterization of neural stem cells on electrospun poly(epsilon-caprolactone) submicron scaffolds: evaluating their potential in neural tissue engineering, *Journal of Biomaterials Science Polymer Edition*, **19**(5), 623–34.

Noh, H. K., Lee, S. W., Kim, J. M., Oh, J. E., Kim, K. H., Chung, C. P. *et al.* (2006), Electrospinning of chitin nanofibers: degradation behavior and cellular response to normal human keratinocytes and fibroblasts, *Biomaterials*, **27**(21), 3934–44.

Pan, H., Jiang, H. and Chen, W. (2006), Interaction of dermal fibroblasts with electrospun composite polymer scaffolds prepared from dextran and poly lactide-co-glycolide, *Biomaterials*, **27**(17), 3209–20.

Panzavolta, S., Gioffre, M., Focarete, M. L., Gualandi, C., Foroni, L. and Bigi, A. (2011), Electrospun gelatin nanofibers: optimization of genipin cross-linking to preserve fiber morphology after exposure to water, *Acta Biomaterialia*, **7**(4), 1702–9.

Pareanteau, N. L., Harden-Young, J. and Ross, R. N. (2000), Skin, in R. Lanza, R. Langer and J. Vacanti (eds), *Principles of Tissue Engineering*, San Diego, Academic Press, 879–90.

Peng, F., Yu, X. and Wei, M. (2011), In vitro cell performance on hydroxyapatite particles/ poly(l-lactic acid) nanofibrous scaffolds with an excellent particle along nanofiber orientation, *Acta Biomaterialia*, **7**(6), 2585–92.

Powell, H. M. and Boyce, S. T. (2008), Fiber density of electrospun gelatin scaffolds regulates morphogenesis of dermal-epidermal skin substitutes, *Journal of Biomedical Materials Research Part A*, **84**(4), 1078–86.

Powell, H. M., Supp, D. M. and. Boyce, S. (2008), Influence of electrospun collagen on wound contraction of engineered skin substitutes, *Biomaterials*, **29**(7), 834–43.

Prabhakaran, M. P., Venugopal, J. R. and Ramakrishna, S. (2009a), Mesenchymal stem cell differentiation to neuronal cells on electrospun nanofibrous substrates for nerve tissue engineering, *Biomaterials*, **30**(28), 4996–5003.

Prabhakaran, M. P., Venugopal, J. R., Chyan, T. T., Hai, L. B., Chan, C. K., Lim, A. Y. *et al.* (2008), Electrospun biocomposite nanofibrous scaffolds for neural tissue engineering, *Tissue Engineering Part A*, **14**(11), 1787–97.

Prabhakaran, M. P., Venugopal, J. and Ramakrishna, S. (2009b), Electrospun nanostructured scaffolds for bone tissue engineering, *Acta Biomaterialia*, **5**(8), 2884–93.

Rainer, A., Centola, M., Spadaccio, C., De Porcellinis, S., Abbruzzese, F., Genovese, J. A. *et al.* (2010), A biomimetic three-layered compartmented scaffold for vascular tissue engineering, *Conference Proceedings of the IEEE Engineering in Medicine and Biology Society*, 839–42.

Rho, K. S., Jeong, L., Lee, G., Seo, B. M., Park, Y. J., Hong, S. D. *et al.* (2006), Electrospinning of collagen nanofibers: effects on the behavior of normal human keratinocytes and early-stage wound healing, *Biomaterials*, **27**(8), 1452–61.

Riesle, J., Hollander, A. P., Langer, R., Freed, L. E. and Vunjak-Novakovic, G. (1998), Collagen in tissue-engineered cartilage: types, structure, and crosslinks, *Journal of Cellular Biochemistry*, **71**(3), 313–27.

Rodriguez, K., Renneckar, S. and Gatenholm , P. (2011), Biomimetic calcium phosphate crystal mineralization on electrospun cellulose-based scaffolds, *ACS Applied Materials & Interfaces*, **3**(3), 681–9.

Ruckh, T. T., Kumar, K., Kipper, M. J. and Popat, K. C. (2010), Osteogenic differentiation of bone marrow stromal cells on poly(epsilon-caprolactone) nanofiber scaffolds, *Acta Biomaterialia*, **6**(8), 2949–59.

Sahoo, S., Ouyang, H., Goh, J. C., Tay, T. E. and Toh, S. L. (2006), Characterization of a novel polymeric scaffold for potential application in tendon/ligament tissue engineering, *Tissue Engineering*, **12**(1), 91–9.

Sahoo, S., Ang, L. T., Cho-Hong Goh, J. and Toh, S. L. (2010), Bioactive nanofibers for fibroblastic differentiation of mesenchymal precursor cells for ligament/tendon tissue engineering applications, *Differentiation*, **79**(2), 102–10.

Salo, J. (2002), Bone resorbing osteoclasts reveal two basal plasma membrane domains and transcytosis of degraded matrix material, Dissertation. University of Oulu Department of Anatomy and Cell Biology, pp. 1–57.

Schnell, E., Klinkhammer, K., Balzer, S., Brook, G., Klee, D., Dalton, P. *et al.* (2007), Guidance of glial cell migration and axonal growth on electrospun nanofibers of poly-epsilon-caprolactone and a collagen/poly-epsilon-caprolactone blend, *Biomaterials*, **28**(19), 3012–25.

Sell, S. A., McClure, M. J., Ayres, C. E., Simpson, D. G. and Bowlin, G. L. (2010), Preliminary investigation of airgap electrospun silk-fibroin-based structures for ligament analogue engineering, *Journal of Biomaterials Science Polymer Edition*, **22**(10), 1253–73.

Sell, S., Barnes, C., Simpson, D. and Bowlin, G. (2008), Scaffold permeability as a means to determine fiber diameter and pore size of electrospun fibrinogen, *Journal of Biomedical Materials Research Part A*, **85**(1), 115–26.

Shalumon, K. T., Anulekha, K. H., Chennazhi, K. P., Tamura, H., Nair, S. V. and Jayakumar, R. (2011), Fabrication of chitosan/poly(caprolactone) nanofibrous scaffold for bone and skin tissue engineering, *International Journal of Biological Macromolecules*, **48**(4), 571–6.

Shanmugasundaram, S., Chaudhry, H. and Arinzeh, T. L. (2011), Microscale versus nanoscale scaffold architecture for mesenchymal stem cell chondrogenesis, *Tissue Engineering Part A*, **17**(5–6), 831–40.

Sharma, P. and Maffiulli, N. (2005), Basic biology of tendon injury and healing. *The Surgeon*, **3**(5), 309.

Shields, K. J., Beckman, M. J., Bowlin, G. L. and Wayne, J. S (2004), Mechanical properties and cellular proliferation of electrospun collagen type II, *Tissue Engineering*, **10**(9–10), 1510–7.

Shim, I. K., Suh, W. H., Lee, S. Y., Lee, S. H., Heo, S. J., Lee, M. C. *et al.* (2009), Chitosan nano-/microfibrous double-layered membrane with rolled-up three-dimensional structures for chondrocyte cultivation, *Journal of Biomedical Materials Research Part A*, **90**(2), 595–602.

Shin, M., Yoshimoto, H. and Vacanti, J. P. (2004), In vivo bone tissue engineering using mesenchymal stem cells on a novel electrospun nanofibrous scaffold, *Tissue Engineering*, **10**(1–2), 33–41.

Smith, M. J., McClure, M. J., Sell, S. A., Barnes, C. P., Walpoth, B. H., Simpson, D. G. *et al.* (2008), Suture-reinforced electrospun polydioxanone-elastin small-diameter tubes for use in vascular tissue engineering: a feasibility study, *Acta Biomaterialia*, **4**(1), 58–66.

Soffer, L., Wang, X., Zhang, X., Kluge, J., Dorfmann, L., Kaplan, D. L. *et al.* (2008), Silk-based electrospun tubular scaffolds for tissue-engineered vascular grafts, *Journal of Biomaterials Science Polymer Edition*, **19**(5), 653–64.

Subramanian, A., Vu, D., Larsen, G. F. and Lin, H. Y. (2005), Preparation and evaluation of the electrospun chitosan/PEO fibers for potential applications in cartilage tissue engineering, *Journal of Biomaterials Science Polymer Edition*, **16**(7), 861–73.

Subramanian, A., Krishnan, U. M. and Sethuraman, S. (2011), Fabrication of uniaxially aligned 3D electrospun scaffolds for neural regeneration, *Biomedical Materials Journal*, **6**(2), 025004.

Sui, G., Yang, X., Mei, F., Hu, X., Chen, G., Deng, X. *et al.* (2007), Poly-L-lactic acid/hydroxyapatite hybrid membrane for bone tissue regeneration, *Journal of Biomedical Materials Research Part A*, **82**(2), 445–54.

Suwantong, O., Waleetorncheepsawat, S., Sanchavanakit, N., Pavasant, P., Cheepsunthorn, P., Bunaprasert, T. *et al.* (2007), In vitro biocompatibility of electrospun poly(3-hydroxybutyrate) and poly(3-hydroxybutyrate-co-3-hydroxyvalerate) fiber mats, *International Journal of Biological Macromolecules*, **40**(3), 217–23.

Teh, T. K., Toh, S. L. and Goh, J. C. (2011), Aligned hybrid silk scaffold for enhanced differentiation of mesenchymal stem cells into ligament fibroblasts, *Tissue Engineering Part C Methods*, **17**(6), 687–703.

Tigli, R. S., Kazaroglu, N. M., Mav, I. S. B. and Gumusderel, I. O. M. (2011), Cellular behavior on epidermal growth factor (EGF)-immobilized PCL/gelatin nanofibrous scaffolds, *Journal of Biomaterials Science Polymer Edition*, **22**(1–3), 207–23.

Timnak, A., Gharebaghi, F. Y., Shariati, R. P., Bahrami, S. H., Javadian, S., Emami, S. H. *et al.* (2011), Fabrication of nano-structured electrospun collagen scaffold intended for nerve tissue engineering, *Journal of Materials Science: Materials in Medicine*, **22**(6), 1555–67.

Venugopal, J. R., Zhang, Y. and Ramakrishna, S. (2006), In vitro culture of human dermal fibroblasts on electrospun polycaprolactone collagen nanofibrous membrane, *The International Journal of Artificial Organs*, **30**(6), 440–6.

Wang, C. Y., Liu, J. J., Fan, C. Y., Mo, X. M., Ruan, H. J. and Li, F. F. (2012), The effect of aligned core-shell nanofibres delivering NGF on the promotion of sciatic nerve regeneration, *Journal of Biomaterials Science Polymer Edition*, **23**(1–4), 167–84.

Wang, C. Y., Zhang, K. H., Fan, C. Y., Mo, X. M., Ruan, H. J. and Li, F. F. (2011a), Aligned natural-synthetic polyblend nanofibers for peripheral nerve regeneration, *Acta Biomaterialia*, **7**(2), 634–43.

Wang, G., Hu, X., Lin, W., Dong, C. and Wu, H. (2011b), Electrospun PLGA-silk fibroin-collagen nanofibrous scaffolds for nerve tissue engineering, *In Vitro Cellular & Developmental Biology – Animal*, **47**(3), 234–40.

Wang, H. B., Mullins, M. E., Cregg, J. M., Hurtado, A., Oudega, M., Trombley, M. T. *et al.* (2009a), Creation of highly aligned electrospun poly-L-lactic acid fibers for nerve regeneration applications, *Journal of Neural Engineering*, **6**(1), 016001.

Wang, H. B., Mullins, M. E., Cregg, J. M., McCarthy, C. W. and Gilbert, R. J. (2010), Varying the diameter of aligned electrospun fibers alters neurite outgrowth and Schwann cell migration, *Acta Biomaterialia*, **6**(8), 2970–8.

Wang, S., Zhang, Y., Wang, H., Yin, G. and Dong, Z. (2009b), Fabrication and properties of the electrospun polylactide/silk fibroin-gelatin composite tubular scaffold, *Biomacromolecules*, **10**(8), 2240–4.

Wang, W., Itoh, S., Konno, K., Kikkawa, T., Ichinose, S., Sakai, K. *et al.* (2009c), Effects of Schwann cell alignment along the oriented electrospun chitosan nanofibers on nerve regeneration, *Journal of Biomedical Materials Research Part A*, **91**(4), 994–1005.

Weiner, S. and Traub, W. (1992), Bone structure: from angstroms to microns, *FASEB Journal*, **6**(3), 879–85.

Woodfield, T. B., Malda, J., de Wijn, J., Peters, F., Riesle, J. and van Blitterswijk, C. A. (2004), Design of porous scaffolds for cartilage tissue engineering using a three-dimensional fiber-deposition technique, *Biomaterials*, **25**(18), 4149–61.

Wu, H., Fan, J., Chu, C. C. and Wu, J. (2010a), Electrospinning of small diameter 3-D nanofibrous tubular scaffolds with controllable nanofiber orientations for vascular grafts, *Journal of Materials Science: Materials in Medicine*, **21**(12), 3207–15.

Wu, L., Li, H., Li, S., Li, X., Yuan, X., Li, X. *et al.* (2010b), Composite fibrous membranes of PLGA and chitosan prepared by coelectrospinning and coaxial electrospinning, *Journal of Biomedical Materials Research Part A*, **92**(2), 563–74.

Wu, L., Li, M., Zhao, J., Chen, D. and Zhou, Z. (2011), Preliminary study on polyvinyl alcohol/wild antheraea pernyi silk fibroin as nanofiber scaffolds for tissue engineered tendon, *Zhongguo Xiu Fu Chong Jian Wai Ke Za Zhi*, **25**(2), 181–6.

Xie, J., MacEwan, M. R., Li, X., Sakiyama-Elbert, S. E. and Xia, Y. (2009), Neurite outgrowth on nanofiber scaffolds with different orders, structures, and surface properties, *ACS Nano*, **3**(5), 1151–9.

Xie, J., Li, X., Lipner, J., Manning, C. N., Schwartz, A. G., Thomopoulos, S. *et al.* (2010), 'Aligned-to-random' nanofiber scaffolds for mimicking the structure of the tendon-to-bone insertion site, *Nanoscale*, **2**(6), 923–6.

Xu, C. Y., Inai, R., Kotaki, M. and Ramakrishna, S. (2004a), Aligned biodegradable nanofibrous structure: a potential scaffold for blood vessel engineering, *Biomaterials*, **25**(5), 877–86.

Xu, C., Inai, R., Kotaki, M. and Ramakrishna, S. (2004b), Electrospun nanofiber fabrication as synthetic extracellular matrix and its potential for vascular tissue engineering, *Tissue Engineering*, **10**(7–8), 1160–8.

Yang, W., Fu, J., Wang, D., Wang, T., Wang, H., Jin, S. *et al.* (2010), Study on chitosan/polycaprolactone blending vascular scaffolds by electrospinning, *Journal of Biomedical Nanotechnology*, **6**(3), 254–9.

Yang, Y., Xia, T., Zhi, W., Wei, L., Weng, J., Zhang, C. *et al.* (2011), Promotion of skin regeneration in diabetic rats by electrospun core-sheath fibers loaded with basic fibroblast growth factor, *Biomaterials*, **32**(18), 4243–54.

Ye, L., Wu, X., Mu, Q., Chen, B., Duan, Y., Geng, X. *et al.* (2011), Heparin-conjugated PCL scaffolds fabricated by flectrospinning and loaded with fibroblast growth factor 2, *Journal of Biomaterials Science Polymer Edition,* **22**(1–3), 389–406.

Yeo, I. S., Oh, J. E., Jeong, L., Lee, T. S., Lee, S. J., Park, W. H. *et al.* (2008), Collagen-based biomimetic nanofibrous scaffolds: preparation and characterization of collagen/silk fibroin bicomponent nanofibrous structures, *Biomacromolecules*, **9**(4), 1106–16.

Yin, Z., Chen, X., Chen, J. L., Shen, W. L., Hieu Nguyen, T. M., Gao, L. *et al.* (2010), The regulation of tendon stem cell differentiation by the alignment of nanofibers, *Biomaterials*, **31**(8), 2163–75.

Yoshimoto, H., Shin, Y. M., Terai, H. and Vacanti, J. P. (2003), A biodegradable nanofiber scaffold by electrospinning and its potential for bone tissue engineering, *Biomaterials*, **24**(12), 2077–82.

Zander, N. E., Orlicki, J. A., Rawlett, A. M. and Beebe, T. P. Jr. (2010), Surface-modified nanofibrous biomaterial bridge for the enhancement and control of neurite outgrowth, *Biointerphases*, **5**(4), 149–58.

Zhang, D., Chang, J. and Zeng, Y. (2008a), Fabrication of fibrous poly(butylene succinate)/wollastonite/apatite composite scaffolds by electrospinning and biomimetic process, *Journal of Materials Science: Materials in Medicine*, **19**(1), 443–9.

Zhang, R. and Ma, P. X. (1999), Poly(alpha-hydroxyl acids)/hydroxyapatite porous composites for bone-tissue engineering. I. Preparation and morphology, *Journal of Biomedical Materials Research Part A*, **44**(4), 446–55.

Zhang, S., Huang, Y., Yang, X., Mei, F., Ma, Q., Chen G. *et al.* (2009a), Gelatin nanofibrous membrane fabricated by electrospinning of aqueous gelatin solution for guided tissue regeneration, *Journal of Biomedical Materials Research Part A*, **90**(3), 671–9.

Zhang, X., Baughman, C. B. and Kaplan, D. L. (2008b), In vitro evaluation of electrospun silk fibroin scaffolds for vascular cell growth, *Biomaterials*, **29**(14), 2217–27.

Zhang, X., Wang, X., Keshav, V., Wang, X., Johanas, J. T., Leisk, G. G. *et al.* (2009b), Dynamic culture conditions to generate silk-based tissue-engineered vascular grafts, *Biomaterials*, **30**(19), 3213–23.

Zhang, Y., Venugopal, J. R., El-Turki, A., Ramakrishna, S., Su, B. and Lim, C. T. (2008c), Electrospun biomimetic nanocomposite nanofibers of hydroxyapatite/chitosan for bone tissue engineering, *Biomaterials*, **29**(32), 4314–22.

Zhu, X., Cui, W., Li, X. and Jin, Y. (2008), Electrospun fibrous mats with high porosity as potential scaffolds for skin tissue engineering, *Biomacromolecules*, **9**(7), 1795–801.

10
Functional nanofibers in lithium-ion batteries

H. QIAO and Q. WEI,
Jiangnan University, China

Abstract: This chapter introduces the topic of lithium-ion batteries, including definitions, the development and work principles of Li-ion batteries and the application of nanofibers as anode materials. The authors successfully prepared porous carbon nanofibers (CNFs) by the electrospinning of a bicomponent polymer solution, followed by thermal treatments under a nitrogen atmosphere. It was found that they are formed with turbostratically disordered graphene sheets and have small pores and large surface areas. The unique structure of these porous CNFs was shown to result in good electrochemical performance such as high reversible capacity and good cycle stability when they are used as an anode for Li-ion batteries.

Key words: electrospinning, nanofibers, anode, electrochemical performance, Li-ion batteries.

10.1 Introduction

A lithium-ion battery is a family of rechargeable battery types in which lithium ions move from the negative electrode to the positive electrode during discharge and back when charging. Chemistry, performance, cost and safety characteristics vary across lithium-ion battery types. Unlike lithium primary batteries (which are disposable), lithium-ion batteries use an intercalated lithium compound as the electrode material instead of metallic lithium.

Lithium-ion batteries are common in consumer electronics. They are one of the most popular types of rechargeable battery for portable electronics, with one of the best energy-to-weight ratios, high open circuit voltage, low self-discharge rate, no memory effect and a slow loss of charge when not in use. Beyond consumer electronics, lithium-ion batteries are growing in popularity for military, electric vehicle and aerospace applications due to their high energy density.

Lithium-ion batteries were first proposed by M. S. Whittingham at Binghamton University in the 1970s.[1] Whittingham used titanium(II) sulfide as the cathode and lithium metal as the anode. The electrochemical properties of lithium intercalation in graphite were first discovered in 1980 by Rachid Yazami *et al.*, who showed the reversible intercalation of lithium into graphite in a lithium/polymer electrolyte/graphite half cell.[2] In 1981, Bell Labs developed a workable graphite anode to provide an alternative to the lithium metal battery. Following cathode research performed by a team led by John Goodenough, in 1991 Sony released the first commercial lithium-ion battery. Their cells used layered oxide chemistry, specifically lithium cobalt oxide. In 1983, Dr Michael Thackeray, Goodenough and coworkers identified manganese spinel as a cathode material.[3] Spinel showed great promise, given low cost, good electronic and lithium ion conductivity and three-dimensional structure, which gives it good structural stability. Although a pure manganese spinel fades with cycling, this can be overcome with chemical modification of the material. A manganese spinel is currently used in commercial cells.

In 1989, Goodenough showed that cathodes containing polyanions – for example, sulfates – produce higher voltages than oxides due to the inductive effect of the polyanion.[4] In 1996, Goodenough, Akshaya Padhi and coworkers identified lithium iron phosphate ($LiFePO_4$) and other phospho-olivines (lithium metal phosphates with olivine structure) as cathode materials.[5] In 2002, Yet-Ming Chiang and his group at MIT showed a substantial improvement in the performance of lithium batteries by boosting the material's conductivity by doping it with aluminum, niobium and zirconium. The exact mechanism causing the increase became the subject of a debate. In 2004, Chiang again increased performance by utilizing iron phosphate particles of less than 100 nm in diameter. This decreased particle density by almost a hundredfold, increased the cathode's surface area and improved capacity and performance.

In recent years, with the rapid development of nanotechnology, nanomaterials are promising candidates for lithium-ion battery electrodes. As lithium-ion battery electrode materials, nanomaterials have some unique physical and chemical properties, such as the large surface area, shorter transport length, high reversible capacity and long cycle life. These properties can significantly improve specific capacity and high-rate performance of lithium-ion batteries.

10.2 Components and materials in lithium-ion batteries

The three primary functional components of a lithium-ion battery are the anode, cathode and electrolyte. The anode of a conventional lithium-ion cell is made from carbon, the cathode is a metal oxide and the electrolyte is a lithium salt in an organic solvent.[6] The most commercially popular anode material is graphite. The cathode is generally one of three materials: a layered oxide (such as lithium cobalt oxide), a polyanion (such as lithium iron phosphate) or a spinel (such as lithium manganese oxide).

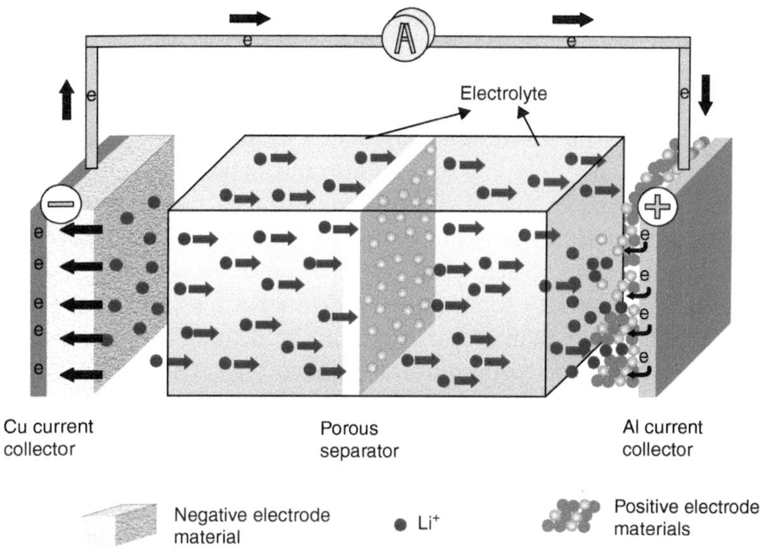

10.1 The work principle of lithium-ion battery.

Both the anode and cathode are materials into which and from which, lithium can migrate. During insertion (or intercalation), lithium moves into the electrode. During the reverse process, extraction (or deintercalation), lithium moves back out. When a lithium-based cell is discharging, the lithium is extracted from the anode and inserted into the cathode. When the cell is charging, the reverse occurs (Fig. 10.1).

The cathode half-reaction (with charging being forward) is:

$$\mathrm{LiCoO_2} \underset{\text{Discharge}}{\overset{\text{Charge}}{\rightleftharpoons}} \mathrm{Li_{1-X}CoO_2} + \mathrm{xLi^+} + \mathrm{xe^-}$$

The anode half-reaction is:

$$6\mathrm{C} + \mathrm{xLi^+} + \mathrm{xe^-} \underset{\text{Discharge}}{\overset{\text{Charge}}{\rightleftharpoons}} \mathrm{Li_xC_6}$$

The overall reaction is:

$$6\mathrm{C} + \mathrm{LiCoO_2} \underset{\text{Discharge}}{\overset{\text{Charge}}{\rightleftharpoons}} \mathrm{Li_{1-X}CoO} + \mathrm{xLi^+} + \mathrm{xe^-}$$

In a lithium-ion battery the lithium ions are transported to and from the cathode or anode, with the transition metal, cobalt (Co), in $\mathrm{Li_xCoO2}$ being oxidized from Co^{3+} to Co^{4+} during charging and reduced from Co^{4+} to Co^{3+} during discharge.

10.3 Applications of nanofibers in lithium-ion batteries

Electrospun polymers, such as polyvinylidenedifluoride (PVDF) and polyacrylonitrile (PAN) and their derivatives, can be used as nanofibers in separators of Li-ion batteries, providing a nanoporous structure leading to increased ionic conductivity of a membrane soaked with liquid electrolyte. Aligned electrospun PVDF fibrous membranes can enhance the tensile strength and modulus of membranes by improving interfiber compaction under hot pressing. Such electrospun membranes may have important applications as battery separators. Uniform electrospun PVDF membrane thickness and fiber diameter can be obtained by using high polymer concentrations and electrospinning at high voltage. This improves mechanical strength and provides the PVDF separator with charge and discharge capacities that exceed commercial polypropylene separators, while resulting in little capacity loss.[7,8] Electrospun PAN nonwoven fibers show higher porosities with lower Gurley values and increased wettabilities compared to conventional separators. Furthermore, cells with separators of PAN nonwovens show enhanced cycle life/performance, higher rate capabilities and smaller diffusion resistance than cells with conventional separators.[9,10]

High surface area conducting nanofibers make electrospun materials attractive as electrodes for batteries. Compared to conventional powder and film electrodes, nanofiber electrodes are expected to have improved electrode properties due to their high surface areas and 1-D nanostructure properties and, therefore should provide higher rates of electron transfer. In addition, their entangled network allows easy access of the ions to the electrode/electrolyte interface. Nanostructured $LiCoO_2$ fiber electrodes with faster diffusion of Li^+ cations prepared by electrospinning afford a higher initial discharge capacity of 182 mAh/g compared with that of conventional powder and film electrodes (140 mAh/g).[11]

In the present commercial lithium-ion batteries, graphite has been widely used as the intercalating anode for its high reversible capacity, even discharge–charge potential profile and low cost. However, graphite does have some disadvantages and cannot satisfactorily meet the performance requirements of some important applications, especially in the safety and rate performance.[12] Therefore, researchers turn to investigate nanostructured materials as alternative anode material for lithium-ion batteries because of their superior performances. In recent years, nanostructured carbon materials, such as CNFs,[13–15] have been employed to construct anodes for lithium-ion batteries because these materials have relatively short lithium-ion pathways, large specific surface area and extremely charming surface activity. However, in order to further improve the performance of carbon materials as anodes for batteries, an effective porous structure that is controllable is needed to provide a desirable surface area and open pore structure, which can achieve larger energy conversion density, higher rate capability and better cycle performance.[16,17] Therefore, activated or porous CNFs with large specific surface area and controlled pore structure could be an ideal candidate to meet these requirements.

Wei and coworkers[18] reported that the reversible specific capacity of the vapor-grown CNFs at a 0.1C rate was 461 mAh/g, and even at a high cycling rate of 10C the reversible capacity is around 170 mAh/g with a 95% Coulombic efficiency. High-purity CNFs with dome-shaped interiors form the noncatalytic thermal decomposition of acetylene over a copper surface at atmospheric pressure, and electrochemical measurements, indicated that the CNFs deliver good cyclability and a reversible capacity of 260 mAh/g at the high specific current of 100 mA/g.[19] Zou *et al.*[20] reported that the CNF electrode which was synthesized by cocatalyst deoxidization process by a reaction between $C_2H_5OC_2H_5$, Zn and Fe powder at 650°C for 10 h, shows a capacity of 220 mAh/g and high reversibility with little hysteresis in the insertion/deintercalation reactions of lithium-ion. Wang and coworkers[21] synthesized porous CNFs by an easy pyrolysis of conducting a polymer in an argon atmosphere, with an average pore diameter of 27.98 nm and a specific surface area of about 74.5 $m^2\ g^{-1}$. The reversible specific capacity of the CNFs at a 0.5C rate is about 400 mAh g^{-1} when used as anode material in lithium-ion batteries. Moreover, they exhibit considerably specific capacity even at a high charge–discharge current; that is, the reversible capacities are around 250 and 194 mAh g^{-1} at a rate of 10 and 20C, respectively. Porous CNFs with large accessible surface areas and well-developed pore structures were prepared by electrospinning and subsequent thermal and chemical treatments.[22–24] The electrochemical performance results show that porous CNF anodes have improved lithium-ion storage ability, enhanced charge–discharge kinetics and better cyclic stability compared with nonporous counterparts. Ji *et al.*[25,26] reported that Mn oxide-loaded porous CNFs were prepared by electrospinning polyacrylonitrile nanofibers containing different amounts of $Mn(CH_3COO)_2$, followed by thermal treatments in different environments. Electrochemical measurements indicated that the resulting composite nanofibers exhibit high reversible capacity, improved cycling performance and elevated rate capability even at high current rates, when used as anodes for rechargeable lithium-ion batteries without adding any polymer binder or electronic conductor. Zhang and coworkers also synthesized Si,[27,28] copper[29] and nickel[30]/carbon composite nanofibers by the same method for use as anodes in lithium-ion batteries. The resultant composite nanofiber anodes exhibited a large accessible surface area, high reversible capacity and relatively good cycling performance. The novel Sn-C composite with a core-shell morphology of Sn nanoparticles encapsulated in porous multichannel carbon microtubes[31] and bamboo-like hollow CNFs[32] was fabricated by a single-nozzle electrospinning technique. The resulting material exhibited excellent characteristics in terms of reversible capacities, cycling performance and rate capability for applications as an anode material in lithium-ion batteries.

10.4 Examples of applications of nanofibers

Porous CNFs are typically in the form of electron-conducting nonwoven webs, providing good mechanical stability. Hence, they can be directly used as anodes for LIBs without adding nonactive materials (e.g., polymer binder and conducting carbon black). The elimination of nonactive materials leads to high energy and power densities, longer cycle life and improved safety of LIBs. In addition, the fully interconnected three-dimensional network pore structure in CNFs facilitates easy access of lithium ions to the inner sites of anodes, which dramatically reduces lithium-ion diffusion distance and improves electron transfer rate. As a result, LIBs using fabricated porous CNFs as anodes offer advantages such as improved energy and power densities, fast-charging capability, long cycle life and low cost. Therefore, the technological impact of designing porous CNFs with controlled structures and enhanced electrochemical performance would be significant and needs further exploration.

With optimized electrospinning conditions, homogeneous and long organic nanofibers in the form of a thin white web were obtained. The conversion of an organic nanofiber into a CNF involves stabilization and thermal-treatment steps. The air stabilization of electrospun nanofibers, which involves complex chemical reactions such as three-dimensional cross linking,[33,34] is accompanied by a color change from white to reddish brown. Finally, by thermally treating the air-stabilized fibers, we obtained a black web consisting of fibrous carbon.

As shown in Fig. 10.2a–c, the electrospun PAN/*polymethyl methacrylate* (PMMA) composite nanofibers exhibit a smooth outer surface, long fibrous morphology and homogeneous diameter distributions in the range of 200–400 nm (see Fig. 10.2a). Air stabilization of the electrospun nanofiber web was then carried out at 280°C for 1 h under controlled air circulation, which is critical to obtain dimensional stability and, hence, a fibrous morphology. The morphology is not disrupted by thermal treatment at 800°C in an inert atmosphere as in the case of conventional PAN-based carbon fibers. The elongated PMMA phase decomposes (or disappears) without carbon residue and the continuous PAN phase is easily transformed into carbon. CNFs exhibited an uneven and wrinkled 'molecular sponge' surface morphology (see Fig. 10.2b). In addition, there are large variations in the fiber diameter of porous CNFs. The pore structure of porous CNFs can be more clearly seen in transmission electron microscope (TEM) images. Figure 10.2c shows a TEM image of porous CNFs made from PAN/PMMA bicomponent

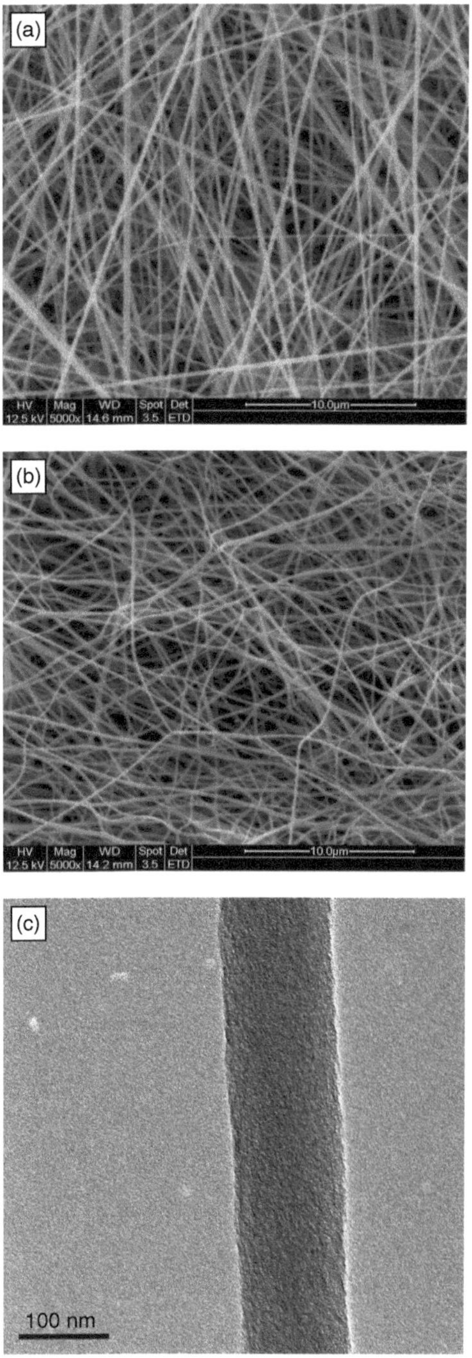

10.2 SEM images of PAN/PMMA composite nanofibers (a), porous carbon nanofibers (b) and TEM image of porous carbon nanofibers (c).

Table 10.1 Surface parameters of as-prepared carbon nanofibers

PAN: PMMA blend ratios	A_{BET} (m^2/g)	V_{BJH} (cm^3/g)	Mean pore diameter (nm)
10:0	2.7	0.0016	134.8
9:1	452.1	0.032	28.1
7:3	572.9	0.053	33.6
5:5	454.9	0.050	37.1

nanofibers. The largely irregular and undulating surface morphology with elongated indents and clear ridges along the fiber surface indicate the presence of relatively large pores (i.e., mesopores with diameters in the range of 2–50 nm). However, the presence of smaller pores (i.e., micropores with diameters less than 2 nm) needs to be studied using other techniques such as the *Brunauer–Emmett–Teller* (BET) nitrogen adsorption method.

The porosity parameters for all samples are summarized in Table 10.1. The total pore volume, BET specific surface area and the fraction and average size of the mesopores (2–50 nm) increase with an increase in PMMA concentration. From these results, we determined that the discontinuous but PMMA phase played a key role in developing mesopores during thermal decomposition. In further work to obtain porous and fibrous carbons in a more controlled way, a systematic kinetic study of the pores created as a result of temperature and heating-rate functions during thermal treatment is critically needed because the complex gas evolution involved (degradation of PMMA) is closely related to pore formation.

Porous CNFs are typically collected in the form of nonwoven porous mats and can be directly used as anodes in LIBs without adding any polymer binder or other additives. The elimination of nonactive additives ensures the high rate of electron conduction and has the potential to improve the electrochemical properties. In addition, anodes made of porous CNFs have more accessible sites, shorter transport distances for both electron and lithium ions as well as larger electrode–electrolyte contact areas resulting from their large surface areas. All of these structural features benefit the improvement of electrochemical performance of porous CNF anodes.

Galvanostatic charge–discharge experiments were carried out at a current density of 50 mA g^{-1} within a voltage window of 0.01–3.0 V to evaluate the electrochemical performance of the porous CNF anode. Figure 10.3 shows the voltage profiles of the porous CNF anode at the first cycle. When lithium ions insert into the anode in the first charge, there are two sloping voltage ranges (1.3 and 0.8 V versus Li^+/Li) that can be discerned. The first slope is attributed to the decomposition of the electrolyte and the formation of the solid electrolyte interphase (SEI) film, while the potential slope in 1.3 V corresponds to the insertion of lithium ions into the porous CNF anode.

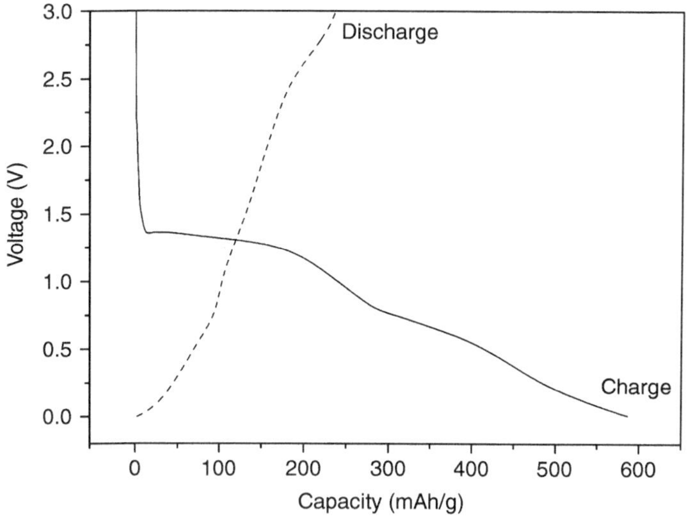

10.3 Initial discharge–charge profile of porous carbon nanofibers electrode at 50 mA g^{-1} within a voltage window of 0.01–3.0 V.

From Fig. 10.3, it is also apparent that the porous CNF anode has a large Li^+ storage capacity. For example, the first charge and discharge capacities are about 586 and 333 mAh g^{-1}, respectively. A large irreversible capacity of about 253 mAh g^{-1} can also be seen at the first cycle, and this leads to a relatively low Coulombic efficiency of about 57%. This irreversible capacity may be caused by the reductive decomposition of electrolyte solution and the subsequent formation of the SEI film.

Figure 10.4 shows the cycling performance of porous CNF anodes. The results further demonstrate that the porous CNF anode delivers good cycle stability and large reversible capacity. Owing to their large surface area for Faradaic reaction, short distance for charge diffusion, low resistance for electron conduction, as well as the added pore space for the volume change accompanying lithium-ion intercalation, porous CNFs offer high lithium storage capacity, fast charge–discharge kinetics and good cyclic stability.

10.5 Future trends

Current advances in electrospinning technology are providing important evidence for new potential roles of electrospun materials in lithium-ion battery applications. The electrospun porous nanofibers used as a separator in Li-ion batteries will improve ion conductivity, thereby enhancing battery efficiency. The use of electrospun porous nanofibers as electrodes in Li-ion batteries will undoubtedly improve the cycle life and increase rate

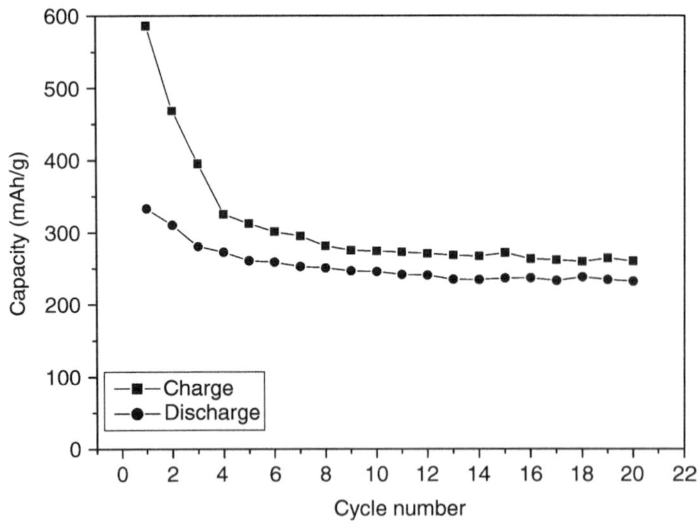

10.4 Cycle performance of porous carbon nanofibers electrode at 50 mA g^{-1} within a voltage window of 0.01–3.0 V.

capabilities and capacitance. Enhancement of the battery performances and cost reduction has been achieved, and now further improvements in performances of the battery are necessary for future development. There is likely to be remarkable improvement in Li-ion batteries in the future so that the Li metal itself will be further explored, and the expectation for nanofibers to play an important role as electrode materials should be strengthened in this area. Taking all these factors into account, nanofiber electrodes as well as the intercalation sciences will greatly be able to contribute to the further development of Li-ion batteries, as second- and third-generation batteries are developed. High performance Li-ion batteries with advanced nanofibers will be required more as the twenty-first century progresses, with its increased use of multimedia and wearable computers, coupled with greater environmental awareness.

10.6 References

1. Whittingham, M. S. (1976), Electrical energy storage and intercalation chemistry, *Science*, **192**, 1126–7.
2. Yazami, R. and Touzain, P. (1983), A reversible graphite-lithium negative electrode for electrochemical generators, *Journal of Power Sources*, **9**, 365–71.
3. Thackeray, M. M., David, W. I. F., Bruce, P. G. and Goodenough, J. B. (1983), Lithium insertion into manganese spinels, *Materials Research Bulletin*, **18**, 461–72.

4. Manthiram, A. and Goodenough, J. B. (1989), Lithium insertion into $Fe_2(SO_4)_3$ frameworks, *Journal of Power Sources*, **26**, 403–08.
5. Padhi, A. K., Nanjundaswamy, K. S. and Goodenough, J. B. (1997), Phospho-olivines as positive-electrode materials for rechargeable lithium batteries, *Journal of the Electrochemical Society*, **144**, 1188–94.
6. Silberberg, M. (2006), *Chemistry: The molecular nature of matter and change*, New York, McGraw-Hill Education.
7. Yang, C. R., Jia, Z. D., Guan, Z. C. and Wang, L. M. (2009), Polyvinylidene fluoride membrane by novel electrospinning system for separator of Li-ion batteries, *Journal of Power Sources*, **189**, 716–20.
8. Gao, K., Hu, X. G., Dai, C. S. and Yi, T. F. (2006), Crystal structures of electrospun PVDF membranes and its separator application for rechargeable lithium metal cells, *Materials Science and Engineering B*, **131**, 100–5.
9. Cho, T. H., Tanaka, M., Onishi, H., Kondo, Y., Nakamura, T., Yamazaki, H., Tanase, S. and Sakai, T. (2008), Battery performances and thermal stability of polyacrylonitrile nano-fiber-based nonwoven separators for Li-ion battery, *Journal of Power Sources*, **181**, 155–60.
10. Cho, T. H., Sakai, T., Tanase, S., Kimura, K., Kondo, Y., Tarao, T. and Tanaka, M. (2007), Electrochemical performances of polyacrylonitrile nanofiber-based nonwoven separator for lithium-ion battery, *Electrochemical and Solid-State Letters*, **10**, A159–62.
11. Gu, Y. X., Chen, D. R. and Jiao, M. L. (2005), Synthesis and electrochemical properties of nanostructured $LiCoO_2$ fibers as cathode materials for lithium-ion batteries, *Journal of Physical Chemistry B*, **109**, 17901–6.
12. Endo, M., Kim, C., Nishimura, K., Fujino, T. and Miyashita, K. (2000), Recent development of carbon materials for Li ion batteries, *Carbon*, **38**, 183–97.
13. Kim, C., Yang, K. S., Kojima, M., Yoshida, K., Kim, Y. J., Kim, Y. A. and Endo, M. (2006), Fabrication of electrospinning-derived carbon nanofiber webs for the anode material of lithium-ion secondary batteries, *Advanced Functional Materials*, **16**, 2393–7.
14. Yoon, S. H., Park, C. W., Yang, H., Korai, Y., Mochida, I., Baker, R. T. K. and Rodriguez, N. M. (2004), Novel carbon nanofibers of high graphitization as anodic materials for lithium ion secondary batteries, *Carbon*, **42**, 21–32.
15. Wang, L., Yu, Y., Chen, P. C., Zhang, D. W. and Chen, C. H. (2008), Electrospinning synthesis of C/Fe_3O_4 composite nanofibers and their application for high performance lithium-ion batteries, *Journal of Power Sources*, **183**, 717–23.
16. Zhou, H. S., Zhu, S. M., Hibino, M., Honma, I. and Ichihara, M. (2003), Lithium storage in ordered mesoporous carbon (CMK-3) with high reversible specific energy capacity and good cycling performance, *Advanced Materials*, **15**, 2107–11.
17. Winter, M., Besenhard, J. O., Spahar, M. E. and Novák, P. (1998), Insertion electrode materials for rechargeable lithium batteries, *Advanced Materials*, **10**, 725–63.
18. Subramanian, V., Zhu, H. W. and Wei, B. Q. (2006), High rate reversibility anode materials of lithium batteries from vapor-grown carbon nanofibers, *Journal of Physical Chemistry B*, **110**, 7178–83.
19. Deng, D. and Lee, J. Y. (2007), One-step synthesis of polycrystalline carbon nanofibers with periodic dome-shaped interiors and their reversible lithium-ion storage properties, *Chemistry of Materials*, **19**, 4198–204.

20. Zou, G. F., Zhang, D. W., Dong, C., Li, H., Xiong, K., Fei, L. F. and Qian, Y. T. (2006), Carbon nanofibers: Synthesis, characterization, and electrochemical properties, *Carbon*, **44**, 828–32.
21. Li, C. C., Yin, X. M., Chen, L. B., Li, Q. H. and Wang, T. H. (2009), Porous carbon nanofibers derived from conducting polymer: Synthesis and application in lithium-ion batteries with high-rate capability, *Journal of Physical Chemistry C*, **113**, 13438–42.
22. Ji, L. W. and Zhang, X. W. (2009), Generation of activated carbon nanofibers from electrospun polyacrylonitrile-zinc chloride composites for use as anodes in lithium-ion batteries, *Electrochemistry Communications*, **11**, 684–7.
23. Ji, L. W., Lin, Z., Medford, A. J. and Zhang, X. W. (2009), Porous carbon nanofibers from electrospun polyacrylonitrile/SiO_2 composites as an energy storage material, *Carbon*, **47**, 3346–54.
24. Ji, L. W. and Zhang, X. W. (2009), Fabrication of porous carbon nanofibers and their application as anode materials for rechargeable lithium-ion batteries, *Nanotechnology*, **20**, 155705–11.
25. Ji, L. W. and Zhang, X. W. (2009), Manganese oxide nanoparticle-loaded porous carbon nanofibers as anode materials for high-performance lithium-ion batteries, *Electrochemistry Communications*, **11**, 795–8.
26. Ji, L. W., Medford, A. J. and Zhang, X. W. (2009), Porous carbon nanofibers loaded with manganese oxide particles: Formation mechanism and electrochemical performance as energy-storage materials, *Journal of Materials Chemistry*, **19**, 5593–601.
27. Ji, L. W. and Zhang, X. W. (2009), Electrospun carbon nanofibers containing silicon particles as an energy-storage medium, *Carbon*, **47**, 3219–26.
28. Ji, L. W. and Zhang, X. W. (2009), Fabrication of porous carbon/Si composite nanofibers as high-capacity battery electrodes, *Electrochemistry Communications*, **11**, 1146–9.
29. Ji, L. W., Lin, Z., Zhou, R., Shi, Q., Toprakci, O., Medford, A. J., Millns, C. R. and Zhang, X. W. (2010), Formation and electrochemical performance of copper/carbon composite nanofibers, *Electrochimica Acta*, **55**, 1605–11.
30. Ji, L. W., Lin, Z., Medford, A. J. and Zhang, X. W. (2009), In-situ encapsulation of nickel particles in electrospun carbon nanofibers and the resultant electrochemical performance, *Chemistry-A European Journal*, **15**, 10718–22.
31. Yu, Y., Gu, L., Zhu, C. B., van Aken, P. A. and Maier, J. (2009), Tin nanoparticles encapsulated in porous multichannel carbon microtubes: Preparation by single-nozzle electrospinning and application as anode material for high-performance Li-based batteries, *Journal of the American Chemical Society*, **131**, 15984–5.
32. Yu, Y., Gu, L., Wang, C. L., Dhanabalan, A., van Aken, P. A. and Maier, J. (2009), Encapsulation of Sn@carbon nanoparticles in bamboo-like hollow carbon nanofibers as an anode material in lithium-based batteries, *Angewandte Chemie International Edition*, **48**, 6485–9.
33. Donnet, J. B. and Bansal, R. C. (1990), *Carbon fibers*, New York, Marcel Dekker.
34. Peebles, L. H. (1995), *Carbon fibers: Formation, structure, and properties*, Boca Raton, FL, CRC.

11
Functional nanofibers in sensor applications

X. F. WANG, B. DING and J. Y. YU,
Donghua University, China

Abstract: Over the last decade, the interest in electrospun nanomaterials and their applications has increased. The fascinating and unparalleled properties of electrospun nanomaterials, such as large surface-to-volume ratios and high open porosity, have opened new and unexpected fields of application, especially in ultrasensitive sensors. By exploiting the inherent physical, electrical and mechanical properties of nanomaterials, it is possible to improve the performance of conventional sensors by increasing their sensitivity, selectivity, portability and power efficiency. In this chapter, the recent progress in the development of electrospun nanomaterials is reviewed. In particular, applications in some predominant sensing approaches, such as acoustic wave, resistive, photoelectric, optical and biological, are discussed.

Key words: electrospinning, nanofibers, nanonets, nanostructures, sensing materials, sensors.

11.1 Introduction: modern electrospinning technology

Electrospinning is a highly versatile technique to process solutions or melts, mainly of polymers, into continuous fibers with diameters ranging from several micrometers down to a few nanometers by applying a high voltage on a polymer solution or melt ejected from a micro-syringe pump. In 1934, Formhals described the operation of electrospinning in a US patent for the first time; but only since the 1990s has it gained substantial attention. Since that time, the process has attracted rapidly growing interest triggered by potential applications of nanofibers in nanoscience and nanotechnology. Several research groups, especially the Reneker group, revived interest in this technology and have demonstrated a possibility to electrospin a wide variety of polymeric fibers (Doshi and Reneker, 1995; Reneker and Chun, 1996; Fong *et al.*, 1999). A number of review articles have been published recently which provide an insight into the most prominent aspects of electrospinning (Huang *et al.*, 2003; Li and Xia, 2004; Ramakrishna *et al.*, 2006;

Teo and Ramakrishna, 2006; Greiner and Wendorff, 2007; Agarwal *et al.*, 2009). To date, over 100 synthetic and natural polymers have been successfully electrospun into fibers with a broad range of applications owing to their low dimensions and large surface areas.

Generally, electrospun fibers are collected as nonwoven membranes with randomly arranged structures, which have greatly limited their applications in electronic devices or biomedical applications (Teo and Ramakrishna, 2006; Agbenyega, 2008; Zhang and Chang, 2008; Liu *et al.*, 2010). In order to fully realize the potential of electrospun fibers, it is important to fabricate fibrous assemblies with controllable microstructures. Recently, several groups have demonstrated that electrospun nanofibers could be collected as uniaxially aligned arrays by using specially designed collectors. For example, Li and coworkers (2003) have demonstrated that nanofibers can be uniaxially aligned by introducing insulating gaps into conductive collectors. Matthews *et al.* (2002) have demonstrated that aligned fibers could be obtained by using a rotating collector at high speed; the mandrel rotation speed and fiber orientation strongly influence the properties of the electrospun nanofibers. Zussman *et al.* (2003) have demonstrated the use of a wheel-like bobbin as the collector to position and align individual polymer nanofibers into parallel arrays. However, because the edge of such a bobbin has to be relatively sharp, this technique does not seem to be feasible for forming well-aligned nanofibers over large areas. In addition, it was possible to obtain various patterned architectures of the electrospun nanofibers by varying the design of electrode pattern (Li *et al.*, 2005). One of the most interesting features associated with this approach is that this technique enables direct integration of nanofibers with controllable configurations into an electrode system such that the nanofibers can be fabricated and aligned simultaneously, which will significantly simplify the production of nanofiber-based devices.

11.1.1 Nanofibers

Different nanofiber morphologies can be obtained via control of the processing conditions, enabling one to produce smooth as well as beaded (Lin *et al.*, 2010) and helical (Kessick and Tepper, 2004) structures (see Fig. 11.1a–c). Recent demonstrations from a number of groups indicate that this technique is also capable of generating nanofibers with ribbon (Koombhongse *et al.*, 2001), necklace-like (Jin *et al.*, 2010), porous (Kanehata *et al.*, 2007; Ding *et al.*, 2008a), core-shell (Sun *et al.*, 2003), hollow (Dror *et al.*, 2007), multichannel tubular (Zhao *et al.*, 2007), nanowire-in-microtube (Chen *et al.*, 2010), multicore cable-like (Kokubo *et al.*, 2007) and tube-in-tube (Mou *et al.*, 2010) structures (see Fig. 11.1d–m).

Compared with inorganic nanorods and carbon nanotubes, the electrospun nanofibers are continuous, which possess them with high axial strength combined with extreme flexibility. Yet, these assemblies would possess excellent

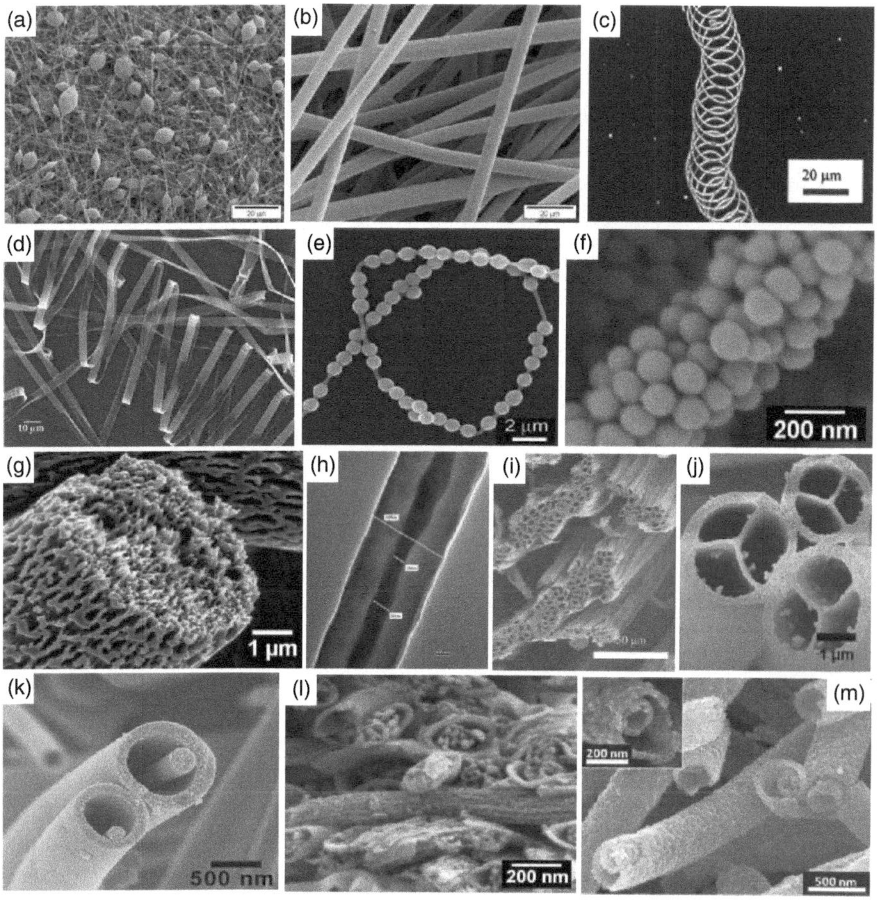

11.1 Different morphologies of electrospun fibers: (a) beaded, (b) smooth (Reprinted with permission from Lin *et al.* (2010). © 2010 American Chemical Society), (c) helical (Reprinted with permission from Kessick and Tepper (2004). © 2004 American Institute of Physics), (d) ribbon (Reprinted with permission from Koombhongse *et al.* (2001). © 2001 Wiley), (e) necklace-like (Reprinted with permission from (Jin *et al.*, (2010). © 2010 American Chemical Society), (f, g) porous (From Ding *et al.* (2008a)), (h) core-shell (Reprinted with permission from Sun *et al.* (2003). © 2003 Wiley), (i) hollow (Reprinted with permission from Dror *et al.* (2007). © 2007 Wiley), (j) multichannel tubular (Reprinted with permission from Zhao *et al.* (2007). © 2007 American Chemical Society), (k) nanowire-in-microtube, (l) multi-core cable-like (Reprinted with permission from Kokubo *et al.* (2007). © 2007 IOP Publishing Ltd) and (m) tube-in-tube (Reprinted with permission from Mou *et al.* (2010). © 2010 American Chemical Society) structured fibers.

structural mechanical properties. Additionally, a multitude of functions can be incorporated into the fibers and an extremely broad range of potential applications exists in which electrospun nanofibers can make major contributions. These include not only textile, ultrafiltration, tissue engineering, catalysis and mechanical reinforcement applications, but also extend to the fabrication of sensors, batteries and other types of devices. It would be meaningful to combine the electrospinning technique with controllable nanofibers assembly for sensors fabrication, as this would provide a simple, rapid and cheap way to construct ultrasensitive devices. Although the use of nanofibers offers the prospect of high sensitivity and rapid detection, the ability to incorporate nanofibers into device architectures is limited by the difficulty in manipulating and locating the nanostructures with respect to the microelectrodes.

11.1.2 Nanonets

Electrospun fibrous membranes typically have fiber diameters in the range 100–500 nm, but properties such as surface area and porosity become more significant when the fiber diameter falls below 20 nm (Huang *et al.*, 2006). The major challenge for developers is to come up with robust methods for manufacturing extremely small nanofibers in large quantities and with a uniform size. We have found a procedure for generating net-like structured nanowires with small diameters (~20 nm) in three-dimensional (3-D) fibrous mats by optimization of various processing parameters during electrospinning (Ding *et al.*, 2006b). This novel process, termed 'electronetting', allows one-step fabrication of ultrathin 'nanonets' with large quantities and uniform size. The electrospun fibers act as a support for the soap-bubble-like structured nanonets comprising interlinked one-dimensional (1-D) nanowires (see Fig. 11.2). The major distribution region (over 70%) of nanowire diameters was in the range of 10–30 nm, which was one order of magnitude less than that of common electrospun nanofibers. The region of pore-width distribution of nanonets ranged from 20 to 550 nm, which was much less than that of pores among electrospun nanofibers. Therefore, nanonets possess great potential for application in filtration systems for the removal of particles or viruses with a size to nanometer ranges. To date, our group has successfully prepared various nanonets based on different polymer systems such as nylon-6 (Ding *et al.*, 2006b), polyacrylic acid (PAA) (Ding *et al.*, 2006b; Wang *et al.*, 2010b), poly(vinyl alcohol) (PVA) (Ding *et al.*, 2008b, 2010) and polyurethane (PU) (see Fig. 11.2). The formation of the nanonets is considered to be due to the electrically forced fast phase separation of the charged droplets which move at high speed between the capillary tip and the collector. The formation, morphology and area density of the nanonets in electrospun fibrous membranes are strongly affected by the solution properties and several parameters in the process of electrospinning.

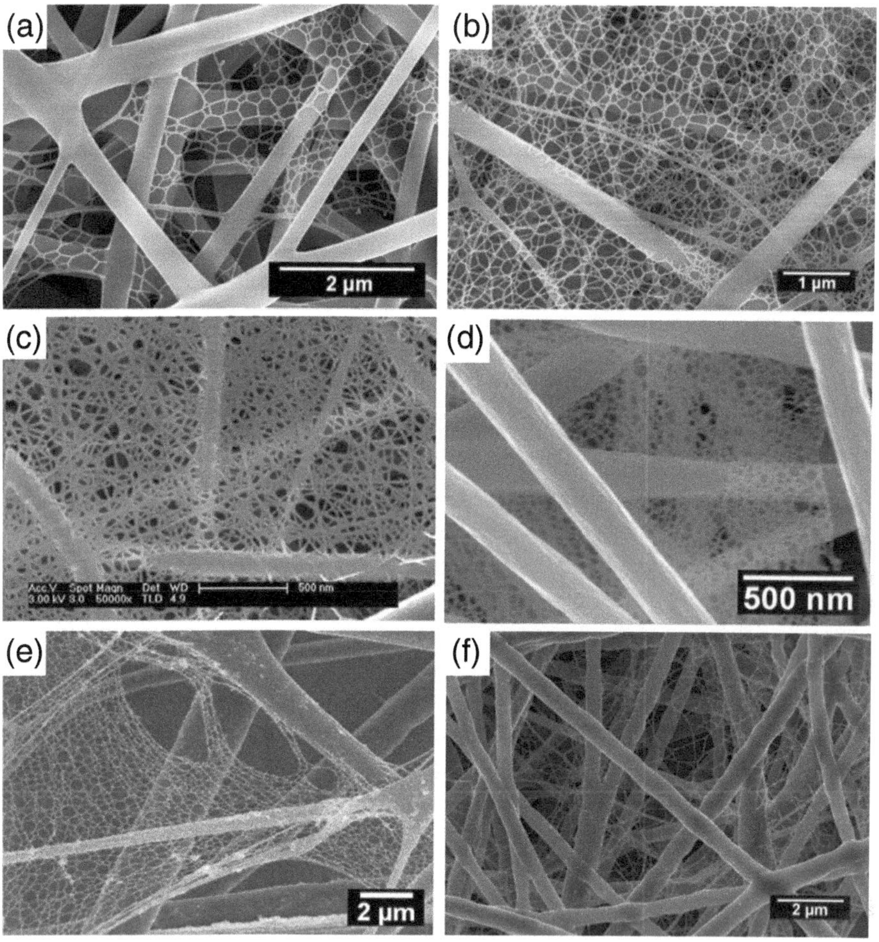

11.2 SEM images of various nanonets based on different polymer systems: (a) PAA (Reprinted with permission from Wang *et al.* (2010b). © 2010 IOP Publishing Ltd), (b) PAA/NaCl (Reprinted with permission from Wang *et al.* (2011). © 2011 Royal Society of Chemistry), (c) nylon-6 (Reprinted with permission from Ding *et al.* (2006b). © 2006 IOP Publishing Ltd), (d) PVA/ZnO (Reprinted with permission from Ding *et al.* (2008b). © 2008 Elsevier Ltd, (e) PVA/SiO_2 (Reprinted with permission from Ding *et al.* (2010). © 2010 Elsevier Ltd), (f) PU/NaCl.

Nanonets exhibit several fascinating characteristics, such as extremely small fiber diameter and pore width, notable specific surface area and high porosity, superior mechanical performances, which make them attractive candidates for use in a range of applications including ultrasensitive sensors and ultrafiltration. Furthermore, nanonets have the potential to be used as ultrafine filters to intercept viruses and bacteria, such as influenza A (H1N1) virus, severe acute respiratory syndrome (SARS) virus, *Escherichia coli*, etc. (Ding *et al.*, 2010). Clearly, there is growing interest in the process, but the results reported to date are primarily focused on the empirical production and the proposed uses of polymer nanofibers. At the same time, thorough understanding of the mechanisms of the breakup of liquid jet and subsequent nanonets formation is needed for the development of robust methods of process control.

11.1.3 Electrospinning and sensors

Development of electrospun nanomaterials, such as nanofibers and nanowebs, provided researchers with an opportunity to construct electronic interfaces with components whose sizes are comparable to the size of molecules, potentially leading to a much more efficient interface. Nanometer cross-sections of nanomaterials gives them enhanced surface sensitivity and allows them to utilize the benefits of size effects, such as quantization and single-molecule sensitivity. The comparatively large surface area and high porosity make electrospun nanomaterials highly attractive candidates for use in a range of devices, including ultrasensitive sensors. This is one of the most desirable properties for improving the sensitivity of sensors because a larger surface area will absorb more analytes and change the sensor's signal more significantly. A survey of open publications related to 'electrospinning and sensors' over the past ten years (at the time of writing) is given in Fig. 11.3.

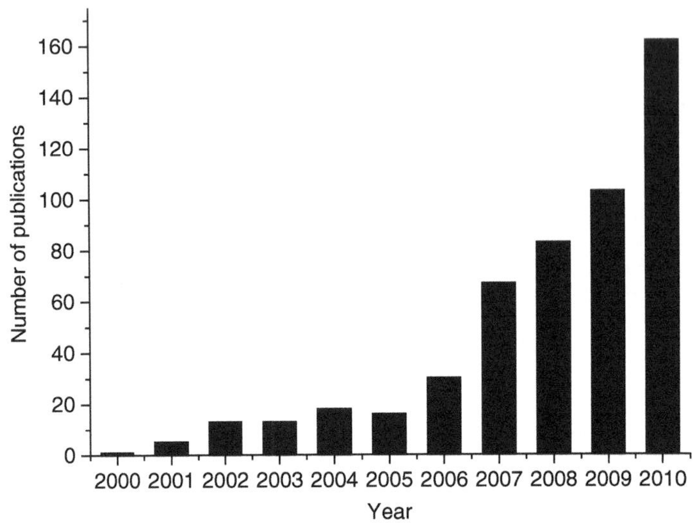

11.3 Comparison of the annual number of scientific publications with the keyword 'electrospinning and sensor', as provided by the search engine of SciFinder Scholar.

The rapid rise in data clearly demonstrates that this subject has attracted increasing attention recently.

11.2 Formation of nanostructured sensing materials

The performance of a sensor is highly dependent on the configuration, morphology, thickness and the composite ratio of the sensing materials. There have been a vast number of methods developed to fabricate nanostructured sensing materials. Either single-component electrospinning or composite/doped electrospinning can be used to produce nanofiber functionalized sensing materials. Within each technique, the structure of the material may be random or ordered. However, up to now, preparation of electrospun sensing materials has always involved complicated polymer syntheses which make the fabrication time consuming; sometimes the leakage of sensing molecules affects the sensing performance. More available systems and new fabrication approaches need exploring. Recently, some research has focused on fibers with secondary porous structures. Moreover, it is a great challenge to immobilize the sensing elements onto the surface of the electrospun porous fibers. This section describes the design and fabrication processes that have been developed to realize electrospinning nanostructured sensing materials.

11.2.1 Single-component electrospinning

Single-component electrospinning can be used to produce nanofibers functionalized sensing materials. An interesting feature of this technique is that more sensing elements can be established on the surface of nanofibers, as well as more absorption sites exposed to analytes. To date, more than 100 types of natural and synthetic polymer have been electrospun into one-component nanofibers, such as poly(ethylene oxide), PVA, polystyrene (PS), polyacrylonitrile (PAN), polylactide, poly(ε-caprolactone), PU, polyamide, cellulose acetate (CA), poly(vinyl acetate) (PVAc) and many more (Huang *et al.*, 2003; Lu *et al.*, 2009); however, only few of them were used as sensing elements (Ding *et al.*, 2005b). Besides polymer-based sensing materials, nanostructured ceramics also have potential applications in nanoscale electronics, optical and sensing devices due to their notable electronic properties. The fabrication of sensing materials can be implemented over single-component electrospinning, but the remaining hindrance is that some polymers or ceramics cannot obtain nanostructured morphology via electrospinning directly (Wang *et al.*, 2010a). A promising approach to solve this problem is to use composite or doped electrospinning and even assemble these materials on the surface of nanomaterials.

11.2.2 Composite or doped electrospinning

Early work on electrospinning mainly dealt with conventional polymers that could be synthesized with sufficiently high molecular weights and could be dissolved in appropriate solvents. In an effort to greatly expand the functionality and thus the scope of applications associated with fibrous structures, composite and doped electrospinning have recently been developed to generate nanofibers with a range of chemical compositions, and therefore various electronic, magnetic, optical and biological properties.

In the past several decades, there have been many reports on the fabrication of composite nanofibers via electrospinning, including polymer/polymer, polymer/inorganic and inorganic/inorganic composites (Wang and Pan, 2008; Lu *et al.*, 2009). Recently, a large number of polymer/metal-oxide composite fibers have been produced by electrospinning in combination with sol-gel processes. In many cases, the composite fibers could be converted into metal-oxide fibers by subsequent calcination. Such composite fibers allow the fabrication of polymer fibers with special functionalities or of precursor fibers. For instance, Wang and coworkers (2006) obtained pure WO_3 nanofibers with controllable diameters of around 100 nm by electrospinning PVAc/tungsten isopropoxide solutions and subsequent calcinations (see Fig. 11.4a, b). The prepared WO_3 ceramic nanofibers have a quick response to ammonia in various concentrations, suggesting potential applications of the electrospun WO_3 nanofibers as a sensor material for gas detection (Fig. 11.4c).

Pure and doped metal-oxide semiconductors (MOS) nanostructured materials, such as ZnO, SnO_2 and other wide band gap metal oxides, have recently attracted considerable interest due to their unique physical properties and a wide range of possible applications (Andersson *et al.*, 1998; Li *et al.*, 2000; Schmidt-Mende and MacManus-Driscoll, 2007). However, most pure MOS are not stable in air and its electrical properties are significantly affected by adsorption of O_2, CO_2, hydrocarbons, S-containing compounds and water (Jimenez-Gonzalez, 1997). Therefore, it is highly desired to prepare 1-D MOS nanostructures doped with a selection of elements, such as Li, Ga, In, N, Al, Sn and P, to enhance and control their mechanical, electrical and optical performance. Electrospinning proved to be a simple, low-cost and reliable technique for dopant incorporation to attain modified structural, electrical and optical properties for these semiconducting nanofibers. Lotus and coworkers (2010) have investigated the effect of dopant addition on morphological, electrical and optical properties of semiconducting oxide nanofibers obtained by electrospinning. Pure SnO_2 and SnO_2 polycrystalline electrospun nanofibers doped with multiwall carbon nanotubes (MWCNTS) are synthesized by Yang *et al.* (2007). The SnO_2/MWCNTS composite nanofibers exhibited enhanced sensitivity compared

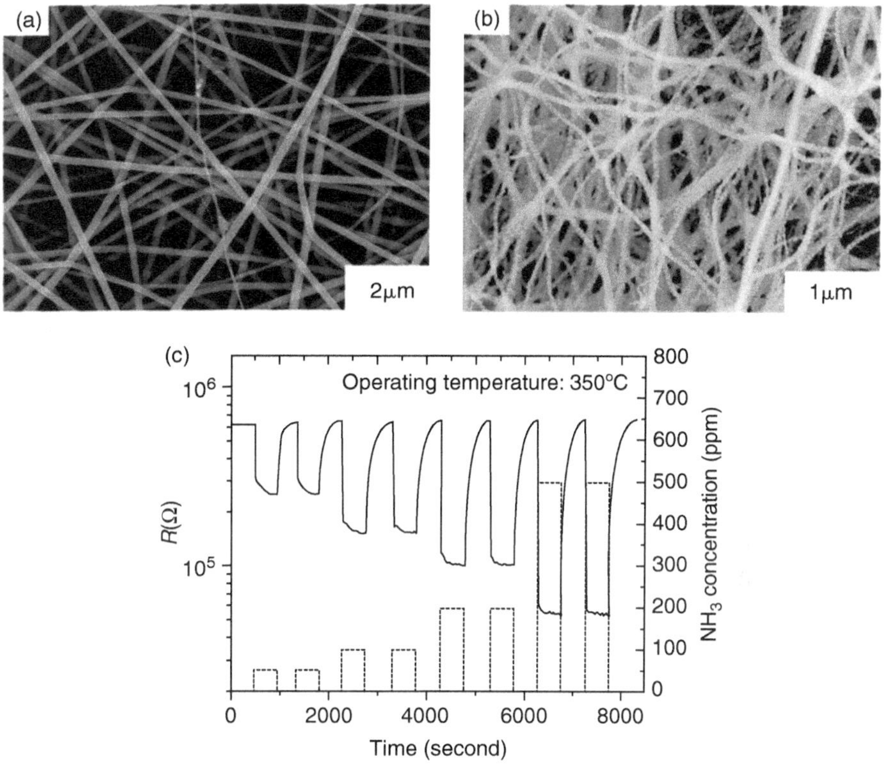

11.4 (a) SEM image of tungsten oxide/PVAc composite fibers, (b) SEM image of the same sample after it had been calcined in air at 500°C for 3 h and (c) response of WO3 nanofibers to ammonia (50–500 ppm). (Reprinted with permission from Wang *et al.* (2006). © 2006 American Chemical Society.)

to pure SnO_2 nanofibers. Doping with materials in the semiconductors may affect the electron transfer, which is considered in their investigation.

11.2.3 Surface assembly of sensing materials

Another area that has been explored for the parallel immobilization of sensing materials over large areas of electrospun nanomaterials is the use of surface assembly. With bifunctional molecules an efficient surface modification can be achieved. A further development of this idea resulted in self-assembled monolayers (SAMs). Between neighboring adsorbed molecules, both hydrogen bonding and weak van der Waals forces are active (Tsukruk *et al.*, 1997). The result is a densely packed layer of sensing molecules standing steeply at the surface. Interfaces functionalized by SAMs can be useful for electron transfer in the direct oxidation of organic

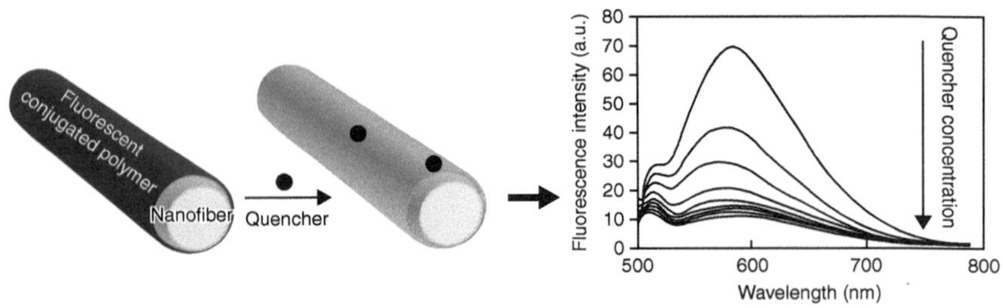

11.5 Assembling fluorescent probes on to electrospun CA nanofibers and fluorescence emission spectra of the sensing film with varying quencher concentration. (Reprinted with permission from Wang *et al.*, (2004). © 2004 American Chemical Society.)

substances, and they may also act as ion-selective surfaces. Assembly of sensing materials can also be driven by electrostatic interactions, resulting in layer-by-layer assembly (LBL) (Decher *et al.*, 1992; Huang *et al.*, 2001; Mamedov *et al.*, 2002; Hammond, 2004). The LBL technique involves the alternate adsorption of anionic and cationic polyelectrolytes on a charged substrate by sequential dipping of the substrate into the appropriate aqueous polyelectrolyte solutions. This yields a highly tailored polymer thin film on the substrate. This technique effectively coated the individual electrospun fibers, thus leaving the inherent high surface area of the electrospun membrane intact (Ding *et al.*, 2004a, 2005a; Ogawa *et al.*, 2007). The LBL assembly technique has been successfully applied to sensor fabrication (Nohria *et al.*, 2006). One of the amazing advantages of this technique is the versatility, in that a vast range of functional groups can be incorporated within the structure of the film. Furthermore, this technique is not limited to planar substrates and has been demonstrated in LBL assembly onto 2- and 3-D architectures (Kato *et al.*, 2002). The combination of electrospinning and LBL assembly with different electrospun substrates and sensing layers or functionalities will afford new properties and tremendous flexibility for fabricating sensors. Wang *et al.* (2004) reported a sensitive optical sensor by assembling fluorescent probes onto electrospun CA nanofibers, which showed fluorescence quenching upon exposure to even an extremely low concentration (ppb) of methyl viologen cytochrome in aqueous solutions (see Fig. 11.5). There are also examples of self-assembling noble metal nanoparticles onto electrospun fibers (Formo *et al.*, 2008; Huang *et al.*, 2008). Li *et al.* (2006) have prepared silver nanoparticles deposited onto the surface of PAN nanofibers by photoreducing the desired metal salt directly. And, the solution route is also a generally simple and attractive approach for the fabrication of noble metal nanoparticles with a great level of control for both the size and dispersity onto the electrospun fibers.

11.3 Sensing mechanisms

11.3.1 Acoustic wave sensors

Quartz crystal microbalances (QCMs), one of a broad class of acoustic wave techniques, whose history dates back to 1880, when Pierre and Jacques Curie discovered the piezoelectric effect, where pressure exerted on a small piece of quartz causes an electrical potential difference between the deformed surfaces (Marx, 2003). The essence of QCM as an analytical tool lies in its ability to detect small mass changes of even smaller than nanogram order deposited on crystal surface and to act as a sensor. Interfacial mass changes can be related to changes in the QCM oscillation frequency by applying Sauerbrey's equation (Sauerbrey, 1959):

$$\Delta f = \frac{-2_o^2 \Delta m}{a(\mu\rho)^{1/2}} \qquad [11.1]$$

where Δf is the measured frequency shift, f_o is the fundamental frequency of a bare QCM chip, Δm is the mass change per unit area, a is the electrode area, ρ is the density of quartz and μ is the shear modulus of quartz crystal.

In the past decade, QCMs have been widely explored as sensors for the detection of biomarkers, drug targets, virus capsids, bacteria and living cells. Up to now, a variety of materials such as metals, ceramics, polymers, self-assembled monolayers, dendrimers, lipids and waxes (Ricco *et al.*, 1998, Grate, 2000) have been employed as sensitive coatings on QCM to improve the sensor sensitivity and selectivity for chemical analytes. It is widely accepted that the sensitivity of QCM sensors toward a specific analyte is enhanced by increasing the specific surface area of the sensing materials (Kolmakov and Moskovits, 2004). Recent efforts have been focused on the development of nanostructured coatings on QCM to improve the sensor sensitivity (Ding *et al.*, 2004b; Chao *et al.*, 2007). We have demonstrated the possibilities of fabricating novel QCM sensors by electrospinning deposition of nanofibrous membranes as sensitive coatings on the electrodes of QCM. Taking advantage of the large specific surface area, high porosity and good interconnectivity of these electrospun fibrous assemblies, we have successfully detected NH_3 (Ding *et al.*, 2004b, 2005b), H_2S (Ding *et al.*, 2006a), formaldehyde (Wang *et al.*, 2010a), moisture (Wang *et al.*, 2010b) and trimethylamine (TMA) (Wang *et al.*, 2011) with new detection limits at room temperature. Sensing experiments were carried out at room temperature in a flow-type gas testing system (Ding *et al.*, 2005b) and a static-type gas testing system (Fig. 11.6a) by measuring the resonance frequency shifts of QCM due to the additional mass loading. The experiments indicated that the sensitivity of the fibrous membrane-coated QCM sensors was several times higher than that of the casting film-coated QCM sensors. Moreover, the fiber-coated QCM sensors with small average fiber diameter had a higher sensitivity. One exciting and challenging aspect of the work of our group lies in the fabrication of a novel TMA sensor by using an electronetting technique to deposit a PAA nanonets on a QCM (Wang *et al.*, 2011). The versatile nanonets created enhanced interconnectivity and additional surface area, and facilitated the diffusion of analytes into the membranes, which significantly boosted the gas diffusion coefficient and sensing properties (Fig. 11.6b). These proof-of-principle experiments demonstrate the possibility of using electrospinning techniques to regulate the structure of membranes and hence obtain high surface activity and gas sensitivity; further experiments using more complex fiber arrangements and structures should expand the capabilities of this platform.

Surface acoustic wave (SAW) devices for gas detection have also attracted growing attention (Janata *et al.*, 1994, Crooks and Ricco, 1998). The change in electrical conductivity or mass of the sensing layer perturbs the velocity of the propagating SAWs due to mechanical and piezoelectric effects. Concentrated efforts have been focused on depositing nanostructured coatings on SAWs via electrospray or electrospinning technique to improve the sensor sensitivity (Sarkar *et al.*, 2006; He *et al.*, 2010; Li *et al.*, 2010). For example, He *et al.* (2010) demonstrated that a novel H_2 sensor could be prepared by electrospinning deposition of poly(vinylpyrrolidone) (PVP)/$LiTaO_3$ composite nanofiber on the electrodes of SAWs. Upon exposure to H_2, the PVP fibers adsorb H_2 molecules due to the hydrogen bond and the mass increase. Therefore, the frequency of the SAW device decreases as the adsorbed molecules reduce the acoustic wave velocity (Fig. 11.7). This indicates that the electrospinning technique can afford a simple approach for the fabrication of composite nanomaterials for sensing technology.

11.3.2 Resistive sensors

Commonly, resistance changes caused by the interaction between electrodes and analytical samples can be obtained by measuring the electrical conductivity (or resistivity) and evaluating to extract analytical results. Significant progress has been achieved in developing highly sensitive MOS resistive sensors using novel 1-D nanostructured architectures such as TiO_2, SnO_2, ZnO, In_2O_3, WO_3 and other wide band gap metal oxides. The fundamental principle of these devices is associated primarily with the adsorption of the chemisorbed species on the surface of the MOS inducing electric charge transport between the two materials, which changes the resistance of the oxide. 1-D sensing architectures provide unparalleled advantages in terms of facilitating fast mass transfer of the analyte

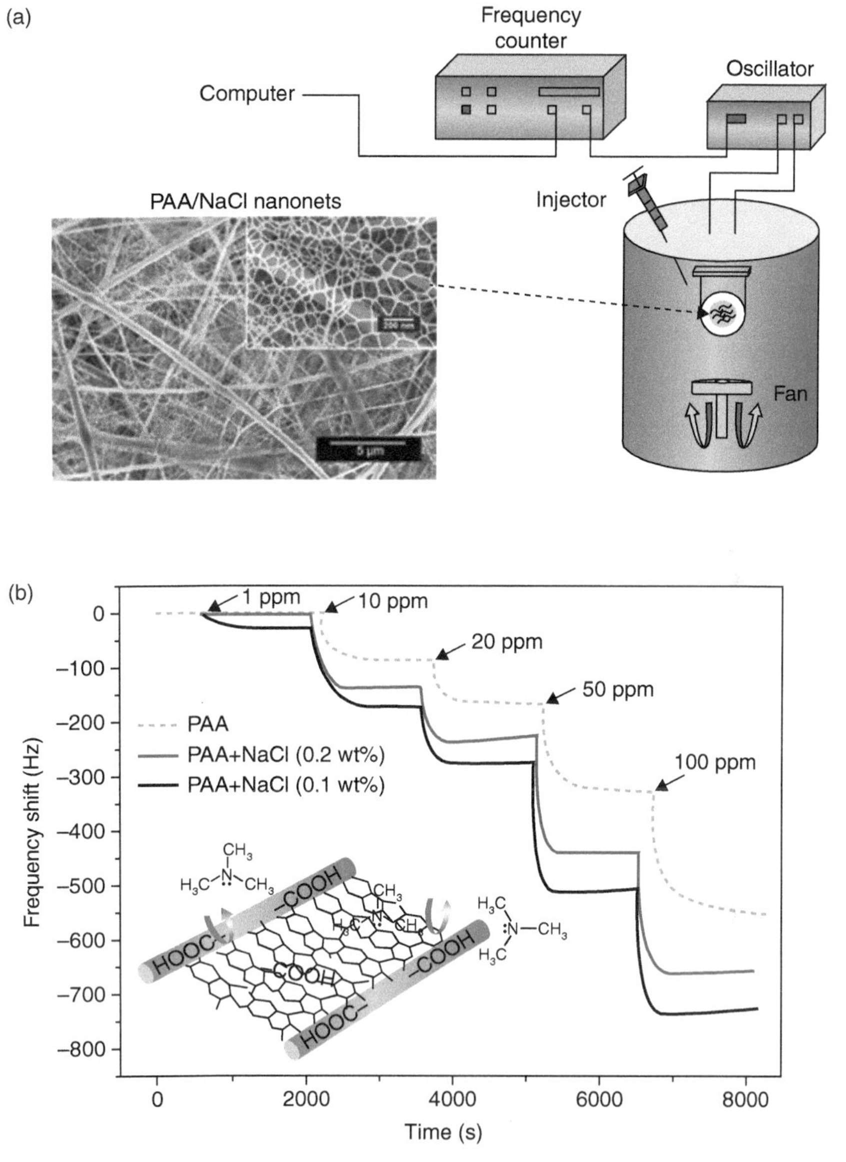

11.6 (a) Schematic of a gas testing system for TMA detection. (b) Response of sensors coated with PAA-NaCl nanonets containing different concentrations of NaCl (0 (dotted line), 0.1 (black line), and 0.2 wt% (gray line) and exposed to increasing TMA concentrations ranging from 1 to 100 ppm. The inset shows the schematic of gas-sensing mechanism between PAA nanonets and TMA. (Reprinted with permission from Wang *et al.* (2011). © 2011 Royal Society of Chemistry.)

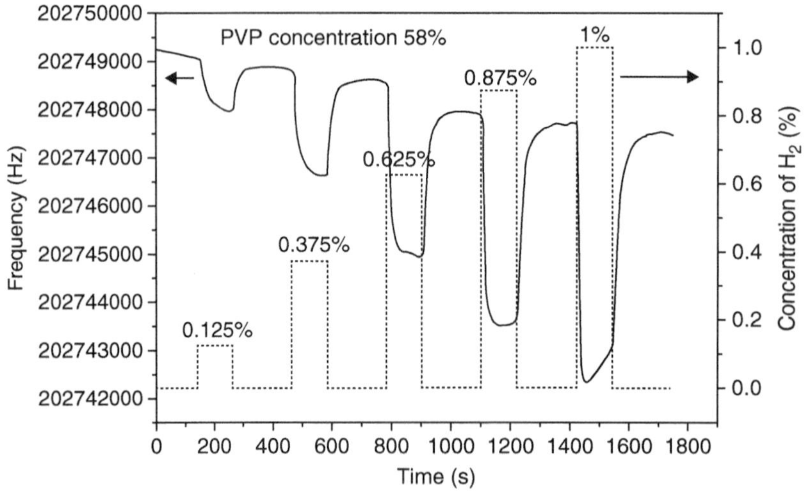

11.7 Dynamic response of PVP/$LiTaO_3$ SAW device toward H_2 at room temperature. (Reprinted with permission from He *et al.* (2010). © 2010 Elsevier Ltd.)

molecules to and from the interaction region, as well as requiring charge carriers to traverse any barriers introduced by molecular recognition events along the entire wire. Among the variety of methods for the preparation of such nanostructured architectures, electrospinning is one of the most simple, versatile and cost-effective approaches, offering the ability to produce long, continuous organic or inorganic nanofibers with potential for alignment and spooling.

So far, many attempts are carried out to prepare ultrasensitive resistive sensors to detect NH_3, H_2S, CO, NO_2, O_2, CO_2, moisture and volatile organic compound (CH_3OH, C_2H_5OH, $C_5H_{10}Cl_2$, $C_6H_5CH_3$, C_4H_8O, $CHCl_3$, $C_2H_2Cl_2$, C_3H_6O, C_3H_7NO, C_2HCl_3, N_2H_4, $(C_2H_5)_3N$, C_6H_{14}, etc.) vapors with new detection limits using electrospun nanofiber functionalized MOS such as TiO_2, ZnO, WO_3, MoO_3, SnO_2, In_2O_3 and ITO (Choi *et al.*, 2009; Ding *et al.*, 2009; Qi *et al.*, 2009; Wang *et al.*, 2009; Zhang *et al.*, 2009). Wang *et al.* (2010c) studied the H_2 sensing properties of p-NiO/n-SnO_2 composite nanofibers (NSNFs) synthesized through an electrospinning method (Fig. 11.8). The sensor response was measured between 260°C and 380°C by comparing the resistance of the sensor in dry synthetic air with that in target gases. Extremely fast response-recovery behavior (~3s) has been obtained at the operable temperature of 320°C, based on the prepared sensor, with the detection limit of approximate 5 ppm H_2. Many applications of electrospun nanofibers could be greatly improved by increasing the surface area and porosity of the fibers.

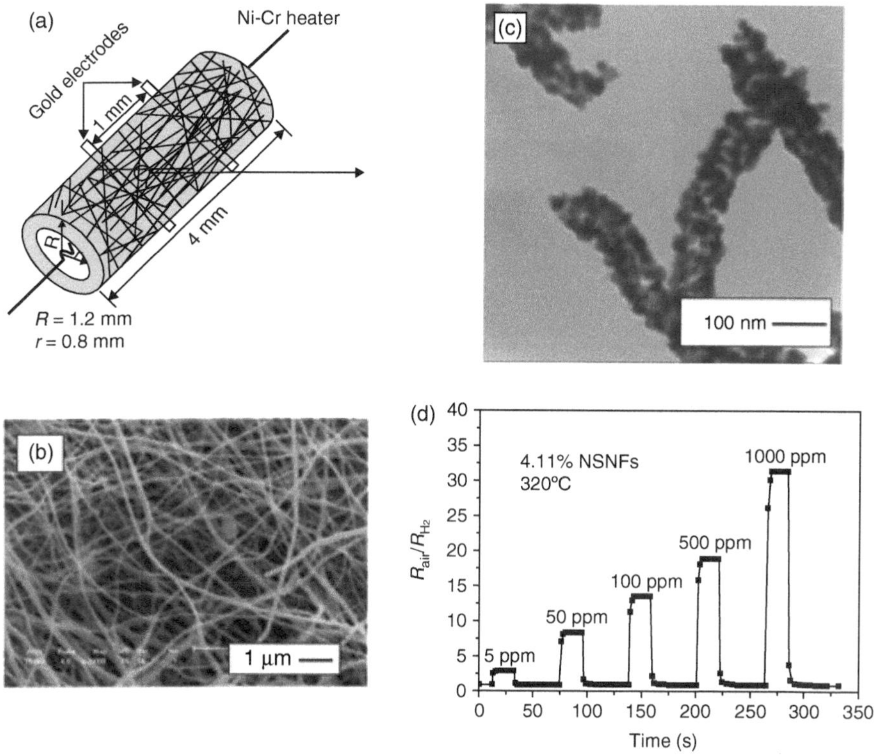

11.8 (a) Schematic illustration of the fabricated device structure for the H_2 sensing measurement and the corresponding TEM image of the 4.11 mol % NSNFs. (b) SEM images of 4.11 mol % NiO–SnO_2 composite nanofibers. (c) Response and recovery characteristic curves of the sensor based on 4.11 mol % NSNFs to H_2 in the range of 5–1000 ppm at 320°C. (Reprinted with permission from Wang *et al.* (2010c). © 2010 American Chemical Society.)

11.3.3 Photoelectric sensors

Recently, metallic nanoparticles encapsulated into dielectric matrices have been considered to have practical applications in photoelectric devices owing to their enhanced third-order nonlinear susceptibility (Haglund *et al.*, 1993). The enhancement of photoresponse behaviors arising from the excitation of the surface plasmon is attributed to the generation of hot electrons. The generated hot electrons drift or diffuse to the oxide barrier and tunnel to the counterelectrode. Generation of hot electrons and electron tunneling through the oxide nanowires may contribute to the enhanced photoresponse behavior in the hybrid nanowire (Hu *et al.*, 2006). Recently, Shi and coworkers (2009) have developed a method to fabricate Au nanoparticles (AuNPs) embedded in silica nanofibers via

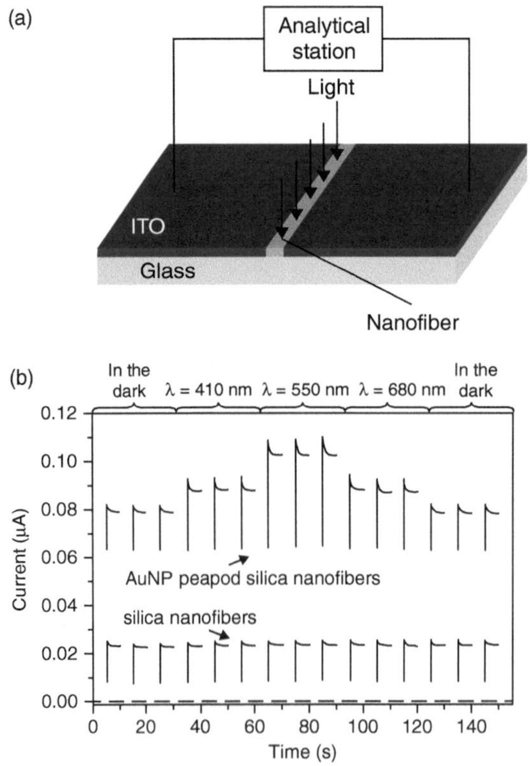

11.9 (a) A schematic of the microreactor for photoelectric response measurements. (b) Photoelectric response measurements of the room-temperature current response as a function of time of light illumination for pure silica nanofibers and AuNPs peapod silica nanofibers. (Reprinted with permission from Shi *et al.* (2009). © 2009 Elsevier Ltd.)

electrospinning and thermal decomposition of hybrid nanofibers, which exhibited an obvious photoelectric response under illumination (Fig. 11.9a). Photoelectric response measurements were conducted under dark condition or under alternate light illuminations of different wavelengths (k = 410, 550 or 680 nm). Data were obtained by an electrochemical analyzer with applied constant voltage of 10 V every 5 min. Upon light illumination, the hybrid peapod nanofibers presented a wavelength-dependent photoelectric current response. Particularly, the current photoresponse reached a maximum value under illumination at a wavelength of 550 nm, which was close to the surface-plasmon-resonance absorption band of the AuNPs nanofibers (Fig. 11.9b). These results show the potential of using gold nanopeapodded silica nanowires as wavelength-controlled optical nanoswitches. The microreactor approach can be applied to the preparation of a range of hybrid metal-dielectric 1-D nanostructures that can be used as functional building blocks for nanoscale waveguiding devices, sensors and optoelectronics.

11.3.4 Optical sensors

Optical sensor-based electrospun nanomaterials can be grouped into three general types: (a) quenching-based fluorescent optical sensors; (b) colorimetric sensors; and (c) Fourier transform infrared (FTIR) spectroscopy optical sensors.

For quenching-based fluorescent optical sensors, utilizing the fluorescence quenching of the sensing material against the targeted chemical molecules, the sensitivity of the device can be dramatically affected by the accessibility of the sensing elements to the quencher or analyte. Electron-deficient metal cations such as Fe^{3+} or Hg^{2+} and nitro aromatic compounds such as 2,4-dinitrotoluene (DNT) and 2,4,6-trinitrotoluene can serve as quenchers for a fluorophore (Wang *et al.*, 2002b, 2002c). The pyrene methanol (PM) (Wang *et al.*, 2002a) and conjugated polymers (Long *et al.*, 2009) were chosen as fluorescent indicators due to their large Stokes shift, high quantum yield, strong absorbance, excellent photostability and lifetime. Here the quenching of fluorescence is due to the interactions of an electron-rich indicator and electron-deficient quencher, and the degree of quenching depends on the amount of analyte. Electrospinning was used as a novel and simple method to fabricate fluorescent optical sensors. Wang and coworkers (2002b) have successfully developed nanofibrous PAA-PM membrane optical sensors for metal ion (Fe^{3+} or Hg^{2+}) and DNT detection using the electrospinning technique. Recently, electrospun nanofibrous PS film doped with a fluorescent conjugated polymer was developed as a sensory device for detection of the explosive DNT (Fig. 11.10a, b) (Long *et al.*, 2009). Polymer acting as a fluorescence probe presented relatively large affinity for the nitroaromatic compound DNT, which is mainly based on photoinduced charge transfer and ϖ–ϖ stacking interaction and could provide a strong driving force for fast fluorescence quenching. Second, the electrospun nanofibrous sensing film possessed remarkable specific surface area, in which more recognition elements were located on the surface.

Lv *et al.* (2010) reported a method for the colorimetric visualization of trace amount of HCl gas using electrospun porphyrinated polyimide (PPI) nanofibrous membrane as schematically illustrated in Fig. 11.10c. The dual chromo- and fluorogenic responses of the nanofibrous membrane upon exposure to HCl gas are interpreted in terms of the out-of-plane distortion of porphyrin macrocycle, which ultimately affects its optical properties. Exposing the PPI nanofibrous membrane to HCl gas for only 10 s generates a fast color change (from reddish russet to green) on the sensor surface, which can be ascribed to the protonation of the neutral porphyrin moieties

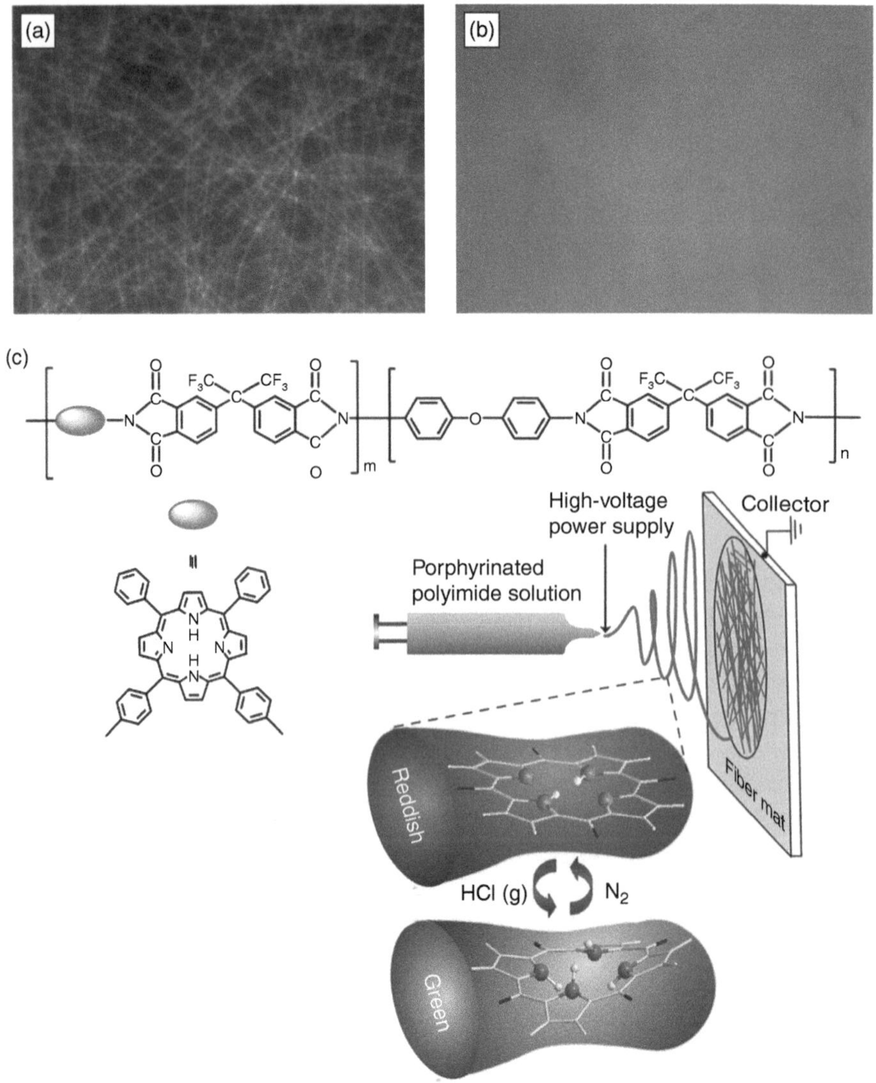

11.10 Fluorescence images of polymer P-Doped PS nanofibrous sensing membranes (a) before and (b) after being exposed to DNT vapor for 30 min. (Reprinted with permission from Long *et al.* (2009). © 2009 American Chemical Society.) (c) Molecular structure of PPI used in this work and the schematic illustration for the whole colorimetric and fluorescent detection of HCl gas.

in PPI . It is notable that the intensity of green color increases with the concentration of HCl gas (5–100 ppm). Additionally, the reusability of this PPI nanofibrous membrane sensor is proved by color recovery after puffing with N_2 for a certain time.

FTIR spectroscopy is another widely used optical method to study the interaction of electromagnetic radiation in the infrared region with chemical compounds. The mechanism of FTIR sensor is based on the fact that different functional groups within a compound absorb radiation at different frequencies and yield unique infrared absorbance spectra. The large band width (400–4500 cm^{-1}) and the distinct absorbance bands render FTIR technology suitable for sensor applications. Hahn *et al.* (Luoh and Hahn, 2006) first utilized the electrospun nanocomposite PAN/oxide fiber mats as optical sensors in conjunction with FTIR spectroscopy to detect CO_2 gas. The absorbance spectra showed a higher sensitivity with a fiber mat, regardless of its type, than without, indicating gas adsorption on the fiber mat.

11.3.5 Biological sensors

Biosensors are becoming essential in the fields of health care, chemical and biological analysis, environmental monitoring and good processing industries (Ahmad *et al.*, 2010). Electrospinning is a versatile method by which to fabricate biocompatible nanomaterials. The novel nanostructure increases the surface area and mass transfer rate, which significantly improves the biochemical binding effect and sensor signal to noise ratio. The past several decades have witnessed big progress on fabricating biological devices for fast and reliable monitoring of biological targets.

Up to now, most biological sensors-based electrospinning techniques were used to detect glucose. Among those devices, amperometric glucose biosensors, with glucose oxidase (GO_x) and without GO_x, have been an intensively investigated research area because a low detection limit can be achieved easily. Ren *et al.* (2006) have developed an amperometric biosensor by electrospinning deposition of nanofibrous GO_x/PVA membranes as sensitive coatings on the surface of the Au electrode. Chronoamperometric measurements demonstrated that electrospun fibrous enzymatic electrodes exhibited a rapid response (1 s) and a higher response current (1 μA level) to glucose in the normal and diabetic level. The linear response range (from 1 to 10 mM) and the lower detection limit (0.05 mM) of the sensor are satisfying. The electrospun method makes it convenient and efficient to prepare the enzymatic electrode for biosensors.

Apart from glucose biosensors, Luo *et al.* (2010) utilized the surface functionalized electrospun nitrocellulose nanofibers as a direct-charge-transfer biosensor to detect *E. coli* O157:H7 and bovine viral diarrhea virus (BVDV) cells. The biosensor was assembled by attaching the three membrane pads (application pad, capture pad and absorption pad) onto the polyvinylidene chloride substrate as shown in Fig. 11.11a. Figure 11.11b shows the pathogen detection principle of the biosensor. The electrospun biosensor exhibits

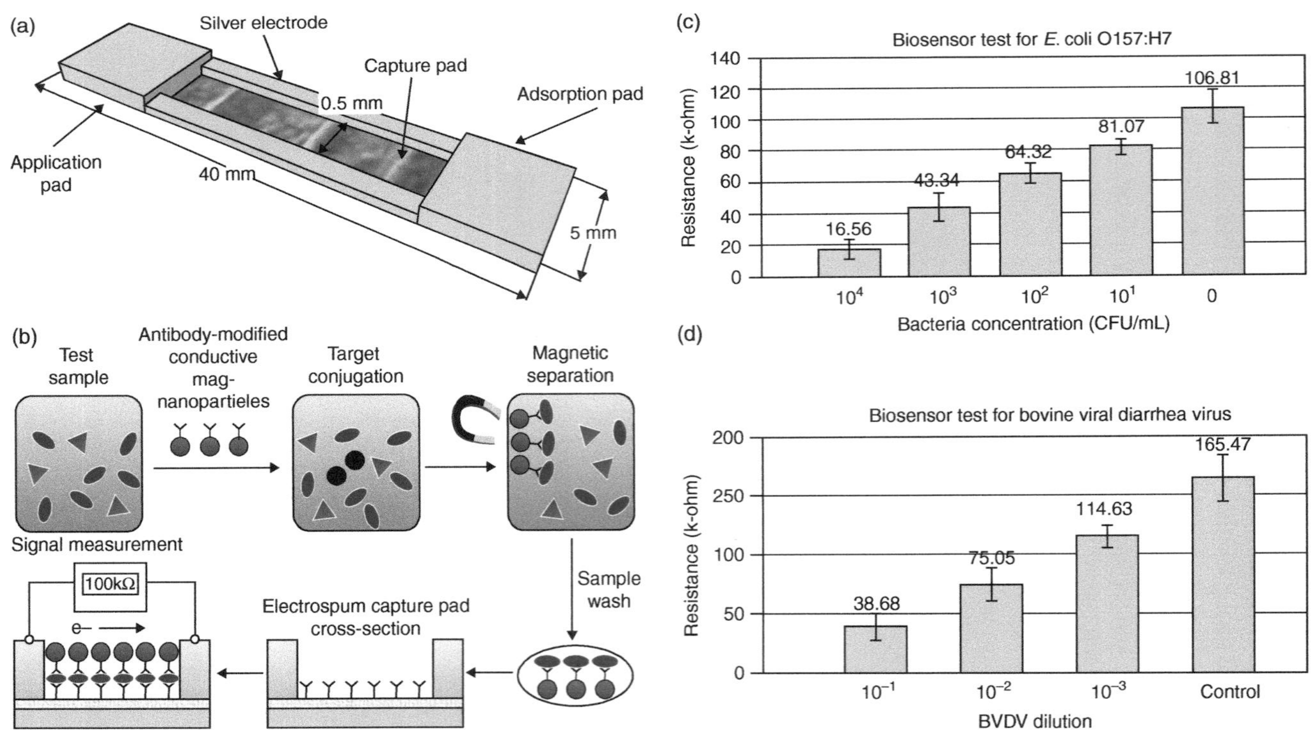

11.11 (a) Schematic of the biosensor structure and membrane assembly consisting of cellulose application and absorption pads and electrospun cellulose nitrate capture pad. (b) Detection scheme of the lateral flow immunosensor based on the antibody functionalized electrospun capture membrane. (c) Biosensor test results for *Escherichia coli* O157:H7 bacteria demonstrate the linear sensor response and lower detection limit of 61 CFU/mL. (d) Biosensor test results for BVDV virus demonstrate the linear sensor response. (Reprinted with permission from Luo *et al.* (2010). © 2010 Elsevier Ltd.)

a linear response to both microbial samples, the *E. coli* O157:H7 and BVDV. The detection time of the biosensor is 8 min, and the detection limit is 61 CFU/mL and 103 CCID/mL for bacterial and viral samples, respectively (see Fig. 11.11c, d). With the advantage of efficient antibody functionalization, excellent capillary capability and relatively low cost, the electrospinning process and surface functionalization method can be implemented to produce nanofibrous capture membranes for different immunodetection applications.

11.4 Future trends and conclusions

Electrospinning is a versatile method of creating high-functional and high-performance nanofibers that can revolutionize the world of structural materials. It is obvious that the current research has focused on seeking the possible applications of these resultant nanofibers with broad functionalities. The utilization of electrospun nanofibers with a large specific surface as sensing materials has received great attention since 2004. The fine structure of electrospun fibers makes them excellent candidates to replace the current widely used solid flat membranes, promising to further increase sensor sensitivity; this opens a new way to fabricate ultrasensitive sensors (Ding *et al.*, 2009). It is important to note that the application of electrospun nanomaterials for sensors still faces many challenges. Preparation of sensing materials in the form of nanostructures may significantly improve their performances in the existing devices or open doors to new types of applications. However, integration of nanomaterials into sensors requires materials of well-controlled orientation, size and other target characteristics, as well as reproducibility locating them in specific positions and orientations. The ability to achieve this, however, remains a major challenge in the field (Frenot and Chronakis, 2003). Fortunately, several approaches have emerged that promise to bridge these gaps and realize the vision of upgraded nanosensors. Given the versatility of electrospinning for generating fibrous membranes with various nanostructures, it is likely that electrospinning will be one of the most significant nanotechnologies in the fabrication of more and more ultrasensitive and practical sensors.

11.5 Acknowledgments

This work is supported by the National Natural Science Foundation of China (No. 50803009), the '111 Project' (No. 111–2-04 and B07024), the Shanghai Committee of Science and Technology (No. 10JC1400600), the National Basic Research Program of China (973 Program, 2011CB606100), the Innovation Program of Shanghai Municipal Education Commission (11ZZ59), the 'Dawn' Program of Shanghai Education Commission (10SG32) and the Fundamental Research Funds for the Central Universities.

11.6 References

Agarwal, S., Greiner, A. and Wendorff, J. H. (2009), Electrospinning of manmade and biopolymer nanofibers-progress in techniques, materials, and applications, *Advanced Functional Materials*, **19**, 2863–79.

Agbenyega, J. (2008), Electrospinning has nanofibers in alignment, *Materials Today*, **11**, 10.

Ahmad, M., Pan, C. F., Luo, Z. X. and Zhu, J. (2010), A single ZnO nanofiber-based highly sensitive amperometric glucose biosensor, *Journal of Physical Chemistry C*, **114**, 9308–13.

Andersson, A., Johansson, N., Broms, P., Yu, N., Lupo, D. and Salaneck, W. R. (1998), Fluorine tin oxide as an alternative to indium tin oxide in polymer LEDs, *Advanced Materials*, **10**, 859–63.

Chao, T. W., Liu, C. J., Hsieh, A. H., Chang, H. M., Huang, Y. S. and Tsai, D. S. (2007), Quartz crystal microbalance sensor based on nano structured IrO_2, *Sensors and Actuators B*, **122**, 95–100.

Chen, H. Y., Wang, N. U., Di, J., Zhao, Y., Song, Y. and Jiang, L. (2010), Nanowire-in-microtube structured core/shell fibers via multifluidic coaxial electrospinning, *Langmuir*, **26**, 11291–6.

Choi, S. H., Ankonina, G., Youn, D. Y., Oh, S. G., Hong, J. M., Rothschild, A. and Kim, I. D. (2009), Hollow ZnO nanofibers fabricated using electrospun polymer templates and their electronic transport properties, *ACS Nano*, **3**, 2623–31.

Crooks, R. M. and Ricco, A. J. (1998), New organic materials suitable for use in chemical sensor arrays, *Accounts of Chemical Research*, **31**, 219–27.

Decher, G., Hong, J. and Schmitt, J. (1992), Buildup of ultrathin multilayer films by a self-assembly process: III. Consecutively alternating adsorption of anionic and cationic polyelectrolytes on charged surfaces, *Thin Solid Films*, **210**, 831–5.

Ding, B., Gong, J., Kim, J. and Shiratori, S. (2005a), Polyoxometalate nanotubes from layer-by-layer coating and thermal removal of electrospun nanofibres, *Nanotechnology*, **16**, 785–90.

Ding, B., Kikuchi, M., Li, C. and Shiratori, S. (2006a), Electrospun nanofibrous polyelectrolytes membranes as high sensitive coatings for QCM-based gas sensors, in Eugene V. Dirote, *Nanotechnology at the leading edge*, New York, Nova Science, 1–28.

Ding, B., Kim, J. H., Miyazaki, Y. and Shiratori, S. M. (2004b), Electrospun nanofibrous membranes coated quartz crystal microbalance as gas sensor for NH_3 detection, *Sensors and Actuators B*, **101**, 373–80.

Ding, B., Kim, J., Kimura, E. and Shiratori, S. (2004a), Layer-by-layer structured films of TiO_2 nanoparticles and poly(acrylic acid) on electrospun nanofibres, *Nanotechnology*, **15**, 913–7.

Ding, B., Li, C. R., Miyauchi, Y., Kuwaki, O. and Shiratori, S. (2006b), Formation of novel 2D polymer nanowebs via electrospinning, *Nanotechnology*, **17**, 3685–91.

Ding, B., Li, C., Du, J. and Shiratori, S. (2008a), Biomimetic super-hydrophobic micro/nanoporous fibrous mat surfaces via electrospinning, in Xiaohua Huang, *Nanotechnology research: New nanostructures, nanotubes and nanofibers*, New York, Nova Science, 131–66.

Ding, B., Ogawa, T., Kim, J., Fujimoto, K. and Shiratori, S. (2008b), Fabrication of a super-hydrophobic nanofibrous zinc oxide film surface by electrospinning, *Thin Solid Films*, **516**, 2495–501.

Ding, B., Wang, M. R., Wang, X. F., Yu, J. Y. and Sun, G. (2010), Electrospun nanomaterials for ultrasensitive sensors, *Materials Today*, **13**, 16–27.

Ding, B., Wang, M. R., Yu, J. Y. and Sun, G. (2009), Gas sensors based on electrospun nanofibers, *Sensors*, **9**, 1609–24.

Ding, B., Yamazaki, M. and Shiratori, S. (2005b), Electrospun fibrous polyacrylic acid membrane-based gas sensors, *Sensors and Actuators B*, **106**, 477–83.

Doshi, J. and Reneker, D. H. (1995), Electrospinning process and application of electrospun fibers, *Journal of Electrostatics*, **35**, 151–60.

Dror, Y., Salalha, W., Avrahami, R., Zussman, E., Yarin, A. L., Dersch, R., Greiner, A. and Wendorff, J. H. (2007), One-step production of polymeric microtubes by co-electrospinning, *Small*, **3**, 1064–73.

Fong, H., Chun, I. and Reneker, D. H. (1999), Beaded nanofibers formed during electrospinning, *Polymer*, **40**, 4585–92.

Formhals, A. (1934), Process and apparatus for preparing artificial threads, US Patent 1.975,504.

Formo, E., Lee, E., Campbell, D. and Xia, Y. N. (2008), Functionalization of electrospun TiO_2 nanofibers with Pt nanoparticles and nanowires for catalytic applications, *Nano Letters*, **8**, 668–72.

Frenot, A. and Chronakis, I. S. (2003), Polymer nanofibers assembled by electrospinning, *Current Opinion in Colloid & Interface Science*, **8**, 64–75.

Grate, J. W. (2000), Acoustic wave microsensor arrays for vapor sensing, *Chemical Reviews*, **100**, 2627–47.

Greiner, A. and Wendorff, J. H. (2007), Electrospinning: A fascinating method for the preparation of ultrathin fibres, *Angewandte Chemie International Edition*, **46**, 5670–703.

Haglund, R. F., Yang, L., Magruder, R. H., Wittig, J. E., Becker, K. and Zuhr, R. A. (1993), Picosecond nonlinear optical response of a Cu:silica nanocluster composite, *Optics Letters*, **18**, 373–5.

Hammond, P. T. (2004), Form and function in multilayer assembly: New applications at the nanoscale, *Advanced Materials*, **16**, 1271–93.

He, X., Arsatb, R., Sadek, A., Wlodarski, W., Kalantar-Zadeh, K. and Jianping, L. (2010), Electrospun PVP fibers and gas sensing properties of PVP/36 YX $LiTaO_3$ SAW device, *Sensors and Actuators B*, **145**, 674–9.

Hu, M. S., Chen, H. L., Shen, C. H., Hong, L. S., Huang, B. R., Chen, K. H. and Chen, L. C. (2006), Photosensitive gold-nanoparticle-embedded dielectric nanowires, *Nature Materials*, **5**, 102–6.

Huang, C. B., Chen, S. L., Lai, C. L., Reneker, D. H., Qiu, H., Ye, Y. and Hou, H. Q. (2006), Electrospun polymer nanofibres with small diameters, *Nanotechnology*, **17**, 1558–63.

Huang, J. S., Wang, D. W., Hou, H. Q. and You, T. Y. (2008), Electrospun palladium nanoparticle-loaded carbon nanofibers and their electrocatalytic activities towards hydrogen peroxide and NADH, *Advanced Functional Materials*, **18**, 441–8.

Huang, Y., Duan, X. F., Wei, Q. Q. and Lieber, C. M. (2001), Directed assembly of one-dimensional nanostructures into functional networks, *Science*, **291**, 630–3.

Huang, Z. M., Zhang, Y. Z., Kotaki, M. and Ramakrishna, S. (2003), A review on polymer nanofibers by electrospinning and their applications in nanocomposites, *Composites Science and Technology*, **63**, 2223–53.

Janata, J., Josowicz, M. and Devaney, D. M. (1994), Chemical sensors, *Analytical Chemistry*, **66**, R207–28.

Jimenez-Gonzalez, A. E. (1997), Modification of ZnO thin films by Ni, Cu, and Cd doping, *Journal of Solid State Chemistry*, **128**, 176–80.

Jin, Y., Yang, D. Y., Kang, D. Y. and Jiang, X. Y. (2010), Fabrication of necklace-like structures via electrospinning, *Langmuir*, **26**, 1186–90.

Kanehata, M., Ding, B. and Shiratori, S. (2007), Nanoporous ultra-high specific surface inorganic fibres, *Nanotechnology*, **18**, 315 602.

Kato, N., Schuetz, P., Fery, A. and Caruso, F. (2002), Thin multilayer films of weak polyelectrolytes on colloid particles, *Macromolecules*, **35**, 9780–7.

Kessick, R. and Tepper, G. (2004), Microscale polymeric helical structures produced by electrospinning, *Applied Physics Letters*, **84**, 4807.

Kokubo, H., Ding, B., Naka, T., Tsuchihira, H. and Shiratori, S. (2007), Multi-core cable-like TiO_2 nanofibrous membranes for dye-sensitized solar cells, *Nanotechnology*, **18**, 165 604.

Kolmakov, A. and Moskovits, M. (2004), Chemical sensing and catalysis by one-dimensional metal-oxide nanostructures, *Annual Review of Materials Research*, **34**, 151–80.

Koombhongse, S., Liu, W. X. and Reneker, D. H. (2001), Flat polymer ribbons and other shapes by electrospinning, *Journal of Polymer Science Part B: Polymer Physics*, **39**, 2598–606.

Li, D. and Xia, Y. N. (2004), Electrospinning of nanofibers: Reinventing the wheel?, *Advanced Materials*, **16**, 1151–70.

Li, D., Ouyang, G., Mccann, J. T. and Xia, Y. N. (2005), Collecting electrospun nanofibers with patterned electrodes, *Nano Letters*, **5**, 913–6.

Li, D., Wang, Y. L. and Xia, Y. N. (2003), Electrospinning of polymeric and ceramic nanofibers as uniaxially aligned arrays, *Nano Letters*, **3**, 1167–71.

Li, Y., Li, P., Yang, M. J., Lei, S., Chen, Y. Q. and Guo, X. S. (2010), A surface acoustic wave humidity sensor based on electrosprayed silicon-containing polyelectrolyte, *Sensors and Actuators B*, **145**, 516–20.

Li, Y., Meng, G. W., Zhang, L. D. and Phillipp, F. (2000), Ordered semiconductor ZnO nanowire arrays and their photoluminescence properties, *Applied Physics Letters*, **76**, 2011–3.

Li, Z. Y., Huang, H. M., Shang, T. C., Yang, F., Zheng, W., Wang, C. and Manohar, S. K. (2006), Facile synthesis of single-crystal and controllable sized silver nanoparticles on the surfaces of polyacrylonitrile nanofibres, *Nanotechnology*, **17**, 917–20.

Lin, J. Y., Ding, B., Yu, J. Y. and Hsieh, Y. (2010), Direct fabrication of highly nanoporous polystyrene fibers via electrospinning, *ACS Applied Materials & Interfaces*, **2**, 521–8.

Liu, Y. Q., Zhang, X. P., Xia, Y. N. and Yang, H. (2010), Magnetic-field-assisted electrospinning of aligned straight and wavy polymeric nanofibers, *Advanced Materials*, **22**, 1–4.

Long, Y. Y., Chen, H. B., Yang, Y., Wang, H. M., Yang, Y. F., Li, N., Li, K. A., Pei, J. and Liu, F. (2009), Electrospun nanofibrous film doped with a conjugated polymer for DNT fluorescence sensor, *Macromolecules*, **42**, 6501–9.

Lotus, A. F., Kang, Y. C., Walker, J. I., Ramsier, R. D. and Chase, G. G. (2010), Effect of aluminum oxide doping on the structural, electrical, and optical properties of zinc oxide (AOZO) nanofibers synthesized by electrospinning, *Materials Science and Engineering: B. Advanced Functional Solid-State Materials*, **166**, 61–6.

Lu, X. F., Wang, C. and Wei, Y. (2009), One-dimensional composite nanomaterials: Synthesis by electrospinning and their applications, *Small*, **5**, 2349–70.

Luo, Y., Nartker, S., Miller, H., Hochhalter, D., Wiederoder, M., Wiederoder, S., Setterington, E., Drzal, L. T. and Alocilja, E. C. (2010), Surface functionalization of electrospun nanofibers for detecting E. coli O157:H7 and BVDV cells in a direct-charge transfer biosensor, *Biosensors and Bioelectronics*, **26,** 1612–7.

Luoh, R. and Hahn, H. T. (2006), Electrospun nanocomposite fiber mats as gas sensors, *Composites Science and Technology*, **66**, 2436–41.

Lv, Y.-Y., Wu, J. and Xu, Z.-K. (2010), Colorimetric and fluorescent sensor constructing from the nanofibrous membrane of porphyrinated polyimide for the detection of hydrogen chloride gas, *Sensors and Actuators B*, **148**, 233–9.

Mamedov, A. A., Kotov, N. A., Prato, M., Guldi, D. M., Wicksted, J. P. and HIRSCH, A. (2002), Molecular design of strong single-wall carbon nanotube/polyelectrolyte multilayer composites, *Natural Materials*, **1**, 190–4.

Marx, K. A. (2003), Quartz crystal microbalance: A useful tool for studying thin polymer films and complex biomolecular systems at the solution-surface interface, *Biomacromolecules*, **4**, 1099–120.

Matthews, J. A., Wnek, G. E., Simpson, D. G. and Bowlin, G. L. (2002), Electrospinning of collagen nanofibers, *Biomacromolecules*, **3**, 232–8.

Mou, F. Z., Guan, J. G., Shi, W. D., Sun, Z. G. and Wang, S. H. (2010), Oriented contraction: A facile nonequilibrium heat-treatment approach for fabrication of maghemite fiber-in-tube and tube-in-tube nanostructures, *Langmuir*, **26**, 15 580–5.

Nohria, R., Khillan, R. K., Su, Y., Dikshit, R., Lvov, Y. and Varahramyan, K. (2006), Humidity sensor based on ultrathin polyaniline film deposited using layer-by-layer nano-assembly, *Sensors and Actuators B*, **114**, 218–22.

Ogawa, T., Ding, B., Sone, Y. and Shiratori, S. (2007), Super-hydrophobic surfaces of layer-by-layer structured film-coated electrospun nanofibrous membranes, *Nanotechnology*, **18**, 165 607.

Qi, Q., Zhang, T., Liu, L. and Zheng, X. J. (2009), Synthesis and toluene sensing properties of SnO_2 nanofibers, *Sensors and Actuators B*, **137**, 471–5.

Ramakrishna, S., Fujihara, K., Teo, W. E., Yong, T., Ma, Z. W. and Ramaseshan, R. (2006), Electrospun nanofibers: Solving global issues, *Materials Today*, **9**, 40–50.

Ren, G. L., Xu, X. H., Liu, Q., Cheng, J., Yuan, X. Y., Wu, L. L. and Wan, Y. Z. (2006), Electrospun poly(vinyl alcohol)/glucose oxidase biocomposite membranes for biosensor applications, *Reactive and Functional Polymers*, **66,** 1559–64.

Reneker, D. H. and Chun, I. (1996), Nanometre diameter fibres of polymer, produced by electrospinning, *Nanotechnology*, **7**, 216–23.

Ricco, A. J., Crooks, R. M. and Osbourn, G. C. (1998), Surface acoustic wave chemical sensor arrays: New chemically sensitive interfaces combined with novel cluster analysis to detect volatile organic compounds and mixtures, *Accounts of Chemical Research*, **31**, 289–96.

Sarkar, S., Levit, N. and Tepper, G. (2006), Deposition of polymer coatings onto SAW resonators using AC electrospray, *Sensors and Actuators B*, **114**, 756–61.

Sauerbrey, G. (1959), The use of quartz oscillators for weighing thin layers and for microweighing, *Z. Phys*, **155**, 206–22.

Schmidt-Mende, L. and Macmanus-Driscoll, J. L. (2007), ZnO-nanostructures, defects, and devices, *Materials Today*, **10**, 40–8.

Shi, W., Lu, W. S. and Jiang, L. (2009), The fabrication of photosensitive self-assembly Au nanoparticles embedded in silica nanofibers by electrospinning, *Journal of Colloid and Interface Science*, **340**, 291–7.

Sun, Z. C., Zussman, E., Yarin, A. L., Wendorff, J. H. and Greiner, A. (2003), Compound core-shell polymer nanofibers by co-electrospinning, *Advanced Materials*, **15**, 1929–32.

Teo, W. E. and Ramakrishna, S. (2006), A review on electrospinning design and nanofibre assemblies, *Nanotechnology*, **17**, R89–106.

Tsukruk, V. V., Rinderspacher, F. and Bliznyuk, V. N. (1997), Self-assembled multilayer films from dendrimers, *Langmuir*, **13**, 2171–6.

Wang, G., Ji, Y., Huang, X. R., Yang, X. Q., Gouma, P. I. and Dudley, M. (2006), Fabrication and characterization of polycrystalline WO_3 nanofibers and their application for ammonia sensing, *Journal of Physical Chemistry B*, **110**, 23, 777–82.

Wang, M. and Pan, N. (2008), Predictions of effective physical properties of complex multiphase materials, *Materials Science and Engineering: R Reports*, **63**, 1–30.

Wang, X. F., Ding, B., Sun, M., Yu, J. Y. and Sun, G. (2010a), Nanofibrous polyethyleneimine membranes as sensitive coatings for quartz crystal microbalance-based formaldehyde sensors, *Sensors and Actuators B*, **144**, 11–7.

Wang, X. F., Ding, B., Yu, J. Y., Si, Y., Yang, S. and Sun, G. (2011), Electro-netting: Fabrication of two-dimensional nano-nets for highly sensitive trimethylamine sensing, *Nanoscale*, **3**, 911–15.

Wang, X. F., Ding, B., Yu, J. Y., Wang, M. R. and Pan, F. K. (2010b), A highly sensitive humidity sensor based on a nanofibrous membrane coated quartz crystal microbalance, *Nanotechnology*, **21**, 055 502.

Wang, X. Y., Drew, C., Lee, S. H., Senecal, K. J., Kumar, J. and Samuelson, L. A. (2002a), Electrospinning technology: A novel approach to sensor application, *Journal of Macromolecular Science, Part A: Pure and Applied Chemistry*, **A39**, 1251–8.

Wang, X. Y., Drew, C., Lee, S. H., Senecal, K. J., Kumar, J. and Sarnuelson, L. A. (2002b), Electrospun nanofibrous membranes for highly sensitive optical sensors, *Nano Letters*, **2**, 1273–5.

Wang, X. Y., Kim, Y. G., Drew, C., Ku, B. C., Kumar, J. and Samuelson, L. A. (2004), Electrostatic assembly of conjugated polymer thin layers on electrospun nanofibrous membranes for biosensors, *Nano Letters*, **4**, 331–4.

Wang, X. Y., Lee, S. H., Drew, C., Senecal, K. J., Kumar, J. and Samuelson, L. A. (2002c), Highly sensitive optical sensors using electrospun polymeric nanofibrous membranes, *Organic Optoelectronic Materials, Processing and Devices*, **708**, 397–402

Wang, Y., Jia, W. Z., Strout, T., Ding, Y. and Lei, Y. (2009), Preparation, characterization and sensitive gas sensing of conductive core-sheath TiO_2-PEDOT nanocables, *Sensors*, **9**, 6752–63.

Wang, Z. J., Li, Z. Y., Sun, J. H., Zhang, H. N., Wang, W., Zheng, W. and Wang, C. (2010c), Improved hydrogen monitoring properties based on p-NiO/n-SnO_2

heterojunction composite nanofibers, *Journal of Physical Chemistry C*, **114**, 6100–5.

Yang, A., Tao, X. M., Wang, R. X., Lee, S. C. and Surya, C. (2007), Room temperature gas sensing properties of SnO_2/multiwall-carbon-nanotube composite nanofibers, *Applied Physics Letters*, **91**, 133 110.

Zhang, D. M. and Chang, J. (2008), Electrospinning of three-dimensional nanofibrous tubes with controllable architectures, *Nano Letters*, **8**, 3283–7.

Zhang, Z. Y., Li, X. H., Wang, C. H., Wei, L. M., Liu, Y. C. and Shao, C. L. (2009), ZnO hollow nanofibers: Fabrication from facile single capillary electrospinning and applications in gas sensors, *Journal of Physical Chemistry C*, **113**, 19, 397–403.

Zhao, Y., Cao, X. Y. and Jiang, L. (2007), Bio-mimic multichannel microtubes by a facile method, *Journal of the American Chemical Society*, **129**, 764–5.

Zussman, E., Theron, A. and Yarin, A. L. (2003), Formation of nanofiber crossbars in electrospinning, *Applied Physics Letters*, **82**, 973–5.

12
Functional nanofibers in clothing for protection against chemical and biological hazards

A. H. NURFAIZEY, University of Canterbury, New Zealand, and Universiti Teknikal Malaysia Melaka, N. TUCKER, New Zealand Institute for Plant and Food Research Limited, New Zealand, J. STANGER and M. P. STAIGER, University of Canterbury, New Zealand

Abstract: Functional nanofibers have great potential to improve protection against chemical and biological warfare agents. This chapter reviews the history of chemical and biological warfare and existing protective technologies. The use of electrospinning to produce nanofibers for protective applications is also introduced. Previous studies demonstrate the applicability of electrospun nanofibers in protective fabric technologies. Advantages of these next-generation protective materials include high penetration resistance, excellent breathability, low basis weight, low pressure drop, the potential for the incorporation of surface chemical functionality and simple processing equipment. Finally, the main development issues and potential research directions of electrospun nanofibers in protective applications are discussed.

Key words: electrospinning, nanofiber, chemical and biological warfare, protective clothing.

12.1 Introduction

Due to their unique properties, electrospun nanofibers have been the subject of much study over the last decade. The advantage of electrospun fibers over other nanoscale fibers, such as carbon nanotubes, is that electrospun fibers have an obvious macroscale dimension – an electrospun nanofiber might only be 80 nm in diameter, but it could be over a kilometer in length. As a result, the high specific surface area functionality of a nanoparticle is retained but without the possibility of any risk to health and safety associated with nanoscale particulates or short fibers. The nanoscale diameter of the fibers means that small particulate and aerosol materials tend to stick to the fiber surface, the method of attraction being a combination of electrostatic and London-van der Waals forces. A wide variety of applications such as filtration, tissue engineering, wound dressings, drug delivery, composite reinforcements, electronics and sensors have been proposed. The one application in which nanofibers have achieved commercial success so far is filtration. Due to the fact that the fibers have nanoscale diameters, membranes constructed from such fibers are generally thin, but have a very high specific surface area coupled with a small pore size. These features are advantageous for producing filtration media with low pressure drop, but high filtration efficiency. In addition, the high surface area of the fiber could be used to provide an effective interface for treatment using additives with specific chemical or biochemical functionalities.

The combination of properties that make nanofibers suitable for active or passive filtration and incorporation into woven and nonwoven fabrics make them particularly suitable candidate materials for protective clothing.

In this chapter the use of functional nanofibers for protective applications, especially in the context of protection against chemical and biological warfare agents, will be discussed.

12.2 Chemical and biological warfare background

The early history of the use of toxic materials as weapons dates back to at least medieval times, when animal and human corpses were projected by catapult into besieged towns in the hope of engendering plague among the defenders. More directly, arrow tips were sometimes poisoned. The majority of deaths resulting from warfare in this period were caused by disease and septicaemia from infected wounds.

In the early nineteenth century, Admiral Lord Thomas Cochrane, 10th Earl of Dundonald (1775–1860), proposed the use of chemical warfare techniques. In 1812 he suggested a method of chemical and biological attack ('stink ships') that was accepted by the UK Admiralty as technically feasible, but rejected on the grounds of inhumanity (Stephenson, 2006). The modern history of the use of chemical warfare agents began about 100 years ago. The German Nobel laureate Fritz Haber is fairly described as the 'father of chemical warfare'. Through his research at Kaiser Wilhelm Physical Institute he produced and directly supervised the use of chlorine gas as a chemical warfare agent. Despite being banned by international convention, chlorine gas was first used by German troops in a large-scale operation at Ypres, Belgium, on April 22, 1914, killing 5000 Allied soldiers (Bajgar *et al.*, 2009). Throughout World War I, several poisonous gases were used by both sides, including phosgene, diphosgene and mustard gas.

Further disregarding treaty obligations, research into more effective warfare agents continued through World War II and beyond. This led to the invention, by Dr. Gerhard Schrader's research group, of a second generation of chemical warfare agents, known as the G-series organophosphorus 'nerve' agents (Tabun, Sarin, Soman and Cyclosarin). An amount of these nerve agents had been created and stockpiled by the Germans in World War II, but they were never used in actual warfare. After World War II, during the Cold War era, the development of chemical and biological warfare agents continued in the United States, the United Kingdom and the Soviet Union, producing the third and fourth generations of nerve agents, including V-series, GV-series and Soviet Novichok-series.

German troops were also the first to use biological warfare in modern times, again during World War I, although their attacks with biological weapons were on a rather small scale and were not particularly successful (Frischknecht, 2003). *Bacillus anthracis* and other disease-causing organisms were used in covert programs intended to infect livestock and animal feed to be exported to the Allied Forces (Hilmas *et al.*, 2009). The development of biological warfare agents also took place in the East, when Japan began a massive biological warfare program before the onset of World War II under Shiro Ishii (known inevitably as the 'father of Japanese biological warfare') (Frischknecht, 2003). During World War II, the Japanese used biological warfare in conflict with China and the Soviet Union.

Although it is difficult to quantify the actual number of deaths caused by modern chemical and biological warfare, it has been estimated that chemical warfare claimed 90 000 lives during World War I (Bajgar *et al.*, 2009), while biological warfare killed several tens of thousands of people during World War II (Frischknecht, 2003). Due to the hazards caused by these weapons (e.g., Gruinard Island off the coast of Scotland was quarantined for 48 years following biological warfare testing using anthrax spores), the need to control the development and production of chemical and biological warfare agents has been at least recognized. According to the United Nations Office for Disarmament Affairs, nuclear, chemical and biological weapons are categorized as weapons of mass destruction (WMD). There are a number of treaties concerning chemical and biological warfare, including the latest Chemical Weapons Convention which came into force on April 29, 1997 and the Biological Weapons Convention which came into force on March 26, 1975. However, it should be noted that the relatively low technological and economic barriers to the development and production of biological and chemical weapons mean that the existence of treaties to limit the production and use of such agents may not be entirely effective (Schneider and Mayer, 1995).

The likely modern scenarios for the use of chemical and biological weapons are terrorist attacks on civil populations rather than tactical battlefield use.

12.2.1 Chemical and biological warfare agents

Chemical warfare agents are highly toxic chemicals intended to harm, kill, incapacitate or control adversaries in warfare (Satoh, 2009). In contrast, biological warfare uses harmful living organisms such as bacteria – for example, *Bacillus anthracis* – which causes anthrax. These weapons are different from conventional weapons in that they do not depend on the percussion of an explosion to produce their terrible effects. Chemical and biological warfare agents can be classified based on the common military groupings of chemical effect (i.e., nerve, vesicant (blister forming), blood, pulmonary, incapacitating and riot control), or biological origin (i.e., bacterial, viral, rickettsial (carried by ticks, fleas or lice), fungal, or bioorigin toxic agents) (Ellison, 2008). The classification of chemical and biological agents is summarized in Fig. 12.1.

Significant quantities of chemical and biological agents have been produced and stockpiled by treaty signatory nations with the intention of using them as weapons. The most well-known agents include organophosphorus nerve agents such as Sarin, Soman and VX agents; pulmonary (choking) agents such as chlorine and phosgene; vesicant (blistering) agents such as mustard gas and lewisite; cellular respiration inhibitors (also known as blood agents) such as hydrogen cyanide; and biological agents such as *Bacillus anthracis*.

Nerve agents are the most toxic proposed man-made chemical warfare agents (Ellison, 2008). These agents are potent anticholinesterase compounds deliberately formulated to disrupt the function of the nervous system by interfering with the enzyme acetylcholinesterase (Ellison, 2008; Watson *et al.*, 2009). Organophosphorus nerve agents include the first generation of nerve agents – that is, G-series (such as Sarin, Soman and Tabun), V-series, GV-series and, more recently, 'новичок' (Novichok or Newcomer) series developed in the USSR specifically to be difficult to detect and to be effective against existing personal protective equipment (PPE). Exposure to solids, liquids or vapors from these agents is hazardous and can result in death within minutes (Ellison, 2008). The most recent attack using a nerve agent was the Aum Shinrikyo Tokyo subway terrorist attack in 1995. In the attack, Sarin gas was used, killing 12 people and inflicting various degrees of injury on thousands of others (Okumura *et al.*, 2009).

Pulmonary (or choking) agents cover a wide variety of chemical structures including halogens, acylhalides, various alkylating agents and metallic and nonmetallic oxides. The majority of these agents are members of the first generation of warfare agents (Ellison, 2008) including chlorine and phosgene, which were used during World War I. Pulmonary agents act on the pulmonary system causing severe irritation and swelling of the nose, throat and lungs (Gulati and Ray, 2009). Vesicant (or blister) agents produce their

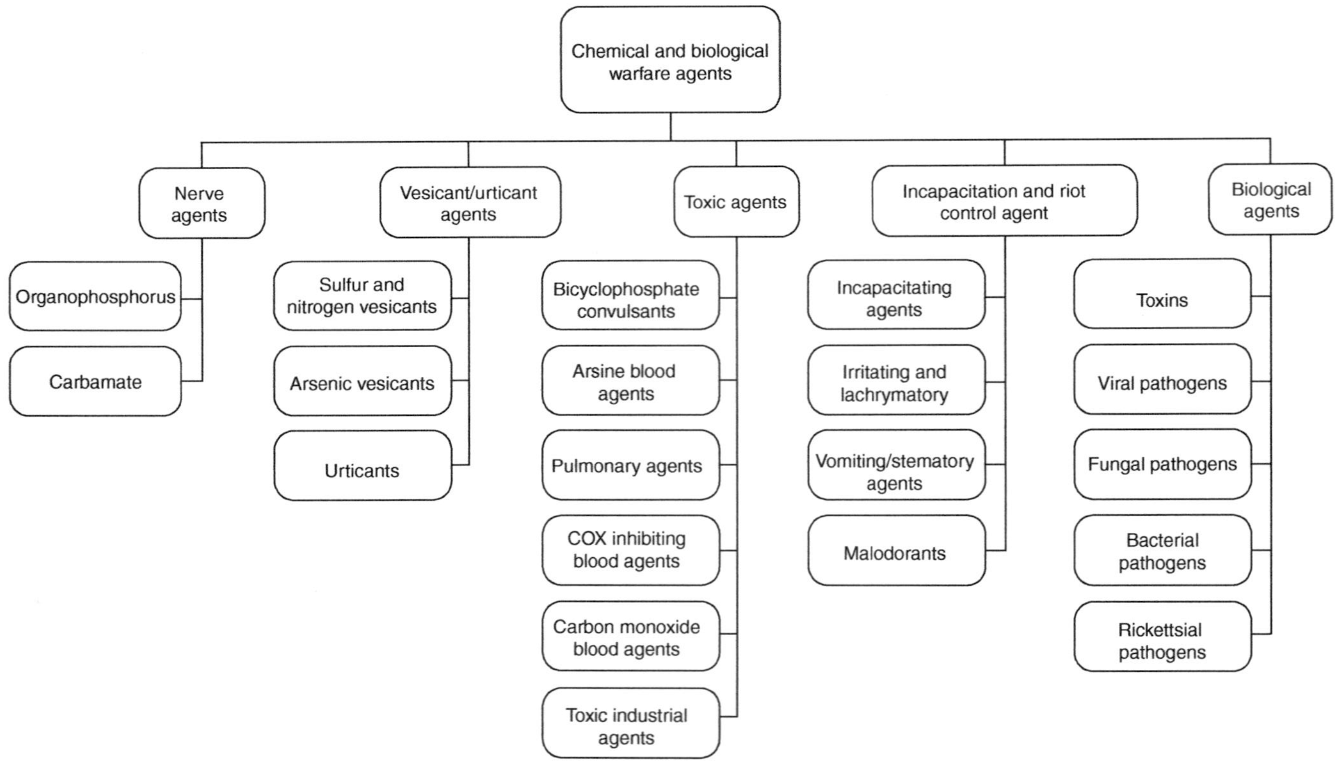

12.1 Classification of chemical and biological warfare agents. (Based on information from Ellison, 2008.)

toxic effects by forming a highly reactive intermediate cyclic structure that alkylates nucleophiles in the cell structure and disrupts the normal function of the affected tissue (Ellison, 2008). These agents act on skin and other epithelial tissues and severely blister the eyes, respiratory tract and internal organs; they also destroy various substances within living cells (Gulati and Ray, 2009). Vesicant agents include sulfur mustard, nitrogen mustard, lewisite and phosgene oxime. Hydrogen cyanide (a so-called blood agent) inhibits the enzyme cytochrome oxidase, preventing the use of oxygen in cellular metabolism (Agency for Toxic Substances and Disease Registry, 2011). As a result, the lack of oxygen rapidly affects all body tissues, especially the central nervous system, producing headaches, dizziness, confusion, stupor, nausea and vomiting (Ellison, 2008).

The rapid (as little as 15 sec; Laurie, 1972) absorption of damaging quantities of these materials means that protective equipment must either be comfortable and practical to wear for long durations or easy to put on following a warning of chemical contamination.

Biological agents are of little use as tactical battlefield weapons. Biological agents act too slowly to be of practical use, and short environmental persistence in active forms and the difficulty in dispersing the pathogen over wide areas has happily limited the use of these materials in combat (Anon., 1969). However, as a weapon of terror, biological agents have found an unfortunate niche. The most well-known biological agent is the bacterium *Bacillus anthracis* which causes the acute disease anthrax. *Bacillus anthracis* is an aerobic, gram-positive, spore-forming and rod-shaped bacteria (Ellison, 2008) causing a virulent, contagious and potentially fatal disease transmitted by inhalational, cutaneous and oral/ingestional routes. The most recent attack using anthrax was in 2001 in the United States. Letters containing anthrax spores were deliberately mailed to a number of locations across the United States killing five people and infecting many others. The chief suspect was an American biodefense researcher (Hilmas *et al.*, 2009).

12.2.2 Protective clothing against chemical and biological warfare agents

Despite the signing of numerous treaties concerning WMD, chemical and biological warfare continues to be a perceived significant threat to the world. The world is in a perennial state of political instability and there continue to be frequent terrorist attacks made with varying degrees of sophistication. In the last three decades, there have been several incidents where potential WMD have been used, either by states – in the 1981–1988 Iran–Iraq war – or by terrorists – the 1995 Aum Shinrikyo Tokyo subway Sarin gas attack. Therefore, as preparation for the possibility of chemical or biological warfare attack, both military and civil forces are equipped with protective clothing.

To adequately protect the body, it must be protected from poisoning by contact, inhalation or ingestion. This is accomplished by the use of respirators and protective clothing in conjunction. The first attempts at developing this technology resulted in simple impervious garments – the gas cape, goggles and face masks used in World War I. These items of PPE were designed to prevent absorption by the skin, and the more rapid absorption by wet membranes such as the eye and the respiratory and gastrointestinal tracts.

If the toxic agent is delivered as a gas, the duration of the toxic effect may be quite short, although the gas may linger in certain confined areas. Chemical agents delivered as aerosols (fine particulates of solid or liquid suspended in the atmosphere) may remain dangerous even after they have precipitated out of the air onto a solid surface. Similarly, biological agents – spores, bacteria or viruses – may remain potent, if not active, for a considerable time. It is therefore desirable that PPE should be capable of being worn for an extended period, and that it should not affect the ability of the wearer to perform operational tasks or the activities that sustain life such as eating, drinking and excretion. The aspirational aim for this type of PPE is that it should be incorporated into standard clothing, and not in any way inhibit the actions of the wearer.

The protective garments used by the American armed forces can provide a prime example of the state-of-the-art PPE technology at any given time in the modern era. The early means of protection against chemical and biological warfare used by the United States Forces was the battledress overgarment (BDO). The BDO liner consisted of an activated charcoal(carbon)-impregnated polyurethane foam, which absorbed ingressive chemical agents, and a water resistant nylon tricot laminate outer shell (The United States Army, 2011). The BDO was worn over clothing and thus reduced the individual's dexterity and also trapped heat, to the discomfort of the user, making prolonged use in hot climates problematic (Stroud, 1999). It should be noted that during the cold war period, the major conflict was expected to be against the Soviet block – such a conflict would have taken place in the cold, temperate climate of northern Europe.

The BDO was replaced by joint service lightweight integrated suit technology (JSLIST) in 1997 (see Fig. 12.2). Instead of layering activated carbon protection into the suit fabric, as in the BDO, the protective carbon is formed into tiny beads and woven into the suit fabric (Stroud, 1999). This new fabric was developed by Blücher GmbH under the trade name of SARATOGA®. The claimed advantage of this technology lies in the presence of tiny beads with diameter of approximately 0.02 inch (0.5 mm) intended to act as efficient chemical absorbers due to their spherical form, which provides the maximum available surface area for adsorption of chemical warfare agents

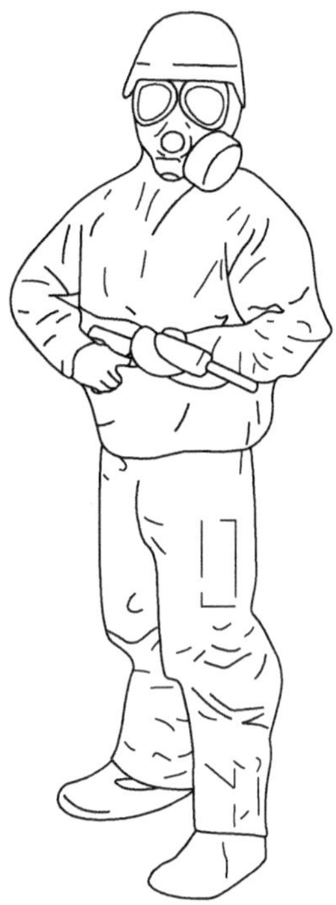

12.2 The Joint Service Lightweight Integrated Suit Technology (JSLIST). (Image from http://www.army.mil/ courtesy of The Office of the Chief of Staff of the Army in coordination with Army Public Affairs.)

and other contaminants (Blücher, 2010). The fabric is also lighter, cooler and provides a higher level of protection without blocking the movement of air and perspiration through the suit (The United States Army, 2011).

Both BDO and JSLIST protective clothing take advantage of the high surface area of the active material to adsorb toxic agents. In the case of activated carbon spheres, the internal surface area is claimed to be between 600 and 1500 m^2/g (Blucher *et al.*, 1985). Activated carbon is a material with a highly porous morphology widely used for purification of liquids and gases (Marsh and Rodriguez-Reinoso, 2006). These adsorbent properties are essentially attributed to the large interparticulate surface area which is made up of a complex network of micropores (diameter less than 2 nm), mesopores (diameter between 2 and 50 nm) and macropores (diameter more than 50 nm) (Bansal and Goyal, 2005). In addition to physical adsorption, activated carbon can also be impregnated with suitable mixtures of metal adsorbant compounds to provide protection against toxic chemical agents by means of chemical adsorption. The principles of the SARATOGA® filter sheet material technology are more fully described by US Patents 4455187, 4510193 and 5277963.

12.3 Electrospun nanofibers

Electrospinning is known as a simple, versatile and scalable method of producing nanofibers (Reneker and Chun, 1996; Ramakrishna *et al.*, 2005). The advantage of this process is that the capital cost is low, and the method is conceptually simple. The process operates at room temperature and does not require coagulation chemistry to form a solid fiber from a liquid precursor. The process of solidification is either by solvent loss if a polymer solution is used, or by freezing if the fiber is spun from a melt. Consequently, the process is ideal for retaining complex chemical functionality in the fibers, as might well be the case in the production of PPE materials. The disadvantages are a low production rate from single-spinneret systems, and certain process control difficulties in coping with variation in feedstock materials. In the process of electrospinning, polymer droplets are charged with high electric potential causing the same charged molecules to repel each other. When the repulsion forces are greater than the surface tension of the polymer droplet, the surface ruptures and a jet of charged molecules emerges and moves toward the nearest grounded or oppositely charged collector. In an equilibrium condition, a conical distortion of the polymer droplet known as the Taylor cone is formed at the starting point of the jets. At a certain distance, the straight jets will undergo a transformation into a superficially chaotic motion known as the bending instability. The fiber rapidly accelerates, being drawn into a nanoscale diameter. In the typical electrospinning process, the fibers are collected in the form of a random-oriented mat. Electrospun fibers are manufactured from molten polymers, or a solution of polymers, in a manner that favors a high degree of molecular orientation, and hence strength. Solidification is achieved by a combination of freezing, drying, cross-linking and orientation by drawing and strain hardening. The strength of such fibers may approach theoretical maxima because the organizing process tends to minimize intramolecular defects and in the final structure the applied loads are taken up directly by oriented molecules.

A typical electrospinning apparatus consists of either a pipette tip with an electrode and polymer solution or a syringe with a metal needle and polymer in the barrel. Various collectors are used at a typical distance of between 2 and 35 cm, with the main common factor for all collectors being a conductive metal component. The polymer is fed to the needle tip via gravity or pressure to form a droplet. This droplet is then charged by applying a potential difference between 5 and 100 kV and a Taylor cone (Taylor, 1969) is formed, followed by a jet of polymer which is accelerated toward the collector by an electrostatic force, stretching the fiber to a submicron diameter. The distance between these electrodes is usually set to give an electric field strength of 100–500 kV/m (Greiner and Wendorff, 2007). The diameter of nanofibers produced by electrospinning is in the range of a few nanometers to one micron (Ramakrishna *et al.*, 2005). Electrospun nanofibers have been proposed for various applications such as filtration, tissue engineering, wound dressing, drug delivery, composite reinforcement, electronics and sensors.

The history of electrospinning contains a number of commercialization attempts to produce woven textiles. The first patent on the process was issued in the UK to Cooley (1900), and by the 1930s a sequence of patents was granted to Formhals (1934–1944). However, the earliest recorded successful attempt at commercializing electrospun nanofibers took place in the Soviet Union. In 1938 N. D. Rozenblum and I. V. Petryanov-Sokolov, working in Prof. N. A. Fuks' group at the Aerosol Laboratory of the L. Ya Karpov Institute in the USSR, generated electrospun fibers which they developed into nonwoven filter materials known as Petryanov filters®. These filters were made by depositing a nonwoven electrospun nanofiber layer onto a macrofiber support substrate (see, for example, Fig. 12.3). The advantage of this method is that it does away with the need to directly handle the fragile nanoscale materials, and allows the nanofibers to be readily incorporated into manufacturing processes. By 1939, this work had led to the establishment of a factory in Tver for the manufacture of electrospun smoke filter elements for gas masks. The material, dubbed 'battlefield filter', was spun from cellulose acetate in a solvent mixture of dichloroethane and ethanol. At the peak of operation the Soviet Union was producing 21 million m^2 per annum of electrospun filtration material (Filatov *et al.*, 2007).

More recently, since the 1980s, electrospun nanofibers have been used to make filter media by the Donaldson Company Incorporated. Known under trade name Ultra-Web®, the material has been successfully used in wide

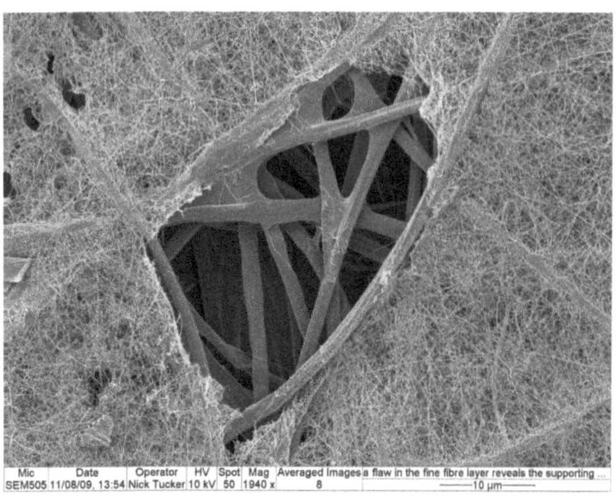

12.3 SEM photo of a commercially available filter membrane showing nanoscale fibers incorporated onto a microscale fiber substrate. (Image courtesy of the New Zealand Institute for Plant and Food Research.)

range of industrial, consumer and defense filtration applications (Graham *et al.*, 2003). In 2006, DuPont™ introduced hybrid membrane technology (HMT), which was targeted toward applications such as air filtration, liquid filtration, energy storage and apparel applications. DuPont™ HMT nanofiber sheets contain continuous polymeric filaments with a typical diameter between 100 nm and 1 micron; these provide filtration efficiency approaching that of microporous membranes (E. I. Du Pont De Nemours and Company, 2011).

12.3.1 The use of electrospun nanofibers for protective applications

Current protective clothing is based either on full hermetic barrier protection, such as hazardous materials suits, or permeable adsorptive protective overgarments, such as the current US forces JSLIST garment (Schreuder-Gibson *et al.*, 2003).

Graham *et al.* (2003) lists the critical performance requirements for chemical protective clothing as including excellent protection against liquid and vapor contaminants, resistance to rain intrusion, air permeability, tear strength, fabric weight requirements and durability. To achieve these requirements, current protective fabric systems consist of several layers of fabrics covered by durable outer shell fabric. The disadvantages associated with this fabric system are bulky size, heavy weight, moisture retention and heat burden (Sundarrajan and Ramakrishna, 2007; Chen *et al.*, 2009). Recently, the direct incorporation of electrospun nanofibers into protective fabrics has been proposed to improve the practicality of this type of protective clothing.

The US Army Natick Soldier Center, based in Natick, Massachusetts, has been studying the application of electrospun nanofibers for the purpose of developing a new generation of protective fabrics. It was reported that electrospun nanofiber layers are extremely efficient at trapping airborne particles (see, for example, Fig. 12.4, showing particulates trapped on an electrospun air filter element) while presenting minimal impedance to the moisture vapor diffusion required for evaporative cooling of the wearer (Schreuder-Gibson *et al.*, 2002).

Lee and Obendorf (2006), from Cornell University, showed that electrospun polypropylene (PP) membranes and laminates exhibited a combination of high protection performance and an acceptable level of air and vapor transport properties compared with currently available PPE materials.

Electrospun nanofibers have smaller fiber diameters and larger specific surface areas than traditional filtration media, and hence significantly increase the filtration efficiency without sacrificing air permeability (Vitchuli *et al.*, 2010). The reason for the effectiveness of small diameter electrospun

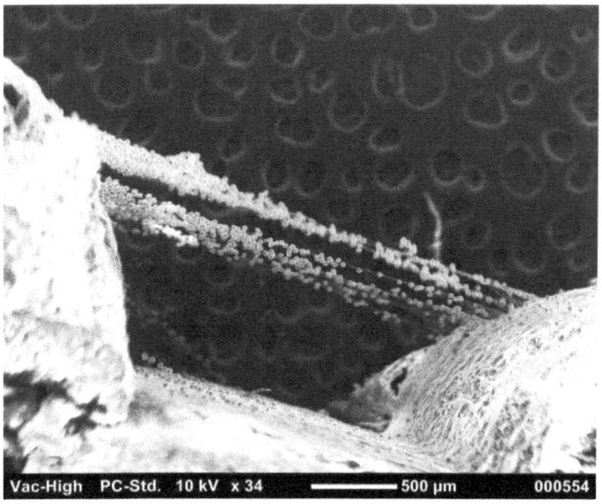

12.4 Spores captured on nanofibers in an air filtration unit. (Image courtesy of Revolution Fibres Ltd, Auckland, New Zealand.)

nanofibers in providing efficient filtration was explained by Graham *et al.* (2002) and Chase (2009). Filtration theory generally relies on an assumption of continuous flow around the fiber with no-slip flow conditions at the fiber surface. However, when the scale of the fiber is very small, the molecular movements of the air molecules are significant in relation to the size of the fibers and the flow field (Graham *et al.*, 2002). A dimensionless number known as the Knudsen number, *Kn*, defined as the ratio of the molecular mean free path length to a representative physical length scale, named after Danish physicist Martin Knudsen (1871–1949), is normally used to determine the nature of the gas flow regime. The Knudsen number can be written as:

$$\mathrm{Kn} = \frac{\lambda}{r_{\mathrm{f}}} \qquad [12.1]$$

where λ is the gas mean free path (in this case, air) and r_{f} is the radius of the fiber (Graham *et al.*, 2002). Depending on certain other thermodynamic conditions (i.e., gas pressure and temperature), the gas flow can be continuous if $\mathrm{Kn} < 10^{-2}$, a slip flow if $10^{-2} < \mathrm{Kn} < 1$, or a free molecule flow if $\mathrm{Kn} > 1$ (Gradon, 1987). The mean free path of air at standard conditions is 0.066 microns; therefore, for nanofibers with diameters smaller than 0.5 microns, slip flow theory must be considered (Graham *et al.*, 2002). Due to the small diameter of nanofibers, fewer air molecules collided at the surfaces than with microfibers. The remaining air molecules that do not collide with the nanofiber surface retain their bulk flow motion and this is referred to as 'slip flow' (Chase, 2009). Slip flow improves the single fiber capture efficiency because

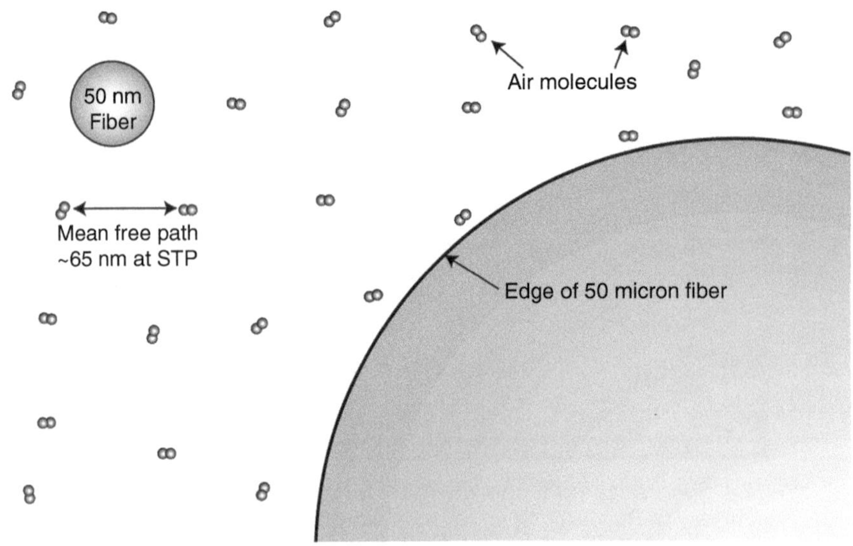

12.5 Slip flow improves the single fiber capture efficiency because the gas flow streamlines pass much closer to the surface of the nanofiber than a microfiber. (After Chase, 2009.)

the gas flow streamlines pass much closer to the surface of the nanofiber than the microfibers – see Fig. 12.5 (Graham *et al.*, 2002; Chase, 2009).

12.4 The advantages of using electrospun nanofibers in protective clothing

The physical characteristics of electrospun nanofibers indicate that the application of nanofibers in protective clothing is likely to be advantageous. Membranes made of submicron electrospun fibers are thin, have very small pore sizes and a high surface area to volume ratio. The application of electrospun nanofibers for protective clothing has the advantages of high penetration resistance of particles, good breathability (evaporative cooling) of the materials, low areal mass and low pressure drop.

In addition, the ability to add chemical or biological functionality to the fiber surface is also likely to improve the performance of the PPE, particularly if it can be coupled with a simple manufacturing process.

12.4.1 High resistance to particulate penetration

One of the advantages of electrospun nanofiber membranes is their high resistance to particulate penetration. As mentioned earlier, the single fiber capture efficiency of the nanofiber is superior to conventional microfibers due to the effect of slip flow phenomena. In addition, layers of randomly oriented electrospun nanofibers exhibit the characteristics of a porous material with very small pore sizes. Gibson *et al.* (2001) demonstrated the high penetration resistance of electrospun fibers against aerosol particles. Standalone electrospun polybenzimidazole (PBI) membranes, polyurethane (PU) membranes and electrospun nylon 6,6 deposited onto activated carbon PU foams were compared with commercially available polytetrafluoroethylene (PTFE) microporous membranes, polyvinylidene fluoride (PVDF) submicron filters and several protective fabrics. The filtration resistance of standalone electrospun membranes was found to be equivalent to PTFE and PVDF filters, showing no aerosol penetration, while electrospun nylon-PU foam filters showed a relationship between coating level and performance (Gibson *et al.*, 2001). It was also noted by Gibson *et al.* (2001) that the penetration resistance of electrospun fibers is superior to fabrics currently used by the US military, such as BDO fabrics and SARATOGA® liners.

Although in general there is a negative relationship between protection performance and air permeability, Lee and Obendorf (2006) showed that electrospun fibers could provide both high protection performance and high air permeability. High air permeability is as important as water vapor transmission in providing thermal comfort for hot and humid working conditions (Lee and Obendorf, 2006). In their study, PP was melt electrospun as a web and a laminate onto a nonwoven substrate prior to testing using pesticide mixtures comprising atrazine (2-chloro-4-ethylamino-6-isopropylamino-1,3,5-triazine) and pendimethalin (N-(1-ethylpropyl)-3,4-dimethyl-2,6-dinitrobenzenamine). ASTM F 2130 'Standard test method for measuring repellency, retention and penetration of liquid pesticide formulation through protective clothing materials' was used to evaluate the performance of the filters. Compared with commercial PPE materials, electrospun PP webs and laminates showed a combination of high protection performance with an acceptable level of thermal comfort (Lee and Obendorf, 2006).

Recently, Lee and Obendorf (2007) demonstrated similar properties in a test using solvent-based polyurethane (PU) electrospun fibers. The notion was that smaller fiber sizes in solvent-based electrospinning would exhibit better barrier performance against pesticide penetration. Similarly, it was noted that the inclusion of electrospun fibers significantly improved barrier performance of the filters while air permeability was still higher than most materials currently in use for protective clothing (Lee and Obendorf, 2007).

12.4.2 Good 'breathability' (evaporative cooling)

Breathability of the fabrics used in protective clothing is very important as it is directly related to wearer comfort. Breathability is the rate at which

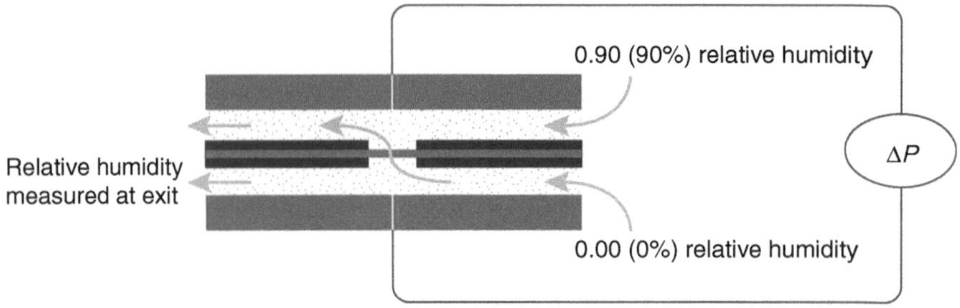

12.6 Dynamic Moisture Permeation Cell (DMPC) test method. (After Gibson *et al.*, 2000.)

water vapor molecules move through a clothing layer. There are a number of laboratory tests for measuring water vapor transport properties of fabrics. This includes 'ASTM E 96 Standard test methods for water vapor transmission of materials', 'ISO 11092 Measurement of thermal and water vapor resistance under steady-state conditions,' and 'BS 7209 Specification for water vapor permeable apparel fabrics.'

Air permeability is a measure of air that flows through the fabric. One of the methods used to measure air permeability is 'ASTM D 737 Standard test method for air permeability of textile fabrics.' In general, materials with high rates of water vapor diffusion and low air permeability are promising candidates for protective clothing applications (Gibson *et al.*, 1998).

A research group from the US Army Natick Research, Development and Engineering Center invented a method to measure the rate of moisture transmission called the dynamic moisture permeation cell (DMPC) (Gibson *et al.*, 2000). Two gas cocurrent streams at specified temperatures and humidities were pumped through two separate tunnels (see Fig. 12.6). A fabric sample was put in a portal between the two tunnels positioned in the tunnel midsection and parallel to the direction of gas flow, acting as a 'gate' between the two tunnels. The flow rates of gas and water vapor transported through the sample can be determined by measuring the temperature, water vapor concentration and flow rates of the gas at both exits. If there is no pressure difference between the two tunnels, water vapor transport is driven by pure diffusion. However, if pressure difference is present, water vapor and gas transport includes convective transport. DMPC is more advantageous than other methods as measurements of both water vapor diffusion and air permeability can be obtained from a single test (Gibson *et al.*, 2000; Schreuder-Gibson *et al.*, 2002). The method was later adopted as 'ASTM F2298 Standard test methods for water vapor diffusion resistance and air flow resistance of clothing materials using the dynamic moisture permeation cell'.

Gibson *et al.* (1998) made a comparison study using DMPC to determine the convective gas flow (air permeability) and moisture vapor transport between electrospun fibers and several commercially available filtration materials. Electrospun polyacrylonitrile (PAN) and PBI nanofiber membranes were used. It was reported that the convective flow resistance of PAN and PBI nanofibers were equivalent to the PTFE microporous membrane but still much higher than normal clothing materials. The moisture vapor transport properties of electrospun fibers were also equivalent to the PTFE membrane and better than other commercial 'breathable' membranes such as SympaTex® and GORE-TEX®. The results suggest that electrospun nanofibers could provide excellent protection against aerosol particles while still allowing evaporative cooling to take place.

Hussain *et al.* (2007), from Texas Tech University, made a comparison study of electrospun nanofiber composite fabrics. The specification for water vapor-permeable apparel fabrics BS 7209 test method was used in this experiment. Polyethylene oxide nanofibers were deposited onto different types of commercial fabrics producing composite Petryanov-type filter membranes. The fabrics used were denim, activated carbon and MicroMax®. Uncoated fabrics samples were prepared as controls. Although the moisture vapor transport rates for all composite membranes were lower than the control samples, the values were not considered to be significantly reduced. The reduction of moisture vapor transport rates were 5% for denim, 3% for activated carbon and 10% for MicroMax®. However, there was no investigation carried out on the particulate capture efficiency of the composite membranes.

12.4.3 Low basis weight and low pressure drop

The small diameter of electrospun membranes also means that effective layers of electrospun nanofibers are correspondingly very thin – just a few hundred microns in thickness. It has been calculated that surface area of nanofiber in a one square meter piece of poly vinyl alcohol nanofiber deposited 10 micron thick would be 40 square meters (Tucker and Stanger, 2010). Due to the porous structure and very small pore size of electrospun random mat materials, there is much evidence that only a small amount of fiber needs to be included to improve the efficiency of a filter. Gibson *et al.* (2001) have shown that even the lightest coating of about 0.001 kg/m^2 of electrospun fibers on polyurethane foam could capture 100% of aerosol particles passing through the foam. Lee and Obendorf (2007) also reported that with only 0.001 kg/m^2 inclusion of electrospun polyurethane fibers onto nonwoven substrate, the penetration of pesticide mixtures reduces from 85% to below 25%.

More recently, Vitchuli *et al.* (2010) studied the relationship between basis weight (the mass of a given area of a sheet material – also known as grammage or areal density), filtration efficiency, pressure drop and air permeability of electrospun nylon 6 nanofiber for protective applications. Nylon 6 in 2,2,2-tri-fluoro ethanol (12 wt%) solution was electrospun at 10 kV producing fibers with average diameter of 150 nm. Fibers were deposited onto a 50:50 nylon/cotton fabric substrate at varying deposition times giving different basis weights for each sample. It was reported that by applying electrospun fibers of about 6.5–8.5 g/m^2 (corresponding to basis weight of less than 3.4%), the filtration efficiency of the filter improved to 99.5% from an initial 38% in the control sample. In addition, the pressure drop and air permeability were not significantly changed compared with the control sample. The results suggest that only a small amount of nanofiber incorporation (low basis weight/areal weight) could provide significant improvement in filtration efficiency without sacrificing pressure drop or air permeability.

12.4.4 Surface functionalization

The ability to incorporate catalysts onto electrospun nanofibers offers interesting advantages in protective applications. Figure 12.7 summarizes how surface functionalization can be incorporated into nanofibers; note that the functional materials may be contained within the precursor material or added during or after fiber manufacture. There is evidence to suggest that selective reactive membranes demonstrate higher efficiencies because agents are adsorbed, not only physically but also chemically. The higher efficiency is due to an increase in net attractive force resulting from the chemical attraction between reactants and an increase in specificity – only reacting species will be chemically attracted to the fiber. Note that the surface may act as a catalytic site detoxifying the agents without being changed by the reaction. Thus the activity of the fiber will not be exhausted by contact with the target material. The functionality may be chemical in nature or biochemical. In the latter case a specific antibody could be bound to the fiber surface.

This chemical functionality also gives an additional benefit in reduction of the risk of secondary contamination related to disposal of used protective clothing. In addition, the superior performance of the functionalized membrane means that highly efficient protective fabrics could be designed using a minimum amount of materials producing lightweight, breathable and comfortable protective clothing.

A study of the use of electrospun nanofiber as a catalyst carrier was conducted by Graham *et al.* (2003). In this study, a catalyst polyoxometallate

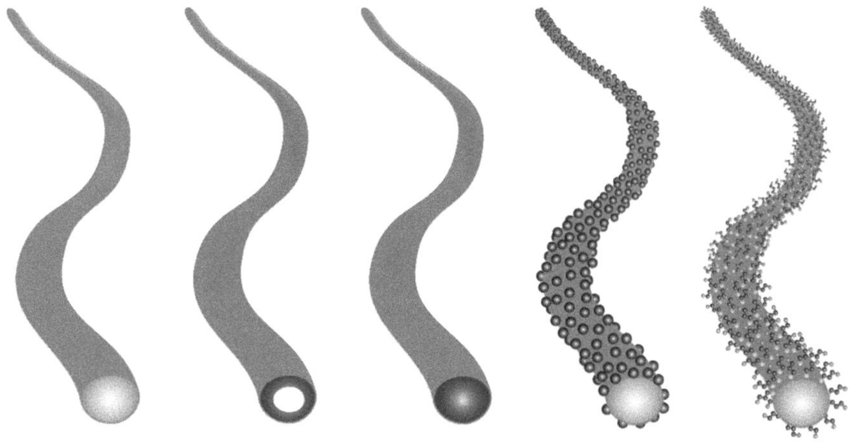

12.7 Methods of incorporation of surface activity into nanofibers. (a) Single component nanofiber; (b) encapsulated nanofiber formed by coaxial electrospinning – functional ingredient may be on the inside for slow release or on the outside surrounding a structural inner core; (c) nanofiber uniformly blended with functional material; (d) functional particulates on nanofiber surface and (e) functional molecules chemically attached to nanofiber surface.

(POM) facilitating the oxidative degradation of sulfur mustard vesicant chemical warfare agent was mixed at 1 wt% content into 10 wt% of polymer in organic solvent. The polymers used were polyurethane (TPU) (Estane 58238, Noveon Inc.) and Donaldson elastomer, FP-10. The final dry electrospun fibers contained 10 wt% POM. The nanofibers were then placed in a solution containing a mustard simulant, 2-chloroethylethylsulfide (CEES). It was reported that the reactive FP-10 nanofibers depleted the most CEES in a 24-h period. The depletion rate was faster than the reactive TPU fibers or the POM alone. Incorporation of catalysts directly into the electrospun nanofibers enhances the reactivity of the catalysts, rather than diluting their effectiveness within the fiber bulk (Graham *et al.*, 2003).

Ramaseshan *et al.* (2006) studied the use of a novel catalyst synthesized from β-cyclodextrin (β-CD) and o-iodosobenzoic acid (IBA) against paraoxon, a model material analogous to the sarin organophosphorus nerve agent. Poly(vinylchloride) (PVC) filter membrane were prepared with the inclusion of β-CD, IBA, a blend of β-CD + IBA and the newly synthesized catalyst. UV spectroscopy measurement was used to measure the degree of hydrolysis of the paraoxon. The results were compared with activated charcoal, which is currently used in protective clothing. It was observed that, among other catalysts, electrospun PVC nanofibers incorporated with the newly synthesized catalyst showed the fastest hydrolysis rate of paraoxon, 11.5 times faster than activated charcoal. The fast hydrolysis reaction of paraoxon was attributed to the presence of (3-carboxy-4-iodosobenzyl)-oxy-β-cyclodextrin molecules which react with paraoxon producing p-nitrophenol and diethyl phosporic acid (Ramaseshan *et al.*, 2006).

Sundarrajan and Ramakrishna (2007) studied the incorporation of magnesium oxide (MgO) and aluminum oxide (Al_2O_3) nanoparticles onto electrospun nanofibers, producing nanocomposite membranes for protection against paraoxon nerve agent. To confirm the suitable carrier for MgO and Al_2O_3, three polymers were used; polysulfone (PSU), poly(vinylidene fluoride-co-hexafluoropropylene) (PVDF copolymer) and PVC. All the functionalized and nonfunctionalized membranes were tested using UV spectroscopy to determine the degree of paraoxon hydrolysis in comparison with conventional activated charcoal. Based on experimental results, the most reactive membrane against paraoxon agent was MgO incorporated into electrospun PSU nanofibers. The rate of hydrolysis was found to be twice that observed with conventional activated charcoal.

12.4.5 Simplicity of the process

From a manufacturing perspective, electrospinning offers attractive advantages over other processing techniques. A number of manufacturers claim to be able to produce fibers in reasonable industrial quantities. Electrospinning is a simple, versatile and scalable process for making nanoscale materials. In its basic form, the electrospinning process consists of two electrodes; one is connected to a high voltage DC power supply and a polymer delivery system while the other acts as a grounded collector. In most cases the process is operated in a normal 'shirt-sleeves' environment, the main health and safety measure being to make sure that operators are insulated from the high voltages used to form the fibers.

The process is also versatile, providing the capacity to 'lace' together different types of polymers, fibers and particles to produce various ultrathin membranes (Gibson *et al.*, 2001). There are considerably more than 50 different polymer solutions which have been successfully electrospun into nanofibers (Huang *et al.*, 2003) with a vast variety of polymer–polymer or polymer–additive combinations available for specific functionalities. The versatility of the process also allows nanofibers to be incorporated onto existing fabric systems producing composite protective fabrics (Graham *et al.*, 2003). The thickness of the electrospun nanofiber membrane is controlled by varying the fiber deposition time. It has been reported by Schreuder-Gibson *et al.* (2002) that air flow resistance and aerosol filtration efficiency can be controlled by varying the thickness of electrospun membrane coatings.

Electrospun nanofibers have also been envisioned to be directly deposited onto three-dimensional forms obtained from 3-D body scanning. By doing this, it is claimed that the seam-sealing problems currently faced by protective clothing manufacturers are eliminated (Gibson *et al.*, 2001).

Compared with other processing methods such as drawing, template synthesis, phase separation and self-assembly, electrospinning is a readily scalable process (Ramakrishna *et al.*, 2005), hence the output could be increased to any required level by increasing the resources or enhancing the techniques. It should be noted that the Soviet industry developed high throughput technologies to address exactly this problem (Filatov *et al.*, 2007). Other advantages of the single-nozzle electrospinning process are repeatability of the process and the ability to control the fiber dimensions (Ramakrishna *et al.*, 2005). However, some high-throughput designs trade throughput against precise control of fiber diameter.

12.5 Issues with development of electrospun fibers

Despite the advantages of electrospinning, there are certain disadvantages associated with electrospun nanofibers. These include low productivity of the process, limited mechanical properties and control of fiber amount and pore size.

One of the main problems with electrospinning is the low productivity of the process. In the typical electrospinning single-nozzle process, the normal feed rate of the polymer is a few milliliters per hour. This means that, for a solution concentration of 0.2 g/mL, the mass of nanofiber collected from a single nozzle in an hour is only 0.3 g (Ramakrishna *et al.*, 2005). Therefore, there is a need to increase the productivity of the process to make it competitive. Currently, there are industrial-scale electrospinning machines available which could raise fiber production levels to acceptably large rates by using either multiple delivery nozzles, as for example the NanoFMG Nanospinner 416 which has an array of 256 individual spinnerets (NanoFMG, 2011), or by needleless designs such as the Elmarco Nano spider (Petras *et al.*, 2010).

Electrospun nanofibers have low mechanical properties on the macroscale – an individual fiber may have a high specific strength due to its highly oriented structure and a low number of mechanical defects, but as an ultrathin fiber, the load to failure will be correspondingly low. For this reason and because of the desirable property of adhesion that would rapidly attract sufficient surface contamination to prevent the operation of the fiber in the manner desired, in an application such as protective clothing, the thin layer of nanofibers cannot operate as a stand-alone filtration material. However, it has been demonstrated that an electrospun membrane could be incorporated onto a substrate which would act as a protective material support – the original 1938 Petryanov filter being a case in point. Graham *et al.* (2003) studied the incorporation of electrospun nylon nanofiber onto the existing layered structure of the JSLIST fabric. The best durability performance was achieved using two layers of nanofibers sandwiched between two layers of spunbond material. This combination protected the nanofibers from direct abrasion and provided a dimensionally stable supporting structure (Graham *et al.*, 2003).

It is the control of pore size which determines the effectiveness and flow-related pressure drop through the membrane. For fibrous filters, Filatov *et al.* (2007) say that these attributes are dependent on the structure of the fiber – that is, fiber diameter, packing density, thickness and uniformity of the fibers. Although small pore size can be achieved by increasing packing density and thickness, it will also inevitably increase the pressure drop across the filter membrane. Ideally it is desirable to achieve small pore size while maintaining an acceptable level of packing density and fiber thickness. One of the solutions to this is controlling the deposition pattern of the nanofiber. Typical electrospinning processes produce random-oriented fibers due to the chaotic nature of the process. Several methods have been used to achieve controlled deposition including the use of moving collectors. However, there are disadvantages associated with the moving collector method due to the limitations imposed by the nature of the moving components. The mass of moving collector systems generally limits the speed of response to below the level required for direct control of precise fiber deposition. This type of system is generally employed to generate even deposition of randomly oriented fibers over large areas. Uneven deposition can result from the mutual repulsion of fibers emitted from multiple-spinneret designs. Passive collector arrays are also used, but these systems tend to produce fibers aligned in single directions. Recently, a nonmoving component method of controlled deposition has been demonstrated using the principle of electric field superposition (Nurfaizey *et al.*, 2012).

12.6 Future trends

Recently electrospun nanofibers have been given a great deal of attention in protective clothing applications. This is mainly attributed to their unique properties of small diameter and high surface area. Due to nanoscale diameter and kilometer range length, electrospun nanofibers are seen as perfect intermediaries between nanoscale and bulk materials. To date, there have been a significant number of feasibility studies related to protective application, some of which have been discussed earlier in this chapter. There is no doubt that in the future this field will continue to advance toward the production of even more efficient protective fabrics. There are still many opportunities to improve, including the incorporation of new active chemicals or biological structures onto the nanofiber surfaces during spinning or through postprocessing techniques.

An interesting example of this technique has been demonstrated by a research group from Cornell University. It is known that Ag nanoparticles have a biocidal effect against both gram-negative and gram-positive bacteria. However, a challenge remains on how to disperse these nanoparticles uniformly on the surfaces of electrospun fiber. Uniform dispersion of Ag nanoparticles on the surface of nanofibers is important to maximize the available area for interaction. Conventionally, these nanoparticles tend to undergo aggregation depending upon the viscosity of the polymer (Lala *et al.*, 2007). Recently, Dong *et al.* (2008) have found a way to overcome this problem. A novel one-step process for uniformly assembling the Ag and other nanoparticles on the surface of electrospun nylon 6 nanofibers has been developed through a pH-controlled assembly method. The basis of the method is to improve the hydrogen bonding between nylon 6 and Ag nanoparticles containing carboxylic acid groups. It was reported that by lowering the pH value of the Ag nanoparticle suspension before the immersion process, at some stage the COONa groups in Ag solution are acidified, becoming COOH. This COOH can be bridged to the amide groups on the surface of electrospun nylon 6 fibers forming intermolecular hydrogen bonds (Dong *et al.*, 2008).

Another step forward on functionalization of electrospun fiber surfaces has been shown by Chen *et al.* (2010). The use of a postelectrospinning assembly method has the advantages of higher surface density of functional groups, enhanced reactivity and ease of implementation compared with the conventional polymer blending method (Chen *et al.*, 2010). Layer-by-layer (LBL) technology was used to electrostatically assemble chemically reactive polyanion, polyhydroxamic acid (PHA) and an antimicrobial polycation, poly-(*N*-vinylguanidine) onto electrospun PAN nanofibers. The advantage of using LBL technology is that one could engineer multifunctional fabrics by selecting appropriate materials for both chemical and biological protection (Chen *et al.*, 2010). The treated electrospun PAN fiber samples were tested using diisopropyl fluorophosphate (DFP) as the chemical agent and a gram-negative strain of *E. coli* and a gram-positive strain of *S. epidermidis* as bacterial contaminants. Test results showed that LBL PHA/PVG-treated nanofiber is 60-fold faster in DFP degradation than untreated samples, while having up to 99.9% killing efficiency against the test bacteria.

It should be noted that current published literature usually establishes the efficacy of functionalized nanofibers by indirect means. The active ingredient is demonstrated to act against the toxic agent. The presence of the active ingredient in the nanofiber is shown, and as a final stage the action of the active ingredient in the nanofiber may be demonstrated by placing a sample onto a culture plate and determining the kill radius as the active ingredient leaches out, in the case of antimicrobial activity. If chemical activity is under scrutiny, then a sample of fabric is placed in solution and change in chemical composition of the solution is measured. It is clear that either of these methods departs from the conditions to be expected in the field. If a vesicant agent biological pathogen is adsorbed onto the active membrane, will it then be neutralized? Will the effect of the human body on the fabric make a difference? The material must remain consistently active if it is positioned in a warm, moist biologically active area such as under the arms, or in a cooler, drier, more exposed area as, say, the back. In summary, little has been published on the efficiency of nanofiber PPE in the field. This is perhaps for reasons of security, or because of the difficulties in assessing performance under nonlaboratory conditions. Nevertheless, for whatever reason, there remains a gap in the knowledge of the actual performance of these materials.

For the future it is likely that the intrinsic physical properties of electrospun nanofibers will be supplemented by the addition of chemical and biochemical functionality. The noninvasive nature of the manufacturing process, taking place at room temperature and not requiring coagulation chemistry, is particularly suited to working with complex chemical moieties. The high specific surface area provides a large reaction site – intrinsically greater than that obtained by embedding particulates into the fiber. If biological polymers are used as the spinnable substrate then it is entirely possible that biochemical functionality can be engineered to be intrinsic to the fiber. The nature of this functionality will naturally include mechanisms that render external toxins and pathogenic organisms harmless to the wearer. In addition to this primary functionality the fiber may also cope with effluvia from the wearer, and make it possible for the garment to be worn for an extended period while remaining in a sanitary condition. It will also be possible to incorporate sensor technology into the garment; this will warn the wearer if toxins are present. It should also be noted that the potential for long nanofiber composites means that a nanofiber battle dress may well provide protection against physical impact. It may also be possible for the nanofiber component to assist in wound control. Nanofiber dressing materials assist the healing process by aiding the mechanisms of coagulation to control bleeding, and by staged delivery of appropriate disinfectants, antibiotics and healing promoters.

12.7 References

Agency for Toxic Substances and Disease Registry (2011), *Medical management guidelines for hydrogen cyanide (HCN)*, Anon., Agency for Toxic Substances and Disease Registry (Online), available from: http://www.atsdr.cdc.gov/mmg/mmg.asp?id=1073andtid=19 (accessed 24 January 2011).

Anon. (1969), CBW: Meselsohn's evidence to Fulbright, *Nature*, **223**, 6.
Bajgar, J., Fusek, J., Kassa, J., Kuca, K. and Jun, D. (2009), Global impact of chemical warfare agents used before and after 1945, in R. C. Gupta (ed.), *Handbook of toxicology of chemical warfare agents*, London: Academic.
Bansal, R. C. and Goyal, M. (2005), *Activated carbon adsorption*, Boca Raton, LA, London, Taylor and Francis.
Blücher GmbH (2010), *JSLIST* (Online), Blücher GmbH, available from: http://www.bluecher.com/en/brands/saratoga/military/jslist/ (accessed 20 November 2010).
Blucher, H., Von Blucher, H. and De Ruiter, E. (1985), *Filter sheet material*, United States patent application 4510193.
Chase, G. (2009), *What is slip-flow and how does it affect filter performance?* (Online). American Filtration and Separations Society, available from: http://www.afssociety.org/education/0810oneminute.html (accessed 24 December 2010).
Chen, L., Bromberg, L., Schreuder-Gibson, H., Walker, J., Hatton, T. A. and Rutledge, G. C. (2009), Chemical protection fabrics via surface oximation of electrospun polyacrylonitrile fiber mats, *Journal of Materials Chemistry*, **19**, 2432–8.
Chen, L., Bromberg, L., Lee, J.A., Zhang, H., Schreuder-Gibson, H., Gibson, P., Walker, J., Hammond, P. T., Hatton, T. A. and Rutledge, G. C. (2010), Multifunctional electrospun fabrics via layer-by-layer electrostatic assembly for chemical and biological protection, *Chemistry of Materials*, **22**, 1429–36.
Cooley, J. F. (1900), *Improved methods of and apparatus for electrically separating the relatively volatile liquid component from the component of relatively fixed substances of composite fluids*, United Kingdom patent application 6385.
Dong, H., Wang, D., Sun, G. and Hinestroza, J. P. (2008), Assembly of metal nanoparticles on electrospun nylon 6 nanofibers by control of interfacial hydrogen-bonding interactions, *Chemistry of Materials*, **20,** 6627–32.
E. I. Du Pont De Nemours and Company (2011), *DuPont™ Hybrid Membrane Technology* (Online), DuPont, available from: http://www2.dupont.com/Separation_Solutions/en_US/assets/downloads/DHM702_HMTSheetRev_me05.pdf (accessed 10 January 2011).
Ellison, D. H. (2008), *Handbook of chemical and biological warfare agents*, Boca Raton, FL, CRC Press LLC.
Filatov, Y., Budyka, A. and Kirichenko, V. (2007), *Electrospinning of micro and nanofibers: Fundamentals and applications in separation and filtration processes*, New York, Begell House Inc.
Formhals, A. (1934), *Process and apparatus for preparing artificial threads*, United States patent application 1975504.
Frischknecht, F. (2003), The history of biological warfare: Human experimentation, modern nightmares and lone madmen in the twentieth century, *Embo Reports*, **4**, S47–52.
Gibson, P., Rivin, D. and Kendrick, C. (2000), Convection/diffusion test method for porous textiles, *International Journal of Clothing Science and Technology*, **12**, 96–113.
Gibson, P., Schreuder-Gibson, H. and Pentheny, C. (1998), Electrospinning technology: direct application of tailorable ultrathin membranes, *Journal of Coated Fabrics*, **28**, 63–72.

Gibson, P., Schreuder-Gibson, H. and Rivin, D. (2001), Transport properties of porous membranes based on electrospun nanofibers, *Colloids and surfaces a-physicochemical and engineering aspects*, **187**, 469–81.

Gradon, L. (1985), Influence of electrostatic interactions and slip effect on aerosol filtration efficiency in fiber filters, *Industrial & Engineering Chemistry Research*, **26,** 306–11.

Graham, K., Gogins, M. and Schreuder-Gibson, H. (2003), Incorporation of electrospun nanofibers into functional structures, 2003 International Nonwovens Technical Conference, Baltimore, Maryland.

Graham, K., Ouyang, M., Raether, T., Grafe, T., McDonald, B. and Knauf, P. (2002), Polymeric nanofibers in air filtration applications, 15th Annual Technical Conference and Expo of the American Filtration and Separations Society, Galveston, Texas.

Greiner, A. and Wendorff, J. H. (2007), Electrospinning: a fascinating method for the preparation of ultrathin fibres, *Angewandte Chemie International Edition*, **46**, 5670–703.

Gulati, K. and Ray, A. (2009), Immunotoxicity, in R. C. Gupta (ed.), *Handbook of toxicology of chemical warfare agents*, London, Academic.

Hilmas, C. J., Katos, A. M., Williams, P. T. and Anderson, J. (2009), Anthrax, in R. C. Gupta (ed.), *Handbook of toxicology of chemical warfare agents*, London, Academic.

Huang, Z. M., Zhang, Y. Z., Kotaki, M. and Ramakrishna, S. (2003), A review on polymer nanofibers by electrospinning and their applications in nanocomposites, *Composites Science and Technology*, **63**, 2223–53.

Hussain, M., Purushothaman, A. and Ramkumar, S. (2007), Breathability and protection studies of nanofiber composite fabrics, 2007 International Nonwovens Technical Conference, Atlanta, GA.

Lala, N. L., Ramaseshan, R., Li, B. J., Sundarrajan, S., Barhate, R. S., Liu, Y. J. and Ramakrishna, S. (2007), Fabrication of nanofibers with antimicrobial functionality used as filters: Protection against bacterial contaminants, *Biotechnology and Bioengineering*, **97**, 1357–65.

Laurie, P. (1972), *Beneath the city streets: The fundamental function of civil defence is the preservation of government*, Harmondsworth, UK, Penguin.

Lee, S. and Obendorf, S. K. (2006), Developing protective textile materials as barriers to liquid penetration using melt-electrospinning, *Journal of Applied Polymer Science*, **102**, 3430–7.

Lee, S. and Obendorf, S. K. (2007), Use of electrospun nanofiber web for protective textile materials as barriers to liquid penetration, *Textile Research Journal*, **77**, 696–702.

Marsh, H. and Rodrìguez-Reinoso, F. (2006), *Activated carbon*, Amsterdam, London, Elsevier.

NanoFMG (2011), *Nanospinner 416* (Online), NanoFMG, available from: http://www.nanofmgroup.com/Nanospinner_eng.html (accessed 17 January 2011).

Nurfaizey, A. H., Stanger, J., Tucker, N., Buunk, N., Wallace, A., and Staiger, M. P. (2012), Manipulation of electrospun fibres in flight: The principle of superposition of electric fields as a control method, *Journal of Materials Science*, **47**, 1156–1163.

Okumura, T., Taki, K., Suzuki, K. and Satoh, T. (2009), The Tokyo subway sarin attack: Toxicological whole truth, in R. C. Gupta (ed.), *Handbook of toxicology of chemical warfare agents*, London, Academic.

Petras, D., Maly, M., Pozner, J., Trdlicka, J. and Kovac, M. (2010), *Rotary spinning electrode*, United States patent application 20100034914.

Ramakrishna, S., Fujihara, K., Teo, W. E., Lim, T. C. and Ma, Z. (2005), *An introduction to electrospinning and nanofibers*, Hackensack, NJ, London, World Scientific.

Ramaseshan, R., Sundarrajan, S., Liu, Y. J., Barhate, R. S., Lala, N. L. and Ramakrishna, S. (2006), Functionalized polymer nanofibre membranes for protection from chemical warfare stimulants, *Nanotechnology*, **17**, 2947–53.

Reneker, D. H. and Chun, I. (1996), Nanometre diameter fibres of polymer, produced by electrospinning, *Nanotechnology*, **7**, 216–23.

Satoh, T. (2009), Foreword, in R. C. Gupta (ed.), *Handbook of toxicology of chemical warfare agents*, London, Academic.

Schneider, B. R. and Mayer, T. N. (1995), *Biological weapons: The poor man's nuke* (Online), Air War College, Maxwell Air Force Base, Alabama, available from: http://handle.dtic.mil/100.2/ADA328852 (accessed 10 January 2011).

Schreuder-Gibson, H. L., Truong, Q., Walker, J. E., Owens, J. R., Wander, J. D. and Jones, W. E. (2003), Chemical and biological protection and detection in fabrics for protective clothing, *MRS Bulletin*, **28**, 574–8.

Schreuder-Gibson, H., Gibson, P., Senecal, K., Sennett, M., Walker, J., Yeomans, W., Ziegler, D. and Tsai, P. P. (2002), Protective textile materials based on electrospun nanofibers, *Journal of Advanced Materials*, **34**, 44–55.

Stephenson, C. (2006), *The Admiral's secret weapon: Lord Dundonald and the origins of chemical warfare*, Woodbridge, Boydell Press.

Stroud, G. (1999), Chemical wear for the EAF, *The CE, Civil Engineer: United States Air Force*, **7**, 12–3.

Sundarrajan, S. and Ramakrishna, S. (2007), Fabrication of nanocomposite membranes from nanofibers and nanoparticles for protection against chemical warfare stimulants, *Journal of Materials Science*, **42**, 8400–7.

Taylor, G. (1969). Electrically driven jets, *Proceedings of the Royal Society of London A: Mathematical, Physical and Engineering Sciences*, **313**, 453–75.

The United States Army (2011), *Joint Service Lightweight Integrated Suit Technology (JSLIST)* (Online), The United States Army, available from: http://www.army.mil/factfiles/equipment/nbc/jslist.html (accessed 10 January 2011).

Tucker, N. and Stanger, J., Vitchuli, N., Shi, Q., Nowak, J., Mccord, M., Bourham, M. and Zhang, X. W. (2010), Electrospun ultrathin nylon fibers for protective applications, *Journal of Applied Polymer Science*, **116**, 2181–7.

Von Blucher, H. and De Ruiter, E. (1984), *Filter sheet material and method of making same*, US patent application 4455187.

Von Blucher, H. and De Ruiter, E. (1994), *Filter sheet material*, US patent application 5277963.

Watson, A., Opresko, D., Young, R., Hauschild, V., King, J. and Bakshi, K. (2009), Organophosphate nerve agents, in R. C. Gupta (ed.), *Handbook of toxicology of chemical warfare agents*, London, Academic.

13
Functional nanofibers in food processing

J. A. LOPES DA SILVA,
University of Aveiro, Portugal

Abstract: Electrospinning enables the production of nano- and microfibers with very large surface-to-volume ratio, high porosity and interconnectivity, with easy chemical functionalization, which makes them very attractive for a wide range of applications. This chapter discusses recent developments on the preparation and application of electrospun nanofibers in food-related areas, including filtration and other separation strategies, enzyme immobilization, enzymatic membrane reactors, sensors and encapsulation of food bioactive compounds and cells.

Key words: electrospinning, micro- and nanofibers, separation techniques, enzyme immobilization, food sensors, encapsulation, bioactive compounds, food industry.

13.1 Introduction

Nanotechnology has assumed increasing importance in food technology and it is anticipated that this importance will grow strongly in coming years. The driving force for this development stems from the enormous potential benefits of nanotechnology in specific areas of food science and technology, both in food processing and safety. Most progress so far achieved has been in food quality monitoring, food packaging and encapsulation and delivery of nutraceuticals. Recent reviews have been published highlighting the latest developments and future directions for the application of nanotechnology in the food area (Weiss *et al.*, 2006; Sozer and Kokini, 2009; Sekhon, 2010; Neethirajan and Jayas, 2011).

Examples of application of nanotechnology in food include the intelligent and bioactive packaging, which are booming research topics all over the world. This will not be directly addressed in this chapter, but the reader can find information on the subject in several excellent recent review articles (e.g., Lopez-Rubio *et al.*, 2006; Sanchez-Garcia *et al.*, 2010; Lagaron and Lopez-Rubio, 2011)

The primary focus of this chapter will be on nano- and microfiber networks produced by the application of high electrostatic potential, a process known as electrospinning, with current or potential application in the food industry. The recent advances in this technique have greatly contributed to the development of functional nanofibers, with applications in diverse areas. The reader may find additional information in some of the excellent reviews that have appeared highlighting prominent aspects of this technique (Frenot and Chronakis, 2003; Burger *et al.*, 2006; Greiner and Wendorff, 2007; Kriegel *et al.*, 2008; Lu *et al.*, 2009) and also in other chapters of this book (Chapters 1 and 2).

In brief, a polymer solution is supplied from a spinneret and forms a droplet at the spinneret exit. Under an applied electrical field, typically of the order of 10^3 V/cm, the electric charges accumulate on the surface of the droplet and cause droplet deformation along the field direction, finally assuming the shape of a cone, known as the Taylor cone. When the field strength overbears the surface tension, a fluid jet emanates under the formed electric field across the spinneret tip and accelerates toward the conductive collector (counter electrode). The high surface charge of the jet under the electric field causes whipping and circular bending motions, a strong elongation of the polymer charged jet and usually a rapid evaporation of the solvent, originating solid thin fibers that are deposited on the collecting target in the form of nonwoven mats. The individual fibers are typically large in length but with small diameters in the range 100 nm to 1 μm, fully interconnected, forming porous three-dimensional membranes.

Nanostructured bi- and tridimensional materials can be obtained with very large surface-to-volume ratio, high porosity and interconnectivity, making them very attractive for applications like filtration, enzyme immobilization, enzymatic membrane reactors, sensors or delivery matrices for nutraceuticals, to refer to those qualities with most interest in the food area. These materials are not only physically versatile, but also different chemical functionalities may be easily introduced through the use of different polymers, polymer blends, nanocomposites made of organic or inorganic materials or by including bioactive ingredients (in emulsions or micellar structures). In addition, post-electrospinning treatments may also be used to functionalize the outer fiber surfaces – for example, by physical adsorption, chemical surface treatments, plasma treatments, surface grafting of different compounds including polymers or a variety of physical coating procedures. Figure 13.1 shows examples of fiber sizes and morphologies that can be produced by electrospinning, to illustrate the diversity of electrospinnable structures, which, associated to the unique characteristics of the electrospun fibers and to the simplicity of the electrospinning fabrication process, make the interest in and potential applications of this technique and resultant materials expand every day.

When considering food applications, an important aspect to consider concerns the toxicity of the obtained materials or solvents used. Despite the versatility of the electrospinning technique regarding the different types of polymers and solvents that may be used, the frequent use of organic solvents and synthetic polymers raises concerns about the risk of toxicity, especially if the materials are intended for contact with food or, of course, if one wishes to formule edible materials. These aspects have received recent interest and research

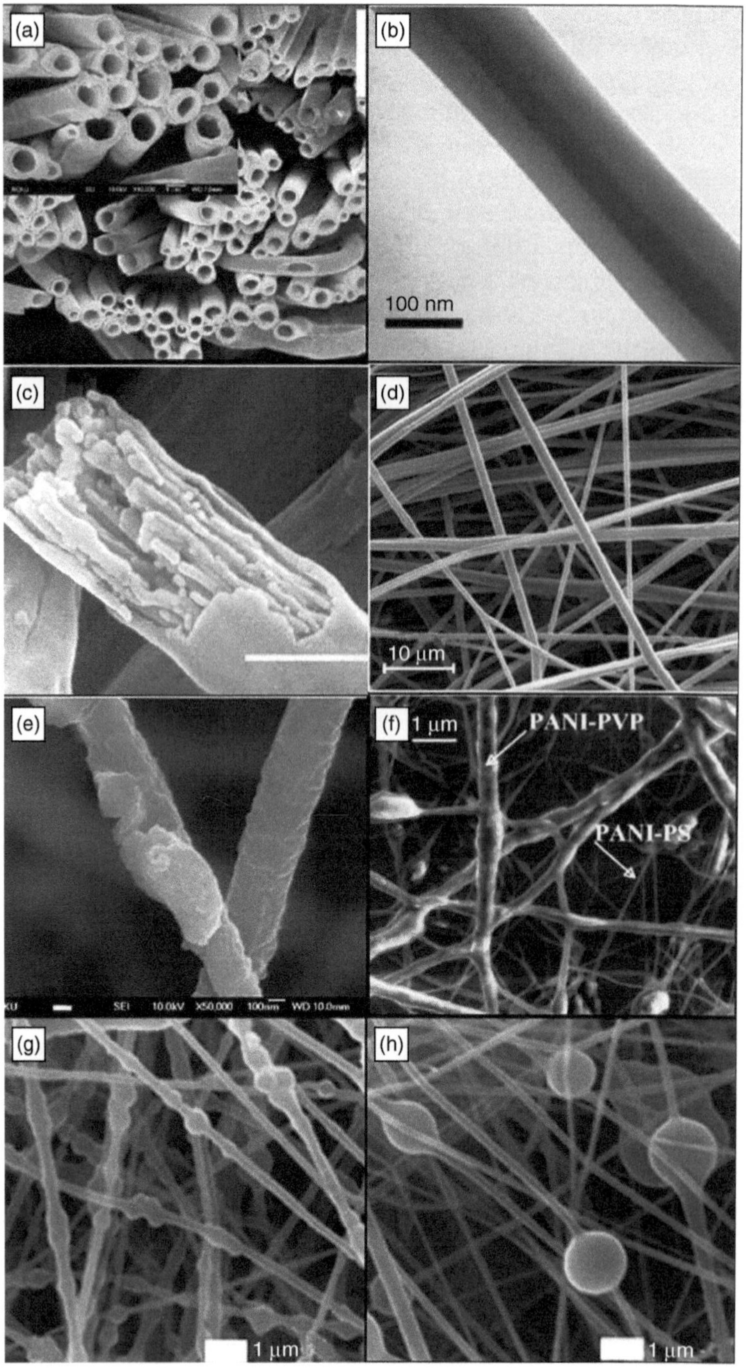

13.1 Examples of fiber sizes and morphologies that can be produced by electrospinning using different polymers, polymer-inorganic

Figure 13.1 (Continued)

composite solutions, spun dope compositions and different electrospinning setups. (a) SEM images of PDLLA/PHB core/shell fibers after extracting the PDLLA core component using DMF (Reprinted with permission from Wang *et al.*, 2010a. Copyright 2010 American Chemical Society), (b) TEM image of a core-sheath polybutadiene/polycarbonate nanofiber (Reprinted from Wei *et al.*, 2006. Copyright 2006, with permission from John Wiley & Sons, Inc.), (c) SEM images of TiO_2/PVAc electrospun nanofibers after calcination at 450°C and removal of the organic component (Reprinted from Kim and Rothschild, 2011. Copyright 2010, with permission from John Wiley & Sons, Inc.), (d) SEM image showing the morphology of hybrid fibers electrospun from PET and chitosan (M_v12 kDa) solutions in trifluoroacetic acid / dichloromethane (TFA/DCM) 80:20, the average fiber diameter was 268 nm (Reprinted with permission from Lopes da Silva *et al.*, 2009. Copyright © 2009 American Scientific Publishers), (e) SEM micrograph of polyacrylonitrile nanofibers with immobilized lipase (magnification 50 000×) (Reprinted from Li *et al.*, 2007b. Copyright 2007, with permission from Elsevier), (f) SEM micrograph of two overlaid nanofibrous mats composed by PANi-PS and PANi-PVP (under- and overlayer, respectively) (Reprinted from Zampetti *et al.*, 2011. Copyright 2010, with permission from Elsevier), (g) SEM images of PET electrospun nanofiber membranes containing 50% of molecularly imprinted nanoparticles (130 nm) and (h) PET electrospun nanofibers containing 50% of nonimprinted microspheres (1.4 μm); the composite nanofibers were proposed to be used as affinity membrane to improve and simplify solid- phase extraction of drug residues in analytical samples (Reprinted from Yoshimatsu *et al.*, 2008. Copyright 2007, with permission from Elsevier).

efforts, including the optimization of modified electrospinning methods and the use of more friendly solvents and polymers. Reference should be made, to cite only a few examples, to the so-called 'green electrospinning' approach advocated by Agarwal and Greiner (2011), the increasing use of polysaccharides and proteins to prepare electrospun nanofiber materials (Stijnman *et al.*, 2011), or the specific preparation of nanofibers from edible biopolymers for the potential encapsulation of bioactive compounds and their use in nutraceutical formulations and in food packaging materials, ensuring the safety and quality of the food product (Fernandez *et al.*, 2009; Wongsasulak *et al.*, 2010).

In the next sections of this chapter, focus will be on four main food areas of application of these functional micro- and nanofibers: separation operations, enzyme immobilization, application in different types of sensor and encapsulation of food bioactive compounds and cells; the most recent developments and future expected trends will be considered.

13.2 Electrospun fibrous mats in food-related separation applications

Separation techniques are day-to-day operations in the food industry, from trivial mechanical separations during processing (centrifugation, membrane filtration, extraction) to more sensitive and specific analytic separations involved in affinity filtrations or solid-phase extractions (SPEs).

Membrane filtration technologies are now widespread in the food industry – even in more traditional ones such as wine making and brewing, where they are used mainly for clarification and concentration purposes. Different types of membranes (hollow fiber, spirally wound, tubular) can be used, not only within operations during production but also in the preparation of food ingredients and in the recycling or treatment of waste streams (Grandison and Lewis, 1996; Skelton, 2000). Examples of applications are also diverse – for instance, microfiltration and ultrafiltration membranes are used for clarification and removal of suspended matter, or even for pasteurization instead of using heat, and for fractionation, concentration and purification, to remove bacteria from skimmed milk, or to concentrate whey protein components, to clarify wine, vinegar and fruit juices. The dairy and fruit juice industries use reverse-osmosis membranes essentially to concentrate liquid products, such as whey or skimmed milk, or to transform normal fruit juices into syrups of a much higher solid content. Ultrafiltration/reverse osmosis combinations and also nanofiltration are largely used to produce ultrapure water, to treat waste water effluents, or for desalting (e.g., to produce low-sodium drinks) and deacidification (e.g., of cheese whey). Other filtration technologies are also important to control airborne and waterborne contaminants, hazardous biological agents, allergens and pollutants, which is of key importance for the food industry.

There are many opportunities for new improvements and the application of nanostructured materials are expected to play an important role in the near future.

13.2.1 Filtration and clarification applications

Electrospun nonwoven fibrous membranes possess characteristics that make them good candidates for filtration applications: a network of continuous interconnected ultrafine fibers with high porosity and small pore sizes (ranging from tens of nanometers to several micrometers) and high specific surface area. In fact, some of the first applications of electrospun nanofiber mats were as filtration membranes, and the first commercialized electrospun materials were in air filters (Barhate and Ramakrishna, 2007).

The good performance of these membranes for air filtration has been shown in numerous research papers, patents and commercialization cases (Graham *et al.*, 2002; Ahn *et al.*, 2006; Barhate and Ramakrishna, 2007). Electrospun nanofibrous membranes have also been studied for pressure-driven liquid filtrations and particulate removal, as micro-, ultra- or nanofiltration membranes (Gopal *et al.*, 2006; Barhate and Ramakrishna, 2007). They were shown to be effective as prefilters to minimize fouling and contamination prior to ultra- or nanofiltration or as filters to pretreat water previous to reverse osmosis (Gopal *et al.*, 2006, 2007; Aussawasathien *et al.*, 2008).

In the case of liquid filtrations, the electrospun nanofibrous membranes are expected to overcome some of the limitations associated to the porous polymeric membranes produced by conventional methods (e.g., low-flux and high-fouling performance) due to their large pore volume (providing high permeability), their highly interconnected pore structure (enabling the fibrous membranes to better withstand fouling) and large surface area to volume ratio. They were shown to replace flux-limiting asymmetric porous ultrafiltration membranes (Wang *et al.*, 2005; Yoon *et al.*, 2006), often exhibiting similar nanoparticle penetration values as commercial filter membranes but at lesser filter mass (Yun *et al.*, 2007); it is possible also that they might exhibit high ability to remove bacteria (by size exclusion) and viruses (by adsorption) simultaneously (Ma *et al.*, 2011).

Electrospun nanofiber membranes have typically limited mechanical properties, which preclude their easy handling, and they often accumulate electrostatic charges during the electrospinning process; this can be advantageous for certain applications like removal of aerosol particles by electrostatic interactions, but in many cases can cause processing difficulties. Deposition on different substrates or preparation of nanofiber composite laminates have been used to alleviate these problems. Nanofiber webs have been applied onto various porous substrates to provide appropriate mechanical properties (allowing easier pleating, filter fabrication and higher durability), as usually used in air filtration (Graham *et al.*, 2002; Fig. 13.2), but also for improved liquid filtration (Ma *et al.*, 2011). Other nanofiber composite laminates have been developed exhibiting high-flux rate, higher rejection

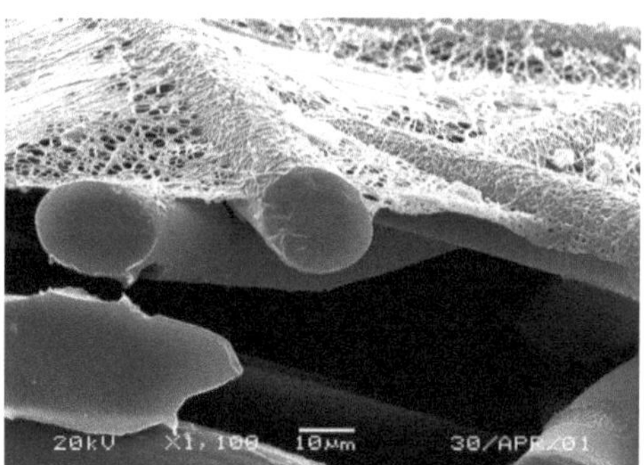

13.2 Electrospun nanofiber web on a polyester substrate developed for air filtration applications; the substrate was chosen to provide improved mechanical properties, while the filtration performance is dominated by the nanoweb. (Reprinted with permission from Graham *et al.*, 2002. Copyright of Donaldson Company, Inc., Minneapolis, MN USA.)

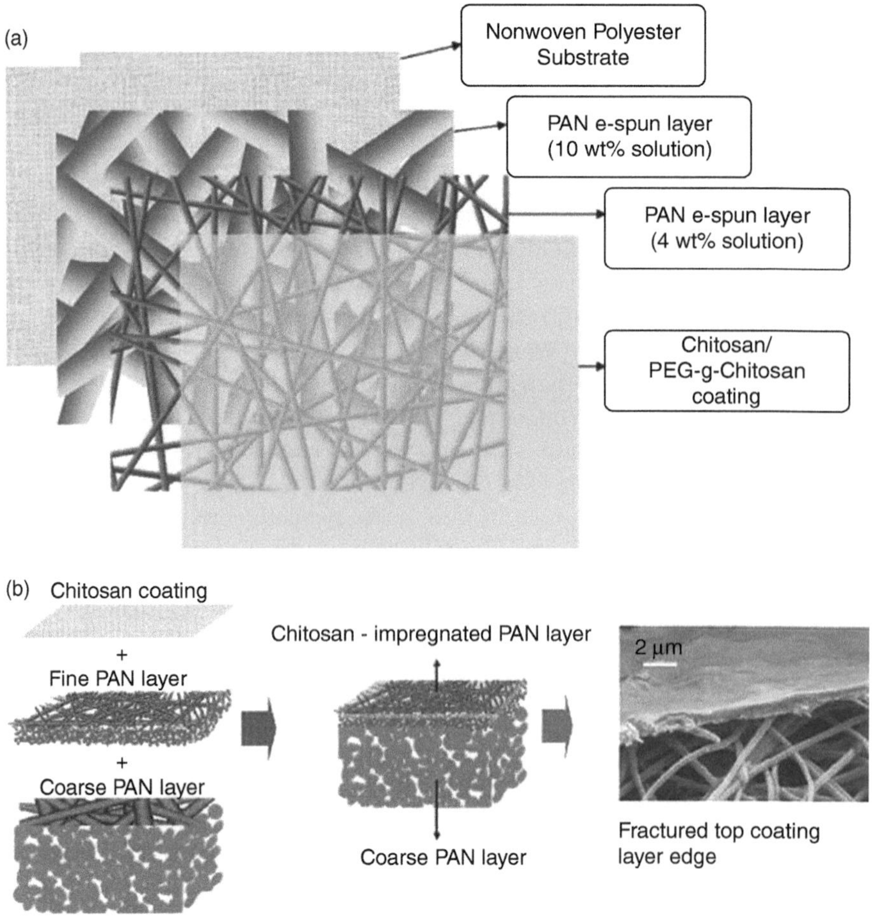

13.3 (a) Schematic diagrams for the assembly of three-layer composite filtration membrane developed as a high-flux and low-fouling filter and (b) fabrication schematics of the electrospun scaffold with a coating layer; the SEM image shows the fractured layer edge between the top layer (chitosan coating) and the mid-layer containing PAN nanofibrous mat (with 4 + 12 wt% sequential electrospinning). (Reprinted from Yoon *et al.*, 2006. Copyright 2006, with permission from Elsevier.)

rates and improved anti-fouling characteristics, besides enabling alternative complementary functionality to be introduced into the nanofiber membrane, such as the preparation of electrospun nanofibrous membranes with top coating layers (Wang *et al.*, 2006a; Yoon *et al.*, 2006), electrospun nanofibrous mats 'sandwiched' between various layers (Wang *et al.*, 2005; Ma *et al.*, 2011) or blended together with micron fibers (Barhate and Ramakrishna, 2007). An example of a three-layer laminate filter is shown in Fig. 13.3, with the mid-layer consisting of the electrospun nanofiber mat.

Similarly to what happens for other filtration materials, the application of these nanofibrous materials are expected to face some constraints related to the need to preserve the peculiar and characteristic product quality and organoleptic attributes. The possible unspecific adsorption of certain food product components may cause some unwanted flavors to be introduced or some wanted flavors and compounds to be lost. These aspects must be taken into consideration when developing a nanofibrous membrane for a particular separation operation. For example, polysaccharides and polyphenols are usually the main culprits in the fouling in wine clarification with membranes, and their removal might be beneficial for technological reasons and to increase the stability of the product. However, the retention of these compounds will affect the organoleptic characteristics of the filtrated wine, so their retention must be controlled.

Studies on the direct use of electrospun nanofibrous membranes for filtration or clarification of liquid food products are rare. One of the few examples – possibly even the only one – was a report by Veleirinho and Lopes da Silva (2009) on the use of electrospun poly(ethylene terephthalate) membranes for clarification of apple juice. The developed nanofibrous membranes enabled a high-flux performance, faster, simple and more economical than the traditional processes. The apple juice obtained from electrospun nanofiber membrane filtration revealed physico-chemical characteristics comparable to that clarified by ultrafiltration or by conventional clarification using filtering aids. The promising results obtained suggest that further developments will soon appear in this field.

Despite all the aforementioned advantages, the use of nanofibers in ordinary separation processes in the food industry still poses challenges that can hinder their applicability. Among them are difficulties to ensure membrane integrity and homogeneity during large-scale production and the fabrication of durable fibers with uniform sizes and morphologies and membranes with controlled shapes and fiber orientation. Nevertheless, more important developments related to the use of electrospun fibrous membranes in food filtration operations are expected to occur for more specific applications.

Due to the versatile scope for creating new functionalities onto electrospun fibers and mats, affinity membranes are especially applealing and are expected to be the focus of further applied-oriented research. Modification of the fiber surface might be achieved by using a diversity of strategies, such as chemical modification, surface grafting of different compounds, including polymers, plasma treatment, or a variety of physical coating procedures. In addition, the simple fiber preparation allows for relatively straightforward incorporation of other components into the electrospun dope and final fibers, including nanoparticles and nanotubes.

Rather than operating purely as a mechanical media to remove particles, macromolecules or microorganisms, these fibrous membranes might be used for separation of many biomolecules, based on a particular selectivity provided by immobilized specific ligands onto the membrane surface. Potential applications would be mainly in the biotechnological industry, intended for the production of highly purified proteins. Examples have been reported of the successful selective separation of bovine serum albumin or bilirubin using regenerated cellulose electrospun nanofiber membranes surface functionalized with Cibacron Blue F3GA (Ma *et al.*, 2005); bovine serum albumin as a model protein, using copolymerized methacrylic acid grafted onto plasma treated electrospun polysulfone fiber mats (Ma *et al.*, 2006); immunoglobulin G using regenerated cellulose electrospun nanofiber membranes surface functionalized with protein A/G (Ma and Ramakrishna, 2008); papain using electrospun hybrid chitosan/nylon-6 nanofibrous mats also surface functionalized with Cibacron Blue F3GA (Zhang *et al.*, 2010), among others.

At the boundary between a filtration affinity/SPE/biosensing membrane, electrospun nanofibrous membranes have been developed with covalently attached molecular recognition elements (e.g., antibodies and peptides) to specifically concentrate and detect the analyte of interest (Senecal *et al.*, 2008) or by using molecularly imprinted nanoparticles encapsulated into the polymer nanofibers also as affinity membrane to simplify and improve SPE (Yoshimatsu *et al.*, 2008). The streptavidin/biotin coupling strategy has also been successfully used in this framework (Guler *et al.*, 2005; Li *et al.*, 2007a; Lu *et al.*, 2008). The ligand specificity and the high surface area of interaction provides fast responses, increased sensitivity and binding efficiency, which might be profitable in sensing applications (Section 13.4). These materials are expected to assist in developing more sensitive methods to detect toxic contaminants and pathogenic microorganisms in food matrices, thus playing an important role in food safety.

Another promising strategy is the use of multifunctional electrospun fiber membranes able to operate not only in separation mechanisms, but also showing additional functionalities – for example, the use of antimicrobial nanofibrous membranes (Section 13.5.1) to improve the microbial safety of beverages (Kriegel *et al.*, 2008); preparation of electrospun chitosan-based fibrous membranes showing also cation binding capacity and antibacterial activity (Desai *et al.*, 2009); or the development of nanofibrous electrospun mats to be used as separation membranes and improved carriers for enzymes (Section 13.3), enabling concomitant filtration and catalytic functionalities, are areas of interest in the food area.

13.2.2 Solid-phase extraction applications

Electrospun nanofibers have attracted recent attention as adsorbents for SPE, especially in water quality, waste water effluents and biological fluid analysis (Kang *et al.*, 2007; Qi *et al.*, 2008a; Liu *et al.*, 2010c; Xu *et al.*, 2010;

Vadukumpully *et al.*, 2011). Electrospun micro-/nanofibers have been successfully used to prepare fiber-based coatings on stainless steel wires for solid-phase microextraction (SPME) or as nonwoven mats to fabricate SPE disk sorbents (Chigome *et al.*, 2010; Zewe *et al.*, 2010).

Again, the high surface-to-volume ratio, high specific surface and number of potential active sites to enable more efficient adsorption make these nanofibrous materials advantageous in this area of application. For example, lower back pressures, allowing for sample processing at higher flow rates, and lower detection limits for targeted analytes are commonly described, in comparison with other available materials. In addition, the easier attachment of the target molecules enables the use of small amounts of nanofibers for the extraction process, which greatly reduces the use of desorption solvents (Xu *et al.*, 2010).

Micro and nanofibers produced by electrospinning are thus a versatile, inexpensive and simple method and an attractive alternative for SPE disk fabrication and for coating SPME fibers. Hence, it is predictable that, through this approach, fabrication of nanofibrous materials for SPE with wide sorbent range capabilities will be possible, and that their application in food analysis will occur and develop in the near future.

Strategies to improve the functionality of the polymer electrospun nanofibers have been studied, and emphasis should be given to the so-called molecularly imprinted SPE proposed by Baggiani *et al.* (2007). Molecularly imprinted polymer microspheres were immobilized within electrospun polymer nanofiber membrane, providing specific recognition capabilities for SPE and filtration applications. The studied analytes included pesticides, drugs and mycotoxins. Screening strategies for whole analyte classes were proposed based on the careful choice of the template structures and on the simultaneous extraction of many related analytes.

13.3 Electrospun microfibers and nanofibers for enzyme immobilization

In industrial applications, immobilized enzymes are favored over free enzymes since they offer straightforward catalyst reuse, simple product purification and continuous operation (Kandimalla *et al.*, 2006; Brady and Jordaan, 2009), with clear advantages in large-scale production operations. Also, the immobilized enzymes show often higher stability, allowing for their application under temperature and pH conditions, and in organic solvent media, that are not practicable when using the free enzyme.

Improvements in current strategies for enzyme immobilization have been recently described, permitting more targeted, multipoint and faster binding, or providing higher volumetric and specific activity, as is the case with self-immobilization techniques (Brady and Jordaan, 2009). However, there is a balance between the advantages and disadvantages of the immobilization process; although far from being well understood as a new application, several drawbacks have been identified. Among them are mass transfer limitations, which reduce the biocatalytic performance. In addition, the structure and dimensions of the carrier material have great importance in the performance of the immobilized enzyme. Nanostructured supports are believed to be able to retain the catalytic activity, as well as to ensure immobilization efficiency and high catalytic performance of the immobilized enzymes. For example, nanoparticles and nanotubes are known to remarkably decrease mass transfer limitations, but their dispersion and reusability are demanding (Wang, 2006).

In this respect, electrospun nanofiber mats have a tremendous potential as novel supports for enzyme immobilization. Due to their large surface area per unit mass and their high porosity, high enzyme loadings are feasible and mass transfer limitations are expected to be reduced, thus allowing high reaction rates to be achieved and providing easier reuse and straightforward implementation of continuous processes. Since a variety of polymers can be electrospun, meeting different requirements as supports for enzyme immobilization, and as engineering of the fiber morphology and structure and subsequent surface functionalization for biomolecule immobilization are also possible, they are highly versatile systems. Furthermore, due to their low resistance to the flow of liquids, they can be used for simultaneous biocatalysis and separation, enabling the potential fabrication of enzymatic membrane-bioreactors.

In most cases, enzymes have been immobilized onto the surface of prefabricated electrospun nanofibers through adsorption and subsequent covalent linkage. Other studies were directed at encapsulating the enzyme into the electrospun nanofibers by mixing the enzyme with the spun dope prior to the electrospinning process, resulting in single-phase fibers with the incorporated enzyme, or by exploring the formation of bicomponent coaxial fibers, with the enzyme included into the core phase.

The first strategy permits better substrate accessibility to the enzyme active sites, but may also originate low enzyme loading due to limited availability of enzyme attachment sites on fiber surface. Moreover, some loss of enzyme activity may occur due to the influence of the binding procedure on enzyme conformation and/or active centers and to the limited mobility and direct exposure to the environment media.

Using the second approach, the immobilized enzyme will be more protected inside the nanofibers, consequently with expected poorer substrate–enzyme interactions. Increasing the fiber's porosity is one of the better strategies to improve accessibility of the substrate to the enzyme active centers. This has been achieved mainly by using a template compound that is removed after electrospinning and fiber formation. For example, Patel *et al.* (2006) removed both the glucose and the polyvinyl alcohol (PVA) polymer templates from electrospun silicate-PVA-glucose-enzyme composite fibers, achieving pore sizes as large as 4 nm into fibers with average diameters between 100 and 200 nm. Another potential problem is the possible release of the enzyme from the fiber matrix due to poor entrapment, as it was observed for certain proteins that were able to diffuse out of the electrospun nanofibers (Jia *et al.* 2002; Jun *et al.* 2005; Dror *et al.*, 2008). Possible strategies to aleviate this problem are the use of polymer coatings on the prefabricated electrospun fibers or the use of core-shell fiber structures with the enzyme entrapped into the inner core (Greiner *et al.*, 2006).

Enzyme immobilization has been performed by optimizing different aspects including appropriated selection of polymers and solvents (to preserve enzyme activity), functional surface groups, spacers, enzyme-binding reactions and adequate fiber composition and structure to allow for enzyme entrapment. An outstanding review on this subject has recently appeared (Wang *et al.*, 2009d) where the reader can find details on the already-explored strategies to immobilize enzymes on the surface or to encapsulate them into electrospun polymer fibers, together with the most successful achievements and drawbacks.

The production of enzyme-based biosensors for food analysis, using electrospun polymer fibers as supporting materials for enzyme immobilization, will be considered in the next section. Here our attention will be directed to bulk phase catalysis that could potentially be scaled up for food-processing applications; we will summarize the information on the work done so far, giving special emphasis to the most recent contributions to the advancement of knowledge in this area.

13.3.1 Immobilization of lipases

Lipases are used in a broad range of applications in several chemical industries, including food, for the catalysis of reactions such as esterifications and transesterifications, aminolysis, etc., mainly in the fats and oils industry, the dairy industry, baking and in production of food ingredients. They can act also as catalysts in biosynthetic reactions with increasing use in food and biotechnological applications.

Lipase immobilization has been widely explored owing to the already-mentioned advantages and to their particular catalytic mechanism, acting at the interface between an aqueous and a nonaqueous phase. Lipases are also among the more studied enzymes for immobilization in/on electrospun polymer fibers. Some studies and results obtained for immobilization of lipases on electrospun nanofibers are shown in Table 13.1.

Table 13.1 Immobilized lipases in/on electrospun polymer micro-/nanofibers

Polymer	Immobilization procedure	Relevant aspects related to the activity of the immobilized enzyme	References
Polyethylene glycol grafted on regenerated cellulose	Covalent binding	40% higher activity at pH 7 than the free enzyme; up to 8–10 times at 60–70°C than the free enzyme	Wang and Hsieh, 2004
Polyacrylic acid grafted on cellulose	Adsorption	–	Chen and Hsieh, 2005
Poly-(acrylonitrile-*co*-maleic acid)	Covalent binding	Retention of 38% activity in relation to the free enzyme	Ye *et al.*, 2006
Poly[acrylonitrile-co-(2-methacryloyloxyethyl phosphorylcholine)] containing phospholipid side moieties	Adsorption	Retention of 73–77% activity in relation to the free enzyme	Huang *et al.*, 2006
Polysulfone/poly(*N*-vinyl-2-pyrrolidone) and polysulfone/poly(ethylene glycol	Adsorption	–	Wang *et al.*, 2006c

Polyacrylonitrile	Covalent binding	Retention of 81% activity in relation to the free enzyme; better activity retention under storage and reuse	Li *et al*, 2007b
Poly(acrylonitrile-co-2-hydroxyethyl methacrylate)	Covalent binding (different methods)	Retention of 32–41% activity in relation to the free enzyme	Huang *et al*., 2008
poly(acrylonitrile-*co*-acrylic acid) surface modified with (a) protein hydrolysate from egg skin; (b) collagen.	Covalent binding	Retention of (a) 20% activity and (b) 62% in relation to the free enzyme	Huang *et al*., 2009
Regenerated cellulose surface modified with Cibacron Blue F3GA (affinity ligand)	Adsorption	Retention of 86% activity in relation to the free enzyme	Lu and Hsieh, 2009a
Electrospun PVA with dispersed organosilicate cages	Entrapment (into the inorganic phase)	Catalytic properties appropriated for flow-through reactors and potential biodiesel production	Sakai *et al*., 2010c
Poly[acrylonitrile-co-(glycidyl methacrylate)]	Covalent binding (several methods)	~10–65% higher activity than the free enzyme; much better thermal stability and reusability	Dai *et al*., 2011

In general, better thermal and solvent stability and reusability were observed for the immobilized lipases. Mention should be made of the wide range of results observed for the activity retention of the immobilized enzyme when compared to free enzyme behavior, depending on the polymers, spacers and immobilizing methods.

Most of the reported results did not progress beyond the lab scale and focused on the study of typical enzyme kinetics. Some studies are worth referring to as representative of a step further, focusing on more defined potential applications for these materials. Li and Wu (2009), working with immobilized lipase onto polyacrylonitrile nanofibrous membranes, with an activity retention of 87% of the free enzyme, have optimized important parameters (oil-to-water ratio, temperature and pH) for the controled hydrolysis of soybean oil. Sakai *et al.* (2010a, b) demonstrated the successful application of immobilized lipase physically adsorbed onto electrospun polyacrylonitrile fibers to transesterification in nonaqueous solvent and on the production of butyl-biodisel; the immobilized lipase showed more than 20-fold higher activity than the free lipase. The same group (Sakai *et al.*, 2010c) developed a flow-through reactor for the continuous production of fatty acid butyl esters based on immobilized lipase (entrapped) in silica cages dispersed into electrospun PVA nanofibers.

13.3.2 Immobilization of proteases

The field of immobilized proteases is becoming increasingly important in food-related areas, including protein removal, separation of racemic aminoacid mixtures, synthesis of peptide sweeteners and – probably one of the most important, now and in the future – the production of biologically active peptides. Proteases have also been immobilized on electrospun nanofibers obtained from various polymers, also using physical encapsulation or covalent binding.

The first work on enzyme immobilization using electrospun nanofibers as supports was performed for a protease, α-chymotrypsin, immobilized on functionalized polystyrene nanofibers (Jia *et al.*, 2002). The prior protection of the enzyme by complexation with an appropriate surfactant allowed the immobilization of α-chymotrypsin into electrospun fibers obtained from a hydrophobic polymer and using organic solvents, with satisfactory retention of the enzyme catalytic activity (Herricks *et al.*, 2005).

The attachment of seed enzymes (α-chymotrypsin) onto nanofibers consisting of a mixture of polystyrene and poly(styrene-co-maleic anhydride), followed by cross-linking of additional enzyme molecules and aggregates onto the covalently attached seed enzyme molecules, was used as a strategy to increase the enzyme loading capacity, resulting in higher activity and stability of the immobilized enzymes (Kim *et al.*, 2005).

Recently, following the same approach, trypsin was immobilized on electrospun polystyrene-poly(styrene-co-maleic anhydride) nanofibers by first

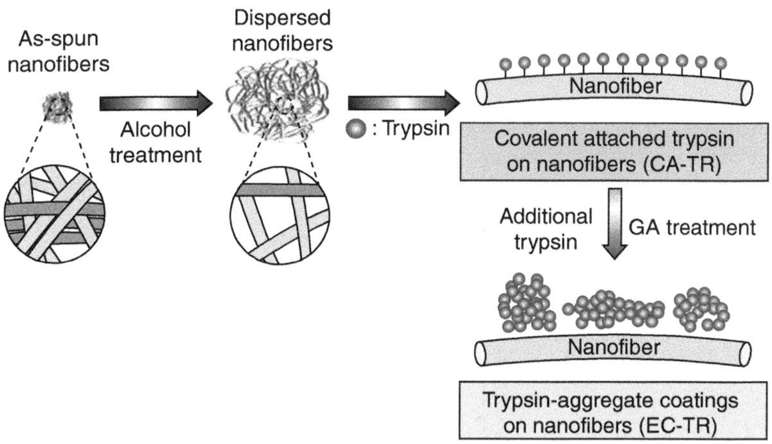

13.4 Schematic diagram for the preparation of dispersed polystyrene/poly(styrene-*co*-maleic anhydride) electrospun nanofibers and trypsin coating. (Reprinted with permission from Jun *et al.*, 2010. Copyright 2010 American Chemical Society.)

conducting the covalent bond of the enzyme to the fiber (monolayer coating) followed by binding of additional enzyme molecules onto the covalent attached trypsin through the action of glutaraldehyde (Ahn *et al.*, 2010; Jun *et al.*, 2010). This interesting strategy allowed for better digestion performance and higher enzyme stability and reusability when compared with the activity of the free enzyme, covalently attached trypsin, or commercial trypsin beads. The schematic diagram of the followed strategy is shown in Fig. 13.4. The main purpose was to use trypsin-immobilized columns for online and automated protein digestion, enabling rapid and efficient protein digestion in proteomics. However, similar strategies may enable the preparation of protease-immobilized nanofibrous membranes for other food-related applications, namely protein removal and production of peptides with potential bioactive properties.

An alternative approach to increase the enzyme loading in the electrospun nanofibers was recently described (Liu *et al.*, 2011), based on multilayer-structured nanofibers composed by polyurethane, a protein coating and the adsorbed enzyme (α-chymotrypsin) and on the high capacity of LiCl embedded into the electrospun polymeric nanofibers to greatly increase the irreversible protein adsorption.

13.3.3 Nano- and microfibers functionalized with other enzymes

Several other enzyme classes have been immobilized in electrospun nano- and microfibers, including other hydrolases and oxireductases (Table 13.2).

Table 13.2 Some examples of oxireductases and hydrolases immobilized in/on electrospun polymer micro-/nanofibers

Enzyme	Polymer	Immobilization procedure	Relevant aspects related to the activity of the immobilized enzyme	References
Oxireductases				
Catalase	Polyacrylonitrile with porphyrin pendants and dispersed carbon nanotubes	Covalent binding	Retention ~ 48% activity in relation to the free enzyme; higher stability to temperature than the free one	Wan *et al.*, 2007
	Cu(II)-chelated poly(vinyl alcohol)	Adsorption (complexation)	Retention ~ 44% activity; higher stability to pH, temperature and storage than the free one	Feng *et al.*, 2011
Glucose oxidase	Polylactide	Entrapment (emulsion electrospinning)	Successful activation of the lactoperoxidase system in milks	Zhou and Lim, 2009
Laccase	Poly(D, L-lactide)	Entrapment (emulsion electrospinning)	Retention of 67% activity in relation to the free enzyme; better reusability	Dai *et al.*, 2010
Peroxidases	Poly(vinyl alcohol)/ tetramethyl orthosilicate/glucose (resulting in silica nanofibers)	Entrapment	Potential use as biosensors	Patel *et al.*, 2006

Hydrolases				
Acetylcholinesterase	Poly(vinyl alcohol)/BSA	Entrapment and cross-linking	Retention of 30–40% activity in relation to the free enzyme; potential usefulness as a biosensor for pesticide determination	Moradzadegan *et al.*, 2010
Cellulase	Poly(vinyl alcohol)	Entrapment and cross-linking (glutaraldehyde)	Retention > 65% activity in relation to the free enzyme	Wu *et al.*, 2005
β-galactosidase	Polycaprolactone (*shell*)/ poly(ethylene oxide)(*core*)	Entrapment (core-shell fiber structures)	Retention ~ 50% activity in relation to the free enzyme	Dror *et al.*, 2008
	Polyacrylamide	Adsorption and entrapment	—	Lu and Hsieh, 2009b

Among them, the glucosidases are largely used in the food industry in starch processing, production of sugars and syrups, baking, brewing, hydrolysis of starch in fruit juices and in vegetable tenderization, hydrolysis of cellulose, conversion of sucrose to glucose and fructose, removal of lactose from milk, clarification of fruit juices and wine, and even in the production of oligosaccharides with nutraceutical properties. However, electrospun nanofiber mats have been poorly explored as immobilizing supports for these enzymes, and one can expect that much more work in this area will appear soon.

Regarding the oxireductases, although they are much less used in the food industries, they are widespread in many other areas, like the textile, pulp and paper industries and in biodegradation and biotransformation of pollutants. Some interesting work has been undertaken on the immobilization of this type of enzyme on electrospun nanofibers, as illustrated in Table 13.2. Approaches to increase the activity and stability of the immobilized enzymes included the use of composite fibers with carbon nanotubes, owing to their electronic properties facilitating the electron transfer between the matrix and immobilized catalases (Wang *et al.*, 2006b; Wan *et al.*, 2007), and the use of nanofibers bearing attached proteins to create a more favorable environment for enzyme catalytic activity (Wang *et al.*, 2009c).

13.4 Electrospun nanofibers in food sensors

Nanofibers, used as the carrier matrix for the biological recognition element of biosensors, or to build nonenzymatic sensors as metal oxide nanofibers, or even as thin polymer composite wires in resistive sensors, have an important role in food analysis and food-processing control.

Different types of sensors have an increasing importance in food analysis, both in the detection and quantification of food constituents, as well as for the tracing of contaminants. Process control in the food industries needs rapid and affordable methods for the detection of several analytes. Also, increasing consumer demand for high-quality and safe food products anticipates the need for more sensitive and selective sensors.

This has led to extensive research efforts related to the development of nanostructured materials for this specific area of application. Nanostructured materials, including nanoparticles, nanofibers, nanowires and nanotubes, are well-known for their unique properties as new materials for the preparation of electrochemical sensors (Wang, 2005; Xiao and Li, 2008). In general, improved sensibility is achieved taking advantage of the large surface areas of the nanoscale structures.

The nanofibers produced by electrospinning have already demonstrated several advantages over other alternative strategies and are expected to increase in importance in the near future. Besides the immobilizing effect, in the case of the incorporation of the sensing compound, thanks to the high surface-to-volume ratio and high porosity of these materials; they can provide large sensing interfaces and increased sensitivity, a more efficient transport of the analytes through the nanofibrous membrane toward the sensor receptor and/or more efficient interfaces between the biological recognition events and the electronic signal transduction, thus originating more rapid responses than conventional systems. In addition, the ability to prepare nanosized fibers from very different polymers and even inorganic fibers after calcination or the organic polymer, the relatively easy preparation of composite fibers and fibers incorporating active compounds provide additional competitive benefits.

A comprehensive review of this subject has recently appeared, focusing on different sensors based on electrospun nanomaterials and several sensing approaches, such as amperometric, resistive, photoelectric and acoustic wave sensors (Ding *et al.*, 2010a). Many of the biosensors based on electrospun materials have been developed for application in several areas not directly related to foodstuffs. However, many of the sensed analytes are also important in food analysis, food quality and process control; we can, therefore, expect that the developed technologies will have direct applications in the food area in the near future.

13.4.1 Role in building improved biosensors

A biosensor can be defined as a device for detecting a substance that combines a biological component with a physical detector component. A biological signal is converted into an electrical output further detected by the transducer element. The transducer (electrochemical, optical, thermal, piezoelectric, etc.) transforms the signal resulting from the interaction of the analyte with the biological element in another signal that can be more easily measured and processed – for example, an electrical signal, temperature or mass change.

The biological recognition element (bioreceptor) may consist of enzymes, other proteins, antibodies, antigens, microorganisms, cells, etc. Typically, the bioreceptor molecules are immobilized on the transducer surface and are expected to bind the analyte molecules selectively. The immobilization of the bioreceptor molecule on the sensor surface is a key factor for the final performance of the sensor. Nanosized materials, including electrospun nanofibers and nanowebs, can make an important contribution in this area of biosensor development and improvement. Many studies have demonstrated that the immobilization of the bioreceptor in/on nanofiber mats in close contact with the transducer contributes to increasing the selectivity and decreasing the response time and analyte detection limits.

Enzymes are still the most employed bio-recognizing systems in the food area, owing to their specific and selective catalytic properties. Unsurprisingly, most of the applications involving electrospun nanofiber materials to construct electrochemical biosensors use these materials as enzyme immobilizing matrices. Typically the enzyme is immobilized onto the fiber surface and the electrospun nanofibrous membrane is then applied as a coating on the electrochemical transducer to build the biosensing device. In Section 13.3 we illustrated the ability of the electrospun nanofibers to immobilize different enzymes providing adequate biocompatible microenvironments for their functionality.

One of the first attempts to use electrospun nanofiber materials to build electrochemical sensors was directed at urea detection, with relevant importance in several areas, including clinical diagnosis and environmental analysis. The urea biosensor was built based on electrospun nanocomposite fibers (7–86 nm) of urease and polyvinylpyrrolidone, exhibiting faster response time, higher sensitivity and a more versatile design when compared with other available methods and technologies (Sawicka *et al.*, 2005). The promising results obtained clearly contributed, in the subsequent years, to the development of the technology in the biosensing area, expanding the studies to other analytes, electrospun polymers and biosensing methods.

Actually, focus is growing on clinical analysis. In fact, glucose is the most explored analyte sensed by this kind of device, owing to its importance in medical diagnostics. The accurate and rapid quantitative determination of glucose and other carbohydrates is also important in the food area, namely in process control in food fermentation industries. The classical determination of glucose is based on monitoring either the consumption of oxygen or the production of hydrogen peroxide, during the oxidation of glucose catalyzed by the enzyme glucose oxidase (GOD). Most enzymatic glucose sensors based on electrospun nanofibers/electrode assemblies are of the amperometric type.

The use of polymer nanofibers to immobilize the enzyme and to improve the contact and electron transfer between the bioreceptor element and the transducer has been shown to mitigate interference and to improve the sensor performance, providing higher sensitivity and faster responses (Ren *et al.*, 2006; Scampicchio *et al.*, 2010). Electrochemical glucose biosensors constructed with polymer nanofibers composites with carbon nanotubes and the immobilized GOD, deposited on the electrode surface, have shown improved performance, owing to the ability of carbon nanotubes to improve electronic conductance and electric communication between the biological element and the transducer (Wang *et al.*, 2009e; Ding *et al.*, 2010b).

Recently, Marx *et al.* (2011) developed another enzymatic amperometric sensor device for fructose based on nanofibers of electrospun gold with immobilized fructose dehydrogenase. The biosensor proved to be accurate and precise in the detection of fructose in both clinical (serum) and food (beverages sweetened with high fructose syrup) areas.

Another high-sensitive enzymatic biosensor relevant in the food area was developed to detect and quantify phenolic compounds, based on tyrosinase immobilized on electrospun nylon-6 nanofibrous membrane deposited on a glass carbon electrode for amperometric detection (Arecchi *et al.*, 2010).

The well-known biotin–streptavidin specific binding was also explored to set up specific biosensors, also based on electrospun immobilizing fiber arrays – for example, for sandwich binding of biotinylated compounds. Indeed, Li *et al.* (2006, 2007a) have shown that polylactic acid-biotin electrospun membranes could be successfully used to capture and detect biological molecules, including streptavidin, biotinylated DNA probes, biotinylated antibodies and a fluorescence labeled secondary antibody. Following the approach of using molecular recognition elements for selective capture of target molecules, covalently attached antibodies and the avidin/biotin coupling strategy were used to functionalize polymer electrospun nanofibers. The developed materials were found to successfully detect and quantify *Staphylococcus* enterotoxin B (Senecal *et al.*, 2008) and *Escherichia coli* O157:H7 and bovine viral diarrhea virus (Luo *et al.*, 2010).

13.4.2 Role in other nonenzymatic sensors

Many other types of sensors have been developed during recent years based on electrospun fibrous materials with demonstrated advantages regarding alternative approaches. Among them are those nonenzymatic sensors for glucose constructed from a diversity of polymers and composite fibers (Manesh *et al.*, 2007; Tiwari *et al.*, 2010) and even metal oxide nanofibers (Wang *et al.*, 2009a; Ding *et al.*, 2010c; Cao *et al.*, 2011).

Recently, Tang *et al.* (2011) have developed a method to produce a carbon paste electrode surface modified with carbon nanofibers, as an amperometric sensor device for xanthine determination, which exhibited high electrocatalytic activity, fast amperometric response, low detection limit and good stability and selectivity. The carbon nanofibers have been obtained by carbonizing polyacrylonitrile electrospun nanofibers. After thermal treatment the carbon nanofibers exhibited diameters ranging from 200 to 400 nm and a rough and porous surface that significantly facilitated the electron transfer and increased the effective electrode surface area. Besides the importance of the xanthine compound in clinical diagnosis, as an indicator of abnormal purine metabolism related to several clinical disorders, it is also used in the food industry as an indicator of fish freshness. Therefore, the developed sensing device may find an important application in food quality control.

Gas and volatile compound sensors have also been developed largely based on electrospun fibers, with the purpose to improve sensor sensitivity and to achieve faster responses to the target gases/vapors. The high surface area and improved reactive gas adsorption, compared with sensors based on flat films or other alternative sensing techniques, have been demonstrated to successfully allow the achievement of these purposes. In fact, the size, morphology and structural characteristics of the electrospun fiber arrays have proved to be quite advantageous for gas sensing.

Electrospun fibrous materials used in this area include semiconductor metal oxides, polyelectrolyte polymers and conducting polymer composites, using different transducer elements and detection techniques such as resistive, acoustic wave, photoelectric and optical techniques (Ding *et al.*, 2009, 2010a). In any case, the electrospun fibers have an important role, providing a high surface interaction with the analyte, adequate electronic conduction by themselves or serving as the matrix to incorporate the conducting elements, or as the template polymer nanofibers for formation of the metal oxide nanofibers after calcinations of the organic component.

High-sensitivity resistive sensors based on electrospun metal oxide nanofibers were successfully developed to detect and quantify O_2, CO_2, CO, NO_2, H_2 and NH_3, among other gases and organic compound vapors, including humidity (e.g., He *et al.*, 2010; Li *et al.*, 2008; Qi *et al.*, 2008b; Wang *et al.*, 2009b; He *et al.*, 2010) and ethanol sensors (e.g., Liu *et al.*, 2009; Qi *et al.*, 2009; Song and Liu, 2009; Wang and Liu, 2009).

As an example of a gas sensor based on electrospun metal oxide nanofibers, Fig. 13.5 shows a diagram of the processing steps used to fabricate metal oxide gas sensors based on TiO_2 or SnO_2 electrospun fibrous mats deposited on oxidized Si (Si/SiO_2) or alumina (Al_2O_3) substrates fitted with interdigitated electrodes (Au or Pt), using poly(vinyl acetate) (PVAc) as the

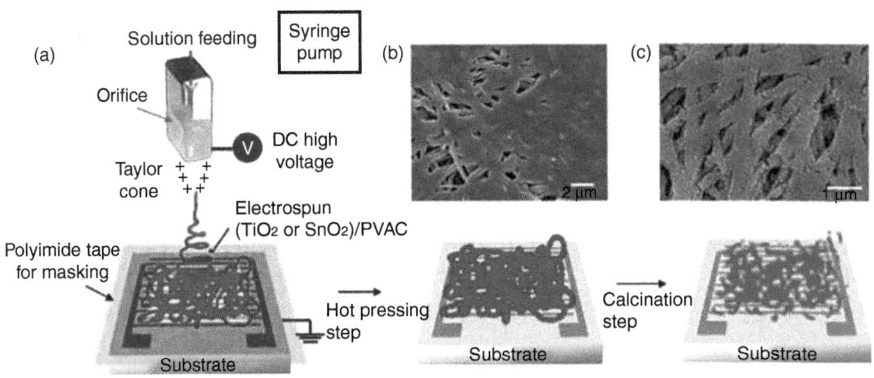

13.5 Schematic diagram of the processing steps (electrospinning, hot pressing, and calcination) used to fabricate metal oxide gas sensors based on TiO_2 or SnO_2 electrospun fibrous and poly(vinyl acetate) as the polymer template. (Reprinted from Kim and Rothschild, 2011. Copyright 2010, with permission from John Wiley & Sons, Inc.)

polymer template and including a hot pressing step to improve the adhesion to the substrate and electrical contacts. This will give better exposition to the gas phase, following the calcination step to decompose the organic component and to oxidize and crystallize the inorganic compound into a metal oxide nanofibrous structure. (Kim and Rothschild, 2011).

Besides the nanofibrous metal oxide sensors, also resistive sensors based on electrospun polyelectrolyte polymers or conducting polymer composites have proved to have advantageous gas/vapor sensing properties (Kessick and Tepper, 2006; Ji *et al.*, 2008; Pinto *et al.*, 2011), including potential for use as humidity (Aussawasathien *et al.*, 2005; Li *et al.*, 2009a) and ethanol sensors (Liu *et al.*, 2010b). The particular fiber size and geometry and the 2-D architecture allow for a rapid and reversible resistance change upon exposure to the target gas, enabling the construction of fast and highly sensitive sensor devices. Several studies also showed the improved sensing properties that can be obtained in gas and vapor detection by using polymer nanosized fibers as the sensitive coatings of acoustic wave sensors, based on the quartz crystal microbalance (Ding *et al.*, 2004, 2005; Sheng *et al.*, 2011; Wang *et al.*, 2010b, 2011).

These sensing devices may find useful applications in food, in the detection and quantification of moisture, ethanol and other volatile organic compound, including volatiles responsible for aroma of many food products. Humidity sensors are important to monitor food storage and shelf-life conditions and food-processing operations where the relative humidity is an important variable. Ethanol sensors are currently in use, or are expected to be in the near future, to monitor a variety of industrial processes, including in food industry to control and manage food fermentation processes – for example, brewing and wine making processes, for medical/breath analysis and environmental monitoring. Also important in the food area is the development of volatile organic compounds sensor systems for potential application as 'electronic noses.'

Electronic noses and tongues, as multisensor devices, have several important applications in food, including: the categorization of alcoholic beverages on the basis of the aromas released; classification of cereal grains; monitoring the freshness of fish, fruits and vegetables; monitoring production processes and final product quality in the dairy industry; and other applications based on monitoring aroma and flavor, ranging from bread and wine to edible oils and even food spoilage by bacteria, as well as in micotoxin analysis.

The outstanding properties of the electrospun nanofibers and mats already discussed are likely to contribute to the development of more sensitive, selective and faster electronic noses and tongues. Some promising results have been obtained with fully integrated chemiresistor microchips – based on electrospun polymer fibers (Zampetti *et al.*, 2011), namely with chemoresistive carbon black–polymer composite sensors, owing to their broad response to diverse volatile organic compounds as detection elements in the electronic nose systems, or based on polyaniline as a conductive polymer due to its thermal stability and sensing performances.

13.5 Electrospun micro- and nanofibers used for encapsulation and controlled release of bioactive compounds

Today, the application of nanotechnology in agro-food products is largely in the area of encapsulation of sensitive nutrients and/or bioactive compounds, designing materials that can protect sensitive food ingredients, as well as the production of encapsulates showing delayed/controlled or targeted delivery of the core materials.

Encapsulation can provide a barrier between the sensitive bioactive compound and the environment (the food matrix, oxygen or temperature, for example). From the standpoint of sensory and consumer acceptance, the encapsulation of certain compounds may allow for masking of unpleasant flavours and odors. Nutritionally, it may allow the passage of certain compounds through the gastrointestinal tract, minimizing their modification until they reach the site of absorption/action, increasing bioavailability. Through proper partitioning, encapsulation can also allow for the inclusion of incompatible compounds in the same food product. Probiotic bacteria are also included within the group of bioactive ingredients; their size and the fact that they must be kept alive offer additional challenges when considering microencapsulation.

A range of techniques can be used to achieve encapsulation of a particular core material/compound; spray drying is the one most used, either at the lab or at industrial scale, although other techniques are also available, including freeze drying, extrusion, coacervation, adsorption, spray-coating and gel-particle technologies. A wide range of materials can also be used for this purpose, originating a diversity of final systems, typically spherical nano- and microparticles, coacervates, or lipid-based systems (nanoliposomes, nanocochleates and archaeosomes). Depending on processes and materials, the compound to be encapsulated will show different stability and release properties. Strategies and recent innovations in encapsulation technologies for biologically active compounds in food systems have been reviewed (Champagne and Fustier, 2007; Quintanilla-Carvajal *et al.*, 2010).

In this section, focus will be on the alternative strategies that can be achieved by using electrospun micro- and nanofibers for encapsulation of sensitive bioactive compounds and/or cells with the purpose of their protection against the influence of adverse environments and/or controlled release under well-defined conditions. So far, antimicrobial compounds, antioxidants, vitamins and probiotic microrganisms have been encapsulated using this approach.

Electrospinning is typically used for the production of continuous nanofiber nonwovens. Long fibers and nanofibrous membranes are not the typical encapsulating materials of choice when intended for internal release. In the food area, the dimensional and morphological nature of the electrospun materials does not call for potential internal applications, by ingestion, in humans, but opens the door for development of novel bioactive packaging strategies, using these nanofibrous matrices on their own or associated with other packaging materials. Multitasking materials with multiple functionalities is an area of increased interest in specific applications and may be another field of interest in which to explore the 'encapsulating' potential of electrospun fibers in foods.

Nevertheless, the easy fabrication, the high surface area and versatility in terms of composition and micro-partitioning within the nanofiber and the typical nonthermal processing involved, makes electrospinning also an attractive technique for the encapsulation and controlled release of bioactive compounds. In fact, the opportunity to encapsulate proteins, preserving their functionality and controlling their release (Chew *et al.*, 2005; Li *et al.*, 2010), the possibility to prepare multiphase nanofibers starting from suspensions or emulsions (Agarwal and Greiner, 2011), either by encapsulating water soluble bioactive agents into biodegradable hydrophobic polymers via emulsion electrospinning (Li *et al.*, 2010 a, b) or by formation of electrospun nanofibers from hydrophilic polymers as carrier vehicles for microemulsions containing solubilized lipophilic functional compounds (Kriegel *et al.*, 2009), the production of core-shell nanofibers by controlling the polymer phase separation during the process (Lopes da Silva *et al.*, 2009) or by using particular electrospinning setups (Yarin *et al.*, 2007), the encapsulation of multiple biological compounds within a single electrospun fiber (e.g. by a combination of electrospinning and the encapsulation of bioactive compounds using nanoparticles) (Dong *et al.*, 2009) and the potential to encapsulate even viable microorganisms and other cells (Salalha *et al.*, 2006; Zussman, 2011) opens routes for the further development of these materials and for their successful application by the food industry.

Therefore, electrospun nanofibers are expected to find increasing numbers of applications as encapsulating materials in food packaging materials, and those obtained from edible polysaccharide and/or protein are expected to find applications also for controlled release of nutraceuticals or pharmaceuticals in the gastrointestinal tract.

13.5.1 Electrospun nanofibrous mats with antimicrobial properties

Preparation of electrospun nanofibrous mats with antimicrobial properties has been described in several works, and the resulting materials have shown properties that validate their beneficial application in diverse areas such as preparation of biomaterials with antibacterial characteristics and formulation of membranes with antimicrobial activity for air and water filtration and packaging.

Among them, electrospun polymer composite fibers prepared with silver nanoparticles have attracted great attention, developed especially for practical medical applications, such as wound-dressing and tissue-engineering scaffolds, but also for antimicrobial nanofiltration applications (Lala *et al.*, 2007; Jeon *et al.*, 2008). Typically these nanocomposite fibers were obtained by electrospinning $AgNO_3$/polymer solutions, followed by a reduction treatment to form the silver nanoparticles – for example, by heating and/or chemical reduction (Xu *et al.*, 2006; An *et al.*, 2009) or by photo-reduction using UV irradiation (Hong *et al.*, 2006; Son *et al.*, 2006; Lala *et al.*, 2007; Xu *et al.*, 2008). Recently, electrospun poly(vinyl pyrrolidone)/nanosilver composite fibers, with high water stability and significant antibacterial efficacy, have been reported by using preformed nanosilver particles added to the polymer solution to be electrospun (Hwang and Jeong, 2011).

In the case of intelligent and bioactive packaging solutions for foods, often the antimicrobial agent is allowed to diffuse into the product to inhibit the proliferation of microorganisms, but in most cases the integrity of the encapsulating matrix should be preserved during storage. Additional efforts are required for the optimization of the materials taking into account the need for use of polymers and solvents that allow contact with food products, the need to ensure the integrity of nanofiber arrays, sometimes in environments with high water activity, and the need for a proper release of the active compound when and wherever desired.

Naturally occuring antimicrobial compounds have been incorporated into electropsun polymer fibers with this purpose in mind. Interesting results have been obtained by Vega-Lugo and Lim (2009) by encapsulating allyl isothiocyanate into soy protein isolate/poly(ethylene oxide) blend and poly(lactic acid) electrospun fibers, observing that the environmental relative humidity might act as the trigger for the bioactive compound release. Dheraprasart *et al.* (2009) encapsulated a food-grade antimicrobial agent, nisin, into gelatin electrospun nanofibers, improving the mechanical properties and water resistance of the fibrous matrices by postcross-linking using glutaraldehyde. Torres-Giner *et al.* (2009) followed a different approach to fabricate water-insoluble fiber mats with efficient biocide properties, by electrospinning blends of zein with the natural antimicrobial chitosan. Preparation of a microemulsion prior to the electrospinning process allowed the incorporation of an antimicrobial lipophilic compound, eugenol, into hydrophilic polymer nanofibers, with satisfactory antimicrobial activity but also a burst release of the encapsulated eugenol and a rapid dissolution of the hydrophilic carrier matrix (Kriegel *et al.*, 2010). Recently, (–)-epigallocatechin-3-gallate, a naturally occurring antimicrobial compound prone to oxidation, was also successfully encapsulated into electrospun PVA nanofibers preserving the antibacterial activity (Sun *et al.*, 2011).

Antimicrobial materials with nonleaching antibacterial agents may have additional advantages in food packaging, avoiding unnecessary migration of the active compound into the food product. An example is the utilization of the contact antimicrobial properties of single-walled carbon nanotubes to prepare electrospun composite polysulfone fibrous mats with antibacterial activity (Schiffman and Elimelech, 2011).

13.5.2 Electrospun nanofibrous encapsulating microorganisms

One of the promising areas is the encapsulation of microorganisms within the electrospun nanofibers, maintaining their viability and functionality. Direct applications in the food area are scarce, but the potential is great, especially in food biotechnology, using membranes for immobilization of microorganisms, biotransformations and production of food ingredients. The use of edible polymers and appropriate solvents may allow extension of this strategy to the encapsulation of probiotics for incorporation in functional foods.

The ability of viruses to maintain their viability within electrospun polymer microfibers was first demonstrated by Lee and Belcher (2004). Zussman and coworkers (Salalha *et al.*, 2006) reported for the first time the successful encapsulation and survival of bacteria (*Escherichia coli*, *Staphylococcus albus*). During the extremely rapid formation of the electrospun thin fibers, under very high elongation rates, typically the solvent rapidly evaporates and the jet dries and solidifies. This process will be accompanied by a quick and significant change in osmotic pressure. Therefore, after electrospinning, those microorganisms more adapted to low water activities and changes in osmotic environment are expected to keep their viability at higher levels than those more sensitive microbes (Gensheimer *et al.*, 2007). This kind of fibrous living membrane can find successful applications in the bioremediation of pollutants in water systems (Klein *et al.*, 2009)

In food technology it is important to maintain the viability of probiotic bacteria from production to consumption, including their passage through the gastrointestinal tract. Microencapsulation is a well-known useful tool for protection and stabilization of these microorganisms for functional food applications (Anal and Singh, 2007). Coaxial electrospun nanofibers were successfully used for this purpose, revealing that encapsulated bifidobacteria could survive both the drastic osmotic changes and electrostatic field, remaining viable for much longer than nonencapsulated bacteria under similar storage conditions (López-Rubio *et al.*, 2009).

Microorganisms immobilized within electrospun polymer fibers have also been used as an alternative bioconversion method. *Pseudomonas nitroreducens* LY was immobilized in PVA electrospun nanofibers and was able to synthesize theanine showing a satisfactory conversion rate of sodium glutamate (Liu *et al.*, 2010a). The survival rate of the bacteria decreased when the electrospinning voltage increased. This means that, besides the influence of all intrinsic and extrinsic factors on the fiber characteristics, the experimental conditions have also to be optimized to guarantee that a satisfactory microbial survival can be achieved. Besides a possible direct damage effect of the electrostatic field upon the cell survival, the voltage has a high influence on the final fiber diameter and the fiber diameter greatly affects the solvent evaporation rate and the osmotic microenvironment of the microorganism.

13.5.3 Encapsulation of other bioactive compounds

Antioxidants are a class of sensitive compounds widely used in the food industry. As food additives and bioactive functional compounds, protection from light and oxygen is needed to ensure their functionality. Recently, Fernandez *et al.* (2009) have reported the successful encapsulation of β-carotene in ultrafine fibers of zein, providing good protection against oxidation and a significant increase in the light stability of the encapsulated component. The same protein has been used to produce electrospun fibers encapsulating (-)-epigallocatechin gallate, a green tea polyphenol (Li *et al.*, 2009b), also by incorporating the antioxidant in the zein fiber-forming solution. Hydrogen bonding, hydrophobic interactions and physical encapsulation have been suggested to explain the immobilization and stabilization of the encapsulated bioactive compound.

Encapsulation of vitamins is another area of interest, although no direct food application has been described so far. Taepaiboon *et al.* (2007) produced all-trans retinoic acid and vitamin E-loaded cellulose nanofibers, showing a gradual and monotonous release of the vitamin. Wu *et al.* (2011) also encapsulated vitamin C and E derivatives into electrospun polyacrylonitrile nanofibers and demonstrated the advantage of the coaxial electrospinning technique to produce encapsulated materials showing better sustained release behavior of the bioactive compounds.

13.6 Future trends and conclusions

Nanostructured materials are becoming increasingly important for the food sector, and among them functionalized nano- and microfibers obtained by electrospinning play an important role, which is expected to become even more relevant in the near future. The importance of these materials is reflected in the diversity of areas where promising results and applications have been obtained already, including food processing, packaging, food analysis and safety.

The simplicity and versatility of the electrospinning technique opens up several avenues for the preparation of fibrous multifunctional materials. We have reviewed currently available information related to the application of functionalized nanofibers in the food area, focusing mainly on separation operations, enzyme immobilization, encapsulation of food bioactive compounds and microorganisms and the fabrication of sensor devices. The continued improvement of nanofiber formulation and functionality is likely to reveal further new areas of application in the food industry.

However, compared with other application areas, the applications of electrospun nanofibers in food and agricultural systems are relatively few. Important in the increased use of these nanofibrous materials in food applications will be the utilization of nontoxic solvents, generally recognized as safe (GRAS) polymers, and other fiber components. Thereafter, it can be foreseen that currently unexplored applications will develop (e.g., as components of 'green' food packaging materials), allowing acceptable and safe food contact applications as food texture modifiers, or as encapsulating edible materials.

The application of these materials in membrane technologies for separation processes in the food industry will likely grow rapidly, as advances are made. Additional experimental and theoretical studies are required to achieve a better control of fiber size and morphology, as well of the homogeneity of the obtained fibrous membranes. A scale-up of the production process, a systematic investigation at the industrial scale of the well-known lab variables influencing the process and cost/benefit studies are still needed.

In the area of enzyme immobilization, deeper knowledge is required on the effects of fiber production process (electrospinning, shear and electrostatic potential) on enzyme activity, better control of fiber structure manipulation to allow better substrate access to the immobilized enzymes and exploitation of alternative strategies using the versatility of electrospinning to incorporate the enzymes and additional functionalities – for example, in the fabrication of multienzymatic bioreactors.

More efforts are needed to develop specific biosensing devices for the food and drink industries. Many developments are expected in the future for the application of functionalized nanofibers in the area of food biosensing, targeting the fabrication of nanosensors with improved detection characteristics (sensitivity and selectivity). Areas where developments are needed and expected include: the development of amperometric microbial sensors for use in the food fermentation industry, taking advantage of the ability to encapsulate living cells into the electrospun nanofibers; the development of multisensor biomimetic devices for more rapid and accurate multicomponent food analysis; and the fabrication of cost effective and portable sensor devices, while still maintaining high sensitivity and selectivity.

13.7 Acknowledgments

The author offers his deepest thanks for the collaboration and contributions of his students, B. Veleirinho, P. Ferreira, A.R. Teles, D. Silva, N. Almeida and S. Pinto. Organizations that provided financial support made possible our work and our continued research interest to develop this promising research area, and they are thoroughly acknowledged – FCT project NanoBioCats (PTDC/CTM-POL/112289/2009), project Nanopacksafer (NANO/NTec-SQA/0033/2007), QOPNA research unit (project PEst-C/QUI/UI0062/2011).

13.8 Sources of further information

Web sources on nanotechnology (general)

- http://www.nanoforum.org/ – European Nanotechnology Gateway
- http://www.observatory-nano.eu/project/ – The ObservatoryNANO provides scientific and economic analysis of nanoscience and nanotechnology developments, including regulation, ethical and societal aspects.

Web sources on food nanotechnology

FAO/WHO (Food and Agriculture Organization of the United Nations/World Health Organization) (2010) FAO/WHO Expert Meeting on the Application of Nanotechnologies in the Food and Agriculture Sectors: Potential Food Safety Implications: Meeting Report, Rome, 130 pp., available from: http://www.fao.org/ag/agn/agns/nanotechnologies_en.asp

FSA (Food Standards Agency). FSA Citizens Forums: Nanotechnology and food. TNS-BMRB Report. JN 219186, April 2011, 39 pp., available from: http://www.food.gov.uk/news/newsarchive/2011/apr/nanoviews

Nanotechnology in agriculture and food, available from: http://www.nanotechproject.org/inventories/agrifood/

Joseph, T. and Morrison, M. (2006), Nanotechnology in agriculture and food, *Nanoforum report*, 14 pp., available from: http://www.nanoforum.org/

Pray, L. and Yaktine, A., *Rapporteurs* (2009), Food Forum; Institute of Medicine, *Nanotechnology in food products: Workshop summary*,

Washington, DC, National Academies Press, available from: http://www.nap.edu/openbook.php?record_id=12633

Web sources on electrospinning

SPINRATI – The electrospinning gateway
A complete source of information about research groups, application news, companies, events and bibliographic resources.
http://www.spinrati.com/index.html

Relevant text books on electrospinning and functional nanofibers

- An introduction to electrospinning and nanofibers, S. Ramakrishna, K. Fujihara, W.-E. Teo, T.-C. Lim and Z. Ma, World Scientific, 2005.
- Electrospinning, Jon Stanger, Nick Tucker and Mark Staiger, Smithers Rapra Technology, 2009.
- Electrospinning and nanofibers, Tao Han, VDM Verlag, 2009.
- New Fibers, T. Hongū and G. O. Phillips (eds), 2nd edn, Woodhead, 1997.
- Polymeric nanofibers, D. H. Reneker and H. Fong, American Chemical Society, Division of Polymer Chemistry, American Chemical SocietyMeeting, 2006.
- Science and technology of polymer nanofibers, A. L. Andrady, John Wiley and Sons, 2008.
- Structure formation in polymeric fibers, D. R. Salem (ed.), Carl Hanser Verlag, 2001.

Selected commercial companies on nanofiber production/electrospinning

Electrospinz
44 Lee Street, Blenheim, New Zealand
http://www.electrospinz.co.nz/

ELMARCO s.r.o.
460 07 Liberec, Czech Republic
– Technology and equipment for the industrial-scale production of nanofibers
http://www.elmarco.com/

eSpin Technologies, Inc.Chattanooga, TN 37416, USA
– Technology to commercially manufacture nanofibers and nanofiber-based products
http://www.espintechnologies.com/

FibeRio Technology Corp.Sugar Road Edinburg, Texas 78539, USA
– Forcespinning™ Technology and Cyclone™ Systems
http://fiberiotech.com/

Finetex Technology Inc.Hampshire 03051–3985, USA

– Nanofibers and nanostructures manufactured on a large scale by electrospinning
http://www.finetextech.com/
MECC Co. LtdFukuoka Japan 838–0137
– Manufacturing and sales of equipment
http://www.mecc.co.jp/en/index.html
NaBond Nanofibers Electrospinning Unit 1001, Fourseas Building, Nathan Road, KL, Hong Kong
– Fiber producers and equipment suppliers
http://www.electro-spinning.com/
Nano – NC Electrospinning and electrospray systems
http://www.nanonc.co.kr/
SNS Nano Fiber Technology LLCHudson, OH 44236, USA
– Producer of specialized nanofiber matrices
http://www.snsnano.com/
The Electrospinning Company Ltd
Chilton, Didcot, Oxon OX11 0QX, UK
– Functionalized nanofibers for the life sciences sector
http://www.electrospinning.co.uk/
TransMIT GmbH – Polynanotec
35394 Gießen, Germany
– Research and development of electrospinning formulations
http://www.polynanotec.com/
YFlowMalaga, Spain
– Nanotechnology company based on electrohydrodynamic microfluidic techniques to generate electrified coaxial microjets of two immiscible liquids – Services and equipment suppliers
http://www.yflow.com/

13.9 References

Agarwal, S. and Greiner, A. (2011), On the way to clean and safe electrospinning – green electrospinning: Emulsion and suspension electrospinning, *Polymers for Advanced Technologies*, **22**(3), 372–8.

Ahn, H. K., Kim, B. C., Jun, S. H., Chang, M. S., Lopez-Ferrer, D., Smith, R. D., Gu, M. B., Lee, S. W., Kim, B. S. and Kim, J. (2010), Robust trypsin coating on electrospun polymer nanofibers in rigorous conditions and its uses for protein digestion, *Biotechnology and Bioengineering*, **107**(6), 917–23.

Ahn, Y. C., Park, S. K., Kim, G. T., Hwang, Y. J., Lee, C. G., Shin, H. S. and Lee, J. K. (2006), Development of high efficiency nanofilters made of nanofibers, *Current Applied Physics*, **6**, 1030–5.

An, J., Zhang, H., Zhang, J., Zhao, Y. and Yuan, X. (2009), Preparation and antibacterial activity of electrospun chitosan/poly(ethylene oxide) membranes containing silver nanoparticles, *Colloid Polymer Science*, **287**, 1425–34.

Anal, A. K. and Singh, H. (2007), Recent advances in microencapsulation of probiotics for industrial applications and targeted delivery, *Trends in Food Science & Technology*, **18**, 240–51.

Arecchi, A., Scampicchio, M., Drusch, S. and Mannino, S. (2010), Nanofibrous membrane based tyrosinase-biosensor for the detection of phenolic compounds, *Analytica Chimica Acta*, **659**, 133–6.

Aussawasathien, D., Dong, J. H. and Dai, L. (2005), Electrospun polymer nanofiber sensors, *Synthetic Metals*, **154**, 37–40.

Aussawasathien, D., Teerawattananon, C. and Vongachariya, A. (2008), Separation of micron to sub-micron particles from water: Electrospun nylon-6 nanofibrous membranes as pre-filters, *Journal of Membrane Science*, **315**, 11–19.

Baggiani, C., Anfossi, L. and Giovannoli, C. (2007), Solid phase extraction of food contaminants using molecular imprinted polymers, *Analytica Chimica Acta*, **591**, 29–39.

Barhate, R. S. and Ramarkrishna, S. (2007), Nanofibrous filtering media: Filtration problems and solutions from tiny materials, *Journal of Membrane Science*, **296**, 1–8.

Brady, D. and Jordaan, J. (2009), Advances in enzyme immobilisation, *Biotechnology Letters*, **31**, 1639–50.

Burger, C., Hsiao, B. S. and Chu, B. (2006), Nanofibrous materials and their applications, *Annual Review of Materials Science*, **36**, 333–68.

Cao, F., Guo, S., Ma, H., Shan, D., Yang, S. and Gong, J. (2011), Nickel oxide microfibers immobilized onto electrode by electrospinning and calcination for nonenzymatic glucose sensor and effect of calcination temperature on the performance, *Biosensors and Bioelectronics*, **26**, 2756–60.

Champagne, C. P. and Fustier, P. (2007), Microencapsulation for the improved delivery of bioactive compounds into foods, *Current Opinion in Biotechnology*, **18**, 184–90.

Chen, H. and Hsieh, Y. L. (2005) Enzyme immobilization on ultrafine cellulose fibers via poly(acrylic acid) electrolyte grafts, *Biotechnology and Bioengineering*, **90**, 405–13.

Chew, S. Y., Wen, J., Yim, E. K. F. and Leong, K. W. (2005), Sustained release of proteins from electrospun biodegradable fibers, *Biomacromolecules*, **6**, 2017–24.

Chigome, S., Darko, G., Buttner, U. and Torto, N. (2010), Semi-micro solid phase extraction with electrospun polystyrene fiber disks, *Analytical Methods*, **2**, 623–6.

Dai, T., Mileti , N., Loos, K., Elbahri, M. and Abetz, V. (2011), Electrospinning of poly[acrylonitrile-*co*-(glycidyl methacrylate)] nanofibrous mats for the immobilization of *Candida Antarctica* lipase B, *Macromolecular Chemistry and Physics*, **212**, 319–27.

Dai, Y., Niu, J., Liu, J., Yin, L. and Xu, J. (2010), In situ encapsulation of laccase in microfibers by emulsion electrospinning: Preparation, characterization, and application, *Bioresource Technology*, **101**, 8942–7.

Desai, K., Kit, K., Li, J., Davidson, P. M., Zivanovic, S. and Meyer, H. (2009), Nanofibrous chitosan non-wovens for filtration applications, *Polymer*, **50**, 3661–9.

Dheraprasart, C., Rengpipat, S., Supaphol, P. and Tattiyakul, J. (2009), Morphology, release characteristics, and antimicrobial effect of nisin-loaded electrospun gelatin fiber mat, *Journal of Food Protection*, **72**(11), 2293–300.

Ding, B., Kim, J., Miyazaki, Y. and Shiratori, S. (2004), Electrospun nanofibrous membranes coated quartz crystal microbalance as gas sensor for NH3 detection, *Sensors and Actuators B: Chemical*, **101**, 373–80.

Ding, B., Wang, M., Wang, X., Yu, J. and Sun, G. (2010a), Electrospun nanomaterials for ultrasensitive sensors, *Materials Today*, **13**, 16–27.

Ding, B., Wang, M., Yu, J. and Sun, G. (2009), Gas sensors based on electrospun nanofibers, *Sensors*, **9**(3), 1609–24.

Ding, B., Yamazaki, M. and Shiratori, S. (2005), Electrospun fibrous polyacrylic acid membrane-based gas sensors, *Sensors and Actuators B: Chemical*, **106**, 477–83.

Ding, Y., Wang, Y. and Lei, Y. (2010b), Direct electrochemistry and electrocatalysis of novel single-walled carbon nanotubes–hemoglobin composite microbelts – towards the development of sensitive and mediator-free biosensor, *Biosensors and Bioelectronics*, **26**, 390–7.

Ding, Y., Wang, Y., Su, L., Bellagamba, M., Zhang, H. and Lei, Y. (2010c), Electrospun Co_3O_4 nanofibers for sensitive and selective glucose detection, *Biosensors and Bioelectronics*, **26**, 542–8.

Dong, B., Smith, M. and Wnek, G. (2009), Encapsulation of multiple biological compounds within a single electrospun fiber, *Small*, **5**(13), 1508–12.

Dror, Y., Kuhn, J., Avrahami, R. and Zussman, E. (2008), Encapsulation of enzymes in biodegradable tubular structures, *Macromolecules*, **41**, 4187–92.

Feng, Q., Xia, X., Wei, A., Wang, X., Wei, Q., Huo, D. and Wei, A. (2011), Preparation of Cu(II)-chelated poly(vinyl alcohol) nanofibrous membranes for catalase immobilization, *Journal of Applied Polymer Science*, **120**, 3291–6.

Fernandez, A., Torres-Giner, S. and Lagaron, J. M. (2009), Novel route to stabilization of bioactive antioxidants by encapsulation in electrospun fibers of zein prolamine, *Food Hydrocolloids*, **23**, 1427–32.

Frenot, A. and Chronakis, I. S. (2003), Polymer nanofibers assembled by electrospinning, *Current Opinion in Colloid and Interface Science*, **8**, 64–75.

Gensheimer, M., Becker, M., Brandis-Heep, A., Wendorff, J. H., Thauer, R. K. and Greiner, A. (2007), Novel biohybrid materials by electrospinning: Nanofibers of poly(ethylene oxide) and living bacteria, *Advanced Materials*, **19**, 2480–2.

Gopal, R., Kaur, S., Feng, C. Y., Chan, C., Ramakrishna, S., Tabe, S. and Matsuura, T. (2007), Electrospun nanofibrous polysulfone membranes as pre-filers: Particulate removal, *Journal of Membrane Science*, **289**, 210–9.

Gopal, R., Kaur, S., Ma, Z. W., Chan, C., Ramakrishna, S. and Matsuura, T. (2006), Electrospun nanofibrous filtration membrane, *Journal of Membrane Science*, **281**, 581–6.

Graham, K., Ouyang, M., Raether, T., Grafe, T., McDonald, B. and Knauf, P. (2002), Polymeric nanofibers in air filtration applications, Fifteenth Annual Technical Conference and Expo of the American Filtration and Separations Society, Conference Proceedings, 2002, 17pp.

Grandison, A. S. and Lewis, M. J. (1996), *Separation Processes in the Food and Biotechnology Industries – Principles and Applications*, Woodhead.

Greiner, A. and Wendorff, J. H. (2007), Electrospinning: A fascinating method for the preparation of ultrathin fibers, *Angewandte Chemie International Edition*, **46**, 5670–703.

Greiner, A., Wendorff, J. H., Yarin, A. L. and Zussman, E. (2006), Biohybrid nanosystems with polymer nanofibers and nanotubes, *Applied Microbiology and Biotechnology*, **71**, 387–93.

Guler, M. O., Soukasene, S., Hulvat, J. F. and Stupp, S. I. (2005), Presentation and recognition of biotin on nanofibers formed by branched peptide amphiphiles, *Nano Letters*, **5**, 249–52.

He, Y., Zhang, T., Zheng, W., Wang, R., Liu, X., Xia, Y. and Zhao, J. (2010), Humidity sensing properties of BaTiO3 nanofiber prepared via electrospinning, *Sensors and Actuators B*, **146**, 98–102.

Herricks, T. E., Kim, S. H., Kim, J., Li, D., Kwak, J. H., Grate, J. W., Kim, S. H. and Xia, Y. N. (2005), Direct fabrication of enzyme-carrying polymer nanofibers by electrospinning, *Journal of Materials Chemistry*, **15**, 3241–5.

Hong, K. H., Park, J. L., Sul, I. H., Youk, J. H. and Kang, T. J. (2006), Preparation of antimicrobial poly(vinyl alcohol) nanofibers containing silver nanoparticles, *Journal of Polymer Science: Part B: Polymer Physics*, **44**, 2468–74.

Huang, X. J., Xu, Z., Wan, L., Innocent, C. and Seta, P. (2006), Electrospun nanofibers modified with phospholipid moieties for enzyme immobilization, *Macromolecular Rapid Communications*, **27**, 1341–5.

Huang, X. J., Yu, A. G. and Xu, Z. K. (2008), Covalent immobilization of lipase from *Candida rugosa* onto poly(acrylonitrile-co-2-hydroxyethyl methacrylate) electrospun fibrous membranes for potential bioreactor application, *Bioresource Technology*, **99**, 5459–65.

Huang, X. J., Yu, A. G., Jiang, J., Pan, C., Qian, J. W. and Xu, Z. K. (2009), Surface modification of nanofibrous poly(acrylonitrile-*co*-acrylic acid) membrane with biomacromolecules for lipase immobilization. *Journal of Molecular Catalysis B: Enzymatic*, **57**, 250–6.

Hwang, S. and Jeong, S. (2011), Electrospun nano composites of poly(vinyl pyrrolidone)/nano-silver for antibacterial materials, *Journal of Nanoscience and Nanotechnology*, **11**(1), 610–3.

Jeon, H. J., Kim, J. S., Kim, T. G., Kim, J. H., Yu, W. R. and Youk, J. H. (2008), Preparation of poly(ε-caprolactone)-based polyurethane nanofibers containing silver nanoparticles, *Applied Surface Science*, **254**, 5886–90.

Ji, S. Z., Li, Y. and Yang, M. J. (2008), Gas sensing properties of a composite composed of electrospun poly(methyl methacrylate) nanofibers and in situ polymerized polyaniline, *Sensors and Actuators B: Chemical*, **133**, 644–9.

Jia, H., Zhu, G., Vugrinovich, B., Kataphinan, W., Reneker, D. H. and Wang, P. (2002), Enzyme carrying polymeric nanofibers prepared via electrospinning for use as unique biocatalyst, *Biotechnology Progress*, **18**, 1027–32.

Jun, S.-H., Chang, M. S., Kim, B. C., An, H. J., Lopez-Ferrer, D., Zhao, R., Smith, R. D., Lee, S.-W. and Kim, J. (2010), Trypsin coatings on electrospun and alcohol-dispersed polymer nanofibers for a trypsin digestion column, *Analytical Chemistry*, **82**(18), 7828–34.

Jun, Z., Aigner, A., Czubayko, F., Kissel, T., Wendorff, J. H. and Greiner, A. (2005), Poly(vinyl alcohol) nanofibers by electrospinning as a protein delivery system and the retardation of enzyme release by additional polymer coatings, *Biomacromolecules*, **6**, 1484–8.

Kandimalla, V. B., Tripathi, V. S. and Ju, H. X. (2006), Immobilization of biomolecules in sol-gels: Biological and analytical applications, *Critical Reviews in Analytical Chemistry*, **36**(2), 73–106.

Kang, X. J., Pan, C., Xu, Q., Yao, Y. F., Wang, Y., Qi, D. J. and Gu, Z. Z. (2007), The investigation of electrospun polymer nanofibers as a solid-phase extraction sorbent for the determination of trazodone in human plasma, *Analytica Chimica Acta*, **587**(1), 75–81.

Kessick, R. and Tepper, G. (2006), Electrospun polymer composite fiber arrays for the detection and identification of volatile organic compounds, *Sensors and Actuators B*, **117**, 205–10.

Kim, B. C., Nair, S., Kim, J., Kwak, J. H., Grate, J. W., Kim, S. H. and Gu, M. B. (2005), Preparation of biocatalytic nanofibres with high activity and stability via enzyme aggregate coating on polymer nanofibres, *Nanotechnology*, **16**, S382–8.

Kim, I. and Rothschild, A. (2011), Nanostructured metal oxide gas sensors prepared by electrospinning, *Polymers for Advanced Technologies*, **22**, 318–25.

Klein, S., Kuhn, J., Avrahami, R., Tarre, S., Beliavski, M., Green, M. and Zussman, E. (2009), Encapsulation of bacterial cells in electrospun microtubes, *Biomacromolecules*, **10**, 1751–6.

Kriegel, C., Arrechi, A., Kit, K., McClements, D. J. and Weiss, J. (2008), Fabrication, functionalization, and application of electrospun biopolymer nanofibers, *Critical Reviews in Food Science and Nutrition*, **48**, 775–97.

Kriegel, C., Kit, K. M., McClements, D. J. and Weiss, J. (2009), Nanofibers as carrier systems for antimicrobial microemulsions. Part I: Fabrication and characterization, *Langmuir*, **25**, 1154–61.

Kriegel, C., Kit, K. M., McClements, D. J. and Weiss, J. (2010), Nanofibers as carrier systems for antimicrobial microemulsions. II. Release characteristics and antimicrobial activity, *Journal of Applied Polymer Science*, **118**, 2859–68.

Lagaron, J. M. and Lopez-Rubio, A. (2011), Nanotechnology for bioplastics: Opportunities, challenges and strategies, *Trends in Food Science and Technology*, **22**, 611–617.

Lala, N. L., Ramaseshan, R., Bojun, L., Sundarrajan, S., Barhate, R. S., Ying-jun, L. and Ramakrishna, S. (2007), Fabrication of nanofibers with antimicrobial functionality used as filters: Protection against bacterial contaminants, *Biotechnology and Bioengineering*, **97**, 1357–65.

Lee, S. W. and Belcher, A. M. (2004), Virus-based fabrication of micro- and nanofibers using electrospinning, *Nano Letters*, **4**(3), 387–90.

Li, D., Frey, M. W., Vynias, D. and Baeumner, A. J. (2007a) Availability of biotin incorporated in electrospun PLA fibers for streptavidin binding, *Polymer*, **48**, 6340–7.

Li, P., Li, Y., Ying, B. and Yang, M. (2009a), Electrospun nanofibers of polymer composite as a promising humidity sensitive material, *Sensors and Actuators B*, **141**, 390–5.

Li, S. F. and Wu, W. T. (2009), Lipase-immobilized electrospun PAN nanofibrous membranes for soybean oil hydrolysis, *Biochemical Engineering Journal*, **45**, 48–53.

Li, S. F., Chen, J. P. and Wu, W. T. (2007b), Electrospun polyacrylonitrile nanofibrous membranes for lipase immobilization, *Journal of Molecular Catalysis B: Enzymatic*, **47**, 117–24.

Li, X. Q., Su, Y., Liu, S. P., Tan, L. J., Mo, X. M. and Ramakrishna, S. (2010), Encapsulation of proteins in poly(L-lactide-co-caprolactone) fibers by emulsion electrospinning, *Colloids and Surfaces B: Biointerfaces*, **75**, 418–24.

Li, Y., Lim, L. T. and Kakuda, Y. (2009b), Electrospun zein fibers as carriers to stabilize (-)-epigallocatechin gallate, *Journal of Food Science*, **74**(3), C233–40.
Li, Z., Zhang, H., Zheng, W., Wang, W., Huang, H., Wang, C., MacDiarmid, A. G. and Wei, Y. (2008), Highly sensitive and stable humidity nanosensors based on LiCl doped TiO2 electrospun nanofibers, *Journal of the American Chemical Society*, **130**, 5036–7.
Liu, B., Li, P., Zhang, C. L., Wang, Y. and Zhao, Y. S. (2010a), Theanine synthesized by immobilizing *Pseudomonas nitroreducens* LY in nanofibrous membranes, *Process Biochemistry*, **45**, 1330–3.
Liu, C. X., Zhang, S. P., Su, Z. G. and Wang, P. (2011), Salt induced irreversible protein adsorption with extremely high loadings on electrospun nanofibers, *Langmuir*, **27**, 760–5.
Liu, L., Zhang, T., Wang, L. and Li, S. (2009), Improved ethanol sensing properties of Cu-doped SnO_2 nanofibers, *Materials Letters*, **63**, 2041–3.
Liu, Y., Zhang, L., Guo, Q., Hou, H. and You, T. (2010b), Enzyme-free ethanol sensor based on electrospun nickel nanoparticle-loaded carbon fiber paste electrode, *Analytica Chimica Acta*, **663**, 153–7.
Liu, Z., Kang, X. and Fang, F. (2010c), Solid phase extraction with electrospun nanofibers for determination of retinol and α-tocopherol in plasma, *Microchimica Acta*, **168**, 59–64.
Lopes da Silva, J. A., Veleirinho, B. and Delgadillo I. (2009), Preparation and characterization of electrospun mats made of PET/chitosan hybrid nanofibers, *Journal of Nanoscience and Nanotechnology*, **9**(6), 3798–804.
Lopez-Rubio, A., Gavara, R. and Lagaron, J. M. (2006) Bioactive packaging: Turning foods into healthier foods through biomaterials, *Trends in Food Science and Technology*, **17**(10), 567–75.
López-Rubio, A., Sanchez, E., Sanz, Y. and Lagaron, J. M. (2009), Encapsulation of living bifidobacteria in ultrathin PVOH electrospun fibers, *Biomacromolecules*, **10**, 2823–9.
Lu, P. and Hsieh, Y. L. (2009a), Lipase bound cellulose nanofibrous membrane via Cibacron Blue F3GA affinity ligand, *Journal of Membrane Science*, **330**, 288–96.
Lu, P. and Hsieh, Y. L. (2009b), Organic compatible polyacrylamide hydrogel fibers, *Polymer*, **50**, 3670–9.
Lu, T., Chen, X., Shi, Q., Wang, Y., Zhang, P. and Jing, X. (2008), The immobilization of proteins on biodegradable fibers via biotin–streptavidin bridges, *Acta Biomaterialia*, **4**, 1770–7.
Lu, X., Wang, C. and Wei, Y. (2009), One-dimensional composite nanomaterials: synthesis by electrospinning and their applications, *Small*, **5**, 2349–70.
Luo, Y., Nartker, S., Miller, H., Hochhalter, D., Wiederoder, M., Wiederoder, S., Setterington, E., Drzal, L. T. and Alocilja, E. C. (2010), Surface functionalization of electrospun nanofibers for detecting *E. coli* O157:H7 and BVDV cells in a direct-charge transfer biosensor, *Biosensors and Bioelectronics*, **26**, 1612–7.
Ma, H., Burger, C., Hsiao, B. S. and Chu, B. (2011), Ultra-fine cellulose nanofibers: New nano-scale materials for water purification, *Journal of Materials Chemistry*, **21**, 7507–10.
Ma, Z., Kotaki, M. and Ramakrishna, S. (2005), Electrospun cellulose nanofiber as affinity membrane, *Journal of Membrane Science*, **265**, 115–23.

Ma, Z., Kotaki, M. and Ramakrishna, S. (2006), Surface modified nonwoven polysulphone (PSU) fiber mesh by electrospinning: A novel affinity membrane, *Journal of Membrane Science*, **272**, 179–87.

Ma, Z. and Ramakrishna, S. (2008), Electrospun regenerated cellulose nanofiber affinity membrane functionalized with protein A/G for IgG purification, *Journal of Membrane Science*, **319**, 23–8.

Manesh, K. M., Santhosh, P., Gopalan, A. and Lee, K. P. (2007), Electrospun poly(vinylidene fluoride)/poly(aminophenylboronic acid) composite nanofibrous membrane as a novel glucose sensor, *Analytical Biochemistry*, **360**, 189–95.

Marx, S., Jose, M. V., Andersen, J. D. and Russell, A. J. (2011), Electrospun gold nanofiber electrodes for biosensors, *Biosensors and Bioelectronics*, **26**, 2981–6.

Moradzadegan, A., Ranai-Siadat, S. O., Ebrahim-Habibi, A., Barshan-Tashnizi, M., Jalili, R., Torabi, S. F. and Khajed, K. (2010), Immobilization of acetylcholinesterase in nanofibrous PVA/BSA membranes by electrospinning, *Engineering in Life Sciences*, **10**, 57–64.

Neethirajan, S. and Jayas, D. S. (2011), Nanotechnology for the food and bioprocessing industries, *Food and Bioprocess Technology*, **4**(1), 39–47.

Patel, A. C., Li, S. X., Yuan, J. and Wei, Y. (2006), In situ encapsulation of horseradish peroxidase in electrospun porous silica fibers for potential biosensor applications, *Nano Letters*, **6**, 1042–6.

Pinto, N. J., Rivera, D., Melendez, A., Ramos, I., Lim, J. H. and Johnson, A. T. C. (2011), Electrical response of electrospun PEDOT-PSSA nanofibers to organic and inorganic gases, *Sensors and Actuators B*, **156**, 849–53.

Qi, D. J., Kang, X. J., Chen, L. Q., Yiyun, Z., Wei, H. and Gu, Z. (2008a), Electrospun polymer nanofibers as a solid-phase extraction sorbent for the determination of trace pollutants in environmental water, *Analytical and Bioanalytical Chemistry*, **390**(3), 929–38.

Qi, Q., Zhang, T. and Wang, L. (2008b), Improved and excellent humidity sensitivities based on KCl-doped TiO2 electrospun nanofiber, *Applied Physics Letters*, **93**, 023105.

Qi, Q., Zhang, T., Liu, L., Zheng, X. and Lu, G. (2009), Improved NH_3, C_2H_5OH, and CH_3COCH_3 sensing properties of SnO_2 nanofibers by adding block copolymer P123, *Sensors and Actuators B*, **141**, 174–8.

Quintanilla-Carvajal, M. X., Camacho-Díaz, B. H., Meraz-Torres, L. S., Chanona-Pérez, J. J., Alamilla-Beltrán, L., Jimenéz-Aparicio, A., and Gutiérrez-López, G. F. (2010), Nanoencapsulation: A new trend in food engineering processing, *Food Engineering Reviews*, **2**(1), 39–50.

Ren, G., Xu, X., Liu, Q., Cheng, J., Yuan, X., Wu, L. and Wan, Y. (2006), Electrospun poly(vinyl alcohol)/glucose oxidase biocomposite membranes for biosensor applications, *Reactive & Functional Polymers*, **66**, 1559–64.

Sakai, S., Liu, Y., Yamaguchi, T., Watanabe, R., Kawabe, M. and Kawakami, K. (2010a), Immobilization of *Pseudomonas cepacia* lipase onto electrospun polyacrylonitrile fibers through physical adsorption and application to transesterification in nonaqueous solvent, *Biotechnology Letters*, **32**, 1059–62.

Sakai, S., Liu, Y., Yamaguchi, T., Watanabe, R., Kawabe, M. and Kawakami, K. (2010b), Production of butyl-biodiesel using lipase physically-adsorbed onto electrospun polyacrylonitrile fibers, *Bioresource Technology*, **101**, 7344–9.

Sakai, S., Antoku, K., Yamaguchi, T., Watanabe, R., Kawabe, M. and Kawakami, K. (2010c), Electrospun PVA fibrous mats immobilizing lipase entrapped in alkyl-silicate cages: Application to continuous production of fatty acid butyl ester, *Journal of Molecular Catalysis B: Enzymatic*, **63**, 57–61.
Salalha, W., Kuhn, J., Dror, Y. and Zussman, E. (2006), Encapsulation of bacteria and viruses in electrospun nanofibres, *Nanotechnology*, **17**(18), 4675–81.
Sanchez-Garcia, M. D., Lopez-Rubio, A. and Lagaron, J. M. (2010), Natural micro and nanobiocomposites with enhanced barrier properties and novel functionalities for food biopackaging applications, *Trends in Food Science and Technology*, **21**(11), 528–36.
Sawicka, K., Gouma, P. and Simon, S. (2005), Electrospun biocomposite nanofibers for urea biosensing, *Sensors and Actuators B*, **108**, 585–8.
Scampicchio, M., Arecchi, A., Lawrence, N. S. and Mannino, S. (2010), Nylon nanofibrous membrane for mediated glucose biosensing, *Sensors and Actuators B: Chemical*, **145**, 394–7.
Schiffman, J. D. and Elimelech, M. (2011), Antibacterial activity of electrospun polymer mats with incorporated narrow diameter single-walled carbon nanotubes, *ACS Applied Materials & Interfaces*, **3**(2), 462–8.
Sekhon, B. S. (2010), Food nanotechnology – an overview, *Nanotechnology, Science and Applications*, **3**, 1–15.
Senecal, A., Magnone, J., Marek, P. and Senecal, K. (2008), Development of functional nanofibrous membrane assemblies towards biological sensing, *Reactive & Functional Polymers*, **68**, 1429–34.
Sheng, L., Dajing, C. and Yuquan, C. (2011), A surface acoustic wave humidity sensor with high sensitivity based on electrospun MWCNT/Nafion nanofiber films, *Nanotechnology*, **22**, 265504.
Skelton, R. (2000), Membrane filtration applications in the food industry, *Filtration + Separation*, April, 28–30.
Son, W. K., Youk, J. H. and Park, W. H. (2006), Antimicrobial cellulose acetate nanofibers containing silver nanoparticles, *Carbohydrate Polymers*, **65**, 430–4.
Song, X. and Liu, L. (2009), Characterization of electrospun $ZnO–SnO_2$ nanofibers for ethanol sensor, *Sensors and Actuators A*, **154**, 175–9.
Sozer, N. and Kokini, J. L. (2009), Nanotechnology and its applications in the food sector, *Trends in Biotechnology*, **27**(2), 82–9.
Stijnman, A. C., Bodnar, I. and Tromp, R. H. (2011), Electrospinning of food-grade polysaccharides, *Food Hydrocolloids*, **25**, 1393–8.
Sun, L., Zhang, C. and Li, P. (2011), Characterization, antimicrobial activity, and mechanism of a high-performance (-)-epigallocatechin-3-gallate (EGCG)-Cu^{II}/polyvinyl alcohol (PVA) nanofibrous membrane, *Journal of Agriculture and Food Chemistry*, **59**, 5087–92.
Taepaiboon, P., Rungsardthong, U. and Supaphol, P. (2007), Vitamin-loaded electrospun cellulose acetate nanofiber mats as transdermal and dermal therapeutic agents of vitamin A acid and vitamin E, *European Journal of Pharmaceutics and Biopharmaceutics*, **67**, 387–97.
Tang, X., Liu, Y., Hou, H. and You, T. (2011), A nonenzymatic sensor for xanthine based on electrospun carbon nanofibers modified electrode, *Talanta*, **83**, 1410–4.
Tiwari, A., Terada, D., Yoshikawa, C. and Kobayashi, H. (2010), An enzyme-free highly glucose-specific assay using self-assembled aminobenzene boronic acid upon polyelectrolytes electrospun nanofibers-mat, *Talanta*, **82**, 1725–32.

Torres-Giner, S., Ocio, M. J. and Lagaron, J. M. (2009), Novel antimicrobial ultrathin structures of zein/chitosan blends obtained by electrospinning, *Carbohydrate Polymers*, **77**, 261–6.

Vadukumpully, S., Basheer, C., Jeng, C. S. and Valiyaveettil, S. (2011), Carbon nanofibers extracted from soot as a sorbent for the determination of aromatic amines from wastewater effluent samples, *Journal of Chromatography A*, **1218**, 3581–7.

Vega-Lugo, A. C. and Lim, L. T. (2009), Controlled release of allyl isothiocyanate using soy protein and poly(lactic acid) electrospun fibers, *Food Research International*, **42**, 933–40.

Veleirinho, B. and Lopes-da-Silva, J. A. (2009), Application of electrospun poly(ethylene terephthalate) nanofiber mat to apple juice clarification, *Process Biochemistry*, **44**, 353–6.

Wan, L. S., Ke, B. B., Wu, J. and Xu, Z. K. (2007), Catalase immobilization on electrospun nanofibers: Effects of porphyrin pendants and carbon nanotubes, *Journal of Physical Chemistry C*, **111**(38), 14091–7.

Wang, C., Yan, K. W., Lin, Y. D. and Hsieh, P. C. H. (2010a), Biodegradable core/shell fibers by coaxial electrospinning: Processing, fiber characterization, and its application in sustained drug release, *Macromolecules*, **43**, 6389–97.

Wang, J. (2005), Nanomaterial-based electrochemical biosensors, *Analyst*, **130**, 421–6.

Wang, P. (2006), Nanoscale biocatalyst systems, *Current Opinion in Biotechnology*, **17**, 574–9.

Wang, W., Zhang, L., Tong, S., Li, X. and Song, W. (2009a), Three-dimensional network films of electrospun copper oxide nanofibers for glucose determination, *Biosensors and Bioelectronics*, **25**, 708–14.

Wang, W., Zhenyu, L., Liu, L., Zhang, H., Zheng, W., Wang, Y., Huang, H., Wang, Z. and Wang, C. (2009b), Humidity sensor based on LiCl-doped ZnO electrospun nanofibers, *Sensors and Actuators B*, **141**, 404–9.

Wang, X., Chen, X., Yoon, K., Fang, D., Hsiao, B. S. and Chu, B. (2005), High flux filtration medium based on nanofibrous substrate with hydrophilic nanocomposite coating, *Environmental Science & Technology*, **39**, 7684–91.

Wang, X., Fang, D., Yoon, K., Hsiao, B. S. and Chu, B. (2006a), High performance ultrafiltration composite membranes based on poly(vinyl alcohol) hydrogel coating on crosslinked nanofibrous poly(vinyl alcohol) scaffold, *Journal of Membrane Science*, **278**, 261–8.

Wang, X., Ding, B., Yu, J., Wang, M. and Pan, F. (2010b), A highly sensitive humidity sensor based on a nanofibrous membrane coated quartz crystal microbalance, *Nanotechnology*, **21**, 055502.

Wang, X., Ding, B., Yu, J., Si, Y., Yang, S. and Sun, G. (2011), Electro-netting: Fabrication of two-dimensional nano-nets for highly sensitive trimethylamine sensing, *Nanoscale*, **3**, 911–15.

Wang, Y. H. and Hsieh, Y. L. (2004), Enzyme immobilization to ultra-fine cellulose fibers via amphiphilic polyethylene glycol spacers, *Journal of Polymer Science, Polymer Chemistry*, **42**, 4289–99.

Wang, Z. G., Xu, Z. K., Wan, L. S., Wu, J., Innocent, C. and Seta, P. (2006b), Nanofibrous membranes containing carbon nanotubes: Electrospun for redox enzyme immobilization, *Macromolecular Rapid Communications*, **27**, 516–21.

Wang, Z. G., Wan, L. S. and Xu, Z. K. (2009c), Immobilization of catalase on electrospun nanofibrous membranes modified with bovine serum albumin or collagen:

Coupling site-dependent activity and protein-dependent stability, *Soft Matter*, **5**, 4161–8.
Wang, Z. G., Wan, L. S., Liu, Z. M., Huang, X. J. and Xu, Z. H. (2009d), Enzyme immobilization on electrospun polymer nanofibers: An overview, *Journal of Molecular Catalysis B: Enzymatic*, **56**, 189–95.
Wang, Z. G., Wang, J. Q. and Xu, Z. K. (2006c), Immobilization of lipase from *Candida rugosa* on electrospun polysulfone nanofibrous membranes by adsorption, *Journal of Molecular Catalysis B: Enzymatic*, **42**, 45–51.
Wang, Z. and Liu, L. (2009), Synthesis and ethanol sensing properties of Fe-doped SnO_2 nanofibers, *Materials Letters*, **63**, 917–19.
Wang, Z. G., Wang, Y., Xu, H., Li, G. and Xu, Z. K. (2009e), Carbon nanotube-filled nanofibrous membranes electrospun from poly(acrylonitrile-*co*-acrylic acid) for glucose biosensor, *Journal of Physical Chemistry C*, **113**, 2955–60.
Wei, M., Kang, B., Sung, C. and Mead, J. (2006), Core-sheath structure in electrospun nanofibers from polymer blends, *Macromolecular Materials and Engineering*, **291**, 1307–14.
Weiss, J., Takhistov, P. and McClements, D. J. (2006), Functional materials in food nanotechnology, *Journal of Food Science*, **71**(9), R107–16.
Wongsasulak, S., Patapeejumruswong, M., Weiss, J., Supaphol, P. and Yoovidhya, T. (2010), Electrospinning of food-grade nanofibers from cellulose acetate and egg albumen blends, *Journal of Food Engineering*, **98**, 370–6.
Wu, L. L., Yuan, X. Y. and Sheng, J. (2005), Immobilization of cellulase in nanofibrous PVA membranes by electrospinning, *Journal of Membrane Science*, **250**, 167–73.
Wu, X. M., Branford-White, C. J., Yu, D. G., Chatterton, N. P. and Zhu, L. M. (2011), Preparation of core-shell PAN nanofibers encapsulated α-tocopherol acetate and ascorbic acid 2-phosphate for photoprotection, *Colloids and Surfaces B: Biointerfaces*, **82**, 247–52.
Xiao, Y. H. and Li, C. M. (2008), Nanocomposites: from fabrications to electrochemical bioapplications, *Electroanalysis*, **20**, 648–62.
Xu, Q., Yin, X., Wang, M., Wang, H., Zhang, N., Shen, Y., Xu, S., Zhang, L. and Gu, Z. (2010), Analysis of phthalate migration from plastic containers to packaged cooking oil and mineral water, *Journal of Agriculture and Food Chemistry*, **58**, 11311–7.
Xu, X., Yang, Q., Wang, Y., Yun, H., Chen, X. and Jing, X. (2006), Biodegradable electrospun poly(L-lactide) fibers containing antibacterial silver nanoparticles, *European Polymer Journal*, **42**, 2081–7.
Xu, X. and Zhou, M. (2008), Antimicrobial gelatin nanofibers containing silver nanoparticles, *Fibers and Polymers*, **9**(6), 685–90.
Yarin, A. L., Zussman, E., Wendorff, J. H. and Greiner, A. (2007), Material encapsulation and transport in core–shell micro/nanofibers, polymer and carbon nanotubes and micro/nanochannels, *Journal of Materials Chemistry*, **17**, 2585–99.
Ye, P., Xu, Z. K., Wu, J., Innocent C. and Seta, P. (2006), Nanofibrous membranes containing reactive groups: electrospinning from poly(acrylonitrile-co-maleic acid) for lipase immobilization, *Macromolecules*, **39**, 1041–5.
Yoon, K., Kim, K., Wang, X. F., Fang, D. F., Hsiao, B. S. and Chu, B. (2006), High flux ultrafiltration membranes based on electrospun nanofibrous PAN scaffolds and chitosan coating, *Polymer*, **47**, 2434–41.

Yoshimatsu, K., Ye, L., Lindberg, J. and Chronakis, I. S. (2008), Selective molecular adsorption using electrospun nanofiber affinity membranes *Biosensors and Bioelectronics*, **23**, 1208–15.

Yun, K. M., Hogan Jr., C. J., Matsubayashi, Y., Kawabe, M., Iskandar, F. and Okuyama, K. (2007), Nanoparticle filtration by electrospun polymer fibers, *Chemical Engineering Science*, **62**, 4751–9.

Zampetti, E., Pantalei, S., Scalese, S., Bearzotti, A., De Cesare, F., Spinella, C. and Macagnano, A. (2011), Biomimetic sensing layer based on electrospun conductive polymer webs, *Biosensors and Bioelectronics*, **26**, 2460–5.

Zewe, J. W., Steach, J. K. and Olesik, S. V. (2010), Electrospun fibers for solid-phase microextraction, *Analytical Chemistry*, **82**, 5341–8.

Zhang, H., Wu, C., Zhang, Y., White, C. J. B., Xue, Y., Nie, H. and Zhu, L. (2010), Elaboration, characterization and study of a novel affinity membrane made from electrospun hybrid chitosan/nylon-6 nanofibers for papain purification, *Journal of Materials Science*, **45**, 2296–304.

Zhou, Y. and Lim, L. T. (2009), Activation of lactoperoxidase system in milk by glucose oxidase immobilized in electrospun polylactide microfibers, *Journal of Food Science*, **74**(2), C170–6.

Zussman, E. (2011), Encapsulation of cells within electrospun fibers, *Polymers for Advanced Technologies*, **22**(3), 366–71.

14
Functional nanofibers in sound absorption, electromagnetic wave attenuation and bioreactor application

Q. WEI and D. GAO,
Jiangnan University, China

Abstract: This chapter reviews the applications of nanofibers in three main areas: optical fields, electromagnetic absorption and bioreactors. The use of nanofibers is discussed in sound-absorbing materials (considering many possible applications – for example in the automobile industry and in products used within the home) and electromagnetic wave attenuating materials. Finally, the chapter discusses nanofibers used in filtration membranes as part of bioreactor materials. Various applications, both in water and air filtration, exhibit scope for considerable development in the future.

Key words: nanofibers, sound absorbing, anti-electromagnetic interference, biosensor.

14.1 Introduction

The unique properties of nanofibers have meant that their use has increased in recent years, especially in areas such as military weapons, medical devices and sensors for many different applications. The manufacture of protective products for use as defense materials is just one application for which nanofibers are suitable. Their light weight and adsorption capacity make nanofibers a key material in this area, as well as having a beneficial effect in the isolation of biochemical weapons due to high specific surface area, porosity, density and other characteristics. In terms of medical applications, nanofibers can be composited with medicine particles, allowing the speed of drug diffusion to be controlled according to specific requirements, such as immediate release, relaxation and delayed release. Nanofibers have also been used in biological engineering as cell scaffolds and as constituents of artificial skin and organs. Function-specific nanomaterials can be composited with polymers to prepare new nanofibers with a variety of features, such as anti-ultraviolet and antibacterial properties, stealth and high wear resistance. Nanofibers have also been used in micro- and nanoelectronic devices, solar cells, LCD devices and sensors, and have great potential in other areas such as sound absorption, electromagnetic absorption and bioreactors.

14.2 Applications in sound absorption

Sound waves are mechanical waves whose propagation velocity is determined by the media through which they travel. Mechanical waves with different frequencies have different propagation speeds in the same media.

As sound absorption of lower frequencies is problematic when using materials made up of coarser fibers, there is a need for development of new, highly efficient sound-absorption materials. This need has been met by nanofiber-based materials. Nanofibers have excellent sound-absorption properties at low frequencies and their absorption of high-frequency sound is also stable (Yang *et al.*, 2011).

Nanofiber materials with absorptive properties are based on a nanofiber composite membrane among the characteristics of which is low frequency oscillation. The level of frequency causing oscillation in nanofiber-based materials is determined by the scale of the nanofibers. Nanofibers can be used in sound-absorbing membranes to decrease the resonance amplitude, accumulate the sound energy and transform it into heat energy. This process is different to the more common soundproofing method of using sound isolation, but it has the advantage that it transforms sound energy into heat energy, which is stored within the nanoporous fibrous materials.

The sound-absorption properties of nanofibers mean that they have great potential in the automotive sector (transportation) as door insulation, headliners and in car boots. They can also be used in home appliances (fridges, dishwashers, washing machines) and in construction of concert halls, theaters, cinemas and large-scale sports stadiums.

14.3 Applications of functional nanofibers as electromagnetic wave attenuating material

Electromagnetic radiation (often abbreviated as EM radiation or EMI) is a form of energy that exhibits wave-like behavior as it travels through space. EMI has both electric and magnetic field components, which oscillate in phase perpendicular to each other and perpendicular to the direction of energy propagation.

Electromagnetic radiation is classified according to its wave frequency. Electromagnetic waves include radio waves, microwaves, infrared radiation, visible light, ultraviolet radiation, X-rays and gamma rays. The eyes of various organisms can sense a small and somewhat variable window of frequencies called the visible spectrum. The photon is the quantum of the electromagnetic interaction, the basic unit of light and all other forms of electromagnetic radiation; it is also the force carrier for electromagnetic force. EM radiation carries energy and momentum that may be imparted to matter with which it interacts.

With the rapid development of radio and radar technology, electronic and communications equipment are becoming more sensitive, intensive and high-frequency, as well as diversifying. This not only leads to electromagnetic interference and pollution, but – more importantly – also leads to leakage of electromagnetic information. For example, electromagnetic radiation from military electronic devices may be picked up and used by enemy reconnaissance. In order to eliminate or reduce electromagnetic interference and leakage in applications such as missile positioning, anti-electromagnetic interference and viability will need to be greatly improved. Therefore, the development of high-performance and wide-band electromagnetic wave absorbing materials is essential for the progress of military applications of electromagnetic radiation.

Absorbent material is an important functional material for military stealth applications. The electromagnetic wave is effectively absorbed by the material and the electromagnetic energy is transformed into heat or other forms of energy (Zhang *et al.*, 2007).

Electromagnetic wave absorbing materials possess two qualities, namely electromagnetic wave impedance matching ability and attenuation characteristics. Electromagnetic wave impedance matching ability means that the media surface has the smallest reflection coefficient possible, which allows the electromagnetic wave to penetrate the media as far as possible. 'Attenuation characteristics' refers to quick absorption of the electromagnetic wave on entering the material. Electric and magnetic loss factors can be used to characterize the electromagnetic loss. Because there is a contradiction between impedance match and strong absorption for single-component absorption, it is necessary to use multiple composite materials in order to adjust the electromagnetic parameters, making it possible to increase the capacity for absorption loss under the best-matching conditions. The loss mechanisms for electromagnetic waves are divided into resistance loss, dielectric loss and magnetic loss. The performance of the absorbent material mainly depends on the absorption loss capacity of the absorbing agent. Therefore, recent research work on absorbing materials has been focused on absorbing agents.

A few studies have been conducted to evaluate the electromagnetic irradiation shielding effect (EMISE) of nanofibers. Yang *et al.* (2005) studied the EMISE of vapor-grown carbon nanofibers (VGCNFs): reinforced liquid crystal polymer (LCP) composites. It was found that EMISE in the frequency range of 0.15–1.5 GHz increased with increasing VGCNF loading. A shielding effect (SE) of 26 dB was reported for a 1.45 mm thick sample made of 15 wt% VGCNF/LCP when used to block an EM wave of 200 MHz. In addition, it was concluded that Simon's model could not be used to predict the SE of nanofiber-filled composites, since it ignores the multiple reflection effect. Experimentally recorded SE over the range of frequencies studied

(15–1500 MHz) was 6–19 dB higher than the values predicted by Simon's model (Simon, 1981; Al-Saleh and Sundararaj, 2009).

Zhang *et al.* (1998) studied the EMISE of VGCNF/polyesterpolyol shape memory polymer (SMP) composites at different VGCNF loadings and EM frequencies. The EMI SE was studied in the K-band (8–26.5 GHz), Q-band (33–50 GHz) and V-band (50–75 GHz) frequency ranges, and was found to increase exponentially in relation to the increasing frequency. For instance, for a 3 mm thick sample made of 6.7 wt% VGCNF/SMP, the EMI SE at 26 GHz was 35 dB, while it was 60 dB at 75 GHz. In addition, it was found in the same study that the EMI SE increased with the thickness of the shielding material. This observation indicated the contribution of the absorptive capacity of the material in shielding the EM waves. For example, EMI SE in the K-band increased from 16 to 35 dB by increasing the thickness of shielding material from 0.5 to 3 mm. The EMI SEs of VGCNF- and MWNT-filled PS in the X-band frequency range (8.2–12.4 GHz) were compared and the EMI SE of VGCNF/PS in the 12.4–18 GHz (Ku-band) was also measured. The researchers found that, for all composites, EMI SE was independent of EM wave frequency. The EMI SE of MWNT/PS composites was considerably higher than that of VGCNF/PS composites. For example, at 5 wt% filler and a frequency of 10 GHz, the EMI SE of VGCNF/PS and MWNT-PS were 7.5 and 24 dB, respectively. Lee and coworkers (2005) evaluated the EMI shielding effectiveness of 40 wt% VGCNF/PVA and compared it to that of 40% CB-filled PVA processed under similar conditions. The VGCNF used in the study was synthesized using different types of catalysts and different feedstocks. They found that the EMI SE of 30 l m thick as-grown VGCNF/PVA film was lower than that of CB/PVA film of similar thickness. However, after heat treating the VGCNF, the EMI SE of the VGCNF/PVA film was higher.

The EMI SE of different VGCNF-filled polymers and polymer blends was also evaluated. The composites were prepared by melt mixing in an internal mixer followed by compression molding. For example, over a frequency range of 50–1500 MHz, an EMI SE of 30 dB was found for a 2 mm thick plate made of 7.5 vol% VGCNF/PE composite.

The authors of this chapter mixed Fe_3O_4 with Polyaniline (PANI), and then formed composited nanofibers by electrostatic spinning. Fe_3O_4/polyaniline nanofiber has conductive and magnetic properties, and has great potential in the field of microwave absorption. The preparation of polyaniline-coated magnetic iron nanoparticles was undertaken as follows: first isooctane and deionized water were mixed with certain proportion, then the O/W emulsion was obtained by adding aerosols and the solution was dispersed by ultrasonic wave. Second, DBSA was added to the aniline aqueous solution until the pH value reached 3. Then, acidic aniline solution was added into the prepared O/W microemulsion. During the next step, a concentration of

30% ammonium sulfate solution was dropped into the solution as it gradually became darker. The emulsion was broken up with acetone, and a dark green precipitate was obtained. Finally, a radar absorbent of a darker green was obtained after drying in vacuum conditions at 45–50°C for about 45–50 h. Figure 14.1 shows the HRTEM images of Fe_3O_4 nanoparticles and PANI/Fe_3O_4 nanocomposites. Fe_3O_4 nanoparticles look spherical with an average diameter of about 10 nm, as presented in Fig. 14.1a. It can be clearly observed that all Fe_3O_4 nanoparticles are covered by PANI, as indicated

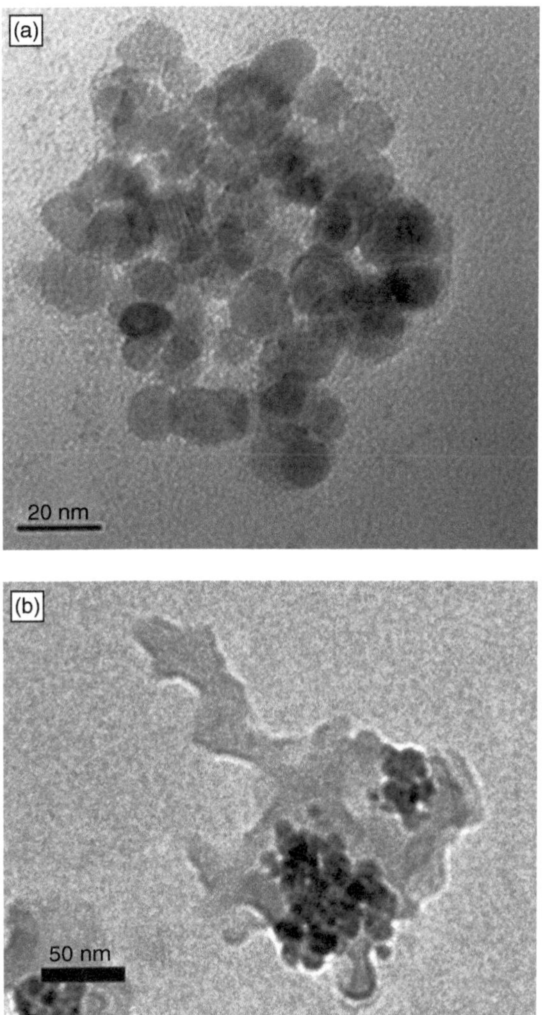

14.1 HRTEM images of (a) Fe_3O_4 nanoparticles and (b) PANI/Fe_3O_4 nanocomposites.

in Fig. 14.1b, and the Fe_3O_4 nanoparticles encapsulated by PANI are also spherical with an average size of about 10 nm, which is similar to the pure Fe_3O_4 nanoparticles shown in Fig. 14.1a. The XRD patterns of PANI, Fe_3O_4 nanoparticles and PANI/Fe_3O_4 nanocomposite with different proportions are shown in Fig. 14.2. As seen in the spectra, the main characteristic peaks of pure PANI are located at $2\theta = 15.07°$, $20.22°$ and $25.18°$, indicating that crystallization occurs in PANI. Moreover, the characteristic peaks of Fe_3O_4 nanoparticles can be observed in the XRD patterns of the PANI/ Fe_3O_4 nanocomposites, thus the existence of Fe_3O_4 nanoparticles in the nanocomposites can be confirmed. On the other hand, the intensity of the main characteristic peak of PANI in the Fe_3O_4 nanoparticles becomes weak compared to the pure PANI, and the downtrend becomes more obvious as the content of Fe_3O_4 nanoparticles increases. It can be seen that the diffraction peaks of PANI disappear in Fe_3O_4/PANI nanocomposite when the content of Fe_3O_4 is 40 wt%, which indicates that the structure of PANI is destroyed with the addition of Fe_3O_4, thus the crystallization degree decreases. The average diameter of Fe_3O_4 nanoparticles in angstroms (D) can be figured out according to Scherer's equation. When the reflecting peak at $2\theta = 34.27°$ is chosen to calculate the average diameter, the estimated average diameter of the Fe_3O_4 nanoparticles are about 10 nm, which is consistent with the HRTEM observation.

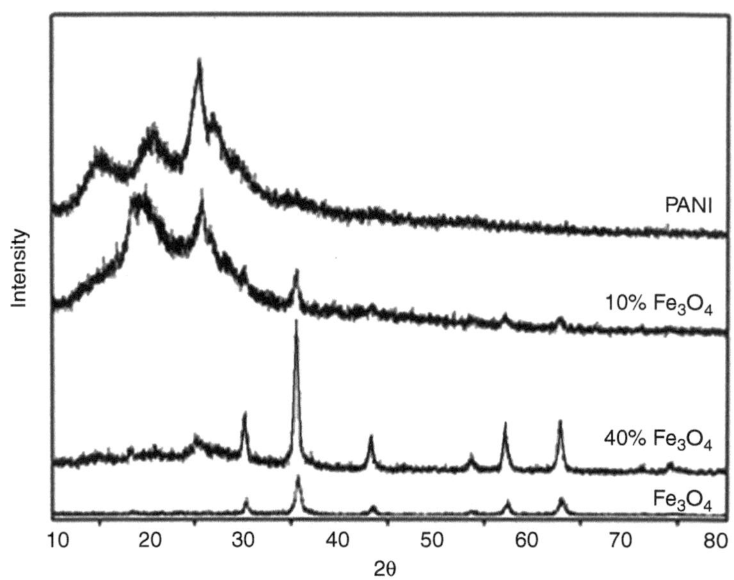

14.2 The XRD patterns of PANI, Fe_3O_4 nanoparticles and PANI/Fe_3O_4 nanocomposite with different proportions.

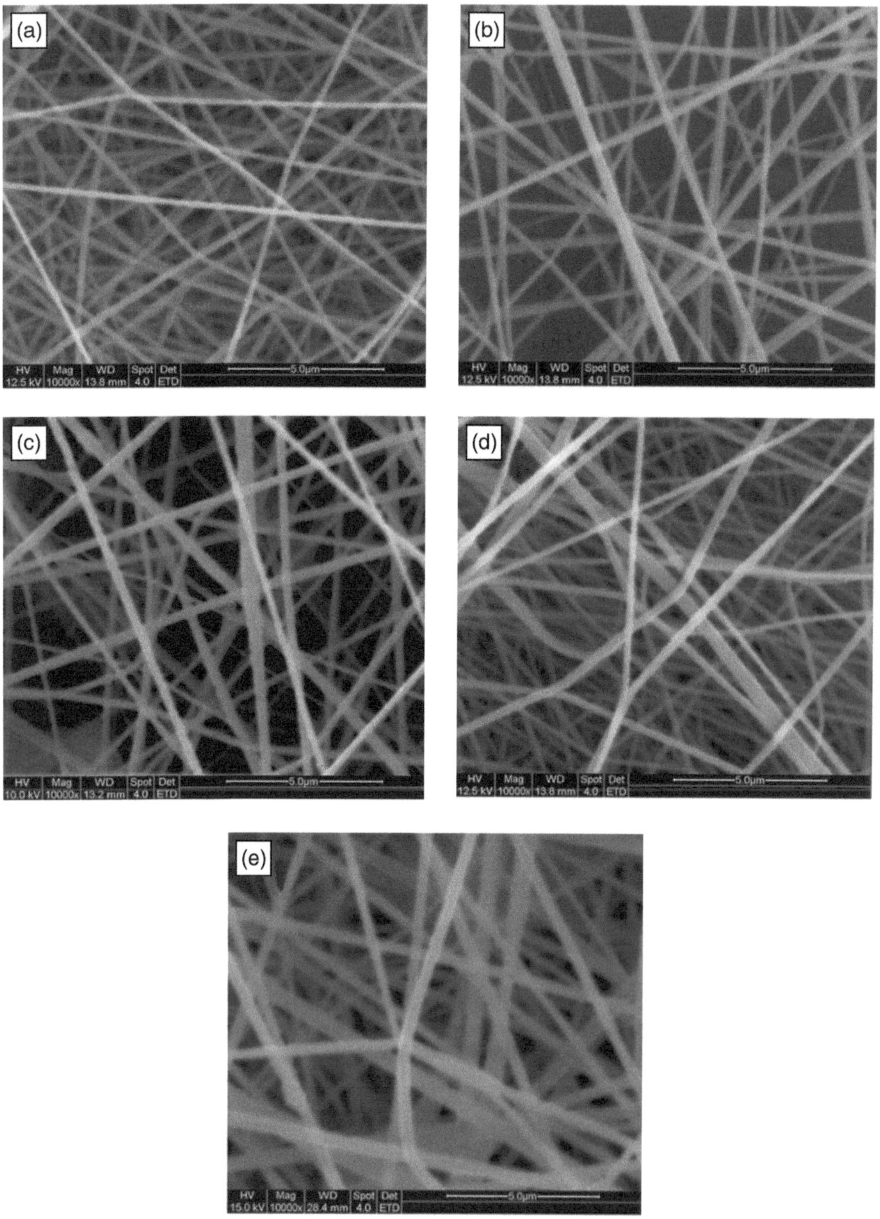

14.3 SEM images of the electrospun PANI and the PANI/ Fe_3O_4–PAN composite nanofibres with different Fe_3O_4 content.

The morphology of the electrospun PANI and the PANI/ Fe_3O_4–PAN composite nanofibers with different Fe_3O_4 content is examined by SEM, as shown in Fig. 14.3. All the fiber mats form very flexible and porous structures in comparison with casting films with the same composition, retaining their green color. From the SEM images, the effect of the content of Fe_3O_4 in PANI on the diameter of the electrospun nanofibers can be clearly observed. Figure 14.3a shows that the PAN and PANI composite nanofibers without Fe_3O_4 have smooth surface morphology; the narrow distribution of fiber diameters can also be seen, demonstrating that PAN and PANI are well blended and single phase. The surface morphology appears rougher when Fe_3O_4 nanoparticles are introduced into the composite, as is shown in Fig. 14.3b, c, d and e, but there are no obvious beads or aggregations observed indicating the homogeneous dispersion of the Fe_3O_4 nanoparticles in the electrospun nanofibers. Figure 14.4 shows the permittivity of the Fe_3O_4, PANI, Fe_3O_4/PANI and Fe_3O_4 /PANI-PAN composite, which was prepared by the *in situ* microemulsion polymerization and electrospinning method. The permittivity of the composites with Fe_3O_4 is lower than that of the composite without Fe_3O_4. The permittivity of the composites containing Fe_3O_4 increases with increasing content of Fe_3O_4.

Several reasons can be associated with such phenomena. First of all, the relative content of PANI in the composite decreases due to the introduction of magnetic material into the composite, which destroys the conjugated

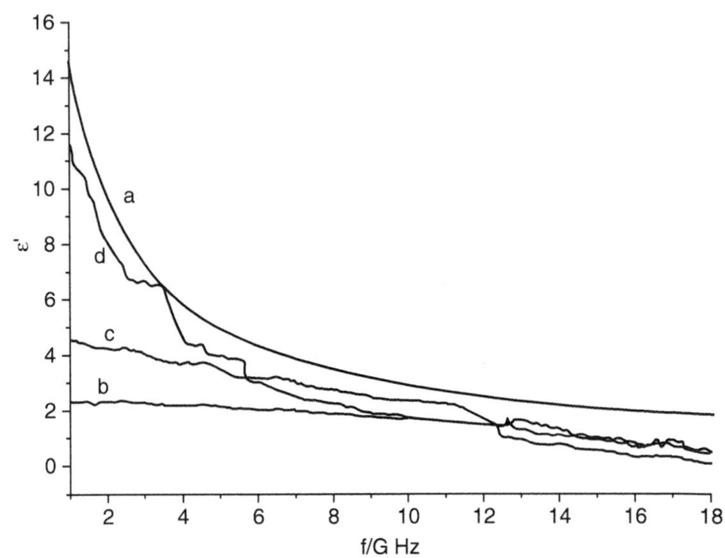

14.4 The permittivity of the Fe_3O_4, PANI, Fe_3O_4/PANI and Fe_3O_4 /PANI - PAN composite corresponding to a,b,c and d, respectively.

structure of PANI to some extent. The conductive properties of the composite are therefore reduced due to structural change. On the other hand, the number of free ion radicals and electric dipoles increases as the Fe_3O_4 content of the composite is increased, which strengthens the ion displacement polarization and orientation polarization of electric dipoles and increases the permittivity of the composite. Moreover, an interface is formed on the core/shell structure between PANI and Fe_3O_4 with the introduction of Fe_3O_4, and the number of two-phase interfaces increases with the content of Fe_3O_4 so that the permittivity of the composites is increased due to the enhancement of interfacial polarization.

Mechanisms of electromagnetic shielding are as follows (Robinson, 1966).

The shielding effectiveness (SE) is typically defined as the ratio of the magnitude of the incident electric field, Ei to the magnitude of the transmitted electric field, Et:

$$SE = \left| \frac{\bar{E}i}{\bar{E}t} \right| \quad [14.1]$$

$$SE(dB) = 20 \cdot \log_{10}\left(\frac{Ei}{Et}\right) \quad [14.2]$$

The above definition of shielding effectiveness is sometimes referred to as the electric field shielding effectiveness (ESE) since it involves the ratio of electric field amplitudes. Shielding effectiveness can also be defined in terms of the ratio of the incident and transmitted magnetic field amplitudes, which is sometimes referred to as magnetic field shielding effectiveness (MSE). For plane waves and the same media (e.g., air) on both sides of the shield, these two definitions are equivalent (i.e., ESE = MSE).

There are several factors that determine the effectiveness of an electromagnetic shield. These factors include the following:

1. frequency of the incident electromagnetic field;
2. shield material parameters (conductivity, permeability and permittivity);
3. shield thickness;
4. type of electromagnetic field source (plane wave, electric field or magnetic field);
5. distance from the source to the shield;
6. shielding degradation caused by any shield apertures and penetrations;
7. quality of the bond between metal shield surfaces.

Electromagnetic fields, in a classical sense, are governed by Maxwell's equations subject to a set of boundary conditions. Schelkunoff, using a transmission line analogy (Schelkunoff , 1955), laid out the basic theory for plane wave attenuation through shields. Using Schelkunoff's formulation (which assumes that the shield is located in the far field of the source), the plane wave shielding effectiveness (in dB) can be expressed as the sum of an absorption loss (A) and a reflection loss (R) plus a multiple reflection correction term (B):

$$\mathrm{SE} = \mathrm{A} + \mathrm{R} + \mathrm{B} \tag{14.3}$$

Absorption loss is incurred as the electromagnetic wave penetrates the shield. The amplitude of the electric and magnetic fields decay exponentially due to induced currents and resultant ohmic losses in the material. The propagation constant of the fields in the metal shield is given by

$$\gamma_s = \sqrt{j\omega\mu_s\sigma_s} \tag{14.4}$$

Where: γ_s = propagation constant of wave inside shield, $j = \sqrt{-1}$, $\omega = 2\pi f$ = radian frequency (in rad/sec), f = frequency (in Hz), μ_s = shield permeability (in H/m) and σ_s = shield conductivity (in mbos/m).

The reflection loss at the interface between two media (e.g., the air and the metal shield) can be thought of as an impedance mismatch loss and is related to the ratio of the wave impedance to the intrinsic impedance of the shield. For plane waves (i.e., radiated electromagnetic waves in the far field of the source), the wave impedance is equal to the characteristic impedance of the propagation medium. The characteristic impedance of free space (and, in close approximation, of air as well) is given by

$$Z_o = \sqrt{\frac{\mu_o}{\varepsilon_o}} = 377\Omega \tag{14.5}$$

where: μ_o = permeability of free space = $4\pi.10^{-7}$ H/m and ε_0 = permittivity of free space = 8.854×10^{-12} F/m. Thus, the wave impedance (i.e., the ratio of the electric field strength to the magnetic field strength) for plane waves propagating in free space is also 377 Ω.

The multiple reflection correction term accounts for the reduction in shielding effectiveness for shields with low absorption loss due to multiple reflections inside the shield. This term is given by

$$B = 20 \cdot \lg\left|1 - \frac{(k-1)^2}{(k+1)^2} e^{\frac{-2t_s}{\delta_s}}\right| \tag{14.6}$$

Since the conductivity of commonly used metallic shield materials, such as aluminum, is very high (typically in the order of 10^7 mhos/m or higher), a metallic barrier theoretically provides a large amount of attenuation. As a rule of thumb, if a continuous metallic barrier has a thickness sufficient to support itself (say 30 mils (30 mils = 0.762 mm) or more), the theoretical shield effectiveness is extremely high (typically on the order of 100 dB or more) at radio frequencies (RF).

In practice, however, the shielding barrier is not continuous but instead contains seams that are fastened (bonded) together with metallic fasteners. These joints often allow electromagnetic fields to leak through the shield, thereby degrading the overall shielding effectiveness. The quality of the electrical bond across the faying surfaces at RF thus, in large part, determines the overall shielding effectiveness of the shielding structure. For example, a shielding effectiveness of the order of 20 dB is typical for riveted aluminum panels with nonconductive sealants.

Providing a direct current (de) path across the faying surface is necessary, but not sufficient, for high shielding effectiveness (riveted joints, in fact, often provide a very low de resistance between bonded panels despite their rather poor shield effectiveness). Instead, a low RF impedance path across the faying surfaces should be provided. RF surface currents flowing across a joint in the skin of an aircraft induce electric fields in the interior of the aircraft. These fields in turn induce currents and voltages on interior wiring. The maximum voltage that can be induced on an interior wire is the voltage drop across the joint given by

$$V_j = J_s Z_{tj} \qquad [14.7]$$

where: V_j = voltage drop across the joint (in V), J_s = surface current density (in A/m) and Z_{tj} = transfer impedance of the joint (in Ω·m). Thus, a good measure of how well a joint is bonded is the transfer impedance of the bonded joint across the frequency band of interest.

14.4 Applications as a bioreactor: filtration membranes

There is growing interest in the use of nanofibers as membrane bioreactors. The immobilization of enzymes on the nanofibers for bioreactor materials is discussed in detail below.

Nanofibers have a small pore size and a large specific surface area compared to traditional nonwoven materials. This, together with their low density and interconnected pore structure, makes the nonwoven nanofiber appropriate for a wide variety of filtration applications such as liquid or gas filtration.

Membrane filtration can be divided broadly into two parts based on the particle sizes filtered. Micro- and ultrafiltration can be used for the removal of larger particles, as membranes operate at low pressures with high productivity. On the other hand, nanofiltration and reverse osmosis (RO) are used to remove salts from water. In nano- and RO membranes, separation takes place mainly by diffusion through the membrane and not by the principle of pore separation. The pressure required to operate nanofiltration and RO is much higher than that required for micro- and ultrafiltration, while productivity is much lower.

14.4.1 Nanofibers for water filtration

Fresh water scarcity is a major problem in some parts of the world. Many technologies are widely employed to get usable water from brackish, waste and salt water. Traditional approaches such as distillation, sand filtration, sedimentation and treatment with chlorine are widely used (Ramakrishna *et al.*, 2010). Filtration using membrane technology is a relatively new approach, with advantages over traditional water purification methods such as scalability, low power consumption and the lack of need for chemicals. Membranes can be operated at ambient temperatures which are favorable for biological, drug and food applications. In general, a membrane acts as a very specific filter that will allow water to pass through while separating out suspended solids and other substances.

14.4.2 Nanofibers for air filtration

The environmental application of nanofiber membranes for air filtration began in the 1980s and has expanded in the last three decades (Wang *et al.*, 2007a). With the advent and development of industrialization, air quality has deteriorated in some areas, requiring regeneration of air through filtration and other processes. Furthermore, stricter emission limits have also contributed to the need for better-quality filtration media. Air filtration spans a wide range of applications, from the removal of particulate materials from work environments, such as high dust loading industrial facilities, to specialized applications, such as providing protection to individuals from toxic agents in the air.

Today, nanofiber mats are used in numerous air filtration applications, either on their own or in combination with other filtration media. The growth in use of nanofibers for air filters has been pioneered by a few companies. Nanofibers are of much smaller dimensions than conventional filtration media: the nanofibers used in filtration can be up to 800 times smaller than conventional filtration fibers. Smaller fibers generally provide better efficiency due to inertial impaction and interception, which makes up for the increased pressure drop when compared to bigger fibers. In addition, when considering nanometer-sized fibers, the slip flow at the fiber surface implies a decrease in drag force on the fiber and therefore a reduced pressure drop. The occurrence of slip flow also results in more contaminants passing near the surface of the fibers, which increases the inertial impaction and interception efficiencies of the fibrous mat. Overall, this results in nanofiber mats that can achieve much higher efficiency at the same pressure drop than conventional fiber mats. The cut-off diameter of the fiber for slip flow consideration at its surface is approximately 500 nm.

14.5 Applications as a bioreactor: immobilization of enzymes

Enzymes are well-known, powerful 'green' catalysts that possess a high degree of specificity and activity (Wang *et al.*, 2009). The specificity involves discrimination between substrates (substrate specificity), similar parts of molecules and optical isomers (Fischbach *et al.*, 2006; Ju and Parales, 2006; Synowiecki and Woosowska, 2006). The mildness and specificity of enzymes endow them with a high efficiency for applications in fine-chemical/pharmaceutical synthesis, food processing, biosensor fabrication, biodegradation, biofuel cells, bioremediation and protein digestion in proteomic analysis (Polat *et al.*, 1997; Akgol and Dinckaya, 1999; Choudhary *et al.*, 2003; Hasan *et al.*, 2006). However, the applications of enzymes are limited by their short active lifetimes and the fact that they are not reusable. Enzyme immobilization is an effective way to overcome these limitations to some extent. First, the multiple-point attachment to the support can restrict the undesirable conformational change of enzyme proteins in unfriendly environments. Second, insoluble supports can be recycled much more easily than soluble enzymes.

The results of immobilization, including the performance of immobilized enzymes, strongly depend on the properties of the supports, which are usually referred to as material types, compositions, and structures, etc. So far, different nanostructured materials have been used as supports such as mesoporous silica, nanotubes, nanoparticles and nanofibers. These stand out among other supports because of their extremely high surface area-to-volume ratios, which can provide large specific surface areas for highly efficient immobilization, as well as stabilizing the enzymes (Kim *et al.*, 2006).

However, some nanostructured materials have disadvantages that are difficult to overcome. For example, mesoporous silica usually confines enzyme molecules on its inner surface, which limits the diffusion of substrate to/off the enzyme and results in lower enzyme activity. Nanoparticles and nanotubes are known to remarkably decrease mass-transfer limitation, and dispersion and recycling of these materials is difficult. On the other hand, electrospun nanofibers have great potential to overcome these problems and may be promising supports for enzyme immobilization. Briefly, the qualification of nanofibers as excellent supports is attributed to: (1) the variety of polymers that can be electrospun and meet different requirements as supports; (2) the high porosity and interconnectivity of electrospun supports, which endows them with a low hindrance for mass transfer; (3) the fact that nanofiber surfaces can be modified to benefit hosting enzymes. The collection of randomly arrayed nanofibers usually forms a nonwoven mesh (or membrane), which can be reused if necessary. From this point of view, the as-spun membranes have also been explored as filters (Barhate *et al.*, 2006; Gopal *et al.*, 2006).

Enzyme-immobilized nanofibrous membranes can deploy their functions of biocatalysis and separation simultaneously, a requirement generally accepted as fundamental for an enzymatic membrane-bioreactor. Applications in biosensors and biofuel cells are also pertinent for these nanofibrous membranes because of their porosity.

14.5.1 Surface attachment of enzymes on nanofibers

Enzyme immobilization on untreated nanofibers

The first example dealing with nanofibers as supports for enzyme immobilization has been described by Jia and colleagues (2002). In their study, polystyrene synthesized with a hydroxyl-containing initiator was electrospun into nanofibers with reactive surfaces. α-chymotrypsin was then covalently attached onto these surfaces, with the amount of loaded enzyme being up to 1.4 wt% of the nanofibers. The study showed that over 27.4% of the external nanofiber surface was covered with a monolayer of enzyme. Specific activity of the immobilized α-chymotrypsin was over 65% of that of the native enzyme in aqueous solution. From this result, it can reasonably be deduced that the immobilized enzyme had a low diffusional limitation. The stability of immobilized α-chymotrypsin to storage and organic solvents (e.g., isooctane and hexane) was much higher than that of the free ones. Untreated silk fibroin (SF) nanofibers have also been used for α-chymotrypsin immobilization (Lee *et al.*, 2005). Interesting results were found with regard to the stabilities of immobilized enzymes on SF nanofibers with different diameters. α-chymotrypsin on SF nanofibers with 205 nm diameter retained more than 90% of the initial activity after 24-h storage in aqueous solution and showed the highest storage stability. However, the highest stability in ethanol was obtained from α-chymotrypsin on SF nanofibers with 320 nm diameter, which retained more than 45% of the initial activity.

Ye *et al.* (2006a) used nanofibers electrospun from poly(acrylonitrile-co-maleic acid) for lipase immobilization. The amount and activity retention of lipase immobilized on the nanofibers were 21.2 mg/g fibers and 37.6%, respectively, while those on the corresponding hollow-fiber membrane were only 2.36 mg/g membrane and 33.9%. Combined with the kinetic parameters, it can be concluded that nanofibers present a low diffusion restriction on substrate. However, low activity retention for the immobilized lipase suggests that loss of enzymatic activity still exists. This can be partly attributed to the immobilization reaction and enzyme-support interactions. Nevertheless, this study provides a simple route to fabrication of nanofibers containing reactive groups for the covalent immobilization of enzymes. To improve the performance of the immobilized enzymes, the reactive groups on the nanofibers have the potential for further modification of the nanofiber surface.

Enzyme loading on a hydrophobic support is often limited because of the reduced accessibility of the enzymes to the support surface. Nair *et al.* (2007) developed a simple method to resolve this problem. PS (polystyrene)/PSMA (containing maleic anhydride) was electrospun into hydrophobic composite nanofibers. These hydrophobic nanofibers were treated with aqueous alcohol solution. Thereafter, the tightly aggregated nanofibers could be dispersed in water to form a loosely entangled structure stable enough for enzyme immobilization. The results indicated that this method increased the amount of enzyme loading up to eight times. It also augmented the steady-state conversion for a continuous-flow reactor filled with the enzyme-loaded nanofibers.

Enzyme immobilization on modified nanofibers

From the above description, it is reasonable to choose nanofibers as the supports for enzyme immobilization with regard to both enzyme loading and enzymatic activity. Depending on the kinds of nanofiber materials, the immobilization approaches and the treatment methods to the supports, the behavior and loading of enzymes can be tailored flexibly (Lee *et al.*, 2005; Ye *et al.*, 2006b). However, loss of enzymatic activity cannot be avoided.

Generally speaking, this can be ascribed to the following factors. First, multipoint attachment to the support restricts the freedom of the immobilized enzyme so that it is difficult for the enzyme protein to adapt to a suitable conformation for catalysis. Second, nonbiospecific interactions tend to occur between enzyme and support, which results in undesired changes in the conformation of enzyme and variation of the microenvironment (Sadana, 1992). Third, especially for redox enzymes, electron transfer involved in the catalysis process may be retarded by the support. These factors are not independent, and all of them can be related to the surface chemistry of support. Therefore, the specific activity and stabilities of immobilized enzymes can be improved through tailoring the surface chemistry of nanofibers.

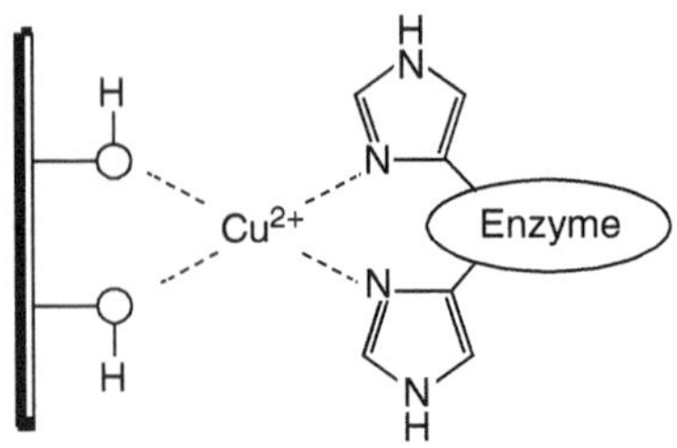

14.5 Schematic of PVA- Cu(II) chetalated nanofibrous membrane immobilized by catalase.

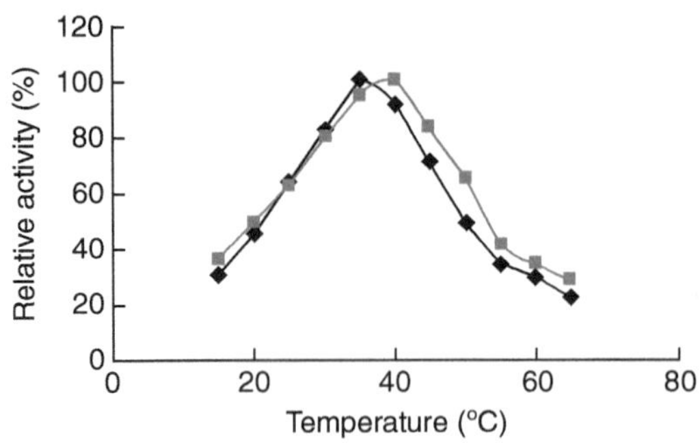

14.6 Effect of temperature on the (■) immobilized and (◆) free catalases.

In our lab, we began by electrospinning PVA nanofibers. Cu(II) was then incorporated into the hydroxyl (OH) groups of the PVA nanofibers via metal chelation, and PVA-Cu(II) chelated nanofibers were prepared. Subsequently, catalases were immobilized onto the metal chelated nanofibers, and the decomposition of H_2O_2 was studied by examining the activity and stability of the immobilized catalases. The Cu(II) chelating scheme is presented in Fig. 14.5. The effect of temperature on the activity of free and immobilized catalases is depicted in Fig. 14.6. The optimum temperatures for the immobilized and free catalase were measured as approximately 40°C and 35°C respectively. The multipoint interactions between enzymes and PVA-Cu^{2+} support might reduce the degree of freedom of the spatial structure of the enzyme, protecting it to some extent from deactivation at high temperature. The effect of pH on the activity of free and immobilized enzymes is shown in Fig. 14.7. According to experimental results, a significant shift in the optimal pH was not observed. The optimal pH value was found to be about 7.0 for free and immobilized enzymes, but the pH profiles of the immobilized enzyme were much broader than those of the free enzyme. The residual relative activity of the immobilized catalase was

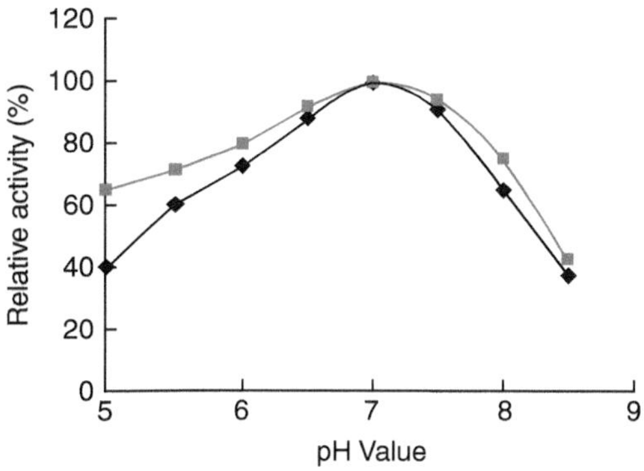

14.7 Effect of pH on the (■) immobilized and (◆) free catalases.

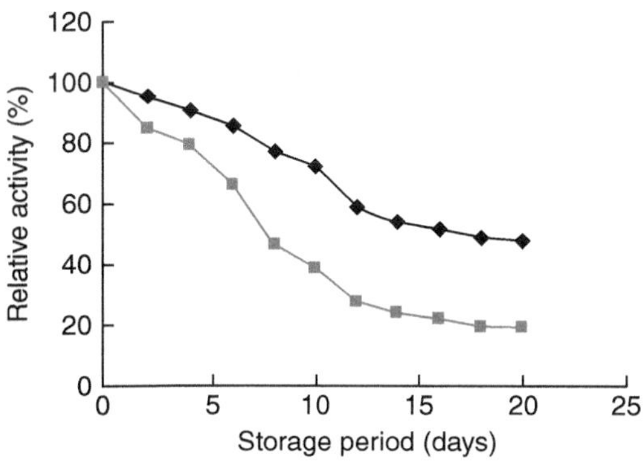

14.8 Storage stability of (◆) immobilized and (■) free catalases.

higher than that of the free catalase in the pH range between 5.0 and 8.5, probably because of the production of oxygen, which formed foams and caused slight diffusional limitation on the surface of the nanofibrous membrane. The storage stability of immobilized and free enzymes is presented in Fig. 14.8. The immobilized catalase lost 52% of its activity within 20 days, and the free catalase lost 81% of its activity during the same period. The result indicated that the storage stability of immobilized catalase was better than that of free catalase. The decrease of the catalase was explained as a time-dependent natural loss. The thermal stability of immobilized and free enzyme is shown in Fig. 14.9. The rate of thermal inactivation of immobilized catalase was lower than that of free catalase. The conformational flexibility of the enzyme could be affected by immobilization of a catalase on the surface of the metal chelated nanofibrous membrane. Multipoint attachment between immobilized enzyme supports is supposed to maintain the tertiary structure of enzyme.

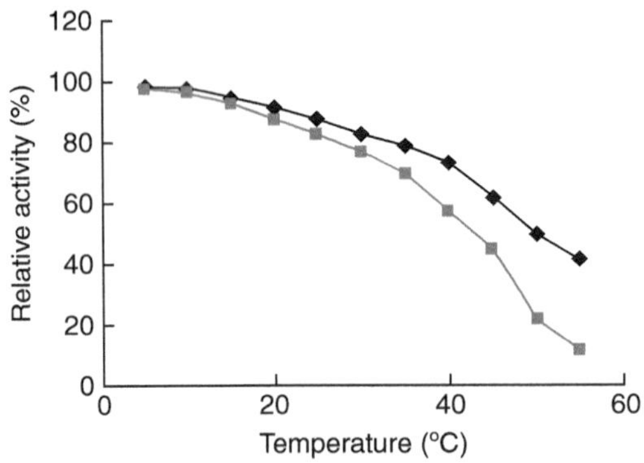

14.9 Thermal stability of (◆) immobilized and (■) free catalases.

Modification toward biocompatible surface

Tailoring the surface chemistry toward biocompatibility is a commonly used method for promoting the activity of immobilized enzymes, which is stimulated by biomimetics methodology. By mimicking the natural mode in living cells where enzymes are mostly present, the biomimetic systems are supposed to be able to stabilize the structure of enzymes and retain their activities. It was mentioned earlier that nanofibers electrospun from poly(acrylonitrile-co-acrylic acid) have been used for lipase immobilization (Ye *et al.*, 2006a). A main reason to choose this polymer is its availability of functional groups, which can react not only with lipase, but also with other molecules (e.g., biomacromolecules). The latter provides a simple route for surface modification that increases the surface biocompatibility of nanofibers. Therefore, the carboxyl-containing nanofibers were modified with chitosan or gelatin to build a dual-layer biomimetic surface (Ye *et al.*, 2006b). Chitosan is the principal derivative of chitin and can be acquired at low cost. As a natural polyaminosaccharide, chitosan offers a unique set of characteristics not provided by common biomaterials, including physiological inertness, antibacterial properties, a remarkable affinity toward proteins and good gel-forming properties, as well as good chelation of heavy metal ions (Krajewska, 2004, 2005).

Gelatin is derived from collagen and shows biological properties almost identical to those of collagen. It is also promising because of its commercial availability at low cost. Tethering these two biomacromolecules on the nanofiber surface can combine their biocompatibility with the mechanical strength of the nanofiber. Moreover, the abundant reactive groups on the backbone of chitosan or gelatin can provide sufficient bonding sites for enzyme immobilization.

Modification toward enzyme mobility

The flexibility of immobilized enzymes can be improved by keeping the enzymes apart from the support surface (Penzol *et al.*, 1998; De Maio *et al.*, 2003; Tamaki *et al.*, 2007). A common method is to introduce spacer arms onto the surface. The flexible spacers can offer the enzyme more freedom of movement and minimize the steric hindrance caused by the solid support, so that the microenvironment for the immobilized enzyme is close to that for the free one. Wang and Hsieh (2004) introduced PEG as the spacer arm onto the alkali-hydrolyzed cellulose nanofibers for the covalent immobilization of lipase. The fibrous structure was retained throughout each process, including alkaline hydrolysis, activation, coupling and activity assays. The immobilized lipase was found to be highly stable in the presence of nonpolar solvents and high temperature. Meanwhile, the results suggested that the molecular structure of PEG made the major difference in the catalysis of lipase, rather than its chain length. The highly hydrophilic PEG layer could offer essential water, which ensured the conformational flexibility of the immobilized enzymes in organic media (Zaks and Klibanov, 1988).

Modification toward electrical conductivity

Electron transfer tends to be involved in reactions catalyzed by redox enzymes. It is apparent that if electron transfer is guaranteed, the catalytic reaction of the enzyme will be sustainable (Vedrine *et al.*, 2003; Fernandes *et al.*, 2003). This is why electrically conductive polymers have traditionally been applied to enzyme immobilization, especially when used in electrochemical biosensors (Chaubey *et al.*, 2000; Ahuja *et al.*, 2008; Gerard *et al.*, 2002). Blending insulating polymers with electrically conductive nanomaterials is also an effective method to fabricate suitable supports for enzyme immobilization (Merkoci *et al.*, 2005; Tkac *et al.*, 2007). Wang and coworkers added multiwalled carbon nanotubes (MWCNTs) to poly(acrylonitrile-co-acrylic acid) nanofibers for the covalent immobilization of catalase (Wang *et al.*, 2006). When the mass ratio of MWCNTs to the polymer was 30%, the activity of the immobilized catalase was increased by 47% without reduction in the enzyme loading. This can be attributed partly to the superb electrical conductivity of MWCNTs, which can also form a charge-transfer complex with polyacrylonitrile. This specificity is thought to assist the electron transfer between the hydrogen peroxide molecules and the intermediate catalase, thus enhancing the activity of the immobilized catalase. Catalase immobilized on the composite nanofibers also showed higher storage stability than that on the untreated nanofibers; this result is associated with the hydrophilization and biocompatibility of MWCNTs (Wang *et al.*, 2007b). This composite nanofibrous support was also used to immobilize another redox enzyme, horseradish peroxidase, whose activity was also obviously enhanced by the filled MWCNTs.

14.5.2 Encapsulation immobilization of enzymes in nanofibers

The encapsulation of enzymes in the nanofibers can be achieved by direct coelectrospinning of enzymes and other components (organic or inorganic materials). Most proteins can only be dissolved in aqueous media. Therefore, in many cases, the polymers to be coelectrospun with enzymes are required to be water-soluble so that they can form homogeneous solution with the enzymes. This can reduce the surface tension of the electrospinning solution, which is necessary for fabricating bead-free nanofibers (Dotti *et al.*, 2007). The commonly used polymers include poly(vinyl alcohol) (PVA), poly(ethylene oxide) (PEO) and poly(N-vinyl-2-pyrrolidone) (PVP) (Xie and Hsieh, 2003; Sawicka *et al.*, 2005; Wu *et al.*, 2005; Wang and Hsieh, 2008). These polymers are commercially available and present good affinity to enzymes. Furthermore, PVA and PEO have dissimilar structures to natural biomacromolecules and the capability to form secondary bonding with proteins, which can dissociate the hydrogen-bonded molecules (protein, chitosan, etc.) and make the coelectrospinning of biomacromolecules easier (Zhang *et al.*, 2001; Frenot *et al.*, 2007; Jia *et al.*, 2007). The coelectrospinning method offers a simple route to the immobilization of enzymes on nanofibers, and the enzyme loading can therefore be substantial (Wang and Hsieh, 2008). Due to these characteristics, the enzyme-immobilized nanofibers were directly applied in the fabrication of enzyme electrodes for biosensors. These biosensors showed some common features, such as a low detection limit and fast response, and provided another advantage in that the electrode could be easily regenerated by peeling away the nanofibrous mesh.

Despite the outstanding characteristics mentioned above, the encapsulation approach also has several disadvantages:

1. The enzyme molecules are not only embedded into the nanofibers, but also reside on the surface, which usually leads to loss of enzymes during measurement and storage.
2. Because most of the enzyme molecules are confined inside the nonporous fibers, the accessibility of the substrate to the enzyme is inhibited.
3. The materials to be electrospun with enzymes are limited to several species only. Even when electrospun from homogeneous solution, nanofibers with beads were still formed (Sawicka *et al.*, 2005; Ren *et al.*, 2006). When the nanofibers are immersed in aqueous medium, the water solubility leads to the swelling and disintegration of nanofibers, resulting in enzyme leakage, thermal instability and poor reusability.

4. Cross-linking of the biocatalytic fibers tends to reduce the activity of immobilized enzymes, though it is normally used for increasing the stability of physically encapsulated enzymes (Wu *et al.*, 2005; Wang and Hsieh, 2008).

On the one hand, cross-linking can reduce the porosity, which limits the accessibility of substrates to the active sites of enzymes. On the other hand, cross-linking itself damages the active sites of the enzymes. Considering these obstacles, one alternative is to use water-insoluble materials for coelectrospinning with enzymes. For that purpose, a coaxial electrospinning setup was used to fabricate poly(α-caprolactone) (PCL)/lysozyme fibers with a core-shell structure, which has often been used for the purpose of controlled release (Jiang *et al.*, 2005; Qi *et al.*, 2006; Zhang *et al.*, 2006). In another example, surfactant was used to stabilize α-chymotrypsin in a PS/PSMA solution, which was then electrospun into biocatalytic nanofibers (Herricks *et al.*, 2005). This two-phase electrospinning method has also been used for fabricating nanofibers with a core-shell structure (Sanders *et al.*, 2003; Qi *et al.*, 2006; Bazilevsky *et al.*, 2007). Silica nanofibers have also been used to encapsulate horseradish peroxidase through electrospinning; no deformation of the fibers and no enzyme leakage were observed in this study. Besides, the fibers were highly mesoporous, which facilitated the diffusion of the substrate to the enzyme. However, a large number of water-insoluble materials (e.g., PS, PSF and PAN) are of poor biocompatibility (Corden *et al.*, 2000; Xu *et al.*, 2003a; Serrano *et al.*, 2004). If they are used to encapsulate enzymes, undesired conformational change of the enzymes will take place.

Two approaches can be used to improve the microenvironment of the encapsulated enzymes. First, additional biocompatible components can be incorporated into the spinning solution to offer a biofriendly microenvironment for enzyme. For instance, these materials included PEG (Jiang *et al.*, 2005) and dextran (Jiang *et al.*, 2006) in the core protein solution for the coaxial electrospinning. Second, the biocompatibility of the polymers is improved by *in situ* polymerization (Wang *et al.*, 2005; Xu *et al.*, 2005) or side chain modification (Huang *et al.*, 2005; Ladmiral *et al.*, 2006; Park *et al.*, 2006), which will also favor the enzyme catalysis.

14.6 Future trends

Electrospun nanofibers have been proven to be excellent supports for enzyme immobilization because they can provide large surface area-to-volume ratios, pore sizes tailored to protein molecule dimensions, functionalized surfaces, multiple sites for interaction or attachment and low mass-transfer limitation. However, the studies of this issue are still limited to a small number as there are problems with their large-scale application.

First, nanofibers continue to be difficult to fabricate in batches, although multiple spinnerets have been developed. Second, very few tools can be used to evaluate the effect of the nanofiber surface on the behavior of the immobilized enzymes. Nevertheless, based on their unique advantages, it is still anticipated that the resultant biocatalytic materials could enable new and expanded uses of enzymes in practical applications such as biosensors, bioremediation, biofuel cells and bioconversions.

Research into nanofibers has been conducted for many years. Institutes and universities in Shanghai, Changchun, Beijing, Suzhou and elsewhere have achieved no small feat in developing the formation and applications of nanofibers, but overall this research remains at laboratory stage. In recent years, many global manufacturers have commercialized nanofiber technology – for example, Elmarco, which has recently marketed nanofiber electrospinning production. This rapid commercialization demonstrates that the development cycle for new material has been significantly shortened in today's world.

With the advances in nanofiber-formation technology and extended research into its application, consumer research should also be put on the agenda. The potential of nanofibers to cause harm to humans is not very clear, and the potential risks of nanotechnology and the options for risk prevention have yet to be fully researched. In the new century, facing the rapid development of new technologies, health and environmental problems created by the development of new materials are not only significant, but also difficult to deal with. It is recommended that employees maintain vigilance against possible harm in the development of nanofiber technology, as the risks cannot be ignored.

Since the 1990s, due to the emergence of new technology, research into nanofibers has focused on the basic forming process, the fiber structure and processing equipment, etc. In recent years, research into the application and development of nanofibers has markedly accelerated, especially the development of functional nanofibers. This change is noteworthy, as the trend of nanofiber development is the same as that of nanotechnology: more and more products are being brought into the market and entering consumer homes and workplaces. It should be clearly understood, however, that the manufacturing process of nanofibers still has low productivity, high processing costs and other disadvantages compared to traditional spinning technology. In addition, the solvents used in electrostatic spinning have implications for the environment and also increase the cost of recycling equipment. It can be said that applied research into nanofibers needs to focus on the challenges of a variety of factors, including the economy of nanofiber technology, solvent recycling, sustainability, human health and research into the packaging and sales of nanofiber products. The trade-offs between the high performance of nanofibers and their high manufacturing cost will be one of the topics about which manufacturers and consumers are most concerned.

14.7 References

Ahuja, T., Mir, I. A. and Kumar, D. (2008), Bimolecular immobilization on conducting polymers for biosensing applications, *Biomaterials*, **28**, 791–805.

Akgol, S. and Dinckaya, E. (1999), A novel biosensor for specific determination of hydrogen peroxide: Catalase enzyme electrode based on dissolved oxygen probe, *Talanta*, **5**, 363–7.

Al-Saleh, M. H. and Sundararaj, U. (2009), A review of vapor grown carbon nanofiber/polymer conductive composites, *Carbon*, **47**, 2–22.

Barhate, R. S., Loong, C. K. and Ramakrishna, S. (2006), Preparation and characterization of nanofibrous filtering media, *Journal of Membrane Science*, **283**, 209–18.

Bazilevsky, A. V., Yarin, A. L. and Megaridis, C. M. (2007), Co-electrospinning of core-shell fibers using a single nozzle technique, *Langmuir*, **23**, 2311–14.

Chaubey, A., Pande, K. K. and Singh, V. S. (2000), Co-immobilization of lactate oxidase and lactate dehydrogenase on conducting polymer films, *Analytical Chimica Acta*, **407**, 7–103.

Choudhary, G., Wu, S. L. and Shieh, P. (2003), Multiple enzymatic digestion for enhanced sequence coverage of proteins in complex proteomic mixtures using capillary LC with ion trap MS/MS, *Journal of Proteome Research*, **2**, 59–67.

Corden, T. J., Jones, I. A. and Rudd, C. D. (2000), Physical and biocompatibility properties of poly-caprolactone produced using in situ polymerisation: A novel manufacturing technique for long-fiber composite materials, *Biomaterials*, **21**, 713–24.

De Maio, A., El-Masry, M. M. and De Luca, P. (2003), Influence of the spacer length on the activity of enzymes immobilised on nylon/polyGMA membranes: Part 2: Non-isothermal conditions, *Journal of Molecular Catalysis B: Enzymatic*, **21**, 253–65.

Dotti, F., Varesano, A. and Montarsolo, A. (2007), Electrospun porous mats for high efficiency filtration, *Journal of Industrial Textiles*, **37**(2), 151–62.

Fernandes, K. F., Lima, C. S. and Pinho, H. (2003), Immobilization of horseradish peroxidase onto polyaniline polymers, *Process Biochemistry*, **38**, 1379–84.

Fischbach, A., Adelt, S. and Müller, A. (2006), Disruption of inositol biosynthesis through targeted mutagenesis in Dictyostelium discoideum: Generation and characterization of inositol-auxotrophic mutants, *Biochemical Journal*, **397**, 509–18.

Frenot, A., Henriksson, M. W. and Walkenström, P. (2007), Electrospinning of cellulose-based nanofibers, *Journal of Applied Polymer Science*, **103**, 1473–82.

Gerard, M., Chaubey, A. and Malhotra, B. D. (2002), Application of conducting polymers to biosensors, *Biosensors and Bioelectronics*, **17**, 345–59.

Gopal, R., Kaur, S. and Ma, Z. (2006), Electrospun nanofibrous filtration membrane, *Journal of Membrane Science*, **281**, 581–6.

Hasan, F., Shah, A. and Hameed, A. (2006), Industrial applications of microbial lipases, *Enzyme and Microbial Technology*, **39**, 235–51.

Herricks, T. E., Kim, S. H. and Kim, J. (2005), Direct fabrication of enzyme-carrying polymer nanofibers by electrospinning, *Journal of Material Chemistry*, **15**, 3241–5.

Huang, X. J., Xu, Z. K. and Wan, L. S. (2005), Novel acrylonitrile-based copolymers containing phospholipid moieties: synthesis and characterization, *Macromolecular Bioscience*, **5**, 322–30.

Jia, H., Zhu, G., Vugrinovich, B. and Wang, P. (2002), Enzyme-carrying polymeric nanofibers prepared via electrospinning for use as unique biocatalysts, *Biotechnology Progress*, **18**, 1027–32.
Jia, Y. T., Gong, J. and Gu, X. H. (2007), Fabrication and characterization of poly (vinyl alcohol)/chitosan blend nanofibers produced by electrospinning method, *Carbohydrate Polymers*, **67**, 403–9.
Jiang, H. L., Hu, Y. and Li, Y. (2005), A facile technique to prepare biodegradable coaxial electrospun nas for controlled release of bioactive agents, *Journal of Controlled Release*, **108**, 237–43.
Jiang, H. L., Zhao, P. C. and Zhu, K. J. (2006), Modulation of protein release from biodegradable core-shell fibers prepared by coaxial electrospinning, *Journal of Biomedical Materials Research Part B*, **79**, 50–7.
Ju, K.-S. and Parales, R. E. (2006), Control of substrate specificity by active-site residues in nitrobenzene dioxygenase, *Applied and Environmental Microbiology*, **72**, 1817–24.
Kim, J., Grate, J. W. and Wang, P. (2006), Nanostructures for enzyme stabilization, *Chemical Engineering Science*, **61**, 1017–26.
Krajewska, B. (2004), Application of chitin- and chitosan-based materials for enzyme immobilizations: A review, *Enzyme Microbial Technology*, **35**, 126–39.
Krajewska, B. (2005), Membrane-based processes performed with use of chitin/chitosan materials: Review, *Separation and Purification Technology*, **41**, 305–12.
Ladmiral, V., Mantovani, G. and Clarkson, G. J. (2006), Synthesis of neoglycopolymers by a combination of 'click chemistry' and living radical polymerization, *Journal of the American Chemical Society*, **128**, 4823–30.
Lee, K. H., Ki, C. S. and Baek, D. H. (2005), Application of electrospun silk fibroin nanofibers as an immobilization support of enzyme, *Fibers and Polymers*, **6**, 181–5.
Merkoci, A., Pumera, M. and Llopis, X. (2005), New materials for electrochemical sensing VI: Carbon nanotubes, *Trends in Analytical Chemistry*, **24**, 826–38.
Nair, S., Kim, J., Crawford, B. and Kim, S. H. (2007), Improving biocatalytic activity of enzyme-loaded nanofibers by dispersing nanofibers into a swollen structure, *Biomacromolecules*, **8**, 1266–70.
Park, J. Y., Acar, M. H. and Akthakul, A. (2006), Polysulfone-graft-poly(ethylene glycol) graft copolymers for surface modification of polysulfone membranes, *Biomaterials*, **27**, 856–65.
Penzol, G., Armisen, P. and Fernández-Lafuente, R. (1998), Use of dextrans as long, inert and hydrophilic spacer arms to improve the performance of immobilized proteins acting on macromolecules, *Biotechnology and Bioengineering*, **60**, 518–23.
Polat, T., Bazin, H. G. and Linhardt, R. J. (1997), Enzyme catalyzed regioselective synthesis of sucrose fatty acid ester surfactants, *Journal of Carbohydrate Chemistry*, **16**, 1319–25.
Qi, H. X., Hu, P. and Xu, J. (2006), Encapsulation of drug reservoirs in fibers by emulsion electrospinning: morphology characterization and preliminary release assessment, *Biomacromolecules*, **7**, 2327–30.
Ramakrishna, S., Jose, R. and Archana, P. S. (2010), Science and engineering of electrospun nanofibers for advances in clean energy, water filtration, and regenerative medicine, *Journal of Materials Science*, **45**, 6283–12.

Ren, G. L., Xu, X. H. and Liu, Q. (2006), Electrospun poly(vinyl alcohol)/glucose oxidase biocomposite-membranes for biosensor applications, *Reactive and Functional Polymers*, **66**, 1559–64.
Robinson, J. G. (1966), Fundamentals of EMI shielding, *Electro-Technology*, **77**, 36–9.
Sadana, A. (1992), Protein adsorption and inactivation on surfaces: Influence of heterogeneities, *Chemical Reviews*, **92**, 1799–818.
Sanders, E. H., Kloefkorn, R. and Bowlin, G. L. (2003), Two-phase electrospinning from a single electrified jet: Microencapsulation of aqueous reservoirs in poly (ethylene-co-vinylacetate) fibers, *Macromolecules*, **36**, 3803–5.
Sawicka, K., Gouma, P. and Simon, S. (2005), Electrospun biocomposite nanofibers for urea biosensing, *Sensors and Actuators B: Chemical*, **108**, 585–8.
Schelkunoff, S. A. (1955), Conversion of Maxwell's equations into generalized telegraphist's equations, *Bell System Technical Journal*, **34**, 995–1043.
Serrano, M. C., Pagani, R. and Vallet-Regì, M. (2004), In vitro biocompatibility assessment of poly(e-caprolactone) films using L929 mouse fibroblasts, *Biomaterials*, **25**, 5603–11.
Simon, H. A. (1981), Information-processing models of cognition, *Journal of American Society for Information Science*, **32**, 364–77.
Synowiecki, J. and Woosowska, S. (2006), Immobilization of thermostable -glucosidase from Sulfolobus shibatae by cross-linking with transglutaminase, *Enzyme and Microbial Technology*, **39**, 1417–22.
Tamaki, T., Ito, T. and Yamaguchi, T. (2007), Immobilization of hydroquinone through a spacer to polymer grafted on carbon black for a high-surface-area biofuel cell electrode, *Journal of Physical Chemistry B*, **111**, 10312–19.
Tkac, J., Whittaker, J. W. and Ruzgas, T. (2007), The use of single walled carbon nanotubes dispersed in a chitosan matrix for preparation of a galactose biosensor, *Biosensors and Bioelectronics*, **22**, 1820–4.
Vedrine, C., Fabiano, S. and Tran-Minh, C. (2003), Amperometric tyrosinase based biosensor using an electrogenerated polythiophene film as an entrapment support, *Talanta*, **59**, 535–44.
Wan, L. S., Xu, Z. K. and Huang, X. J. (2005), Copolymerization of acrylonitrile with n-vinyl-2-pyrrolidone to improve the hemocompatibility of polyacrylonitrile, *Polymer*, **46**, 7715–23.
Wang, H., Zheng, G. and Sun, D. H. (2007a), Electrospun nanofibrous membrane for air filtration, 7th IEEE International Conference on Nanotechnology, 1244–7.
Wang, Y. and Hsieh, Y. L. (2008), Immobilization of lipase enzyme in polyvinyl alcohol (PVA) nanofibrous membranes, *Journal of Membrane Science*, **209**, 73–81.
Wang, Y. H. and Hsieh, Y. L. (2004), Enzyme immobilization to ultra-fine cellulose fibers via amphiphilic polyethylene glycol spacers, *Journal of Polymer Science Part A*, **42**, 4289–99.
Wang, Z., Wan, L. and Liu, Z. (2009), Enzyme immobilization on electrospun polymer nanofibers: An overview, *Journal of Molecular Catalysis B: Enzymatic*, **56**, 189–95.
Wang, Z. G., Ke, B. B. and Xu, Z. K. (2007b), Covalent immobilization of redox enzyme on electrospun nonwoven poly(acrylonitrile-co-acrylic acid) nanofiber mesh filled with carbon nanotubes: A comprehensive study, *Biotechnology and Bioengineering*, **97**, 708–20.

Wang, Z. G., Xu, Z. K. and Wan, L. S. (2006), Nanofibrous membranes containing carbon nanotubes: Electrospun for redox enzyme immobilization, *Macromolecular Rapid Communications*, **27**, 516–21.

Wu, L., Yuan, X. and Sheng, J. (2005), Immobilization of cellulase in nanofibrous PVA membranes by electrospinning, *Journal of Membrane Science*, **250**, 167–73.

Xie, J. and Hsieh, Y. L. (2003), Ultra-high surface fibrous membranes from electrospinning of natural proteins: casein and lipase enzyme, *Journal of Material Science*, **38**, 2125–33.

Xu, H., Yan, F. and Monson, E. E. (2003a), Room-temperature preparation and characterization of poly(ethylene glycol)-coated silica nanoparticles for biomedical applications, *Journal of Biomedical Materials Research Part A*, **66**, 870–9.

Xu, Z. K., Kou, R. Q. and Liu, Z. M. (2003b), Incorporating α-allyl glucoside into polyacrylonitrile by water-phase precipitation copolymerization to reduce protein adsorption and cell adhesion, *Macromolecules*, **36**, 2441–7.

Yang, S., Lozano, K. and Lomeli, A. (2005), Electromagnetic interference shielding effectiveness of carbon nanofiber/LCP composites, *Composites Part A*, **36**, 691–7.

Yang, Y., Liu, X. and Jia, H. (2011), How do vapor grown carbon nanofibers nucleate and grow from deoiled asphalt, *Materials Chemistry and Physics*, **126**, 424–31.

Ye, P., Xu, Z. K. and Wu, J. (2006a), Nanofibrous membranes containing reactive groups: Electrospinning from poly(acrylonitrile-co-maleic acid) for lipase immobilization, *Macromolecules*, **39**, 1041–5.

Ye, P., Xu, Z. K. and Wu, J. (2006b), Nanofibrous poly(acrylonitrile-co-maleic acid) membranes functionalized with gelatin and chitosan for lipase immobilization, *Biomaterials*, **27**, 4169–76.

Zaks, A. and Klibanov, A. M. (1988), Enzymatic catalysis in nonaqueous solvents, *Journal of Biological Chemistry*, **263**, 3194–201.

Zhang, C., Ni, Q. and Fu, S. (2007), Electromagnetic interference shielding effect of nanocomposites with carbon nanotube and shape memory polymer, *Composites Science and Technology*, **64**, 2973–80.

Zhang, C., Yi, X. S. and Yui, H. (1998), Morphology and electrical properties of short carbon fiber-filled polymer blends: high-density polyethylene poly(methyl methacrylate), *Journal of Applied Polymer Science*, **69**, 1813–19.

Zhang, L., Huang, Y. and Chen, J. Y. (2001), The mode III full-field solution in elastic materials with strain gradient effects, *Journal of the Mechanics and Physics of Solids*, **49**, 979–93.

Zhang, Y. Z., Wang, X. and Feng, Y. (2006), Coaxial electrospinning of fitcBSA encapsulated PCL nanofibers for sustained release, *Biomacromolecules*, **7**, 1049–57.

15
Functional nanofibers for water purification

H. MA, B. CHU and B. S. HSIAO,
Stony Brook University, USA

Abstract: Based on breakthroughs in fundamental concepts in membrane design, novel thin-film nanofibrous composite (TFNC) membranes with hierarchical structures have been developed and demonstrated for various water purification processes. Benefits such as high throughput, high rejection, high durability, low energy consumption and low cost can be achieved simultaneously through this technology. Performance advantages of TFNC membranes over commercially available filtration membranes and additional challenges are illustrated in detail. This chapter is intended to serve as a reference for TFNC membranes that can significantly advance the future development of energy-efficient membranes.

Key words: thin-film nanofibrous composite membrane, microfiltration, ultrafiltration, nanofiltration, reverse osmosis, electrospinning, polysaccharide, water purification.

15.1 Introduction

Water, as an essential component in life, is indispensable for the survival of human beings. There are abundant amounts of water on earth, as 71% of the earth's surface is covered by the sea. Unfortunately, we cannot use salt water to meet our common needs. Thus, one of the major challenges in the twenty-first century is finding solutions to get fresh water using less energy (Park *et al.*, 2008; Baird, 2010). The issues of energy and water are closely related to each other.

To be more specific, 97% of the total water on earth (~1.36 billion km^3) is saline water in the oceans (Fig. 15.1). Of the remaining 3% 'fresh' water, ice caps and glaciers constitute 2.06%, leaving only about 0.94% of 'fresh' water, of which 0.90% is groundwater that cannot be readily accessed. Therefore, there is only about 0.04% of accessible 'fresh' water, which also includes swamps and other undesirable sources, leaving only about 0.01% of fresh water from rivers, lakes (and rain) that we can access without consuming too much energy (Gleick, 1996, p. 817). Unfortunately, the scarce fresh water sources are distributed unevenly on earth and are often contaminated

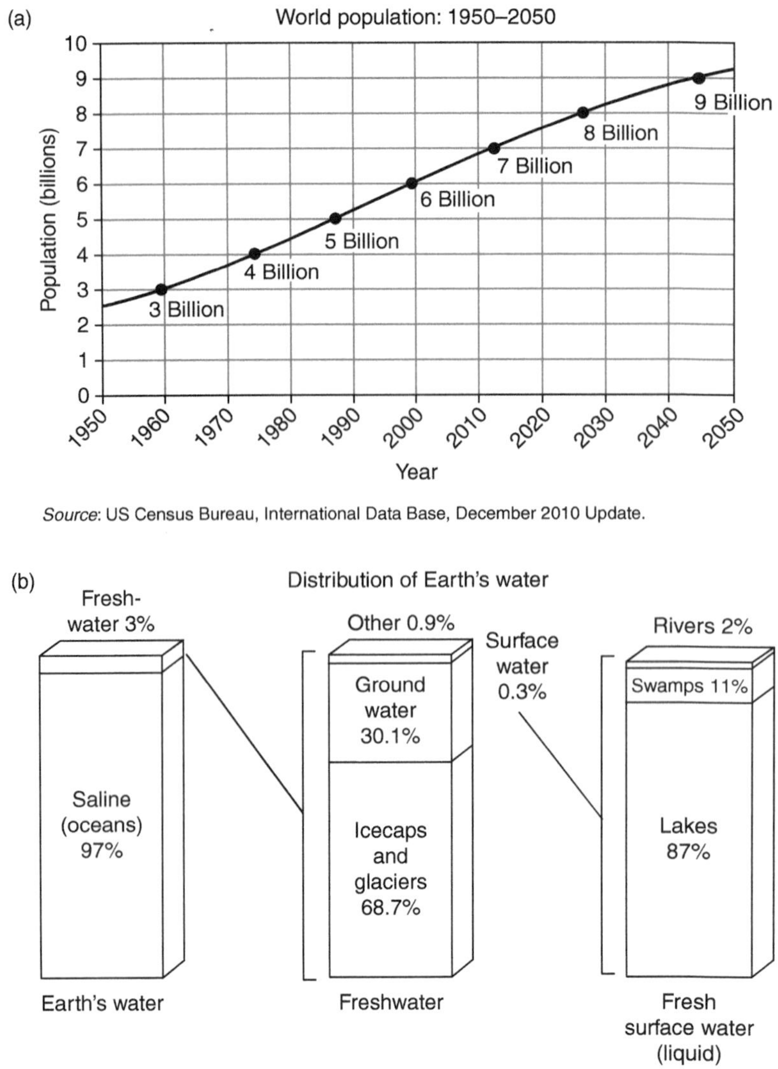

15.1 Growing population and distribution of usable water on Earth. (From Gleick, 1996; Anon, 2010.)

with bacteria and viruses, which can cause diseases and kill millions of people each year. With the rapid growth in population (Anon., 2010; also shown in Fig. 15.1) and mounting water pollution problems, many countries have already faced grave water shortages. Other countries may have an abundant amount of fresh water but lack the means to deliver it to the people in a cost-effective way (Doll *et al.*, 2002, p. 27). As a result, there is a dire need for the development of energy-efficient water purification technologies.

15.1.1 Membrane technology for water purification

Existing technologies for water purification include distillation (Bousfield, 1905; Sulivan, 1979; Tomaszewska, 2000), ion-exchange (Sivetz and Scheibelhut, 1955; Xu, 2005; Nagarale *et al.*, 2006), chemical treatment, such as hypochlorite (Otowa *et al.*, 1997; Crump *et al.*, 2004) and ferrates (Sharma *et al.*, 2005; Sharma, 2008), ultraviolet treatment (Mone, 2011; Yencho, 2011) and filtration (Baker, 2004). The applications of distillation, ion-exchange and chemical treatment, have often been limited by their relatively high energy cost and intrinsic problems (Amjad, 1993; Sagle and Freeman, 2004). Filtration, on the other hand, is a more energy-efficient (compared to distillation and ion-exchange) and safer (compared to chemical treatment) technology, which can overcome several aforementioned disadvantages. Therefore, many new filtration technologies have been developed and their impact on the water purification industry has become more noticeable (Lonsdale, 1982; Van der Bruggen and Vandecasteele, 2002).

Typically, the filtration process can be divided into four classes: microfiltration (MF), ultrafiltration (UF), nanofiltration (NF) and reverse osmosis (RO) (Ulbricht, 2006), based on the pore size of the membranes used. MF membranes possess a pore size larger than 50 nm, which has been widely used for drinking water purification. These membranes can remove particles and waterborne bacteria. The operating pressure of these membranes is typically very low (e.g., sometimes less than 2 psi or by gravity). UF membranes have smaller pore sizes ranging between 2 and 50 nm, which have mainly been used for industrial and community wastewater treatment (e.g., bilge water). The typical operating pressure of these membranes is 10–30 psi. The pore sizes of NF membranes are normally smaller than 2 nm, which can be best used to eliminate multivalent ions (e.g., lead, chromium, magnesium and calcium), surfactants or dye molecules from industrial wastewater. The typical operating pressure for NF membranes is under 100 psi. RO membranes for desalination of brackish water or seawater use higher operating pressures (e.g., 150–1000 psi), depending on the salt content and separation needs. All filtration membranes have been fabricated from inorganic (ceramic) (Tavolaro and Drioli, 1999), organic (polymer) (Ulbricht, 2006) or hybrid materials (combination of ceramic and polymer materials) (Kumar and Fellner, 2003). There is no universal rule, except for the concern of cost effectiveness, which is the central issue of water purification.

15.1.2 Challenges in membrane technology

The development of membrane science has greatly accelerated advances in water filtration technology. However, with the rapid population increase and growing environmental concerns, the need to find new water production technology with significant economic and energy benefits has become a pressing global issue. In spite of the tremendous progress made in the last three decades, membrane technology currently faces two critical challenges (Shannon *et al.*, 2008; Ma *et al.*, 2011a). The first challenge is to identify or to create new materials (e.g., high durability, low cost and fewer environmental concerns) that can fabricate more cost-effective membranes. The second challenge is to design and to optimize structure and morphology in existing membranes that can lead to a substantial increase in filtration efficiency (e.g., high permeation flux, high retention and low pressure drop). These two challenges can be explained in more detail as follows.

Only a limited selection of materials, such as ceramic (Bein, 1996) and synthetic polymers (e.g., polyacrylonitrile (PAN), polysulfone (PSU), polyethersulfone (PES), polyvinylidene fluoride (PVDF), cellulose acetate (CA) and cross-linked polyamide (PA)) (Petersen, 1993; Ulbricht, 2006), have been successfully used for commercial membranes. All aforementioned materials have their advantages and limitations with regard to their applicability for different water purification processes. Ceramic membranes are mechanically durable and chemically inert. However, the high cost has seriously affected their usage, especially in developing countries. Polymers, such as PAN, PSU, PES and PVDF, are routinely used for the fabrication of MF and UF membranes, which have good chemical resistance, mechanical properties and processibility, and are cost-effective. However, their hydrophobic properties often induce fouling problems, drastically decreasing the water permeability. Membranes made of cellulose derivatives are pH sensitive and can only be used within a narrow pH range, while membranes containing cross-linked PA have low tolerance for chlorine (Sagle and Freeman, 2004). Often, volatile and toxic solvents have to be used for processing these materials during membrane fabrication, causing serious environmental concerns. Although there are some new synthetic materials, exhibiting excellent properties for water filtration, their complex preparation schemes, high cost and environmental concerns make them difficult to scale up for immediate commercialization (Holt *et al.*, 2006; Kumar *et al.*, 2007; Zhou *et al.*, 2007; Brady-Estevez *et al.*, 2008; Peng *et al.*, 2009).

With respect to the design and optimization of structure and morphology to enhance the filtration performance of existing membranes, it is necessary to briefly review the current state-of-the-art membrane technology. Thin-film composite (TFC) membranes have been proven as a practical model for the fabrication of polymer membranes (Mulder, 1991; Petersen, 1993). Typically, TFC membranes consist of a two- or three-layer structure (Chu and Hsiao, 2009a). The bottom layer, which primarily provides mechanical support to the membrane, could be a commercial, tough, nonwoven substrate of fibrous materials (e.g., polyethylene terephthalate (PET) nonwoven mats). The barrier layer, typically produced by the phase inversion method, has relatively low porosity and an asymmetric porous structure (i.e., ~ 50% bulky porosity and 20% surface porosity in the barrier layer) (Deniz, 2006; Ma *et al.*, 2010a). Often, the unconnected pores in the barrier layer can be easily plugged by feeding substances, thus decreasing the permeation flux. A barrier layer with very high porosity, interconnected pore structure and directional water channels has been recognized as a way to substantially increase the permeation flux without sacrificing the selectivity and rejection ratio.

In this chapter, we will review the recent development of nanofiber technology, which has opened new doors for the fabrication of highly permeable, energy-saving and durable membranes for water purification (Burger *et al.*, 2006; Chu *et al.*, 2008; Botes and Cloete, 2010; Ramakrishna *et al.*, 2010). This development includes the use of inorganic nanofibers (e.g., TiO_2 nanowire) (Rao *et al.*, 2003), carbon nanofibers (Mauter and Elimelech, 2008; Upadhyayula *et al*,. 2009) and electrospun polymeric nanofibers (Burger *et al.*, 2006; Reneker *et al.*, 2006; Thavasi *et al.*, 2008). The combination of the optimized TFC structure and the newly developed thin-film nanofibrous composite (TFNC) concept has provided a new pathway for the fabrication of highly efficient MF, UF, NF and RO membranes (Wang *et al.*, 2005; Yoon *et al.*, 2006, 2008; Chu *et al.*, 2008; Ma *et al.*, 2011b).

15.2 A new class of thin-film nanofibrous composite (TFNC) membranes

The newly demonstrated TFNC membranes consist of a two-layered (Ma *et al.* 2011a) or a three-layered hierarchical structure (Chu and Hsiao 2009a), as schematically shown in Fig. 15.2. In both structures, the bottom layer is composed of a nonwoven substrate (e.g., PET or polyethylene/polypropylene nonwoven web), which provides the overall mechanical strength to the membrane. The typical thickness of the bottom substrate is in the range of a few hundred microns, having fibers with diameters of about 10–30 microns. This nonwoven substrate has a relatively high porosity (about 40–60%), the Young's modulus up to 500 MPa, the ultimate tensile strength up to 40 MPa and the elongation-to-break ratio around 20% (which provides sufficient elastic property to the membrane (Ma *et al.* 2010a)).

It should be noted that a variety of MF membranes for the removal of bacteria and viruses from contaminated water has been demonstrated by direct surface modification of such nonwoven webs, based on polyester (Yeh *et al.*, 2002), cellulose (Ostreicher, 1992) or glass (Anon., 2000) microfibers. Positively charged polymers, including polyethylenimine (PEI) and polyamino-polyamide (cross-linked by epichlorohydrin/multi-epoxy groups), have been used to modify the surface of these fibers. As a result, the

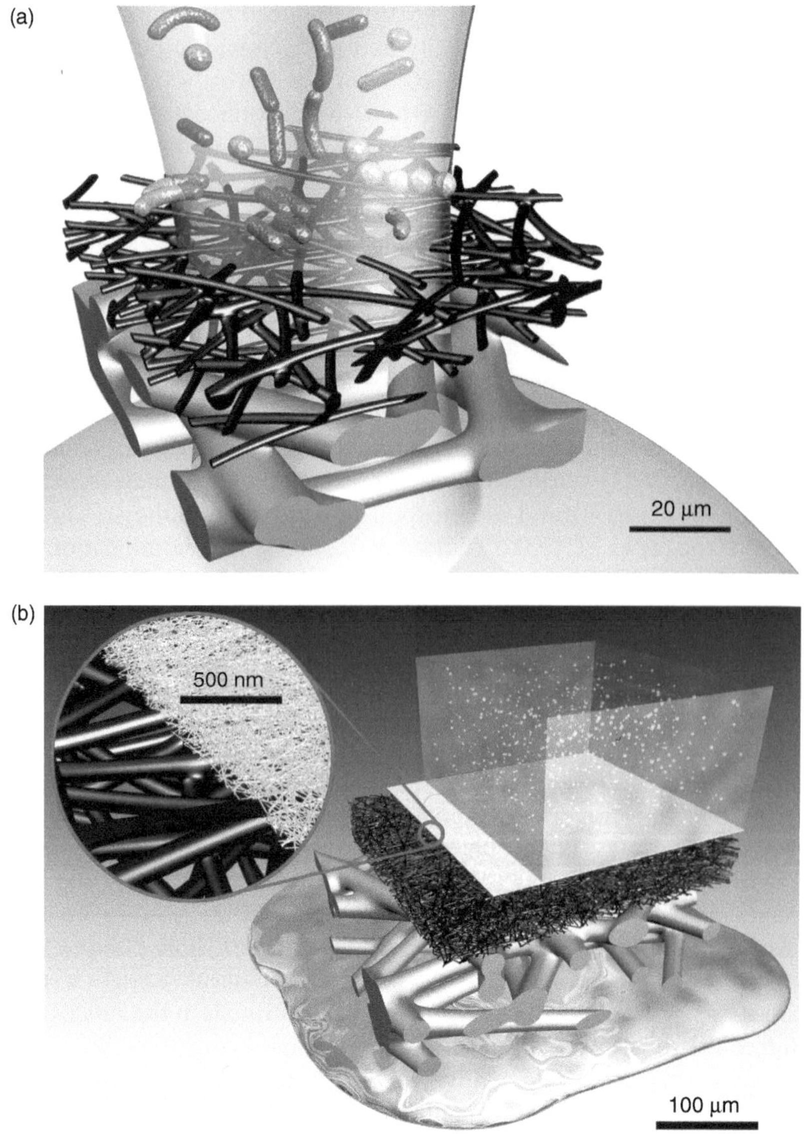

15.2 Hierarchical structures of TFNC membranes. (From Ma *et al.*, 2011a.)

charged microfibrous filter can adsorb negatively charged viruses and bacteria via electrostatic interactions (Hou *et al.*, 2003). However, the adsorption capacity of such microfilters is often limited by the relatively small surface-to-volume ratio and the total amount of active sites (Chu *et al.*, 2010a).

Based on the new concept of TFNC membranes, several breakthroughs in filtration performance were made as a result of (1) the unique properties of nanofibers (e.g., very large surface-to-volume ratio and interconnected three-dimensional porous structure) in the mid-layer support or the barrier layer and (2) the implementation of directed water channels created through the interface between the nanofiber and the polymer matrix in the barrier layer. For example, the first breakthrough performance was observed in microfiltration, where the electrospun nanofibrous scaffold was used directly as the barrier layer (the layer thickness ranged from 20 to 100 μm). The advantages of these MF membranes are due to the high porosity, interconnected three-dimensional structure and small fiber diameter in the barrier layer (Chu *et al.* 2008, 2010b; Chu and Hsiao 2009a), and can be understood by considering the following factors. First, the typical porosity for an electrospun scaffold is above 80% with both the surface porosity and the bulky porosity being essentially the same (the surface porosity of the electrospun scaffold is about three to four times higher than that of conventional membranes prepared by the phase-inversion method (Deniz, 2006)). The higher surface porosity would lead to an increase in the permeation flux (by about three to four times). Second, the three-dimensional nonwoven structure of the electrospun scaffold provides good mechanical property for a porous structure, implying that the layer thickness can be relatively thin. Third, all pores in an electrospun scaffold, as defined by stacking of nanofibers, are interconnected, which, to a limited extent, alleviates the fouling problem. Fourth, the typical fiber diameter is of the order of 100–1000 nm, which is significantly smaller than that of the microfibrous substrate (10~30 μm), implying a one to two order of magnitude larger surface-to-volume ratio. Further increase in the surface-to-volume ratio can be achieved by infusing even finer nanofibers (e.g., ultrafine polysaccharide nanofibers of 5~10 nm in fiber diameter) (Samir *et al.*, 2005; Siro and Plackett, 2010) into the electrospun scaffolds, which will be discussed in detail later (Chu *et al.*, 2010c; Ma *et al.*, 2012). The high surface-to-volume ratio is also of benefit to increase the adsorption capacity of surface modified membranes. Finally, the mean pore size and pore size distribution, based on the small diameter of electrospun nanofibers (e.g., about 100 nm), in the scaffold have been shown to be able to remove most of the waterborne bacteria with sizes above 200 nm through size exclusion (Ma *et al.*, 2011a). We note that there should be a partial integration of structures between the nonwoven PET substrate and the electrospun scaffold in order to provide good adhesion between the two layers for practical applications (Tang *et al.*, 2009a).

The second breakthrough performance of TFNC membranes is due to the unique nanocomposite structure of the barrier layer, where part of the electrospun scaffold has been incorporated (Chu *et al.*, 2008; Chu and Hsiao, 2009a) in the barrier matrix. The thickness of the barrier layer in this structure is typically of the order of 50–500 nm, depending on the fabrication process through physical (coating, Yoon *et al.*, 2006; Ma *et al.*, 2011b) or chemical (interfacial polymerization, Yoon *et al.*, 2009a; Yung *et al.*, 2010) means. The key feature of this structure is that as the embedded nanofibers are interconnected, the interface between the nanofiber and the barrier matrix can be used to generate water channels for more efficient water transportation. In other words, there is a structural integration between the barrier layer and the top electrospun nanofibrous scaffold layer, creating a nanocomposite barrier layer with capability to generate directed water channels. The nanocomposite structure can further withstand higher operating pressures and back-flushing treatments. These directed water channels could be of great benefit in increasing the water permeation flux. It should, nevertheless, be noted that in order to generate a thin barrier layer, the structure requires an asymmetrical mid-layer support with the top-most layer having fiber diameters sufficiently smaller than the thickness of the barrier layer (in the range of 30–100 nm). Second, the interface between the barrier matrix and the incorporated nanofibers must be able to facilitate water flow in order to form directed water channels. Although we do not know the nature (hydrophobic versus hydrophilic) of the optimal interface to transport water molecules, some demonstrated studies have clearly indicated the potential benefits of directed water channels in the barrier layer – for example, using the barrier layer of polyvinyl alcohol (Wang *et al.*, 2005; Ma *et al.*, 2010b), cellulose (Ma *et al.*, 2010a) and chitosan (Yoon *et al.*, 2006) on top of electrospun PAN scaffolds. Recently, the use of some very small nanofibers (with diameters in the range of a few nanometers), such as ultrafine cellulose nanofibers (Ma *et al.*, 2011b) and oxidized carbon nanotubes (Wang *et al.*, 2005), has further confirmed the possibility of generating a thinner and more efficient barrier layer with high water permeability. This will be the subject of future studies.

In addition to the above two breakthroughs, there are some practical advantages of TFNC membranes involving the use of hydrophilic materials as the barrier layer, including cross-linked PVA (Wang *et al.*, 2006; Yoon *et al.*, 2009b), cellulose (Ma *et al.*, 2010a), chitosan (Yoon *et al.*, 2006), chitin (Ma *et al.*, 2011c) and polysaccharide ultrafine nanofibers (Ma *et al.*, 2011b). The increase in hydrophilicity of the barrier layer was found to enhance the antifouling properties of TFNC membranes. It is interesting to note that most of the commercial membranes are made of hydrophobic materials, such as PAN, PES, PVDF, etc., because these polymers are durable under long-term operation in aqueous solutions. However, the use of hydrophobic polymers as the barrier layer can result in fouling and thereby reduce the water permeability during the filtration process (Yoon *et al.*, 2009b; Ma *et al.*, 2010a). The reasons for fouling, including reversible and irreversible types, are complex and can be associated with both properties of the membrane and the nature of filtration solution. Often, fouling is caused by the interactions between the surface of the membrane and the fluid to be filtered, where the adsorption is also associated with the surface energy of the filtration membrane (Hamza *et al.*, 1997). Typically, the hydrophobic surface has a lower surface energy, which would lead to the adsorption of lower surface energy materials, such as oil molecules, and thereby reduce the water permeability during a long duration of filtration (Ju *et al.*, 2009; Yu *et al.*, 2009). Thus, for filtration of oily wastewater, the hydrophilic membrane surface will be of benefit to maintain high permeation flux. The TFNC membranes using PVA, cellulose, chitosan, chitin and polysaccharide nanofibers as the barrier layers usually have a hydrophilic surface (for instance, the cellulose nanofiber based TFNC membrane has a water contact angle of 10.9°), which has been shown to possess very good antifouling properties during oily wastewater filtration (Ma *et al.*, 2011b).

Based on the above breakthroughs and added advantages in filtration performance, functional nanofibers have exhibited great promise for use in the fabrication of high flux MF, UF, NF and RO membranes for water purification; from treating contaminated drinking water to desalination. As the fabrication of TFNC membranes is relatively simple, 'green' and scalable, the nanofiber technology may play an important role in the future development of more energy-efficient and cost-effective water purification technology (Ma *et al.*, 2011b). Some specific examples of these applications are described as follows.

15.3 TFNC membranes for microfiltration

In the MF application for drinking water, the primary target is to exclude large particles, bacteria and viruses in contaminated fresh water (Cheryan, 1998). One sixth of people in the world have to struggle with waterborne diseases, and about 80% of the diseases are caused by contaminated food and water (Tiwari *et al.*, 2008). Microfibers, including synthetic polyesters, cellulose and glass nonwoven fibers, have been used for microfiltration for several decades (Sobsey and Glass, 1984; Borrego *et al.*, 1991). The mean pore size of microfilters made with nonwoven fibers is far greater than that of typical waterborne bacteria (0.2 to a few microns) (Anon., 2011) and viruses (a few tens of nanometers) (Krug and Wagner, 2011). To effectively remove bacteria and viruses from contaminated water, those microfibers have to be modified with functional groups, which can interact with the target impurities. It is well known that both bacteria and viruses can form charged species at certain pH values in suspension (as associated with their surface property or isoelectric point (pI)). Therefore, the fibrous filter with opposite charges located on the fiber surface will be able to adsorb bacteria or viruses (Yang and Tong, 1997). The major interactions between filter fiber surface and target impurities include electrostatic interactions (Zerda *et al.*, 1985), hydrophobic interactions (Lukasik *et al.*, 2000)and specific coordination complex formation (Hu *et al.*, 2008). Correspondingly, there are three means to modify the fibrous filters: surface physical coating (Ding *et al.*, 2005), surface chemical hydrolyzation (Yang and Tong, 1997; Chu *et al.*, 2010b) and surface chemical grafting of ligands (Ma *et al.*, 2006; Ma and Ramakrishna, 2008; Jia *et al.*, 2010). However, it should be noted that the adsorption capacity of the conventional microfibrous membrane is limited by the surface-to-volume ratio and the relatively large mean pore size/pore size distribution. On the other hand, the new microfiltration membrane fabricated from nanofibers can effectively overcome or improve the aforementioned disadvantages.

15.3.1 Novel microfiltration membranes fabricated by electrospinning technology

Nanofibers have been used for microfiltration soon after the rejuvenation of the electrospinning technology (Chu *et al.*, 2004; Agarwal *et al.*, 2008; Kaur *et al.*, 2008). The fabricated submicron-diameter fibers from natural and synthetic polymers have been applied widely to different applications, such as tissue engineering, drug delivery, sensors and electrodes in electronics, gas storage and air purification (Agarwal *et al.*, 2008; Thavasi *et al.*, 2008). The principle of electrospinning can be briefly described as follows. During the electrospinning process, when the applied voltage is strong enough to overcome the surface tension of the (charged) polymer solution at the tip of the spinneret, a fine fluid jet stream can be ejected. As this jet travels from the spinneret to the collector, the solvent in the jet stream is evaporated and the solution jet stream is stretched. The combined stretching and evaporation process results in the formation of a nonwoven nanofibrous scaffold at the collector (Reneker *et al.*, 2006). The fiber diameter and the scaffold morphology are affected by both processing and materials parameters. Electrospun membranes have a high porosity (~ 80%) with fine diameters (0.1~0.5 μm), yielding a very high specific surface area and an interconnected-pore structure. One representative application of electrospun scaffolds (self-supported or supported with a nonwoven substrate, e.g., PET) in water purification is the use as a microfilter for removal of nanoparticles (Barhate and Ramakrishna, 2007), which can be regarded as a model for waterborne bacteria – for example, *E. coli* and *B. diminuta*.

15.3.2 The role of nanofibers and microfibers in microfiltration membranes

The major difference between the role of nanofibers and that of microfibers in microfiltration is the filtration mechanism (Chu *et al.*, 2010c). As the mean pore size of the electrospun nanofibrous scaffold is 1–2 orders of magnitude smaller than that of conventional nonwovens, the former allows the removal of bacteria by size exclusion instead of by adsorption only. In recent studies of electrospun scaffolds, the mean pore size and pore size distribution of nanofibrous scaffolds based on PVDF (Gopal *et al.*, 2006), PAN (Barhate *et al.*, 2006; Chu *et al.*, 2010b), PSU (Gopal *et al.*, 2007), polyvinyl alcohol (PVA) (Chu *et al.*, 2010b) and regenerated cellulose scaffolds (Ma *et al.*, 2005) have been reported, and they are suitable for the removal of bacteria. Typically, the diameter of waterborne bacteria is larger than 0.2 μm – for example, the dimensions of *B. diminuta* are 0.3 μm × 0.9 μm and those of *E. coli* are 0.5 μm × 2.0 μm (Anon., 2011). Thus, it becomes essential that the mean pore size of the electrospun membrane should, at least, be less than 0.3 μm. In addition, the pore size distribution should be relatively narrow in order to achieve a high retention value for bacteria (e.g., a 6 log reduction value, LRV). The average pore size of electrospun membranes was found to be associated with the average fiber diameter at constant porosity as described next. To approach the 0.3 μm mean pore size in the membrane, the mean fiber diameter of the electrospun layer should be of the order of, and preferably less than, the 100 nm range. TFNC membranes based on electrospun PVA or PAN scaffolds on the PET nonwoven support have been demonstrated to reach this goal (Chu *et al.*, 2010c).

15.3.3 Experimental relationship between pore size and fiber diameter in nanofibrous membranes

We have established empirical relationships between the maximum pore size, the mean pore size and the average fiber diameter of electrospun nanofibrous membranes (Ma *et al.*, 2011a). The maximum pore size of a nanofibrous membrane is 8~12 times higher than the average fiber diameter, while the average pore size is 2~4 times the mean fiber diameter when the porosity of the membrane is kept at around 80%. The relationships are less affected by the thickness of the membrane (when it reaches a threshold value of about 50 μm). Therefore, the pore size of the electrospun membrane could be adjusted by tuning the average diameter of the nanofibers. The finer the fiber diameter, the smaller the pore size. To obtain high retention value and high surface-to-volume ratio, the nanofiber diameter should be as small as possible. Experimentally, the nanofiber diameter can be affected by many parameters, including materials parameters such as polymer, solvent, concentration and viscosity, as well as processing parameters such as applied voltage, distance between the spinneret and the collector, solution flow rate, temperature, humidity, etc. (Zong, 2002; Huang *et al.*, 2003; Pham *et al.*, 2006).

15.3.4 Nanofibrous microfiltration membranes for removal of bacteria

Among the above parameters, one major factor to control the fiber diameter is polymer concentration. In practice, nanofibers with small diameters could be fabricated only with polymer solutions at low enough polymer concentrations. However, to be able to carry out electrospinning, the polymer solution concentration must be higher than the overlap concentration of the polymer. Thus, one tends to use polymers with sufficiently high molecular weights. For example, PAN with a molecular weight of 150 kDa could be dissolved in N,N-dimethylformamide (DMF). At concentrations of 6, 8, 12 wt%, the fiber diameters of the electrospun scaffolds prepared from these solutions were 110~1280 nm, respectively (Yoon *et al.*, 2009b; Chu *et al.*, 2010c). All electrospun PAN membranes had a high porosity of about 80–85% and the pure water permeability was ~ 2500 L/m^{2}h/psi. The water contact angle of PAN electrospun membranes changed from 57.6° to 5.4° during 8 sec due to the highly porous structure.

The nanofibrous scaffold prepared from the 6 wt% solution (the mean fiber diameter was 110 ± 10 nm), deposited on the PET nonwoven substrate, yielded the maximum pore size of about 0.8 μm and the mean pore size of about 0.3 μm. This membrane was tested using 0.2 μm-particles in aqueous suspension at a concentration of 250 ppm and an *E. coli* phosphorous buffer solution (PBS, pH = 7.2, 10^6 colony forming unit (cfu)/mL), where higher than 95% retention ratio for 0.2 μm-particle removal and 6 LRV for *E. coli* removal had been achieved. It is important to note that the membrane exhibited 2~3 times higher pure water flux than typical commercial MF membranes with a similar mean pore size of 0.22 μm. Recently, the mass production capability of making electrospun scaffolds with a mean fiber diameter of 110 ± 10 nm has been successfully demonstrated from a custom designed continuous electrospinning device with 8 wt% of PAN (M_w = 150 kDa) in DMF. The electrospun membranes fabricated with this apparatus typically have high pure water permeability (about 1700 ± 50 L/m^{2}h/psi) and 6 LRV for *B. diminuta* (10^6 cfu/mL) removal (Chu *et al.*, 2010c).

PVA electrospun nanofibrous scaffolds were fabricated from an aqueous solution and the resulting mean fiber diameter could easily reach 100 ± 10 nm (Chu *et al.*, 2010b). After cross-linking with glutaraldehyde (GA) in acetone, the water-resistant nanofibrous membrane was challenged with 0.2 μm-particles (250 ppm) where higher than 95% retention ratio was observed. Moreover, the nanofibrous scaffolds exhibited high Young's modulus and high tensile strength due to the formation of cross-linked networks, which could overcome the wear and tear as well as water resistant requirements under typical cross-flow conditions.

PES electrospun nanofibrous scaffolds were fabricated from a mixed solvent of DMF/N-methyl-pyrrolidone (NMP) at different composition ratios (Yoon *et al.*, 2009c). The fiber diameter increased from 550 to 760 nm with increasing NMP content (from 0 to 50 wt%) at a fixed PES concentration of 26 wt%. Ammonium persulfate (APS) was used to oxidize PES electrospun nanofibers, which changed the water contact angle of the nanofibrous membrane from 127° to 28°, making the modified PES scaffold more suitable for water purification. The water permeability of the modified electrospun PES membrane could approach a value of 2.91×10^3 L/m^{2}h/psi, with high Young's modules of 280 MPa (measured by tensile experiments). PES electrospun nanofibrous scaffolds were also prepared from lower polymer concentrations (i.e., 20~22 wt%) by using a solvent mixture of DMF/NMP = 60/40. This condition represented a more optimized one compared to the first condition, considering the adhesion of PES nanofibers with PET, as well as the fiber diameter and uniformity (Tang *et al.*, 2009a).

PVDF (Gopal *et al.*, 2006), PAN (Barhate *et al.*, 2006) and PSU (Gopal *et al.*, 2007) were also used to fabricate self-standing nanofibrous MF membranes (without the PET nonwoven support), which had been comprehensively characterized with SEM and capillary flow porometry (CFP), and tested with aqueous suspensions of spherical particles of different sizes. The fiber diameter of these scaffolds ranged from 200 to 1000 nm and the mean pore size ranged from 1.5 to 5.0 μm. It was noted that the porosity of these membranes was reported to be about 95%, which was much higher than the typically measured value of 80%. Impressive water permeability value was obtained due to the highly porous structure of these membranes, while their particle retention ratios were still higher than 90% when the particle size was larger than 1.0 μm at a concentration of 100 ppm. Smaller particles in the range of 0.1~0.5 μm diameter, however, could pass through the membrane with only a low retention ratio (14~47%), which limited the application of the membrane for removal of small bacteria such as *B. diminuta*.

15.3.5 Nanofibrous microfiltration membranes for removal of viruses

Although nanofibrous MF membranes exhibit clear advantages over conventional MF membranes, smaller particles such as viruses with sizes of a few tens of nanometers cannot be eliminated from contaminated water by using nanofibrous MF membranes. To overcome this shortcoming, surface modifications of electrospun nanofibers have been carried out to expand the application of nanofibrous MF filters by taking advantage of the high surface-to-volume ratio of the nanofibrous scaffold. Viral filtration using an MF filter involves a different mechanism, known as depth filtration or adsorption. The principle of this mechanism involves three types of interactions between the active groups on the filter surface and the target molecules: electrostatic interactions, hydrophobic interactions and coordination complex formation. Thus, the two strategies for modifying the nanofiber surface are: physical or/and chemical means. The physical means involve the immobilization of functional polymers/oligomers/monomers on the surface of nanofibers by complex formation or charge interactions (Chu *et al.*, 2010a). The physically modified nanofibrous membranes thus could be hydrophilic, hydrophobic, positively charged or negatively charged. The chemical means involve the surface grafting of functional polymers, oligomers or monomers, or chemical etching to produce functional groups/ligands, which could interact with the target molecules. The two methods can also be combined together: for example, the nanofibers can first be chemically etched to produce charged surface, which are subsequently complexed with oppositely charged polyelectrolytes (as in the process of layer-by-layer fabrication) as demonstrated earlier (Decher, 1997).

Selected examples of modified nanofibrous scaffolds for MF application are as follows. In one study, PAN electrospun nanofibrous scaffolds were modified to possess positively charged surfaces for the removal of negatively charged viruses. By simply dip-coating the electrospun PAN scaffold in a positively charged monomer (e.g., 1-(1-vinylimidazolium)ethyl-3-vinylimdazolium dibromide (VEVIMIBr), sodium acrylate (SA)/ethoxylated (9) trimethylolpropane triacrylate (SR502), 1-vinyl-3-butylimidazolium bromide (VBIMIBr)/SR502) or polymer (e.g., branched polyethylenimine (PEI)/ethylene glycol diglycidyl ether (EGdGE)/glycidyltrimethylammonium chloride (GTMACl)) (Chu *et al.*, 2010a) solutions, the resulting membrane demonstrated high permeation flux (~ 192 $L/m^2h/psi$) and low pressure drop (~ 2 psi) for MF application and 4 LRV against the removal of MS2, a bacteriophage as a model virus. The positively charged PAN membrane also exhibited an excellent capacity, verified by the adsorption of negatively charged dye, metanil yellow (i.e., the capacity was about 10 mg/g of membrane and the rejection was 97% when 2000 mL of 1.5 ppm metanil yellow and 1 $mL/min/cm^2$ flow rate were used). The capacity of the modified PAN electrospun membrane was many times better than that of the commercial Cuno membranes. The modified electrospun nanofibrous membranes could be used over a wide pH range (3–9) and exhibited good mechanical properties (sufficiently strong to be used as a self-standing MF membrane for water purification).

In another study, functional electrospun PSU nanofibrous membrane was modified by chemical linking of the Cibacron blue F3GA ligand to absorb albumin (i.e., bovine serum albumin (BSA)) (Ma *et al.*, 2006). Such modified membranes exhibited high adsorption capacity (22 mg/g membrane) and good durability (no loss of adsorption capacity during three cycles of regeneration). Similar modification was also applied to cellulose (CA) nanofibrous membranes by hydrolysis in an alkali aqueous solution (Ma *et al.*, 2005). The modified cellulose nanofiber membrane was tested by adsorption of BSA and bilirubin, where a relatively high capacity (i.e., 13 and 4 mg/g membrane, respectively) was also obtained. The permeability of this membrane was around twice as high as that of commercial Millipore 0.22-μm MF membrane. An alternative modification on the cellulose (CA) nanofibrous membrane through hydrolysis was also demonstrated by generating aldehyde groups on the surface of cellulose fibers using the oxidation reaction with sodium periodate ($NaIO_4$) under neutral conditions (Ma and Ramakrishna, 2008). The aldehyde groups located on the surface of the oxidized cellulose nanofibers could further react with the amino groups in proteins (BSA, protein A/G). The functional nanofibrous membrane modified by this approach could be used for purification of immunoglobulin (IgG).

Carbon nanotubes (CNTs), including single-wall carbon nanotubes (SWCNT) (Brady-Estevez *et al.*, 2008) and multi-wall carbon nanotubes (MWCNT) (Brady-Estevez *et al.*, 2010), were used to fabricate microfilters for the removal of *E. coli* and MS2 from contaminated water. The carbon nanotubes were deposited onto PVDF/PTFE microporous substrates by a sonication/filtration procedure using CNTs/dimethylsulfoxide (DMSO) suspensions. Up to 3 LRV against MS2 and antibacterial property of this filter was obtained (the original concentration of MS2 was ~ 10^6 pfu/mL). Heavy loading of CNT would dramatically decrease the permeation flux, although higher LRV of 5~7 could be achieved. Some salts, such as NaCl, could be added to help increase the retention of MS2 (a LRV value of 6.6 was achieved). While the CNT filters had impressive advantages over other MF membranes, there were some problems for practical applications. First, the CNT MF filter has a shorter usage time compared to that of the commercial MF membrane (e.g., Cuno) (Chu *et al.*, 2010a). Second, the CNT MF filter has to be stored in the wet state because the dry membrane has lower permeation flux. Finally, the cost of CNTs is still relatively too high to be used for the fabrication of such specific MF filters.

15.3.6 Nanofibrous microfiltration membranes containing ultrafine polysaccharide nanofibers

A novel TFNC-based MF membrane has been demonstrated by infusing ultrafine polysaccharide nanofibers into a PAN electrospun nanofibrous scaffold deposited on the PET nonwoven substrate (Chu *et al.*, 2010c; Ma *et al.*, 2012). Ultrafine polysaccharide nanofibers, such as cellulose nanofibers (Saito *et al.*, 2006; Eichhorn *et al.*, 2010), cellulose nanowhiskers (Habibi

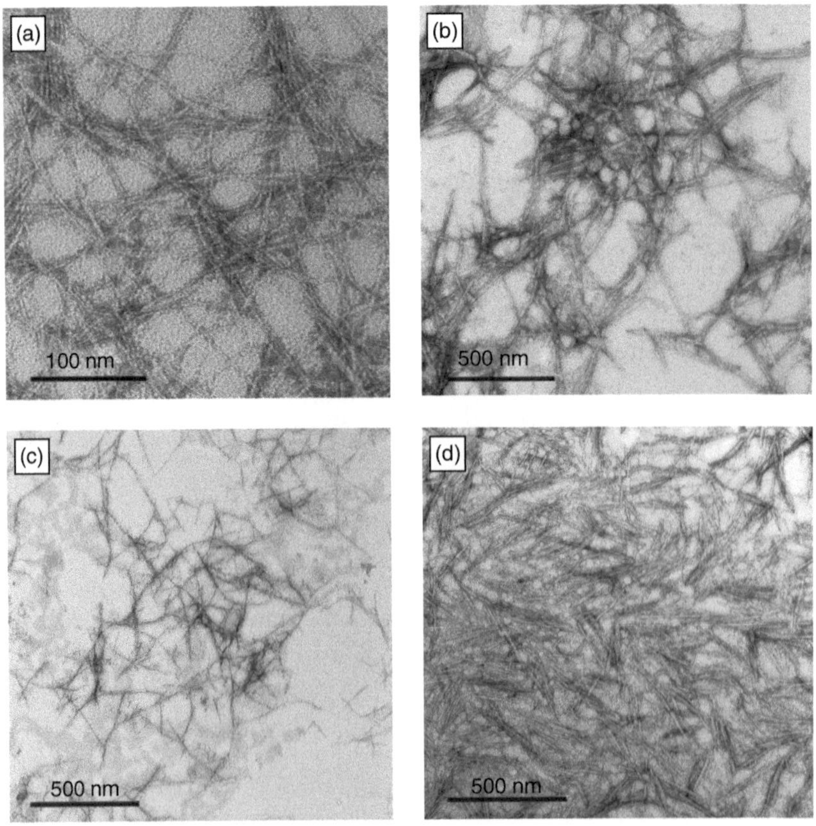

15.3 (a) TEM images of cellulose nanofiber from wood pulp, (b) chitin nanofibers from crab shell, (c) cellulose nanowhiskers from cellulose microcrystalline and (d) cellulose nanofibers from cotton linter pulp. (From Ma *et al.*, 2011b).

et al., 2010) and chitin nanofibers (Fan *et al.*, 2008a, 2008b), are new nanoscale materials which can be used to enhance the adsorption capacity of nanofibrous MF filters. These ultrafine nanofibers can be prepared from biomasses, such as cellulose (e.g., wood or cotton pulps), chitin or tunicates by using the TEMPO/NaBr/NaClO (aq.) oxidation/mechanical treatment method (Isogai *et al.*, 2011). The typical diameter of the final polysaccharide nanofibers is 5~20 nm with a length of a few microns. The TEM images of polysaccharide nanofibers fabricated from different natural sources are shown in Fig. 15.3.

The properties of polysaccharide nanofibers are depicted in Table 15.1. There are several advantages of ultrafine polysaccharide nanofibers when compared with electrospun nanofibers or carbon nanotubes (Ma *et al.*, 2011b). First, the surface-to-volume ratio of polysaccharide nanofibers is 20~40 times higher than that of a typical electrospun nanofiber. Second, the surface of

Table 15.1 Properties of polysaccharide nanofibers

Nanofiber	Diameter (nm)[a]	Length (μm)	Carboxylate (mmol/g oxidized cellulose)	Aldehyde (mmol/g oxidized cellulose)	Thermal Stability (°C)[b]	Crystallinity (%)[c]	Crystal Size (nm)[d]	Compatibility[e]	
								H_2O	DMF
Cellulose Nanofiber	5~8	1~3	0.70	0.25	240	63.6	4.3	√	√
Chitin Nanofiber	15~20	0.3~0.5	0.67	0.35	290	91.2	6.5	√	√
Cellulose nanowhisker	5~10	0.3~0.5	0.83	0.28	262	83.1	4.5	√	√

[a] Determined from TEM images using the Leika software.
[b] Determined from thermal gravimetric analysis (TGA) analysis.
[c] Determined from wide-angle X-ray diffraction (WAXD) measurements carried out at the X27C beam line, the National Synchrotron Light Source, Brookhaven National Laboratory.
[d] Estimated from WAXD profile using the Scherer equation (Warren 1969).
[e] Determined by UV spectroscopy.
'√': good compatibility.

polysaccharide nanofibers is covered with hydroxyl, carboxylate and aldehyde groups, which are formed from the oxidation of primary hydroxyl groups, implying a high density of negative charges. Thus, the surfaces of such ultrafine nanofibers can be highly hydrophilic, and the highly functional polysaccharides offer great potential for further modifications. Third, the high crystallinity, high chemical resistance and high thermal stability of polysaccharide nanofibers offer opportunities for many diverse applications with different materials requirements. Finally, ultrafine polysaccharide nanofibers are easily fabricated from many cellulose sources, implying lower cost and fewer environmental concerns. In our laboratory, polysaccharide nanofibers have been used, for the first time, to modify electrospun nanofibrous scaffolds for the removal of bacteria and viruses for drinking water purification (Ma *et al.*, 2012). However, it should be noted that cellulose, being a natural product, exhibits different properties from different sources and one should be aware of the source in controlling the quality and specifications of the final product.

In one study, cellulose nanowhiskers were infused into the PAN electrospun scaffold and immobilized onto the fiber surface, especially at the cross-point of the nanofibers, by cross-linking of cellulose nanowhiskers through heat treatment (Ma *et al.*, 2012). The pore size became smaller and the pore size distribution became narrower for the TFNC MF membranes after modification. The modified TFNC MF membrane had a negatively charged cellulose surface, serving as an effective adsorption filter for viruses (e.g., MS2), and a size exclusion filter for bacteria (e.g., *E. coli* and *B. diminuta*) removal. Specifically, the cellulose nanowhiskers modified TFNC MF membrane exhibited 2 LRV against MS2 (a commercial MF filter typically has 1 LRV or less) and 6 LRV against *B. diminuta* under an operating pressure of only 2 psi. This property could be attributed to the very high surface-to-volume ratio of infused cellulose nanowhiskers with a typical fiber diameter of 5~7 nm. The pressure drop of TFNC membrane increased slightly due to the infusion of cellulose nanowhiskers. However, it remained sufficiently low to be used as an MF membrane. The mechanical properties of modified TFNC MF membrane were improved significantly due to the presence of cross-linked cellulose nanowhiskers, acting as trestles. Moreover, any nanofibrous membranes not limited to electrospun nanofibrous scaffolds and not limited to PAN could be functionalized using this strategy (Chu *et al.*, 2010c). The modification procedure was quite simple and straightforward, with water being used as the only suspending medium without imposing additional environmental concerns.

15.4 TFNC membranes for ultrafiltration (UF)

The TFNC UF membrane, composed of a three-tier structure, is shown in Fig. 15.4 (Chu and Hsiao, 2009a). The bottom layer could be a PET non-woven substrate, providing mechanical strength to the membrane. The

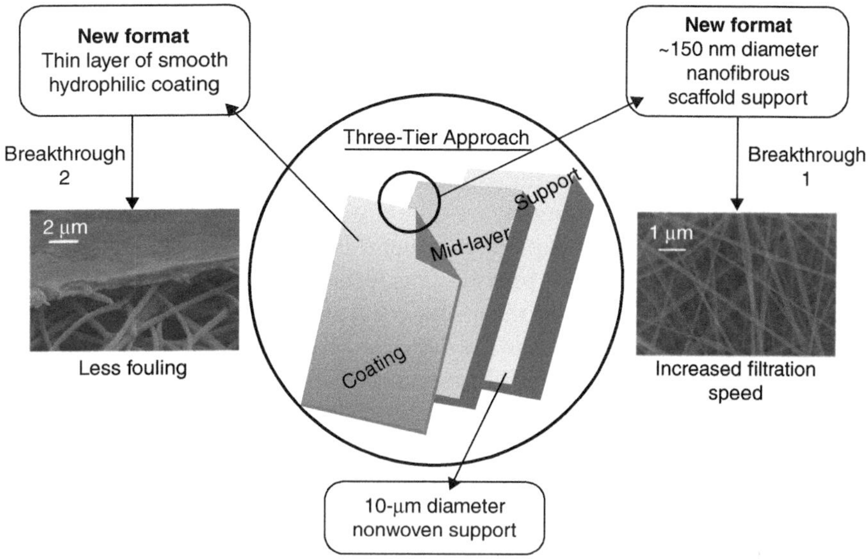

15.4 Schematic representation of TFNC UF membrane. (From Chu and Hsiao, 2009a.)

middle layer is composed of electrospun nanofibers with high porosity and interconnected pores, instead of the conventional microporous membrane prepared by the phase-inversion method. Typically, PAN, cross-linked PVA, PES and other polymers could be employed to fabricate electrospun nanofibrous scaffolds with a typical layer thickness of ~ 50 μm. The middle nanofibrous layer serves as a support for the top barrier layer in the three-tier structure (the previously described MF filter only has a two-tier structure). The thickness of the barrier layer is typically between 100 and 1000 nm, with the pore size ranging from 5 to 50 nm, depending on its applications. Hydrophilic polymers, such as PVA cross-linked with glutaraldehyde (GA) (Wang *et al.*, 2005), regenerated cellulose (Ma *et al.*, 2010a) and ultrafine polysaccharide nanofibers (Ma *et al.*, 2011b) have been used successfully to fabricate the barrier layer to alleviate the fouling problem. Specifically, TFNC UF membranes have been demonstrated successfully for the filtration of oily wastewater – for example, bilge water.

Bilge water is a type of wastewater that comes from ships and is a source of major pollution in the ocean (Ma *et al.*, 2010b). Therefore, there is a need to develop highly efficient (i.e., high-flux and low-fouling) UF membranes to tackle this environmental problem. The major ingredients of bilge water are diesel, detergent and other organic compounds, such as acetone, chlorobenzene, hydrazine, phenol, etc. (Chu *et al.*, 2010b). The requirements for the membrane used for the purification of bilge water include good chemical resistance, high permeation flux and low fouling. More importantly, the key features should be cost effective, exhibiting a low environmental footprint, robust performance and low energy use. To test the bilge water filtration, emulsified oily water with droplet sizes in the 70~1000 nm range can be used as a model contaminant (Li *et al.*, 2006; Chen *et al.*, 2009; Masuelli *et al.*, 2009). The emulsified oily water is prepared with 1350 ppm soybean oil, 150 ppm surfactant Dow Corning 193 and de-ionic water using a homogenizer. The average size of the droplets is about 110 nm, which falls within the range of UF application. The rejection performance in this study was determined by the UV-vis method using the adsorption peak located at 240 nm (Wang *et al.*, 2005). TFNC UF membranes were demonstrated to have high filtration efficiency, high chemical resistance and suitable pore size – therefore making them good candidates for development of an energy-efficient UF system for purification of bilge water. Some examples of the TFMC UF membranes are as follows.

15.4.1 TFNC UF membranes containing cross-linked PVA barrier layer

Cross-linked PVA was used to fabricate the barrier layer of TFNC UF membrane for filtration of oil/water emulsion (Wang *et al.*, 2005; Yoon *et al.*, 2009b; Ma *et al.*, 2010b). PVA is water soluble and it cannot be used for water purification without effective cross-linking. Di/tri-acylchloride, dianhydride, diacid, diepoxy and dialdehyde have been used to cross-link PVA (Bolto *et al.*, 2009); however, only dialdehydes (e.g., GA) and diepoxy can be used in aqueous solution (GA has been tested more extensively). In this study, the nonwoven PET substrate was used as the bottom supporting substrate, while PVA electrospun nanofibers cross-linked in the GA/hydrochloric acid/acetone solution formed the middle layer. The fiber diameter of the cross-linked PVA electrospun nanofibrous scaffolds was 100~300 nm. The fabrication of barrier layer was carried out by casting of 2.0 wt% of PVA aqueous solution, pre-cross-linked with GA (0.06 mole/mole PVA unit), on the top surface layer of cross-linked PVA electrospun scaffold. The three-tier PVA-based TFNC UF membrane with a barrier layer thickness of 1.8 μm was employed to filter 1350 ppm of oil/water emulsion, while another TFNC UF membrane having polyether-b-polyamide (PEBAX1074) as the barrier layer (0.8 μm barrier thickness) was also used for comparison (Wang *et al.*, 2006). The filtration test was carried out at 100 psi and 30~35°C. Stable permeability values of PVA and PEBAX TFNC membranes were observed up to 25 h of operation. The permeability of the PVA TFNC membrane was 1.30 L/m^{2}h/psi, which was about twice as high as that of the PEBAX TFNC membrane with a permeability of 0.57 L/m^{2}h/psi. Both membranes exhibited a similar rejection ratio of above 99.5%, as calculated with the following formula:

$$R\% = \frac{C_f - C_p}{C_f} \times 100\% \qquad [15.1]$$

where C_f and C_p are the concentrations of the feed solution and of the permeation solution, respectively (the concentrations were determined by UV at a wavelength of 240 nm). The success of the above TFNC UF membranes could be attributed partially to the unique characteristics of electrospun scaffold with high porosity and interconnected pores and the utilization of a highly hydrophilic barrier layer (e.g., PVA).

The permeability of TFNC UF membranes can be dramatically increased by introducing oxidized MWCNTs as a means to induce water channels in the PVA or PEBAX barrier layer (Wang *et al.*, 2005). For example, the addition of 8~10 wt% of oxidized MWCNT could increase 2~3 times permeability of PVA or PEBAX-based TFNC UF membranes when compared with that of TFNC membranes without oxidized MWCNTs. With the similar rejection ratio of 99.8% to oil/water emulsion after 30 h of operation, the stable permeability of the PVA-MWCNT TFNC UF membrane was 3.30 L/m^{2}h/psi, which was about twice as high as 1.61 L/m^{2}h/psi of PEBAX-MWCNT TFNC UF membrane. One possible reason for the higher permeability of the nanocomposite barrier layer is the formation of directed water channels between the interface of the oxidized MWCNT and PVA matrix. This was also confirmed indirectly by the evidence revealed by differential scanning calorimeter (DSC) measurements (Ma *et al.*, 2011b). It was found that the melting point of water confined in the nanoscale channels formed by interactions of functional groups – for example, carboxyl or hydroxyl groups of oxidized MWCNTs – shifted from 9.4 to –5.5°C, when the content of MWCNT in water increased to 66.7 wt%.

The permeability of PVA-based TFNC UF membrane could be further improved by decreasing the thickness of the barrier layer (Yoon *et al.*, 2009b). Saturated boric acid aqueous solution was used to wet the PAN electrospun scaffolds, which significantly reduced the penetration of PVA solution to the electrospun nanofibrous support. A GA cross-linked PVA/PAN TFNC membrane with a barrier layer thickness of 0.5 μm was fabricated and characterized. The maximum pore size and the average pore size calculated from the molecular weight cut-off (MWCO) measurements were 8.9 and 4.5 nm, respectively. The PVA/PAN TFNC UF membrane with such a relatively thin barrier layer could withstand pressures operating at up to 130 psi without loss of the rejection ratio for filtration of oil/water emulsion. The antifouling property of PVA/PAN TFNC UF membranes was tested over a 24-h period with the permeation flux remaining relatively stable while those of two commercial membranes (PAN10 and PAN400) declined by 64.4% and 52.3% of the original flux, respectively. A long-term fouling test of the PVA TFNC membrane for oil/water emulsion was carried out at 90 psi for 190 h. When compared with those of PAN10 and PAN400, the permeability of the PVA-based TFNC UF membrane was 2.56 L/m^{2}h/psi, which was about 12 times higher than that of PAN10 (0.22 L/m^{2}h/psi) and three times higher than that of PAN400 (0.83 L/m^{2}h/psi). It should be noted that the rejection ratio of the PVA TFNC UF membrane was 99% after 190 h of operation, whereas those of PAN10 and PAN400 were 99% and 98%, respectively.

Ultrafine cellulose nanofibers (CNs) and oxidized MWCNT were used as additives to enhance the water permeability of the PVA barrier layer in the TFNC membrane, where the additives could form directed water channels (Ma *et al.*, 2010b). The TFNC UF membranes were suitable for operation at low operating pressures – for example, 30 psi. The PVA-based TFNC UF membrane containing cellulose nanofibers (1.3 wt% to PVA) or MWCNT (10.0 wt% to PVA) showed pure water permeability around twice as high when compared with that of the PVA TFNC membrane without the additives. It is very interesting to note that cellulose nanofibers appeared to be more effective than oxidized MWCNTs. Even with only one-tenth of the amount of cellulose nanofibers used (i.e., 1.3 wt% of CN and 10.0 wt% of MWCNT in the PVA barrier layers), the pure water permeability of the corresponding membrane was higher than that of MWCNT. This could be attributed to the fact that cellulose nanofibers have higher density functional groups than those of oxidized MWCNTs, thus resulting in a larger amount of effective directed water channels in the barrier layer. Moreover, the PVA-cellulose nanofiber TFNC membrane was more durable than those of the PVA-MWCNT TFNC membrane during UF application due to the better compatibility between cellulose nanofibers and the PVA matrix (Ma *et al.*, 2010b). The stable permeability of the PVA-cellulose nanofiber (1.3 wt% to PVA) TFNC UF membrane was 11.67 L/m^{2}h/psi, which was about 8 times higher than that of PAN10 (0.83 L/m^{2}h/psi) with the same rejection ratio and 1.7 times higher than that of the pure PVA TFNC membranes (also with the same rejection ratio) at 30 psi/35°C for 24 h of operation. The PVA-cellulose nanofiber TFNC UF membrane can withstand up to 250 psi pressure during the filtration process.

An alternative PVA TFNC UF membrane with a two-tier structure (without the PET nonwoven support) was prepared by melting the deposited PVA electrospun nanofibers on the PAN electrospun membrane (unsupported) with water vapor (Wang *et al.*, 2010). The barrier layer with a thickness of 0.6 μm was formed from the partially melted electrospun PVA nanofibrous scaffold that was subsequently cross-linked in a GA/HCl/water/acetone mixture for 2 h. The membrane was used for the filtration of oil/water emulsion. A dead-end filtration cell with stirring was employed to test the membrane, where the operating pressure was 45 psi. A rejection ratio of 99.5% and a permeability of 4.67 L/m^{2}h/psi were observed at the beginning of the filtration.

Modified PVA-based TFNC UF membrane was also fabricated by coating of vinyl group-modified PVA followed by radical cross-linking with potassium persulfate (thermal initiator)/Darocur1173 (photo initiator) as an initiator (Tang *et al.*, 2009b). Specifically, PVA was reacted with acryloyl chloride in NMP and purified by precipitation in methanol. The vinyl group modified PVA was dissolved in water and pre-polymerized to a certain degree in order to avoid heavy penetration of the PVA-containing solution to the electrospun PAN nanofibrous scaffold. The pre-treated and highly viscous PVA solution could be applied to form a coating on the PAN scaffold surface. The cast membrane was subsequently cross-linked by heating/ UV treatments. The UV-cured PVA-based TFNC UF membrane was used to purify oil/water emulsion, where about twice the permeability (1.83 L/ m^{2}h/psi) of PAN10 was observed with a similar rejection ratio (99.5%).

While PVA-based TFNC membranes have exhibited permeability between twice and ten times as high, with a similar rejection ratio when compared with those of commercial UF membranes for UF of oil/water emulsions, there are some critical issues associated with the practical application in its present state of development. First of all, the coating process must be carefully controlled due to the use of the GA cross-linking reaction. The coating solution has to be pre-cross-linked to avoid the heavy penetration into the highly porous electrospun nanofibrous scaffold. This approach sets more stringent requirements for processing and therefore may hinder the low-cost scaleup production. Second, GA and other organic compounds (e.g., acetone) were used to cross-link PVA, which may cause environmental concerns. Finally, the PVA-based TFNC UF membranes must be stored in a wet state before use. The swelling and the degree of cross-linking have not been optimized to ascertain whether such a filter can retain its high performance for practical applications. Therefore, new materials and simpler processes are necessary to advance the development of TFNC UF membranes.

15.4.2 TFNC UF membranes containing regenerated polysaccharide barrier layer

Regenerated cellulose has been used recently as an alternative material for the fabrication of barrier layers in TFNC membranes for water purification, especially when one can use ionic liquids as solvents to dissolve native cellulose (Swatloski *et al.*, 2002; Pinkert *et al.*, 2009; Chu *et al.*, 2009b; Ma *et al.*, 2010a). Ionic liquids are organic molten salts with an asymmetrical structure and low lattice energy. Therefore, ionic liquids have relatively low melting points, often below 150°C (Welton, 1999; Dupont *et al.*, 2002). The typical composition of an ionic liquid includes both a cation, such as imidazolium, pyridium, ammonium or phosphonium, and an anion, such as chloride, bromide, tetrafluroboric or hexafluorophosphate. The ionic liquid with chloride as the anion can dissolve cellulose due to the hydrogen-bonding formation between chloride ions and the hydroxyl groups, thereby disrupting the hydrogen bonds between cellulose chains (Moulthrop *et al.*, 2005; Remsing *et al.*, 2006). This innovation addresses a key processing problem associated with cellulose and other polysaccharides, such as chitin (Xie *et al.*, 2006) – that is, they cannot be dissolved in common solvents. In a study, two ionic liquids, 1-butyl-3-methylimidazolium chloride (BMIMCl) and 1-ethyl-3-methylimidazolium acetate (EMIMAc), were employed to prepare cellulose solution with concentrations in the range of 0.1 wt% (Ma *et al.*, 2010a). The cellulose solution was homogeneous, as confirmed by small-angle X-ray scattering (SAXS). A high-flux TFNC membrane was fabricated by casting regenerated cellulose as a barrier layer on the PAN electrospun nanofibrous scaffolds on the PET nonwoven substrate. The thickness of the barrier layer could be as thin as ~0.5 μm with PAN electrospun nanofibers partially imbedded, as seen in the SEM image. The thin barrier layer formation could be attributed to the formation of a gel-like cellulose layer at the interface of water (being held in the electrospun scaffold) and ionic liquid, which limited the penetration of the cast solution into the electrospun scaffold. The cellulose-based TFNC UF membrane with a relatively thin barrier layer of 0.3 μm was obtained by using low viscosity (and fortunately low-toxic) EMIMAc as the solvent. The pure water permeability of the resulting TFNC membrane was six times higher than that of PAN-based commercial UF membranes – for example, PAN10 (Sepro). The cellulose-based TFNC UF membrane with a barrier thickness of 0.3 μm has the permeability of 9.30 $L/m^2h/psi$ for oil/water emulsion, which was eight times higher than that of PAN10 (1.18 $L/m^2h/psi$) after 24 h of operation at 30 psi. The rejection ratios of cellulose-based TFNC UF and PAN10 membranes were 99.8% and 99.9%, respectively. The average pore size of the cellulose-based TFNC membrane was 20 nm with a maximum pore size of 50 nm according to the MWCO test. The high permeability of the cellulose-based TFNC membrane could be attributed to the thin barrier layer and the hydrophilic surface, as confirmed by a relatively low water contact angle (23.7°) that implies a high surface energy and the decreasing tendency for oily fouling. Other regenerated polysaccharides, such as chitin, microcrystalline cellulose and their blends, were also employed to prepare the barrier layer of the TFNC UF membrane by using the electrospun PAN scaffold on a nonwoven PET support (Ma *et al.*, 2011c). All regenerated cellulose, chitin and cellulose/chitin blend barrier layers exhibited lower crystallinity when compared with those of native polysaccharides, as determined by wide-angle X-ray diffraction

(WAXD). The cellulose/chitin blend TFNC UF membrane exhibited the permeability of 7.65 L/m^{2}h/psi, which is ten times higher than that of PAN10 when tested in filtration of oil/water emulsion at 30 psi over an 100 h period. The rejection ratios of both membranes remained about the same (above 99.5%). The pore size and pore size distribution were also investigated by using the MWCO test. The average pore size and the maximum pore size of the cellulose/chitin-based TFNC UF membrane were 14 and 83 nm, respectively. With chitin being incorporated into the barrier layer, the cellulose/chitin or chitin TFNC UF membrane could possess good antibacterial properties (Singh and Ray, 2000), whereby applications such as a membrane bioreactor (MBR) could be feasible. The used ionic liquids could be extracted with ethanol in the coating process and recycled by distillation of ethanol. The recycled ionic liquids could be reused to dissolve cellulose without obvious changes and therefore the process could be considered a cost-effective and environmentally friendly one (Ma *et al.*, 2010a, 2011c).

Chitosan, not a natively charged polysaccharide but a semi-synthetic polymer, was also used to fabricate the barrier layer of the TFNC membranes containing electrospun PAN scaffold and the PET nonwoven substrate (Yoon *et al.*, 2006). The mid-layer was slightly modified by first depositing coarse electrospun nanofibers (from 10 wt% PAN/DMF solution, 750 nm mean fiber diameter) on the PET substrate, followed by depositing finer electrospun nanofibers on the top surface (from 4 wt% PAN/DMF solution, 124 nm mean fiber diameter). The cast chitosan solution was based on an acetic acid/water mixture and the PAN electrospun scaffold was also wetted with an aqueous solution of sodium hydroxide to limit the penetration of the cast solution. The chitosan-based TFNC UF membrane with a barrier layer thickness of 1.0 μm showed the permeability of 1.00 L/m^{2}h/psi from the oil/water emulsion, while that of NF270, as a comparison, was 0.05 L/m^{2}h/psi under 130 psi for 24 h of filtration. It is noted that the rejection ratio of chitosan TFNC UF membrane was 99.9%, which appeared to be slightly higher than that of NF270 (99.4%). Furthermore, the chitosan barrier layer could be modified for other applications using established chitosan chemistry (Kumar *et al.*, 2004).

15.4.3 TFNC UF membranes containing barrier layer made of ultrafine polysaccharide nanofibers

TFNC UF membranes with top barrier layers made of cellulose nanofibers, cellulose nanowhiskers and chitin nanofibers were also demonstrated (Chu *et al.*, 2010b; Ma *et al.*, 2011b). For example, ultrafine cellulose nanofibers with 5-nm mean fiber diameter were prepared from wood pulps by using the TEMPO/NaBr/NaClO oxidation process, followed with the fibrillate

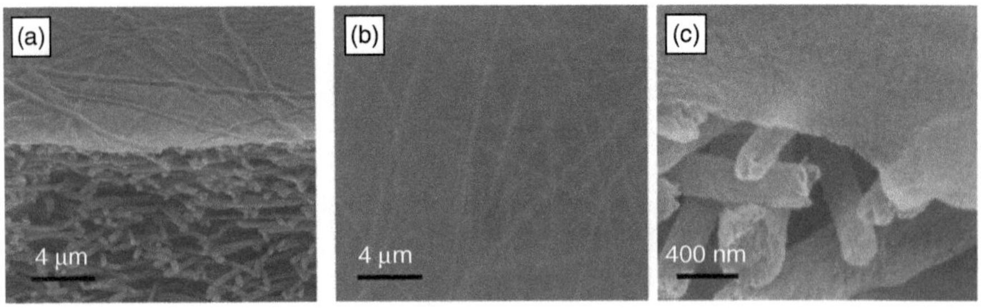

15.5 SEM images of (a) cross-sectioned view of TFNC membrane with cellulose nanofiber barrier layer; (b) top view of TFNC membrane and (c) magnified cellulose nanofiber barrier layer of TFNC membrane. (From Ma *et al.*, 2011b.)

treatment (Saito *et al.*, 2007). The 0.05 wt% of cellulose nanofibers in aqueous suspension behaved as a Newtonian fluid and could be used to prepare the barrier layer of the TFNC UF membrane based on electrospun PAN scaffold and PET nonwoven substrate. The electrospun PAN scaffold was wetted with an acidic aqueous solution (pH = 2) first, where cellulose nanofibers formed a gel at the interface and limited the penetration of the cast solution. The cellulose nanofiber barrier layer thickness was in the range of 0.1 μm, as seen by SEM (Fig. 15.5). The relatively thin barrier layer thickness rendered high water permeability.

The average pore size of the barrier layer based on ultrafine cellulose nanofibers was 21 nm, while the maximum pore size was 55 nm, as confirmed from MWCO data. The permeability of cellulose nanofiber-based TFNC UF membrane was 10.4 $L/m^2h/psi$, which was 12 times higher than that of PAN10 (0.90 $L/m^2h/psi$) at 30 psi for 48 h of filtration of oil/water emulsion. The rejection ratio of the cellulose nanofiber-based TFNC membrane was 99.6% after 48 h of filtration, which was similar to that of PAN10. PAN400 is another high-flux commercial UF membrane with a lower rejection ratio of 98.2% for filtration of oil/water emulsion. Nevertheless, the permeability of 4.5 $L/m^2h/psi$ was three times lower than that of the cellulose nanofiber-based TFNC membrane. The high permeability of cellulose nanofiber-based TFNC UF membranes could also be partially attributed to the formation of directed water channels on the surface of those nanofibers. In addition, the hydrophilic surface of the cellulose nanofibers (the water contact angle was 10.9°) might be very useful in alleviating fouling for filtration of oil/water emulsions. Ultrafine chitin nanofibers or cellulose nanowhiskers could also be used to fabricate the barrier layer in TFNC the membrane. High permeability values for filtration of oil/water emulsion (i.e., 7.2 and 9.1 $L/m^2h/psi$, respectively) were obtained by these two TFNC membranes after 48 h of filtration performance at 30 psi.

It is worthwhile to note that the cellulose nanofiber-based TFNC UF membrane exhibited excellent chemical resistance, which could be partially attributed to the high crystallinity of α-cellulose in the cellulose nanofibers, instead of the lower crystallinity in regenerated β-cellulose. The carboxylate groups located on the surface of cellulose nanofibers have antibacterial properties, whereby oxidized cellulose has been typically used as hemostat in medical applications (Dineen, 1976). In these cellulose nanofiber-based TFNC UF membranes, the applicability in a broad range of pH values from 4 to 10 as well as anti-hypochlorite property were also observed. Various types of TFNC UF membranes with different compositions and performance are summarized in Table 15.2.

Based on the experimental results obtained thus far, the following summary can be made for ultrafine polysaccharide nanofiber-based TFNC UF membranes. These membranes clearly offer high filtration efficiency (i.e., high flux and high retention), as well as low energy consumption and, potentially, low manufacturing cost. The production process does not raise major environmental concerns as the coating process uses mainly water as the solvent or suspending medium.

15.5 TFNC membranes for nanofiltration and reverse osmosis

As mentioned before, the most abundant water source is the ocean. Therefore, desalination has been regarded as one of the essential processes to address the shortage of drinking water on a global scale. RO and NF membranes are considered as two critical components for this application (Vainrot *et al.*, 2008). Two types of commercial NF/RO membranes are widely used in the water industry today: cellulose acetate membranes made by phase-inversion (Strathmann *et al.*, 1971) and polyamide membranes made by interfacial polymerization (Petersen, 1993). The cellulose acetate RO membrane was first created in the 1960s, where the materials included not only cellulose acetate but also cellulose triacetate and their mixture. The asymmetrical structure in cellulose acetate membranes fabricated by the phase inversion method typically offers good filtration efficiency and durability, but can only be used in a relatively narrow pH range due to the hydrolysis of cellulose ester. The polyamide barrier layer was developed after cellulose acetate, and quickly became the preferred polymer for manufacturing RO membranes for desalination. The typical monomers used in interfacial polymerization are trimesoyl chloride (TMC) and m-phenylenediamine (MPD) (Wu *et al.*, 2006). A cross-linked polyamide barrier layer is formed by diffusion of MPD from water to the immiscible organic phase (e.g., hexane), where TMC was located, and polycondensation rapidly occurs at the interface. Aliphatic diamine, such as piperazine (PIP), can also used

Table 15.2 TFNC UF membranes

Membranes	Top layer (thickness: nm)	Middle layer	Bottom layer	MWCO(kDa)	Permeability ($L/m^2h/psi$)	Rejection ratio (%)	Applied pressure (psi)	Ref.
PVA-TFNC	PVA (1800)	PVA e-spun	PET	-	1.30	> 99.5	100	Wang *et al.* 2006
PVA-TFNC	PVA (500)	PAN e-spun	PET	40	2.56	99.0	90	Yoon 2009b
PVA-TFNC	PVA (600)	PAN e-spun	PET	10	6.85	99.7	30	Ma *et al.* 2010b
PVA-TFNC	PVA-MWCNT (1800)	PAN e-spun	PET	-	3.30	99.8	100	Wang *et al.* 2005
PVA-TFNC	PVA-UFCN (850)	PAN e-spun	PET	17.5	11.67	99.5	30	Ma *et al.* 2010b
PVA-TFNC	PVA-UV (ND)	PVA e-spun	PET	-	1.83	99.5	30	Tang *et al.* 2009b
PVA-TFNC	PVA (400)	PAN e-spun	N/A	-	4.67	99.5	45	Wang *et al.* 2010
Cellulose-TFNC	Cellulose (300)	PAN e-spun	PET	2000	9.30	99.8	30	Ma *et al.* 2010a
Cellulose-TFNC	Microcrystal cellulose (600)	PAN e-spun	PET	5000	5.50	99.8	30	Ma *et al.* 2011c
Chitin-TFNC	Chitin (600)	PAN e-spun	PET	5000	8.00	99.6	30	Ma *et al.* 2011c
Cellulose/chitin-TFNC	Cellulose-chitin (600)	PAN e-spun	PET	5000	7.70	99.7	30	Ma *et al.* 2011c
Chitosan-TFNC	Chitosan (1000)	PAN e-spun	PET	-	1.00	99.9	130	Yoon 2006
UFCN-TFNC	Cellulose Nanofiber (100)	PAN e-spun	PET	2000	10.40	99.6	30	Ma *et al.* 2011b

UFCN-TFNC	Cellulose Nanowhisker (200)	PAN e-spun	PET	2000	9.10	99.7	30	Ma *et al.* 2011b
UFCN-TFNC	Chitin Nanofiber (450)	PAN e-spun	PET	2000	7.20	99.9	30	Ma *et al.* 2011b
PEBAX-TFNC	PEBAX (800)	PVA e-spun	PET	–	0.57	> 99.5	100	Wang *et al.* 2006
PEBAX-TFNC	PEBAX-MWCNT (800)	PVA e-spun	PET	–	1.61	99.8	100	Wang *et al.* 2005
PAN10	PAN (N/A)	N/A	PET	70	0.87	99.9	30	Ma *et al.* 2010a
PAN400	PAN (N/A)	N/A	PET	2000	4.50	98.2	30	Ma *et al.* 2010a
NF270	Polyamide (N/A)	PSU	PET	–	0.05	99.4	130	Yoon *et al.* 2006

in interfacial polymerization to fabricate NF membranes with a larger pore size in the barrier layer (Lin *et al.*, 2007; Liu *et al.*, 2008). The fabrication of RO and NF membranes through interfacial polymerization often employs porous UF membranes, prepared by the phase inversion method, as the support. The membrane with this composite structure is termed thin-film composite (TFC) membrane (Petersen, 1993). TFC NF/RO membranes have good filtration efficiency (flux and rejection) for inorganic salts, including magnesium sulfate ($MgSO_4$), magnesium chloride ($MgCl_2$), sodium sulfate (Na_2SO_4) and sodium chloride (NaCl) in aqueous solutions, when compared to that of cellulose ester-based RO/NF membranes. Nevertheless, the anti-chlorine property of polyamide membranes was less robust due to the tendency of nitrogen oxidation in the amido linkage (Iborra *et al.*, 1996; Sagle and Freeman, 2004; Park *et al.*, 2008). Alternative monomers used in the interfacial polymerization were toluene-2,4-diisocyanate (TDI), di/triamine, diacid and their derivatives, while cross-linked polyurea and polyurethane could be formed as the barrier layer of the TFC membranes (Baker, 2004).

Polyamide-based TFNC membranes involving nanofibrous scaffolds opened a new pathway to improve the filtration efficiency of brackish water or seawater. Both PAN (Yoon *et al.*, 2009a) and PES (Yung *et al.*, 2010) electrospun nanofibrous scaffolds were employed as the support layer, instead of the UF support by the phase inversion method. The advantages of highly porous electrospun nanofibrous support have been discussed previously for MF and UF applications and some (e.g., directed water channels) could also be expected to work in NF/RO. In the studies of Yoon and Yung, the polyamide barrier layer was prepared from PIP and TMC monomers, where bipiperazine (BP), BMIMCl and 1-octyl-3-methylimidazolium chloride (OMIMCl) were used as additives to tune the polycondensation reaction. The permeability of polyamide TFNC membranes for filtration of 2000 ppm $MgSO_4$ was 36 % higher than that of NF270 which was the representative commercial NF membrane with a similar barrier layer composition. Both membranes (TFNC and NF270) exhibited a similar rejection ratio of 97.8% and 97.4%, respectively. With a similar rejection ratio, the polyamide TFNC membrane fabricated with 30% of BP and 70% PIP exhibited three times higher permeability than that of NF90. The polyamide TFNC membrane could withstand up to 200 psi without loss of retention capability.

Cellulose acetate electrospun fiber mat was employed as the scaffold where polyelectrolytes, such as chitosan (CHI), sodium alginate (SA) and poly(styrene sulfonate) sodium salt (PSS), were coated on the surface by means of a layer-by-layer (LbL) approach (Ritcharoen *et al.*, 2008), which has been regarded as an efficient method for fabrication of NF membranes (Malaisamy and Bruening, 2005). The high charge density on the surface of the polyelectrolyte membrane was beneficial to reject targeted ions based on the Gibbs-Donnan effect (Krasemann and Tieke, 2000). The rejection ratio of nanofibrous composite membranes containing polyelectrolyte multilayers against the filtration of 2000 ppm NaCl was in the range of 6–15%, which increased with an increasing layer number of CHI/SA polyelectrolytes. The corresponding water permeability of these membranes decreased from 60 to 40 L/m^{2}h under the same operating condition.

The TFNC membranes developed for the applications of NF and RO in water purification are rarely reported in the literature, which will be an active subject for future study. In addition to NF/RO, TFNC membranes may also be very useful for forward osmosis (FO), where the key requirement for the membrane is the high permeability and open structure in the support.

15.6 Conclusions

Highly permeable TFNC membranes with hierarchical fibrous structures have been fabricated, characterized and applied to a wide range of water purification processes. The typical structure of TFNC MF membrane contains a PET nonwoven substrate as the bottom support and a layer of cross-linked PVA, PAN or PES electrospun scaffolds as the barrier layer. The highly porous and interconnected structure of the nonwoven layers in TFNC MF membranes offers 2~5 times higher water permeability than that of commercial MF membranes, while maintaining high retention capability. This feature was achieved simply by the fact that the mean pore size and pore size distribution of the barrier layer by using the electrospun scaffold could be correlated with the fiber diameter. With small fiber diameters in the 100 nm range, which can be achieved by the electrospinning process, the mean pore size and the maximum pore size are suitable for fabricating a cost-effective filter to exclude most known bacteria by size exclusion, down to the reduction of one part per million. Surface modifications of TFNC MF membranes with charged species, such as PEI and polysaccharide nanofibers, have further expanded the application for the removal of viruses and heavy metal ions (e.g., Cr, As and Pb) (Sun *et al.*, 2009) from contaminated drinking water and yet maintaining high retention capability and low pressure drop.

TFNC UF membranes were composed of a three-tier structure with PET nonwoven as the bottom support layer; electrospun scaffold (e.g., cross-linked PVA, PAN and PES) as the middle layer; and cross-linked PVA, chitosan, cellulose or ultrafine polysaccharide nanofibers as the barrier layer, which could be integrated with the mid-layer support to increase the mechanical stability and to introduce directed water channels. For water purification, all materials in the barrier layer should be hydrophilic, mechanically durable and have antifouling characteristics against oily wastewater and proteins. Two to ten times higher water permeability with the same retention capability was achieved when compared with that of corresponding commercial UF membranes. Moreover, the polysaccharide barrier layer in TFNC membranes could be further functionalized for adsorption of bacteria, viruses and heavy metal ions to enhance the removal capability (Araki *et al.*, 2001; Hasani *et al.*, 2008; Roy *et al.*, 2009).

The use of nanofibers offers a platform that can also be used to fabricate highly permeable NF and RO membranes. The three-tier structure in TFNC NF/RO membranes would be the same as that in TFNC UF membranes, except that the former barrier layer should have much smaller pore sizes and narrower size distributions than the latter. Selected TFNC NF/RO membranes have been demonstrated using cross-linked polyamide by interfacial polymerization, and polyelectrolyte by the LbL method (Picart *et al.*, 2002). The key feature to enhance the permeability of the barrier layer in TFNC membranes is due to the introduction of directed water channels, which would enhance the flux but not lose the retention capability. As only limited materials (e.g., cellulose acetates and polyamides) have been used to fabricate conventional TFC membranes, the approach of TFNC NF/RO membranes may lead to a new pathway to fabricate more efficient membranes for desalination.

There is an urgent need to develop more cost-effective and energy-efficient membranes using new and innovative approaches (Interlandi, 2010). The development of TFNC membranes represents a promising attempt to reach this goal, where the resulting membranes have exhibited high flux, high retention, low fouling and low cost. In particular, the use of nonwoven fibrous structures with smaller and smaller diameters, especially with the combined applications of electrospun nanofibrous scaffold (diameter around 100 nm) and ultrafine cellulose nanofibers (diameter around 5 nm), can provide many unique advantages over the conventional membranes for water purification.

15.7 Acknowledgment

Financial support of this work was provided by the National Science Foundation (DMR-1019370) and the Office of Naval Research (N000140310932).

15.8 References

Agarwal, S., Wendorff, J. H. and Greiner, A. (2008), Use of electrospinning technique for biomedical application, *Polymer*, **49**, 5603–21.

Amjad, Z. (1993), Reverse osmosis: Membrane technology, water chemistry, and industrial applications, Van Nostrand Reinhold, New York.

Anon. Kimberly-Clark Worldwide Inc. (2000), Cationically charged coating on glass fibers, WO/0037385.

Anon. (2010), US Census Bureau, Population Division, International Data Base, December 2010 update, available from: http://www.census.gov/ipc/www/idb/worldpopgraph.php, accessed on 17 March 2011.

Anon. (2011), Water treatment solution, Lenntech, available from: http://www.lenntech.com/library/diseases/

Araki, J., Wada, M. and Kuga, S. (2001), Steric stabilization of a cellulose microcrystal suspension by poly(ethylene glycol) grafting, *Langmuir*, **17**, 21–7.

Baird, G. M. (2010), The aging water infrastructure dilemma, Water Utility Infrastructure Management, available from: http://www.uimonline.com/index/webapp-stories-action?id=461

Baker, R. W. (2004), *Membrane technology and applications* (2nd edn), John Wiley & Sons, Chichester.

Barhate, R. S. and Ramakrishna, S. (2007), Nanofibrous filtering media: Filtration problems and solutions from tiny materials, *Journal of Membrane Science*, **296**, 1–8.

Barhate, R. S., Loong, C. K. and Ramakrishna, S. (2006), Preparation and characterization of nanofibrous filtering media, *Journal of Membrane Science*, **283**, 209–18.

Bein, T. (1996), Synthesis and applications of molecular sieve layers and membranes, *Chemistry of Materials*, **8**, 1636–53.

Bolto, B., Tran, T., Hoang, M. and Xie, Z. (2009), Crosslinked poly(vinyl alcohol) membranes, *Progress in Polymer Science*, **34**, 969–81.

Borrego, J. J., Cornax, R., Preston, D. R., Farrah, S. R., McElhaney, B. and Bitton, G. (1991), Development and application of new positively charged filters for recovery of bacteriophages from water, *Applied and Environmental Microbiology*, **57**, 1218–22.

Botes, M. and Cloete, T. E. (2010), The potential of nanofibers and nanobiocides in water purification, *Critical Reviews in Microbiology*, **36**, 68–81.

Bousfield, W. R. (1905), The purification of water by continuous fractional distillation, *Journal of the Chemical Society*, **87**, 740–7.

Brady-Estevez, A. S., Kang, S. and Elimelech, M. (2008), A single-walled-carbon-nanotube filter for removal of viral and bacterial pathogens, *Small*, **4**, 481–4.

Brady-Estevez, A. S., Schnoor, M. H., Vecitis, C. D., Saleh, N. B. and Elimelech, M. (2010), Multiwalled carbon nanotube filter: Improving viral removal at low pressure, *Langmuir*, **26**, 14975–82.

Burger, C., Hsiao, B. S. and Chu, B. (2006), Nanofibrous materials and their applications, *Annual Review of Materials Research*, **36**, 333–68.

Chen, W., Su, Y., Zheng, L., Wang, L. and Jiang, Z. (2009), The improved oil/water separation performance of cellulose acetate-graft-polyacrylonitrile membranes, *Journal of Membrane Science*, **337**, 98–105.

Cheryan, M. (1998), *Ultrafiltration and microfiltration handbook*, CRC Press, Lancaster, PA.

Chu, B., Hsiao, B. S. and Yoon, K. (2008), Nanofiber and nanocomposite-fiber technology for environmental applications, *AATCC Review*, **8**, 31–3.

Chu, B. and Hsiao, B. S. (2009a), The role of polymers in breakthrough technologies for water purification, *Journal of Polymer Science Part B: Polymer Physics*, **47**, 2431–5.

Chu, B., Hsiao, B. S. and Fang, D. (The Research Foundation of State University of New York) (2004), Apparatus and methods for electrospinning polymeric fibers and membranes, US6713011.

Chu, B., Hsiao, B. S. and Ma, H. (The Research Foundation of State University of New York) (2010a), Functionalization of nanofibrous microfiltration membranes for water purification, Fill in Technology Disclosure, 12/22/2010, R-8327.

Chu, B., Hsiao, B. S. and Ma, H. (2009b), High flux fluid separation membranes comprising a cellulose or cellulose derivative layer, US 2009/0078640 A1; WO 2009/025900 A2.

Chu, B., Hsiao, B. S. and Ma, H. (The Research Foundation of State University of New York) (2010b), High flux high efficiency nanofiber membranes and methods of production thereof, WO 2010/042647 A2.

Chu, B., Hsiao, B. S., Ma, H. and Wang, R. (The Research Foundation of State University of New York) (2010c), High flux microfiltration membranes with viruses adsorption capability for water purification, US patent application, 61/375,965.

Crump, J. A., Okoth, G. O., Slutsker, L., Ogaja, D. O., Keswich, B. H. and Luby, S. P. (2004), Effect of point-of-use disinfection, flocullation and combined flocullation-disinfection on drinking water quality in western Kenya, *Journal of Applied Polymer Science*, **97**, 225–31.

Decher, G. (1997), Fuzzy nanoassemblies: Toward layered polymeric multicomposites, *Science*, **277**, 1232–7.

Deniz, S. (2006), Characteristics of polycarbonate membranes with polyethylene glycol prepared via dry-wet phase inversion method, *Desalination*, **200**, 42–3.

Dineen, P. (1976), Antibacterial activity of oxidized regenerated cellulose, *Surgery, Gynecology & Obstetrics*, **142**, 481–6.

Ding, B., Fujimoto, K. and Shiratori, S. (2005), Preparation and characterization of self-assembled polyelectrolyte multilayered film on electrospun nanofibers, *Thin Solid Films*, **491**, 23–8.

Doll, P., Lehner, B. and Kaspar, F. (2002), Global modeling of groundwater recharge, in G. H. Schmitz (ed.), Proceedings of Third International Conference on Water Resources and the Environment Research, Technical University of Dresden, Germany, 1, 27–31.

Dupont, J., de Souza, R. F. and Suarez, P. A. Z. (2002), Ionic liquid (molten salt) organometallic catalysis, *Chemical Reviews*, **102**, 3667–91.

Eichhorn, S. J., Dufresne, A., Aranguren, M., Marcovich, N. E., Capadona, J. R., Rowan, S. J., Weder, C., Thielemans, W., Roman, M., Renneckar, S., Gindl, W., Veigel, S., Keckes, J., Yano, H., Abe, K., Nogi, M., Nakagaito, A. N., Mangalam, A., Simonsen, J., Benight, A. S., Bismarck, A., Berglund, L. A. and Peijs, T. (2010), Review: Current international research into cellulose nanofibers and nanocomposites, *Journal of Materials Science*, **45**, 1–33.

Fan, Y., Saito, T. and Isogai, A. (2008a), Chitin nanocrystals prepared by TEMPO-mediated oxidation of α-chitin, *Biomacromolecules*, **9**, 192–8.

Fan, Y., Saito, T. and Isogai, A. (2008b), Preparation of chitin nanofibers from squid pen β-chitin by simple mechanical treatment under acid conditions, *Biomacromolecules*, **9**, 1919–23.

Gleick, P. H. (1996), Water sources, in S. H. Schneider (ed.), *Encyclopedia of climate and weather*, Oxford University Press, New York, vol. 2, 817–23, available from: http://ga.water.usgs.gov/edu/waterdistribution.html

Gopal, R., Kaur, S., Ma, Z., Chan, C., Ramakrishna, S. and Matruura, T. (2006), Electrospun nanofibrous filtration membrane, *Journal of Membrane Science*, **281**, 581–6.

Gopal, R., Kaur, S., Feng, C. Y., Chan, C., Ramakrishna, S., Tabe, S. and Matruura, T. (2007), Electrospun nanofibrous polysulfone membranes as pre-filters: particulate removal, *Journal of Membrane Science*, **289**, 210–19.

Habibi, Y., Lucia, L. A. and Rojas, O. (2010), Cellulose nanocrystals: Chemistry, self-assembly, and applications, *Chemical Reviews*, **110**, 3479–500.

Hamza, A., Pham, V. A., Matsuura, T. and Santerre, J. P. (1997), Development of membranes with low surface energy to reduce the fouling in ultrafiltration applications, *Journal of Membrane Science*, **131**, 217–27.

Hasani, M., Cranston, E. D., Westman, G. and Cray, D. G. (2008), Cationic surface functionalization of cellulose nanocrystals, *Soft Matter*, **4**, 2238–44.

Holt, J. K., Park, H. G., Wang, Y., Stadermann, M., Artyukhin, A. B., Grigoropoulos, C. P., Noy, A. and Bakajin, O. (2006), Fast mass transport through sub-2-nanometer carbon nanotubes, *Science*, **312**, 1034–7.

Hou, K. C., Bretl, D. S. and Hembree, R. D. (Procter & Gamble) (2003), Microorganism filter and method for removing microorganism from water, US6565749 B1.

Hu, M., Wan, L., Liu, Z., Dai, Z. and Xu, Z. (2008), Fabrication of glycosylated surface on microporous polypropylene membranes for protein recognition and adsorption, *Journal of Materials Chemistry*, **18**, 4663–9.

Huang, Z., Zhang, Y., Kotaki, M. and Ramakrishna, S. (2003), A review on polymer nanofibers by electrospinning and their applications in nanocomposites, *Composites Science and Technology*, **63**, 2223–53.

Iborra, M. I., Lora, J., Alcaina, M. I. and Arnal, J. (1996), Effect of oxidation agents on reverse osmosis membrane performance to brackish water desalination, *Desalination*, **108**, 83–9.

Interlandi, J. (2010), The new oil: Should private companies control our most precious natural resource?, *Newsweek*, October 8.

Isogai, A., Saito, T. and Fukuzumi, H. (2011), TEMPO-oxidized cellulose nanofibers, *Nanoscale*, **3**, 71–85.

Jia, W., Wu, Y., Huang, J., An, Q., Xu, D., Wu, Y., Li, F. and Li, G. (2010), Poly(ionic liquid) brush coated electrospun membrane: A useful platform for the development of functionalized membrane systems, *Journal of Materials Chemistry*, **20**, 8617–23.

Ju, H., McCloskey, B. D., Sagle, A. C., Kusuma, V. A. and Freeman, B. D. (2009), Preparation and characterization of crosslinked poly(ethylene glycol) diacrylate hydrogels as fouling-resistant membrane coating materials, *Journal of Membrane Science*, **330**, 180–8.

Kaur, S., Gopal, R., Ng, W. J., Ramakrishna, S. and Matsuura, T. (2008), Next-generation fibrous media for water treatment, *MRS Bulletin*, **33**, 21–6.

Krasemann, L. and Tieke, B. (2000), Selective ion transport across sefl-assembled alternating multilayers of cationic and anionic polyelectrolytes, *Langmuir*, **16**, 287–90.

Krug, R. M. and Wagner, R. R. (2011), Virus, *Encyclopedia Britannica*, available from http://www.britannica.com/EBchecked/topic/630244/virus

Kumar, B. and Fellner, P. (2003), Polymer-ceramic composite protonic conductors, *Journal of Power Sources*, **123**, 132–6.

Kumar, M., Grzelakowski, M., Zilles, J., Clark, M. and Meier, W. (2007), Highly permeable polymeric membranes based on the incorporation of the functional water channel protein Aquaporin Z, *Proceedings of the National Academy of Science*, **104**, 20719–24.

Kumar, M. N. V. R., Muzzarelli, R. A. A., Muzzarelli, C., Sashiwa, H. and Domb, A. J. (2004), Chitosan chemistry and pharmaceutical perspectives, *Chemical Reviews*, **104**, 6017–84.

Li, H., Cao, Y., Qin, J., Jie, X., Wang, T., Liu, J. and Yuan, Q. (2006), Development and characterization of anti-fouling cellulose hollow fiber UF membranes for oil-water separation, *Journal of Membrane Science*, **279**, 328–35.

Lin, S. W., Sicairos, S. P. and Navarro, R. M. F. (2007), Preparation, characterization and salt rejection of negatively charged polyamide nanofiltration membranes, *Journal of the Mexican Chemical Society*, **51**, 129–35.

Liu, M., Yu, S., Zhou, Y. and Gao, C. (2008), Study on the thin-film composite nanofiltration for the removal of sulfate from concentrated salt aqueous: preparation and performance, *Journal of Membrane Science*, **310**, 289–95.

Lonsdale, H. K. (1982), The growth of membrane technology, *Journal of Membrane Science*, **10**, 81–181.

Lukasik, J., Scott, T. M., Andryshak, D. and Farrah, S. R. (2000), Influence of salts on virus adsorption to microporous filters, *Applied and Environmental Microbiology*, **66**, 2914–20.

Ma, H., Burger, C., Hsiao, B. S. and Chu, B. (2011a), Ultra-fine cellulose nanofibers: New nano-scale materials for water purification, *Journal of Materials Chemistry*, **21**, 7507–10.

Ma, H., Yoon, K., Rong, L., Mao, Y., Mo, Z., Fang, D., Hollander, Z., Gaiteri, J., Hsiao, B. S. and Chu, B. (2010a), High-flux thin-film nanofibrous composite ultrafiltration membranes containing cellulose barrier layer, *Journal of Materials Chemistry*, **20**, 4692–704.

Ma, H., Burger, C., Hsiao, B. S. and Chu, B. (2011b), Ultra-fine polysaccharide nanofibrous membranes for water purification, *Biomacromolecules*, **12**, 970–6.

Ma, H., Burger, C., Hsiao, B. S. and Chu, B. (2012), Nanofibrous microfiltration membrane based on cellulose nanowhiskers, *Biomacromolecules*, **13**, 180–6.

Ma, H., Yoon, K., Rong, L., Shokralla, M., Kopot, A., Wang, X., Fang, D., Hsiao, B. S. and Chu, B. (2010b), Thin-film nanofibrous composite ultrafiltration membranes based on polyvinyl alcohol barrier layer containing directional water channels, *Industrial & Engineering Chemistry Research*, **49**, 11978–84.

Ma, H., Burger, C., Hsiao, B. S. and Chu, B. (2011c), Thin-film nanofibrous composite membranes containing cellulose or chitin barrier layers fabricated by ionic liquids, *Polymer*, **52**, 2594–9.

Ma, Z., Masaya, K. and Ramakrishna, S. (2006), Immobilization of Cibacron blue F3GA on electrospun polysulfone ultra-fine fiber surfaces towards developing an affinity membrane for albumin adsorption, *Journal of Membrane Science*, **282**, 237–44.

Ma, Z. and Ramakrishna, S. (2008), Electrospun regenerated cellulose nanofiber affinity membrane functionalized with protein A/G for IgG purification, *Journal of Membrane Science*, **319**, 23–8.

Ma, Z., Kotaki, M. and Ramakrishna, S. (2005), Electrospun cellulose nanofiber as affinity membrane, *Journal of Membrane Science*, **265**, 215–23.

Malaisamy, R. and Bruening, M. L. (2005), High-flux nanofiltration membranes prepared by adsorption of multilayer polyelectrolyte membranes on polymeric supports, *Langmuir*, **21**, 10587–92.

Masuelli, M., Marchese, J. and Ochoa, N. A. (2009), SPC/PVDF membranes for emulsified oily wastewater treatment, *Journal of Membrane Science*, **326**, 688–93.

Mauter, M. S. and Elimelech, M. (2008), Environmental applications of carbon-based nanomaterials, *Environmental Science & Technology*, **42**, 5843–59.

Mone, J. (2011), Everything you need to know about ultraviolet water purification, available from: http://www.harvesth2o.com/uv.shtml

Moulthrop, J. S., Swatloski, R. P., Moyna, G. and Rogers, R. D. (2005), High-resolution ^{13}C NMR studies of cellulose and cellulose oligomers in ionic liquid solutions, *Chemical Communications*, **12**, 1557–9.

Mulder, M. (1991), *Basic principles of membrane technology*, Kluwer Academic, Dordrecht, The Netherlands.

Nagarale, R. K., Gohil, G. S. and Shahi, V. K. (2006), Recent developments on ion-exchange membranes and electro-membrane processes, *Advances in Colloid and Interface Science*, **119**, 97–130.

Ostreicher, E, A. (Cuno Incorporated), 1992, Use of cationic charge modified filter media, US5085780.

Otowa, T., Nojima, Y. and Miyazaki, T. (1997), Development of KOH activated high surface area carbon and its application to drinking water purification, *Carbon*, **35**, 1315–19.

Park, H. B., Freeman, B. D., Zhang, Z. B., Sankir, M. and McGrath, J. E. (2008), Highly chlorine-tolerant polymers for desalination, *Angewandte Chemie International Edition*, **47**, 6019–24.

Peng, X., Jin, J., Nakamura, Y., Ohno, T. and Ichinose, I. (2009), Ultrafast permeation of water through protein-based membranes, *Nature Nanotechnology*, **4**, 353–7.

Petersen, R. J. (1993), Composite reverse osmosis and nanofiltration membranes, *Journal of Membrane Science*, **83**, 81–150.

Pham, Q. P., Sharma, U. and Mikos, A. G. (2006), Electrospinning of polymeric nanofibers for tissue engineering applications: a review, *Tissue Engineering*, **12**, 1197–211.

Picart, C., Mutterer, J., Richert, L., Luo, Y., Prestwich, G. D., Schaaf, P., Voegel, J. C. and Lavalle, P. (2002), Molecular basis for the explanation of the exponential growth of polyelectrolyte multilayers, *PNAS*, **99**, 12531–5.

Pinkert, A., Marsh, K. N., Pang, S. S. and Staiger, M. P. (2009), Ionic liquids and their interaction with cellulose, *Chemical Reviews*, **109**, 6712–28.

Ramakrishna, S., Ma, Z. and Matsuura, T. (2010), *Polymer membranes in biotechnology: Preparation, functionalization and application*, Imperial College Press, London.

Rao, C. N. R., Deepak, F. L., Gundiah, G. and Govindaraj, G. (2003), Inorganic nanowires, *Progress in Solid State Chemistry*, **31**, 5–147.

Remsing, R. C., Swatloski, R. P., Rogers, R. D. and Moyna, G. (2006), Mechanism of cellulose dissolution in the ionic liquid 1-n-butyl-3-methylimidazolium chloride: A ^{13}C and $^{35/37}Cl$ NMR relaxation study on model systems, *Chemical Communications*, **12**, 1271–3.

Reneker, D. H. and Fong H. (2006), *Polymeric Nanofibers*, ACS Symposium Series; American Chemical Society: Washington, DC.

Ritcharoen, W., Supaphol, P. and Pavasant, P. (2008), Development of polyelectrolyte multilayer-coated electrospun cellulose acetate fiber mat as composite membranes, *European Polymer Journal*, **44**, 3963–8.
Roy, D., Semsarilar, M., Guthrie, J. T. and Perrier, S. (2009), Cellulose modification by polymer grafting: A review, *Chemical Society Reviews*, **38**, 2046–64.
Sagle, A. and Freeman, B. (2004), Fundamentals of membranes for water treatment, available from: http://texaswater.tamu.edu/readings/desal/Membranetechnology.pdf
Saito, T., Nishiyama, Y., Putaux, J. L., Vignon, M. and Isogai, A. (2006), Homogeneous suspensions of individualized microfibrils from TEMPO-catalyzed oxidation of native cellulose, *Biomacromolecules*, **7**, 1687–91.
Saito, T., Kimura, S., Nishiyama, Y. and Isogai, A. (2007), Cellulose nanofibers prepared by TEMPO-mediated oxidation of native cellulose, *Biomacromolecules*, **8**, 2485–91.
Samir, M. A. S. A., Alloin, F. and Dufresne, A. (2005), Review of recent research into cellulosic whiskers, their properties and their application in nanocomposite field, *Biomacromolecules*, **6**, 612–26.
Shannon, M. A., Bohn, P. W., Elimelech, M., Georgiadis, J. G., Marinas, B. J. and Mayes, A. M. (2008), Science and technology for water purification in the coming decades, *Nature*, **452**, 301–10.
Sharma, V. K. (2008), Ferrates: Synthesis, properties, and applications in water and wastewater treatment, 985, ACS Symposium Series, American Chemical Society.
Sharma, V. K., Kazama, F., Hu, J. and Ray, A. K. (2005), Ferrates (iron (VI) and iron (V)): environmentally friendly oxidants and disinfectants, *Journal of Water Health*, **3**, 45–58.
Singh, D. K. and Ray, A. R. (2000), Biomedical application of chitin, chitosan, and their derivatives, *Journal of Macromolecular Science: Reviews of Macromolecular Chemistry and Physics*, **C40**, 69–83.
Siro, I. and Plackett, D. (2010), Microfibrillated cellulose and new nanocomposite materials: A review, *Cellulose*, **17**, 459–94.
Sivetz, M. and Scheibelhut, C. H. (1955), Application of mixed bed ion-exchange to water purification in nuclear reactors, *Industrial & Engineering Chemistry Research*, **47**, 1020–2.
Sobsey, M. D. and Glass, J. S. (1984), Influence of water quality on enteric virus concentration by microporous filter methods, *Applied Environmental Microbiology*, **47**, 956–60.
Strathmann, H., Scheible, P. and Baker, R. W. (1971), A rationale for the preparation of Loeb-Sourirajan-type cellulose acetate membranes, *Journal of Applied Polymer Science*, **15**, 811–28.
Sulivan, J. A. (1979), Distillation and purification of water at sea, *Mariners Mirror*, **65**, 161–2.
Sun, X., Peng, B., Ji, Y., Chen, J. and Li, D. (2009), Chitosan (chitin)/cellulose composite biosorbents prepared using ionic liquid for heavy metal ions adsorption, *AIChE Journal*, **55**, 2062–9.
Swatloski, R. P., Spear, S. K., Holbrey, J. D. and Rogers, R. D. (2002), Dissolution of cellulose with ionic liquids, *Journal of the American Chemical Society*, **124**, 4974–5.
Tang, Z., Qiu, C., McCutcheon, J. R., Yoon, K., Ma, H., Fang, D., Lee, E., Kopp, C., Hsiao, B. S. and Chu, B. (2009a), Design and fabrication of electrospun

polyethersulfone nanofibrous scaffold for high-flux nanofiltration membranes, *Journal of Polymer Science, Part B: Polymer Physics*, **47**, 2288–300.

Tang, Z., Wei, J., Yung, L., Ji, B., Ma, H., Qiu, C., Yoon, K., Wan, F., Fang, D., Hsiao, B. S. and Chu, B. (2009b), UV-cured poly(vinyl alcohol) ultrafiltration nanofibrous membrane based on electrospun nanofiber scaffolds, *Journal of Membrane Science*, **328**, 1–5.

Tavolaro, A. and Drioli, E. (1999), Zeolite membranes, *Advanced Materials*, **11**, 975–96.

Thavasi, V., Singh, G. and Ramakrishna, S. (2008), Electrospun nanofibers in energy and environmental applications, *Energy & Environmental Science*, **1**, 205–21.

Tiwari, D. K., Behari, J. and Sen, P. (2008), Application of nanoparticles in waste water treatment, *World Applied Sciences Journal*, **3**, 417–33.

Tomaszewska, M. (2000), Membrane distillation-examples of applications in technology and environmental protection, *Polish Journal of Environmental Studies*, **9**, 27–36.

Ulbricht, M. (2006), Advanced functional polymer membranes, *Polymer*, **47**, 2217–62.

Upadhyayula, V. K. K., Deng, S., Mitchell, M. C. and Smith, G. B. (2009), Application of carbon nanotube technology for removal of contaminants in drinking water: A review, *Science of the Total Environment*, **408**, 1–13.

Vainrot, N., Eisen, M. S. and Semiat, R. (2008), Membranes in desalination and water treatment, *MRS Bulletin*, **33**, 16–20.

Van der Bruggen, B. and Vandecasteele, C. (2002), Distillation vs. membrane filtration: overview of process evolutions in seawater desalination, *Desalination*, **143**, 207–18.

Wang, X., Chen, X., Yoon, K., Fang, D., Hsiao, B. S. and Chu, B. (2005), High flux filtration medium based on nanofibrous substrate with hydrophilic nanocomposite coating, *Environmental Science & Technology*, **39**, 7684–91.

Wang, X., Fang, D., Yoon, K., Hsiao, B. S. and Chu, B. (2006), High performance ultrafiltration composite membrane based on poly(vinyl alcohol) hydrogel coating on crosslinked nanofibrous poly(vinyl alcohol) scaffold, *Journal of Membrane Science*, **278**, 261–8.

Wang, X., Zhang, K., Yang, Y., Wang, L., Zhou, Z., Zhu, M., Hsiao, B. S. and Chu, B. (2010), Development of hydrophilic barrier later on nanofibrous substrate as composite membrane via a facile route, *Journal of Membrane Science*, **356**, 110–16.

Warren, B. E. (1969), *X-ray diffraction*, Addison-Wesley, Reading, MA, 251–4.

Welton, T. (1999), Room temperature ionic liquids: Solvents for synthesis and catalysis, *Chemical Reviews*, **99**, 2071–83.

Wu, C., Zhang, S., Yang, D., Wei, J., Yan, C. and Jian, X. (2006), Preparation, characterization and application in wastewater treatment of a novel thermal stable composite membrane, *Journal of Membrane Science*, **279**, 238–45.

Xie, H., Zhang, S. and Li, H. (2006), Chitin and chitosan dissolved in ionic liquids as reversible sorbents of CO_2, *Green Chemistry*, **8**, 630–3.

Xu, T. (2005), Ion exchange membranes: State of their development and perspective, *Journal of Membrane Science*, **263**, 1–29.

Yang, M. C. and Tong, J. H. (1997), Loose ultrafiltration of proteins using hydrolyzed polyacrylonitrile hollow fiber, *Journal of Membrane Science*, **132**, 63–71.

Yeh, E. B., Ostreicher, E. A., Sale, R. D. and Bucholz, M. B. (2002), Method for charge-modifying polyester, US 0155225 A1.

Yencho, S. A. (Water of Life, LLC) (2011), Ultraviolet water purification system, US7862768B2.

Yoon, K., Hsiao, B. S. and Chu, B. (2008), Functional nanofibers for environmental applications, *Journal of Materials Chemistry*, **18**, 5326–34.

Yoon, K., Kim, K., Wang, X., Fang, D., Hsiao, B. S. and Chu, B. (2006), High flux ultrafiltration membranes based on electrospun nanofibrous PAN scaffolds and chitosan coating, *Polymer*, **47**, 2434–41.

Yoon, K., Hsiao, B. S. and Chu, B. (2009a), High flux nanofiltration membranes based on interfacially polymerized polyamide barrier layer on polyacrylonitrile nanofibrous scaffolds, *Journal of Membrane Science*, **326**, 484–92.

Yoon, K., Hsiao, B. S. and Chu, B. (2009b), High flux ultrafiltration nanofibrous membranes based on polyacrylonitrile electrospun scaffolds and crosslinked polyvinyl alcohol coating, *Journal of Membrane Science*, **338**, 145–52.

Yoon, K., Hsiao, B. S. and Chu, B. (2009c), Formation of functional polyethersulfone electrospun membrane for water purification by mixed solvent and oxidation processes, *Polymer*, **50**, 2893–9.

Yu, H., Cao, Y., Kang, G., Liu, J., Li, M. and Yuan, Q. (2009), Enhancing antifouling property of polysulfone ultrafiltration membrane by grafting zwitterionic copolymer via UV-initiated polymerization, *Journal of Membrane Science*, **342**, 6–13.

Yung, L., Ma, H., Wang, X., Yoon, K., Wang, R., Hsiao, B. S. and Chu, B. (2010), Fabrication of thin-film nanofibrous composite membranes by interfacial polymerization using ionic liquids as additives, *Journal of Membrane Science*, **365**, 52–8.

Zerda, K. S., Gerba, C. P., Hou, K. C. and Goyal, S. M. (1985), Adsorption of viruses to charge-modified silica, *Applied Environmental Microbiology*, **49**, 91–5.

Zhou, M., Nemade, P. R., Lu, X., Zeng, X., Hatakeyama, E. S., Noble, R. D. and Gin, D. L. (2007), New type of membrane material for water desalination based on a cross-linked bicontinuous cubic lyotropic liquid crystal assembly, *Journal of the American Chemical Society*, **129**, 9574–5.

Zong, X., Kim, K., Fang, D., Ran, S., Hsiao, B. S. and Chu, B. (2002), Structure and process relationship of electrospun bioabsorbable nanofiber membranes, *Polymer*, **43**, 4403–12.

16
Functional nanofibers in microelectronics applications

T. S. NATARAJAN,
Indian Institute of Technology Madras, India

Abstract: The market demand for new devices and gadgets is ever increasing; this applies enormous pressure on researchers all over the world who are engaged in materials science to come up with new materials such as polymers and ceramics, as alternatives to the conventional inorganic materials like silicon and gallium arsenide. In addition to this, recent developments in device geometries have been moving toward using even smaller materials, namely from 2-D thin films to 1-D fibers, tubes, rods and even zero-dimensional quantum dots, etc. In this context, this chapter attempts to review the recent research trends in the fabrication of fibers by a specialized but simple process of electrospinning, and the applications of such functionalized nanofibers in the area of new electronic devices.

Key words: electrospinning, polymer nanofiber, coaxial fibers, nanowire, Schottky diode, FET, electroluminescence, LED, gas sensor, actuators.

16.1 Introduction: polymers in electronics

The importance of the field of electronics and its applications in modern times hardly needs emphasizing. However, such rapid developments in electronics demand newer and better materials with very attractive properties for the fabrication of various devices and components in exotic and innovative applications (Roth *et al.*, 1989; Gurunathan *et al.*, 1999; Nalwa, 2001; Fréchet, 2005; Dodabalapur, 2006; Ahn *et al.*, 2011). It is, therefore, not surprising that we have come a long way from the conventional inorganic semiconductors, such as silicon and germanium, which require very sophisticated and expensive schemes of preparation when it comes to device fabrication. Large-volume applications in new areas, such as photovoltaics for solar energy conversion, ultra-sensitive sensors, thin-film transistors and electroluminescent display devices, have ushered in polymers and even certain ceramics as useful alternate materials for fabrication of electronic devices (Cho *et al.*, 1995; Möller *et al.*, 2003; Riul, 2003; Dutta and Narayan, 2004; Kim *et al.*, 2005c; Dudhe *et al.*, 2010; Xu *et al.*, 2010c). These new materials have several attractive properties compared to their inorganic counterparts.

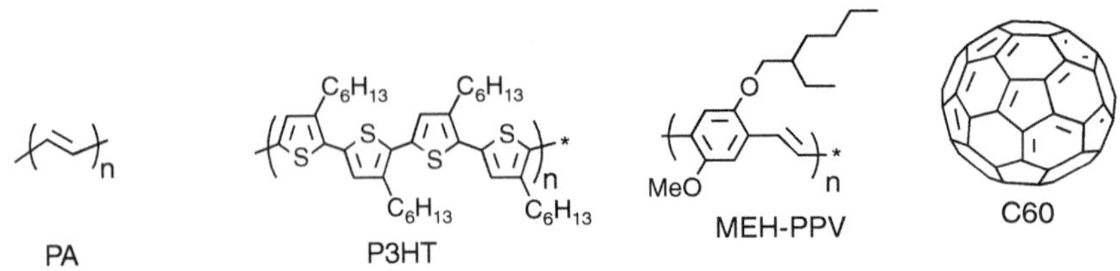

16.1 Conducting polymers useful for opto-electronic applications.

Apart from being mechanically robust and thermally processable, they do not require any special fabrication technologies. Large-area devices with these low-cost materials can be very easily prepared using very simple and modest schemes such as spin coating, ink-jet printing, etc. Flexible plastic electronics are becoming a reality already in many of our cell phones, tablet PCs and e-book readers. Printed electronics which significantly reduce the cost of production of sensors and displays have indeed revolutionized electronic applications (Eom *et al.*, 2009; Krebs *et al.*, 2010).

The electrical and electronic industries have been using polymers or plastics extensively as insulators of metallic conductors, photoresists, etc. However, since the discovery by Shirikawa and others, in 1977, that poly(acetylene) (Fig. 16.1) can be chemically doped to yield a polymer with enhanced conductivity as high as 11 orders of magnitude compared to the pristine polymer, several laboratories around the world have initiated research in this exciting area of 'electro-active' or 'conducting' polymers (Shirakawa *et al.*, 1977; Sirringhaus *et al.*, 1998; Schweitzer and Bassler, 2000). Initially, the focus was mainly to enable highly conducting materials as a replacement of conventional metals, which are relatively expensive, but later the efforts shifted the focus on developing several semiconducting polymers.

The problem with semiconducting polymers, such as poly(acetylene), and unsubstituted poly(heterocycles), such as poly(thiophene), is that, invariably, they are insoluble and untreatable. Therefore, research was aimed at synthesizing polymers with enhanced performance exhibiting far greater solubility and manufacturability, such as MEH-PPV and P3HT (Fig. 16.1). Thus, during the last couple of decades, a large number of polymers have been synthesized from new polymerizable monomeric structures, such as C_{60} (Fig. 16.1), having physical and chemical properties useful for a variety of electronic and opto-electronic devices. Examples of these applications include: conducting elements, field-effect transistors (FET) (Hangarter *et al*,. 2010), light-emitting diodes (LED) (Burroughes *et al.*, 1990; Braun and Heeger, 1991; Tessler *et al.*, 1998; Howard and Prache, 2001), photovoltaic cells (PV) (O'Connor *et al*,. 2008), sensors (Luoh and Hahn, 2006), photodetectors (Wu *et al.*, 2009) and organic circuits for integration into display technologies and low-cost electronics. The greatest advantage of working with such polymer materials is the fact that they demand inexpensive materials processing and high throughput device assembly, mainly via printing methodologies, and may mostly be processed via solution techniques, thereby eliminating the need for expensive lithography and vacuum deposition steps necessary in silicon-based microelectronics. Low-temperature solution processing enables processing options such as spin casting, photolithography (Lee, 1980; Xu *et al.*, 2010b), inkjet printing (Pekkanen *et al.*, 2010), soft lithography (Barink *et al.*, 2011), screen printing (Lee *et al.*, 2002) and micromolding (Park *et al.*, 2006) onto almost any type of substrate, including flexible ones. Thus, unlike silicon-based electronics, plastic electronics is useful in areas such as large-area rollable electronic displays, efficient lighting and low-cost and long-life solar cells. Plastic devices and electronics have thus carved a niche market for themselves in spite of the fact that some of their characteristics may not measure up to their inorganic semiconductor counterparts.

In addition, in the very recent past, there has been feverish activity among researchers all over the world to reduce the dimensions of these devices from 2-D films to 1-D rods, tubes and fibers, and even zero-dimensional quantum dots (Vayssieres, 2006; Lu *et al.*, 2009; Jie *et al.*, 2010). Though some of these developments are still under investigation at the laboratory scale, they have demonstrated already unparalleled performance characteristics at these nanoscale dimensions.

In this chapter we will attempt to review the exciting research work done in the area of applications of one-dimensional (1-D) polymer fibers in microelectronics with special emphasis on the simple and versatile technique of electrospinning. We will confine ourselves to discussions on the current trends and applications in the area of polymer fiber based electronic components and devices. We make only brief mention of the developments in allied areas, like sensor design, photovoltaic devices, etc., as these aspects are dealt with in detail in other chapters in this book.

The discussions here are broadly divided into two sections. The first section elaborates on the technique of electrospinning and its variants briefly to provide an independent background for the various applications discussed later in the chapter. The second section concentrates on the recent applications of such electrospun fibers in microelectronics, with brief mention about other related applications to put these in context.

16.2 Electrospinning of nanofibers

Though some of the ideas of electrospinning (Formhals, 1934, 1939, 1940; Rutledge and Fridrikh, 2007) were available even as early as 1934 with the patents of Formhals, it was not until early 1990s that the method became popular when several researchers started applying the technique with

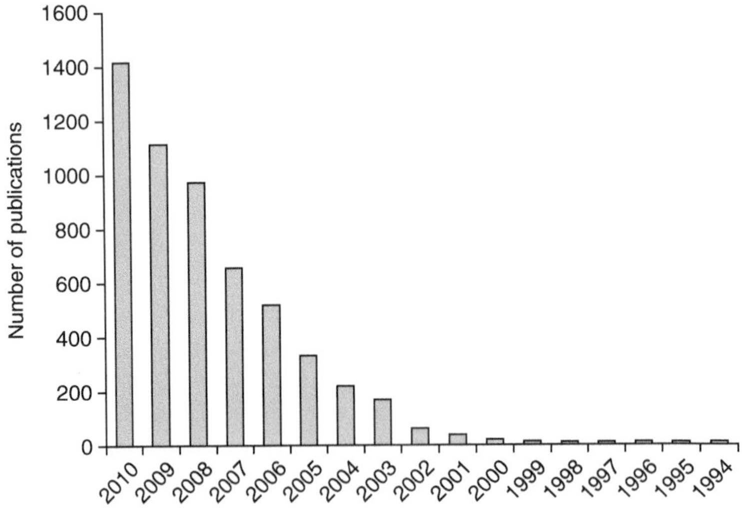

16.2 The increasing number of publications on electrospinning over the years is a clear indication of the importance of the area of research.

different kinds of polymers (Taylor, 1964; Reneker and Chun, 1996; Deitzel *et al.*, 2001, 2002; Dersch *et al.*, 2003; Huang *et al.*, 2003; Zussman *et al.*, 2003; Li and Xia, 2004; Subbiah *et al.*, 2005; Teo and Ramakrishna, 2006; He *et al.*, 2007).

If the number of papers published annually in this area is a fair indicator of the importance that the technique has acquired over the years, the field has indeed grown by leaps and bounds. Figure 16.2 shows the exponential growth of the number of publications based on electrospinning over the years.

Perhaps, one of the striking aspects of electrospinning is its simplicity as it demands only a very modest laboratory facility. If one has a high-voltage supply in some corner of the lab which can go up to 10 kV (say), one can start spinning polymer nanofibers almost immediately. If, in addition, one can also get hold of a syringe pump, then it makes it easier to spin continuous fibers for a long time. Since the technology itself is the subject matter of a separate chapter in this book, we will describe only briefly the process and some interesting variations of the method.

Electrospinning, basically, requires the following: (1) a polymer solution in preferably low-boiling-point solvent (e.g., poly(styrene) (PS) or poly(methylmethacrylate) (PMMA) in chloroform or tetra hydrofuran (THF)) having a concentration of, say, 15% w/w taken in a syringe with a needle; (2) a syringe pump to control the flow rate of the polymer solution during electrospinning; (3) a variable DC power supply in the range 10–30 kV.

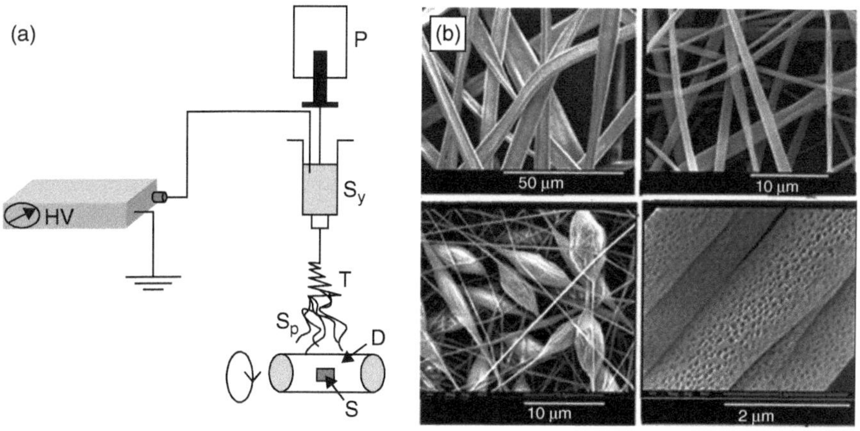

16.3 (a) A simple schematic of electrospinning setup to prepare aligned polymer fibers. HV–high voltage power supply (0–30 kV), P–computer controlled syringe pump to regulate the flow rate, Sy–syringe containing polymer solution, T–Taylor cone, Sp–splaying, D–rotating drum, S–a substrate which can be a metal foil (counter electrode) wrapped around the drum to collect the fibers. (b) Different morphologies of fibers obtained by electrospinning, such as beaded structures, flat ribbon-like and highly porous structures.

16.2.1 The principle of electrospinning

A simple schematic is shown in Fig. 16.3a. Initially a pendent droplet of the solution is formed at the tip of the needle which is connected to the positive terminal of a variable high-voltage power supply. The negative terminal is connected to a metal plate kept at some distance below the tip of the needle. As the voltage is increased, the tear-shaped droplet at the tip of the needle becomes an inverted cone due to the electrostatic field overcoming the surface tension of the solution. This is called the 'Taylor cone' (Fong *et al.*, 1999). On further increasing the voltage, an electrified jet emanates from the needle toward the counter metal electrode. Once the jet comes out into the atmosphere, the low-boiling-point solvent evaporates, leaving behind the solidified polymer strands all charged to the same polarity. The jet splits at this stage into multiple jets due to mutual electrostatic repulsion among the various polymer strands. This phenomenon is called 'splaying' and it continues until the strands reach the counter electrode. After collecting the deposition on the counter electrode by allowing this process for a few minutes, one can observe the electrode under a powerful microscope; random coils of micron or submicron fibers will have formed on the metal plate. Since the whole scheme is carried out under normal temperature and pressure in the laboratory, it is rather simple and easy to set up.

Electrospinning actually depends on several parameters which can be broadly divided into system parameters and solution parameters. The system parameters are: electrical potential, flow rate and concentration of the polymer solution, distance between the tip of the capillary and the collecting screen and other ambient parameters like temperature, humidity and the air velocity in the chamber. In addition, it is dependent on a few solution parameters like, molecular weight, molecular weight distribution and architecture (branched or linear) of the polymer, solution properties such as viscosity, surface tension and conductivity (Spivak *et al.*, 2000; Theron *et al.*, 2004; He *et al.*, 2005; Rutledge and Fridrikh, 2007; Thompson *et al.*, 2007; Uyar and Besenbacher, 2008; Xu, 2009; Luo *et al.*, 2010).

Though understanding the fundamental process of electrospinning is indeed a challenging effort, as it involves the combined principles of electrodynamics and hydrodynamics, there are many reports of attempts to model the process theoretically or through simulation (Katta *et al.*, 2004; Xu *et al.*, 2010a; Laudenslager and Sigmund, 2011). However, it is not easy to optimize the parameters to accurately predict the final diameter of the fibers or to achieve a certain level of control on the formation of a specific pattern of the fibers on the target electrode. However, there are a number of papers published which bring out the effects of various parameters when they alone are changed selectively, keeping all other parameters constant. The polymer fiber obtained comes with several defects and morphologies. Some of these defects, such as beaded structures, flat ribbon-like structures and highly porous structures are shown in the Fig. 16.3b. They can be controlled to a large extent by optimizing some of the parameters such as viscosity, surface tension of the solution, etc.

16.2.2 Applications of nanofibers

Nanofibers find a wide variety of applications, mostly in biomedical areas such as wound dressing (Thakur *et al.*, 2008), tissue engineering (Vasita and Katti, 2006), artificial blood vessels (Chang and Zhang, 2011), etc. They can also be very useful in filtering (Sambaer *et al.*, 2010; Yun *et al.*, 2010), sensors (Li *et al.*, 2006), protective clothing (Huck, 1988) for military applications, as well as in nanoelectronics as nanowires in interconnects, supercapacitors (Yoon and Ko, 2008; Wee *et al.*, 2010), fuel cells (Shabani *et al.*, 2010), etc. A brief overview of the application areas is given in Fig. 16.4.

Having looked at how electrospinning is carried out, and at the applications of nanofibers, it is time to take a look at some of the impressive work that has already been done in the area of electronics based on these electrospun fibers.

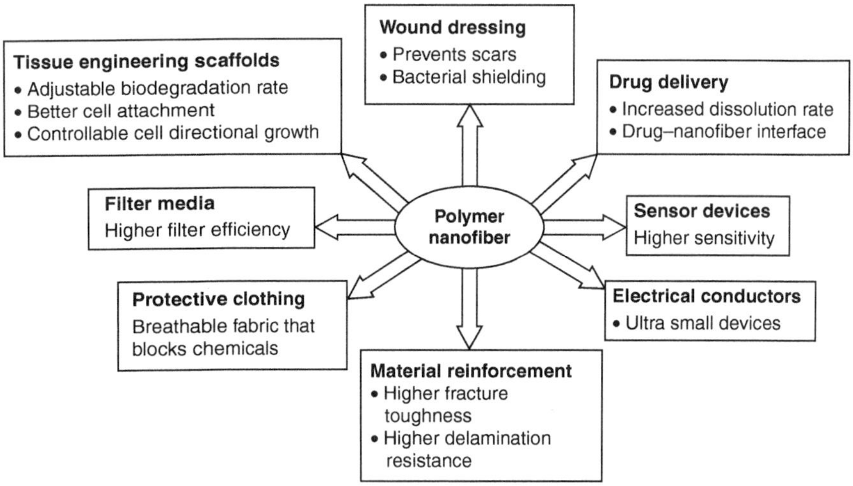

16.4 Nanofibers find very wide areas of applications. Figure shows in a nutshell the ways in which nanofibers of polymers can find useful applications in filtering, tissue engineering, sensors, protective clothing, etc.

16.3 Nanofibers in microelectronics

Plastic electronics, sometimes called printed electronics or organic electronics, is becoming more and more popular and it has got its own wide market. Unlike silicon-based electronics, plastic electronics is useful in areas such as large-area rollable electronic displays, efficient lighting and low-cost and long-life solar cells. It has a great advantage over conventional electronics, in the sense that the circuits can be produced at relatively low cost by ink-jet printing onto surfaces, whether rigid or flexible. In addition, the basic materials are polymers and other organic substances; these do not require very expensive processes – unlike silicon-based semiconductor electronics, which is identified as a high-cost and high-tech process. Plastic devices and electronics thus have carved a niche market for themselves, in spite of the fact that some of the characteristics may not measure up to their inorganic semiconductor counterparts.

In this context, nanofibers and membranes produced by electrospinning show improved ionic conductivity, electrochemical stability, low internal resistance and improved charge–discharge characteristics compared with those prepared by conventional schemes. Nanowires and nanotubes can function as miniaturized devices as well as electrical interconnects. Nanodevices such as field-effect transistors (FETs) (Ishii *et al.*, 2011), Schottky diodes (Rivera and Pinto, 2009), electrical switches (Pinto *et al.*, 2008) and memories (Ling *et al.*, 2007) have thus far been demonstrated. Electrospinning has been shown to be very promising in achieving 1-D nanostructures with the advantages of reduction of the size of devices and compact component density in circuits (Miao *et al.*, 2010). Electrodes made of electrospun polymer fibers in Li-ion batteries exhibit excellent rate capability, high reversible capacity and good cycling performance (Toprakci *et al.*, 2011). High surface area of electrospun activated carbon nanofibers improves supercapacitor energy density (Frackowiak and Beguin, 2002). Electrospun conducting nanocomposite wires having quasi-1-D structures show high electrical conductivity and have been used in ultra-sensitive chemical sensors, optoelectronics and catalysis. They are also used in flexible electrodes. Again, electrospun fiber-based actuators show improved performance since they exhibit high strain and fast response (Wang *et al.*, 2009a).

16.3.1 Nano wires

That the physical properties of materials are critically dependent on their dimensions has been amply demonstrated in several investigations on two-dimensional (2-D) (thin-film) and three-dimensional (3-D) (nanocrystalline) samples. This has increased the interest in 1-D nanostructures such as nanorods, nanowires, nanotubes, nanobelts and nanofibers since their distinctive geometries might have a great advantage. Indeed, they exhibit novel physical and chemical properties, and have potential applications in nanodevices and show size- and shape-dependent optical, magnetic and electronic properties (Xie *et al.*, 2002; Zeng *et al.*, 2007; Zhu, 2010; Zou *et al.*, 2010; Doudou *et al.*, 2011).

There are many schemes for preparing 1-D nanomaterials, such as the template-directed method (Ratnamala *et al.*, 2008), vapor-phase approach (Swihart, 2003), vapor–liquid–solid (VLS) growth (Bae *et al.*, 2008), solvothermal synthesis (Demazeau *et al.*, 1997) and self-assembly. Impressive research efforts have been focused toward synthesizing carbon nanotubes and semiconductor nanowires. Nanowires exhibit several amazing characteristics, such as very large surface area-to-volume ratios, flexibility in surface functionalities and superior mechanical performances. In a similar fashion, polymers also have excellent mechanical and thermal properties; they can be easily drawn in the form of thin sheets, wires, etc., or molded into different shapes and sizes. However, if we can make these polymers, which have such wonderful mechanical and thermal properties to also conduct electricity, then we can have the best of both worlds and we can avoid the expensive metallic components wherever possible and replace them with such conducting plastics. Thus, these polymers can indeed be made electrically conductive by two distinct ways. One way is by combining the polymers with conducting metal or carbon powders to make them reasonably conducting and molding them into different objects. Here the polymer acts as a matrix in which the conducting metal particles are held in position to conduct electricity. The advantage is that we need only a very small 'loading' of the metal particles to achieve reasonable levels of electrical conduction, at the same time without deteriorating the attractive properties of the host polymer matrix – namely its mechanical and thermal properties. The conductivity can also be tuned to some extent by changing the amount of 'loading' of the metal powder. The second method to obtain conducting polymers is to intrinsically modify their structures so that they can conduct electricity. Nobel Prize in Chemistry in the year 2000 was awarded to three scientists, namely MacDiarmid, Shirikawa and A. J. Heeges (2000), for demonstrating for the first time that poly(acetylene) can be made conducting intrinsically.

Thus, the insulating polymer fibers obtained by electrospinning can also be developed as conducting nanowires by suitably dispersing nanoparticles of silver, gold, graphite, carbon nanotubes (CNTs), etc., in the polymer solution before electrospinning (Ajayan *et al.*, 2000; Pantano *et al.*, 2008; Jacob *et al.*, 2009; Sanada *et al.*, 2009). Figure 16.5 shows the current-voltage characteristics of single-conducting nanowire obtained by electrospinning. The polymer in this case is PMMA and dispersed with 1% MWCNT (Sundaray *et al.*, 2006). In addition to such conducting metal particles and CNTs, several other nanoparticles such as ceramic oxides, ferroelectric materials and nanoparticles of compounds, such as ZnO, TiO_2, $BaTiO_3$, etc., were also added to prepare such functionally attractive nanofibers and nanowires by electrospinning (He, 2005; Kobayashi *et al.*, 2008; Tang *et al.*, 2010). One can also get rid of the polymers from the composite nanowires by controlled heat treatment when the polymers decompose and escape, leaving behind

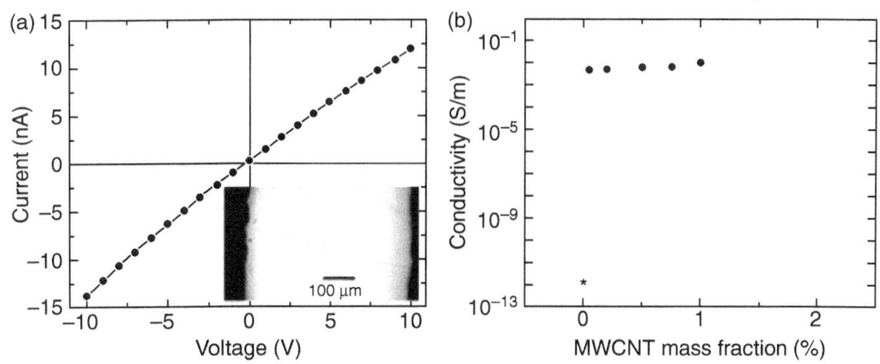

16.5 (a) The *I–V* characteristics of a single fiber. The resistance of the fiber from the slope of the line is about 780 milliohm. The inset shows the single fiber obtained between the gold electrodes. (b) Variation of the conductivity of the single composite fibers with a different mass fraction of MWCNT.

nanowires of these exotic materials. Thus, the preparation of such functionalized nanowires of complex oxides, which remained a great challenge for some years, has found a very simple method in electrospinning. These ceramic oxide fibers can also be prepared directly by using the combined techniques of sol-gel and electrospinning (Li *et al.*, 2003). The interesting properties of ferroelectric oxides, such as spontaneous polarization, high dielectric permittivity piezo- and pyroelectricity, make ferroelectric 1-D nanostructures an extremely interesting class of material for research, as well as for useful applications. Interestingly, necklace-like single-crystalline lead titanate (PbTiO3) nanowires have been reported in a large-scale area using electrospinning (Lu *et al.*, 2006).

16.3.2 Schottky diode

The Schottky diode, also known as hot carrier diode, is a semiconductor diode with a low forward voltage drop and a very fast switching action. A metal-semiconductor junction is formed between a metal and a semiconductor, creating a Schottky barrier instead of a semiconductor–semiconductor junction as in conventional diodes. The semiconductor would typically be N-type silicon and typical metals used are molybdenum, platinum, chromium or tungsten. The metal side acts as the anode while the N-type semiconductor acts as the cathode of the diode. When a current flows through a silicon diode there is a small voltage drop between 0.6 and 1.7 volts across it. In contrast, in the case of a Schottky diode the voltage drop is between 0.15 and 0.45 volts approximately. This lower voltage drop can provide higher switching speed and better system efficiency. Figure 16.6a shows the circuit symbol of a Schottky diode and Fig. 16.6b shows the typical *I–V* characteristics of such a diode and a resistor to compare the two devices.

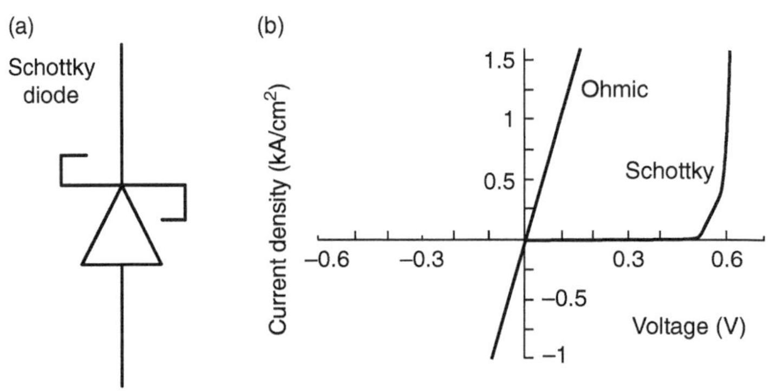

16.6 (a) Symbol of a Schottky diode and (b) *I–V* characteristics of a resistor (linear graph) and that of a diode (non-linear) for comparison.

Thus, a diode is an electronic device that allows current to pass through only in one direction, with a small voltage drop. This device is very useful in several applications such as rectifiers, filters, etc. and in electronic circuits.

In a typical design, the Schottky diode is prepared by using an *n*-doped Si wafer <111>, having sheet resistance 0.1–1.0 Ω cm with a 200 nm thermally grown oxide layer and electrospun poly(aniline) (PANI). Schottky diodes have been fabricated by electrospinning to make 1-D organic conducting polymer over an inorganic n-doped semiconductor (Pinto *et al.*, 2006). This makes them attractive for the fabrication of low-power, highly sensitive and rapid-response sensors and rectifiers. The device performance and the role of surface states on barrier height and charge transport was studied by varying the fiber diameter, and also by varying the fiber doping level at a fixed diameter. Clear rectification is observed for diodes fabricated from thick fibers with significantly reduced rectification ratios for diodes fabricated from thinner ones.

Handling single nanofibers in well-organized patterns is somewhat difficult. Hence Lotus *et al.* (2009) attempted to fabricate diodes using twisted fibers in the form of yarns (Rivera and Pinto, 2009). Again, since the yarn is more robust than single nanofibers it can easily be transferred to any substrate for fabricating devices. They successfully fabricated p-n junction diodes using twisted electrospun nanofibers of ZnO and NiO.

Figure 16.7a shows the optical microscope image of the twisted nanofiber p-n junction yarn obtained after annealing at 873 K. Figure 16.7b shows the SEM pictures of the diode. Both the n-type (ZnO) yarn, shown in white, and the p-type NiO yarn, shown in black, are made up of many fibers. Figure 16.7c shows the current-voltage (*I–V*) characteristics of the ZnO/NiO nanofiber hetero structure in diode configuration.

16.3.3 Field-effect transistor (FET)

Technologically, the most important polymer device fabricated and studied widely is the field-effect transistor, since it forms the central building block in switches, displays and logic circuits. A transistor is a linear semiconductor device that can control a larger current through one of its terminals with a low power applied electrical signal, which could be either a current, as in the case of a bipolar junction transistor (BJT), or a voltage, when it is called a field-effect transistor. Thus, there are two types of transistors: current-controlled bipolar transistor whose operation involves conduction by two types of charge carriers, namely, electrons and holes; and voltage-controlled field-effect transistor which is unipolar, since the main current through it is either electrons or holes. Circuit symbols of the two transistors are shown in Fig. 16.8a and b.

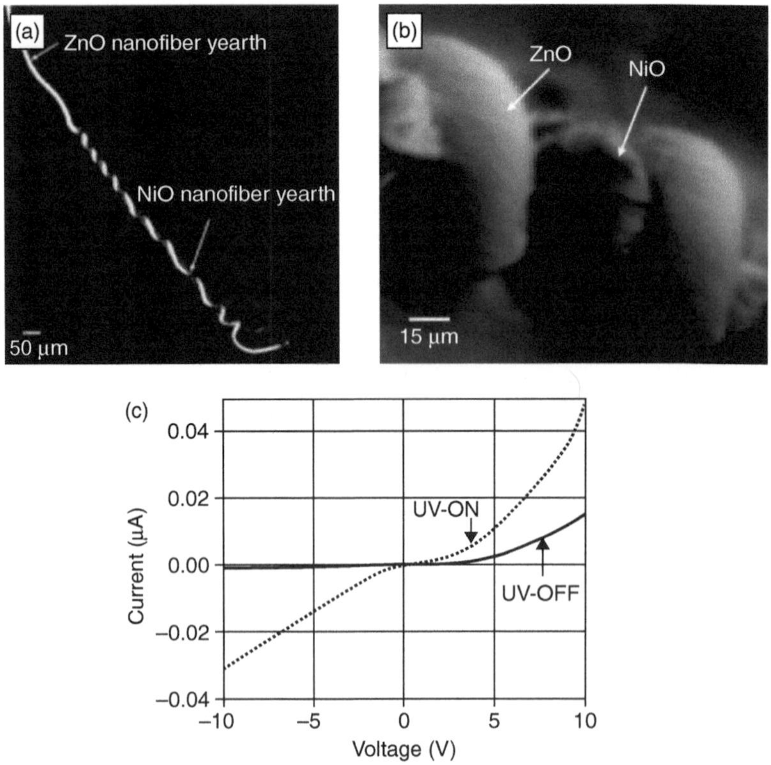

16.7 (a) Optical microscope image of the twisted pn-junction NiO/ZnO nanofiber yarns. (b) SEM image of the yarns annealed at 873 K. (c) *I–V* characteristics of the p-NiO and n-ZnO nanofiber yarn based pn-junction diode (Lotus *et al.*, 2009).

Bipolar transistors are 'normally-off' devices, since, if there is no current through the base, no current through the collector or the emitter. FETs, on the other hand, are 'normally-on' devices: no voltage applied to the gate allows maximum current through the source and drain. Also the amount of current allowed through an FET is determined by a voltage signal rather than a current signal, as against the BJT. In fact, with the gate-source PN junction reverse-biased, there should be nearly zero current through the gate connection. Hence, an FET is classified as a voltage-controlled device, and the bipolar transistor as a current-controlled device. Thus, in the FET device structure shown in Fig. 16.8c, the current flow between the source and the drain electrodes is controlled by the application of two independent biases, one between the source and the drain (V_d) and another between the source and the gate (V_g). The latter bias controls the carrier density in the FET channel, hence it determines whether the device is switched 'off' ($V_g = 0$ V, low carrier density) or switched 'on' ($V_g = 0$ V, high carrier density). Figure

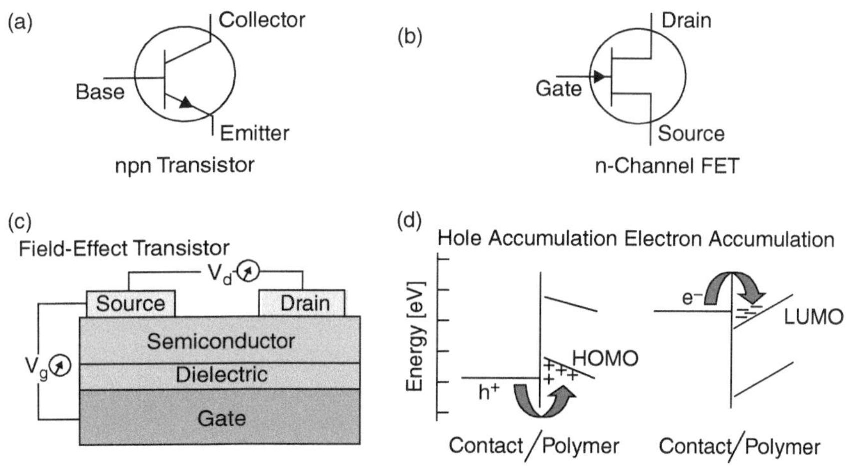

16.8 Circuit symbols of two types of transistors, namely, (a) BJT and (b) FET, showing the different terminals and their designations. (c) Shows the schematic of the fabrication of FET and (d) shows the energy level diagram to understand the charge transport in the FET through electron and hole accumulation at the interface between the metal contact and the polymer.

16.8d shows the energy-level diagram to illustrate the charge transport in the FET through electron and hole accumulation at the interface between the metal contact and the polymer.

Among the organic semiconductor materials used in FETs, poly (thiophenes), because of their chemical/electrochemical stability and ready functionalization, are among the most extensively investigated (Lotus *et al.*, 2009). One of the first reports of a polyheterocycle-based FET utilized a poly(thiophene) (Lee *et al.*, 2005b) and poly(3-hexyl thiophene) (P3HT) (Nakada *et al.*, 2004). Since then, the synthesis and evaluation of a large number of heterocycle polymers have been accomplished by chemical modification(s) of the conjugated core, variations of ring-to-ring connectivity and substitution pattern. Interestingly, majority of polymeric semiconductors for FETs can efficiently transport hole (p-channel semiconductors) but the electron-transporting polymers (n-channel semiconductors) are rare (Könenkamp *et al.*, 2004). Among the p-channel polymers, P3HT is the most investigated and the FET performance strongly depends on polymer purity, regioregularity, doping, molecular weight (Pearton *et al.*, 2004), the solvent used for film deposition (Nomura *et al.*, 2003), film morphology (Ryu *et al.*, 2005), film thickness and the fabrication process (Tsumura *et al.*, 1986).

Field-effect transistors based on individual CP nanofibers have been reported extensively as mentioned above. Pinto *et al.* (2003) reported an electrospun poly(aniline)/poly(ethylene oxide) nanofiber field-effect transistor.

Liu *et al.* (2005b) reported a single nanofiber field-effect transistor from electrospun poly(3-hexylthiophene). The transistor exhibited a hole field-effect mobility of 0.03 cm^2/Vs in the saturation regime, and a current on/off ratio of 103 in the accumulation mode. An ultra-short poly(3-hexylthiophene) field-effect transistor with effective channel length down to 5–6 nm and width ~2 nm was also reported by Qi *et al.* (2004). In addition, Alam *et al.* (2005) reported electrolyte-gated transistors based on conducting poly(aniline), poly(pyrrole) and poly(3,4-ethylenedioxythiophene) (PEDOT) nanowire junction arrays. In the presence of a positive gate bias, the poly(aniline) nanowire-based transistors exhibit a large on/off current ratio of 978, which can vary according to the acidity of the gate medium. Lee *et al.* (2009) also reported electrolyte-gated conducting poly(aniline) nanowire field-effect transistors. Besides nanofiber field-effect transistors, nanostructured polymer/semiconductor (metal) diodes were also reported. Guo *et al.* (2008, 2009) reported an organic/inorganic p-n junction nanowire consisting of poly(pyrrole) and CdS fabricated using an Al_2O_3 template, which displays a strong photo dependent rectifying effect. It was also reported that single Au-poly(pyrrole)-Cd-Au nanorods exhibit diode behavior with rectifying ratio ~200 at ±0.6 V at room temperature.

Pinto *et al.* (2006, 2008) reported a Schottky diode using an n-doped Si/SiO2 substrate and an electrospun fully doped poly(aniline) nanofiber. Liu *et al.* (2006) demonstrated that Au/template-synthesized poly(pyrrole) nanofiber devices show rectifying behavior and that can used as nanorectifiers.

The metal semiconductor field-effect transistor (MESFET) is similar to a junction FET except the gate is a Schottky diode instead of a junction diode. Typical field-effect transistors are fabricated as thin-film transistors with large channel length and width. Merlo and Frisbie (2004) have reported an FET with a mobility value of 0.02 cm^2/Vs out of entangled webs of nanofibers of regioregular poly(3-hexylthiophene) (RRP3HT), whose thin-film counterpart has reported the highest field-effect mobility of 0.1 cm^2/Vs (Sirringhau *et al.*, 1999). In this scenario, the field-effect characteristics of single RRP3HT nanofiber of diameter 100 nm scale reported by Liu *et al.* (2005) shows field-effect mobility comparable to those observed in thin-film transistors. The nanofiber FET structure, designed with a semiconductor cylinder over a planar dielectric layer, is approximated by a coaxial capacitor. When a negative voltage is applied, an accumulation layer of positive charges are formed at the interface, resulting in an increase in conductivity between source and drain. Its current-voltage characteristics exhibited behavior expected for a p-channel organic transistor operating in the accumulation mode. A plot of drain current (I_{DS}) to drain voltage (V_{DS}) shows a linear behavior with V_{DS} initially and then saturating at higher V_{DS}, which is an expected behavior of a p-channel organic FET. A current modulation is a ratio of the current in the accumulation mode to the current in the zero-gate mode (I_{on}/I_{off}). When measured for single fiber RRP3HT, the current modulation is 10^3, which is is comparatively less than its thin-film counterpart because of the relatively large off current (10–10A) due to contamination during electrospinning (Liu *et al.* 2005b). Babel *et al.* (2005) fabricated an FET with bottom contact geometry using a heavily doped Si as the gate electrode and a SiO_2 as the gate dielectric, as shown in Fig. 16.9. On the top of the device, a layer of electrospun blend fibers of MEH-PPV/PHT is deposited through a coaxial two-capillary spinneret. The device shows typical p-channel output characteristics, with saturation of drain current at higher drain voltages when operated in the accumulation mode. The composition-dependant saturation region field-effect mobility is observed in these blends as shown in Fig. 16.9b. The hole mobility of the mat of nonwoven nanofiber blend is less than that of its thin-film counterpart because of the highly porous nature and the smaller area covered in the active channel region compared to that of the cast film. By estimating the effective area covered by the blend nanofibers from the SEM images, it is concluded that the effective area is only 10% of the actual channel length, hence the effective field-effect mobility of holes in these blend nanofibers are one order of magnitude higher.

Electrospinning offers a simple means of fabricating 1-D field-effect transistors. Regioregular poly(3-hexylthiophene) (RRP3HT) has shown field-effect mobility as high as 0.1 cm^2/Vs (Chang *et al.*, 2004). Many reports of thin-film transistors based on this polymer have been published. Pinto *et al.* first reported fabrication of an electrospun hybrid nanofiber FET using doped poly(aniline) andpoly(ethylene oxide) (2003, 2006). Liu *et al.* (2005a) have reported field-effect characteristics of a single RRP3HT nanofiber with 100 nm diameter and found field-effect mobility comparable to thin-film counterparts. Their transistor was fabricated on a low resistivity (3–5 mΩ) single side polished-type Si<100> wafer which acted as the gate electrode. The details are shown in Fig. 16.10. Over this a 150 nm thick gate dielectric (SiO_2) was thermally grown in a dry oxidation process at 1100°C. By selectively removing the oxide by wet etching, 30 nm thick Al and 120 nm thick Au contacts were deposited sequentially followed by rapid thermal annealing at 450°C in 5% N_2/H_2. Figure 16.10a shows the configuration and Fig. 16.10b shows the I_{DS} versus V_{DS} characteristics of the fabricated RRP3HT nanofiber FET.

Copper oxide (CuO) is a p-type semiconductor with a narrow band gap of 1.2 eV; it has been extensively studied. To prepare 1-D nanowires of CuO several techniques are used, such as wet chemical methods (Wang *et al.*, 2002), sonochemical methods (Kumar *et al*,. 2001), thermal decomposition method (Wang *et al.*, 2003b) and template method (Hsieh *et al.*, 2003). Recently Jiang *et al.* (2002) have grown CuO by direct oxidation of metallic copper. Yao *et al.* (2005) prepared CuO nanofibers by directional aggregation and crystallization of CuO particles. However, it remained a great challenge to prepare long nanofibers of CuO until Wu *et al.* (2006)

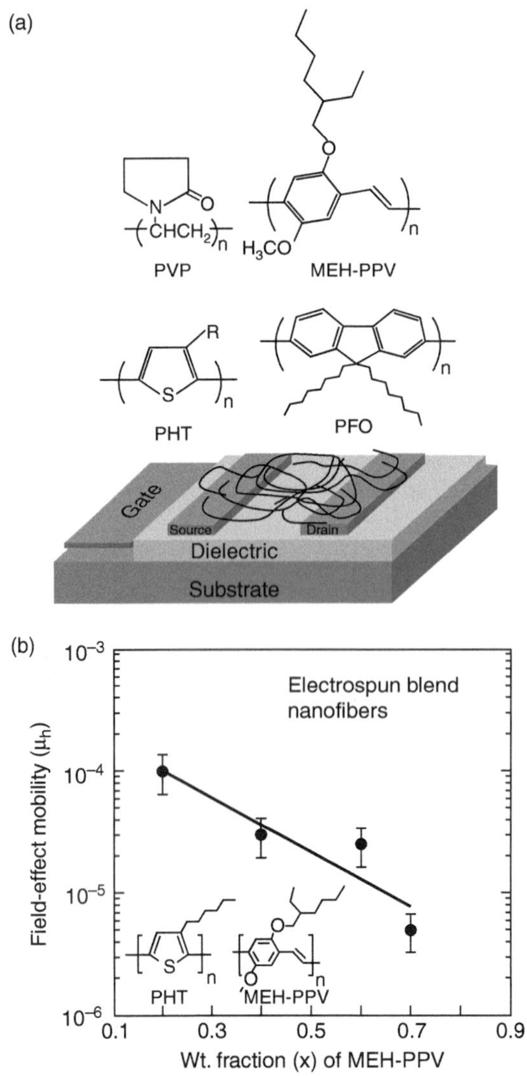

16.9 (a) Molecular structures of polymers used for electrospinning and schematic of field-effect transistor. (b) Compositional dependence of the field-effect mobility in mats of MEH-PPV/PHT blend nanofibers (Babel *et al.*, 2005).

prepared continuous copper oxide nanofibers by electrospinning. They also assembled a single CuO nanofiber into a field-effect transistor.

The procedure they followed was to prepare composite fibers of precursor solution with copper acetate along with poly(vinyl acetate) through electrospinning and then the fiber mat obtained was calcined at 500°C in air for 4 h to obtain CuO nanofibers at a heating rate of 10°C/min. The diameters of the fibers

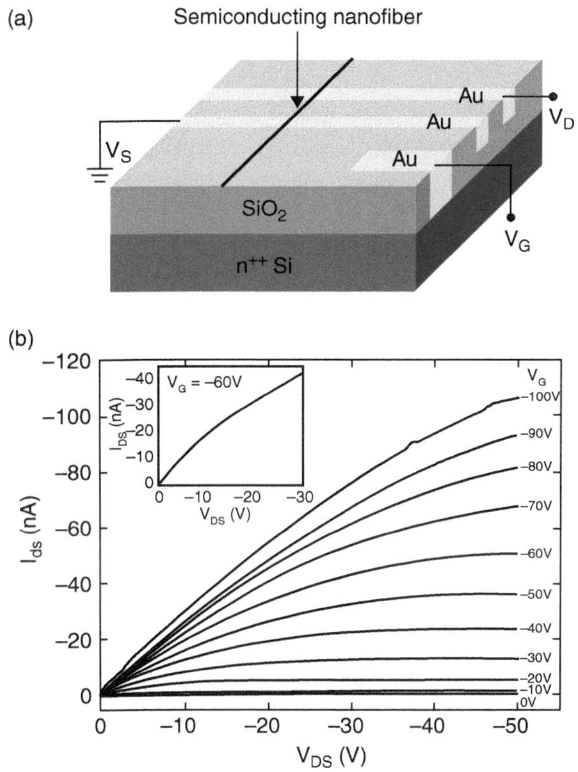

16.10 (a) Schematic of the nanofiber field effect transistor and (b) I_{DS} versus V_{DS} characteristics of a RRP3HT nanofiber FET, with effective channel length of 10 nm and diameter of 180 nm, showing the accumulation mode operation when different negative gate bias applied. Inset: The forward and reverse sweeps exhibited a negligible hysteresis of the device (Liu *et al.*, 2005b).

were in the range 262 ± 25 nm. The average diameter of CuO fibers obtained after calcination was 62 ± 8 nm, as seen in the FE-SEM image at the bottom inset of Fig. 16.11a. They prepared oriented CuO fibers bridging two electrodes which were set on heavily p-doped Si wafer capped with a 500 nm oxide layer with the underlying conducting silicon, used as a back gate. The gate dependent *I–V* characteristics of a single CuO nanofiber contacted with Ag electrodesis are shown in Fig. 16.11a and their enlarged plot is shown in Fig. 16.11b. The inset of Fig. 16.11b shows the current (I_{DS}) versus gate voltage (V_g) at $V_{DS} = 0$ V.

16.3.4 Light-emitting nanofibers

A light-emitting diode (LED) emits light when an electric current passes through it. It consists of a p-n-junction diode with semiconducting materials

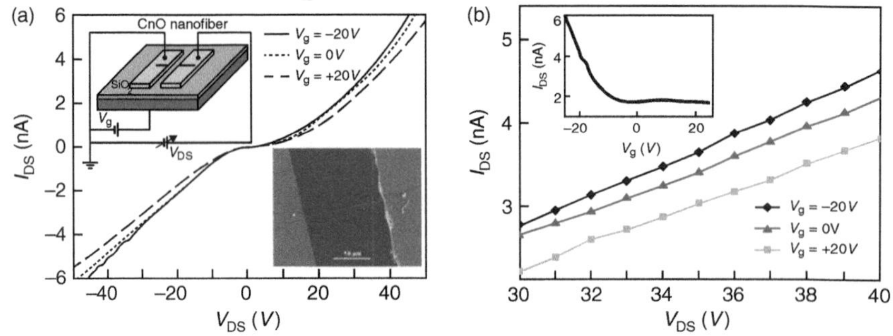

16.11 (a) Gate dependent *I–V* characteristics of a single CuO nanofiber contacted with Ag electrodes. Measurements were done at room temperature. Upper inset: schematic view of nanofiber FET configuration. The source serves as the back gate. Bottom inset: FE-SEM image of a 60 nm diameter CuO nanofiber with a length of ~100 μm suspending across the silver electrodes. (b) Enlarged plot of current (I_{DS}) versus V_{DS} for different gate voltage (V_g); inset shows I_{DS} versus V_g for V_{DS} = 0 V (Wu *et al.*, 2006).

with a direct band gap with energies corresponding to near-infrared, visible or near-ultraviolet light. In an LED, when a current flows from p- to n-side, the charge carriers flow into the junction from electrodes maintained at different voltages. At the junction, when an electron meets a hole, it falls into a lower energy level, and releases energy in the form of a photon. The Cambridge group, led by Richard Friend (Burroughes *et al.*, 1990), reported in 1990 the emission of light from a semiconducting polymer sandwiched between two contacts and connected to a battery. Thus, the discovery of electroluminescence (EL) in conjugated polymers – that is, the emission of light upon electron-hole recombination due to flow of an electric current – has provided a new impetus for the development of polymer light-emitting diodes (PLEDs). In a PLED, and its small molecule-based equivalent OLED (organic LED), the holes and electrons injected from the device contacts recombine and produce luminescence with a wavelength (color) corresponding to the energy difference between the molecular excited and ground states.

The color of the LED depends on the bandgap of the material. In an OLED, the electroluminescent material or the emissive layer is replaced with an organic semiconducting compound with delocalized ϖ-electrons. For the majority of conjugated polymers conventional polymers are more easily oxidized than reduced. Hence an electron injection is more difficult than a hole injection, leading to an unbalanced charge injection into the device. To solve this problem we use metals with a low work function (e.g., calcium) as the cathode, as well as other interfacial technology using electron-injection layers. However, since calcium is highly susceptible to

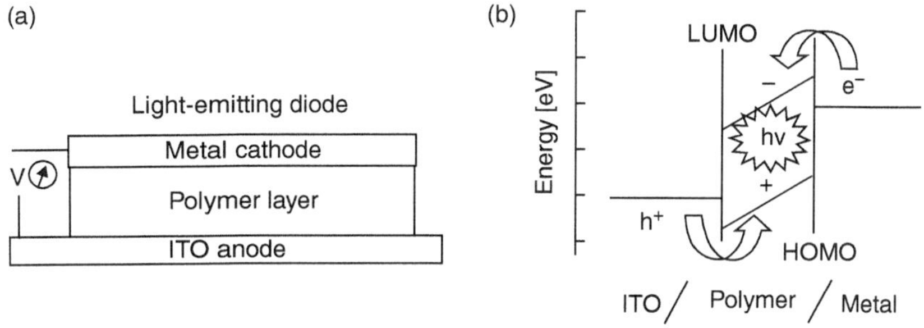

16.12 (a) Shows the simple configuration of polymer LED while (b) shows the energy level diagram showing the hole and electrons injected from the electrodes combine in the active region of the suitable polymer to emit visible radiation.

atmospheric degradation, this contact is covered with a metal that is not sensitive to oxygen and moisture, like aluminum. Thus, PLEDs exhibiting high external quantum efficiencies have been obtained, with efficiencies that are comparable with the best EL devices based on inorganic materials. Turn-on voltages of 5 V or below have also been achieved by the use of charge-transporting layers.

Polymer-based electroluminescent displays provide a good alternative to the well-established display technologies based on cathode-ray tubes and liquid-crystal displays (LCDs) with respect to processability and viewing angle. For the application in large-area displays and flexible displays, in particular, for which the conventional methods are not well suited, polymer light-emitting diodes offer great advantages. In addition, light-emitting sources with constrained dimensions will become useful because of sensing and lab-on-a-chip applications. The reduced size of an illumination source, coupled with on-chip detection mechanisms, can enable the development of fully packaged, functional devices that operate without the need of external optics. This can, in turn, enhance the sensitivity and improve the signal-to-noise ratio of detected fluorescence as the number of stages through which the emitted light has to propagate is now limited. Standard lab-on-a-chip devices use external illumination sources, such as lasers or light-emitting diodes (LEDs). Even though inexpensive submillimeter LEDs are available already, they are still much larger than the typical micro- or nanofluidic device and have to be mounted outside of the sensing region. Thus, fiber-based nano LEDs can be very useful in such applications.

In the recent past, PLEDs (Granström *et al.*, 1995, 1996), field emission (Wang *et al.*, 2001; Kim *et al.*, 2003, 2005a; Grimsdale *et al.*, 2009) and electrochromic displays (Cho *et al.*, 2005a, 2005b, 2007; Kim *et al.*, 2005b; Yan *et al.*, 2006) using well-aligned poly(aniline), poly(pyrrole) and PEDOT nanofibers/tubes have been reported. Granström *et al.* (1996) prepared micrometer- and nanometer-sized polymeric light-emitting diodes based on hard template synthesized PEDOT nanowires. Again, arrays of nanoscale conjugated-polymer light-emitting diodes were fabricated by Boroumand *et al.* (2005). Poly(aniline), PEDOT and poly(pyrrole) nanofibers/tubes can be used as nanotips for field-emission displays. It was found that poly(aniline) nanofibers within hard template membrane showed stable field-emission behavior with low threshold voltage of 5–6 V/m and high emission current density of 5 mA/cm^2 (Wang *et al.*, 2001). However, for template-synthesized PEDOT nanowires (Kim *et al.*, 2003; Yan *et al.*, 2006), the turn-on field was 3.5–4 V/m, the current density was 0.1 mA/cm^2 and the field-enhancement factor of nanotips was about 1200, which is comparable to those of carbon nanotubes. However, the stability of organics into such high fields and the resulting increase in temperature are the main concerns.

The ability of some of the conducting polymers to change color under an applied potential have made them useful as an active layer in electrochromic devices (Cho *et al.*, 2005a, 2005b, 2007; Kim *et al.*, 2005b, 2008). For instance, a fast color switching electrochromic device based on nanotubular PEDOT was reported by Cho *et al.* (2005b). Another electrochromic device from nanostructured PEDOT grown on vertical Si nanowires was also reported (Ostermann *et al.* 2006).

In addition to polymers and organic molecules, some of the inorganic oxides also can be used for electroluminescent applications. Thus, the ferroelectric 1-D nanostructures are extremely interesting materials due to the wide range of properties of ferroelectric oxides, such as spontaneous polarization, high dielectric permittivity, as well as piezo- and pyroelectricity. Hongquan *et al.* (2009) prepared singly distributed 1-D nanostructures of Y_2O_3, YVO_4 and YBO_3 doped with europium by electrospinning. These nanofibers and tubes exhibited good luminescent properties. They observed the fuorescence life times $^5D_0$ level for Eu^{3+} in Y_2O_3: Eu nanowires, YVO_4: Eu nanowires and YBO3: Eu nanotubes, which were determined to be 560 and 493 μs and 1.63 ms, respectively.

Among the class of organic electroluminescent materials, ionic transition metal complexes (iTMCs) can be useful in the fabrication of efficient, single-layer light-emitting devices. These have attractive features due to the operational mechanism of iTMC light-emitting devices, which is similar to that of polymer light-emitting electrochemical cells. In these materials, due to the ionic space charge effects, apart from the enhancement of electronic charge injection, the light-emission profile is also confined to a thin layer of the active region between the electrodes. These characteristics make iTMC devices attractive for on-chip light sources.

Moran-Mirabal *et al.* (2007) produced point illumination sources by depositing on interdigitated electrodes the fibers containing electroluminescent iTMCs imbedded in poly(ethylene oxide) (PEO) through electrospinning technique. They obtained oriented fibers over a wide area by employing a rotating disc on which the substrate was mounted, as shown in Fig. 16.13. The light emission from the electrospun fibers was detected by CC camera with an applied voltage around 3.2 V.

Figure 16.14a shows multiple interelectrode gaps across a single fiber. In Fig. 16.14b the emission recorded at 4 V is shown and in Fig. 16.14c its pseudocolor rendition to compare the intensity contrast among the gaps is shown.

Xia and co-workers reported that metal nanoparticles of different shapes could be deposited on electrospun titania nanofibers (Ostermann *et al.*, 2006). Based on this Wang *et al.* (2007) reported the synthesis of highly photoluminescent composite fibers of TiO_2 nanoparticles and conjugated polymer, namely, poly (phenylene vinylene) (PPV), by combining the sol-gel and electrospinning methods.

The diameters of the hybrid fibers ranged from 100 to 300 nm and the average size of TiO_2 nanoparticles within the fibers was 10–60 nm. Figure 16.15a shows the FE-SEM images of pure PPV fibers while Fig. 16.15b shows the TiO2/PPV hybrid fibers. The corresponding fluorescence photos of the above two types of fibers are shown in Fig. 16.15c and 16.15d. They observed from FT-IR analysis that a new band around 1632 cm^{-1} assigned

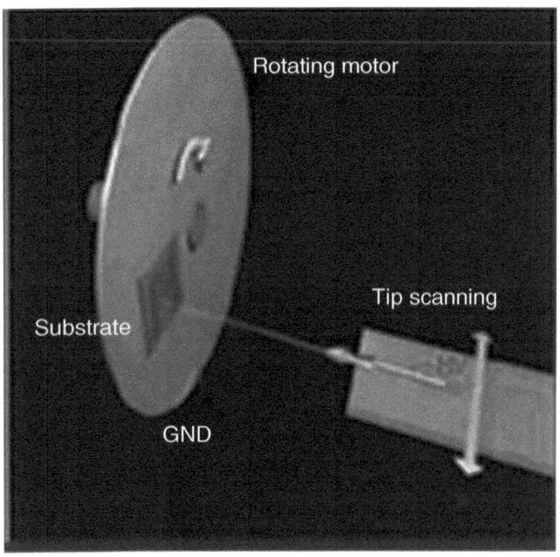

16.13 Electroluminescent nanofibers can be easily produced with electrospinning. The figure shows the schematic of the electrospinning setup used to deposit the fibers on the substrate. Fibers are oriented by the rotating motion of the substrate. Oriented fibers are obtained over a wide area by scanning the tip sideways (Moran-Mirabal *et al.*, 2007).

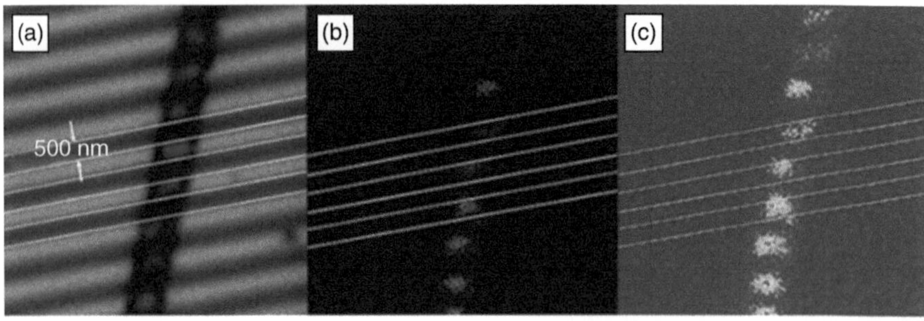

16.14 Light emission from multiple interelectrode gaps can be observed for a single fiber (a) Bright field image of an electrospun fiber spanning several 500 nm interdigitated electrode (IDE) gaps. Electrode gaps have been outlined. (b) Emission from fiber imaged in (a) at 4V. (c) Pseudocolour rendition of (b) for comparison of intensity in individual gaps (Moran-Mirabal *et al.*, 2007).

to the Ti–O–C vibration appeared, which resulted in the stronger luminance of the fluorescence of the TiO_2/PPV hybrid fibers compared to free-standing PPV nanofibers.

Thus, the relationship between the microstructure of the inorganic/polymer and the optical properties of the fibers was investigated by employing different characterization methods. The incorporation of TiO_2 nanoparticles into conjugated polymer matrices could be very useful, as it could further enhance the semiconducting polymer PPV's excellent photoluminescent (PL) and electroluminescent (EL) properties, as well as its photovoltaic (PV) (Huang *et al.*, 1997) and nonlinear optical properties.

16.3.5 Optoelectronic devices

In recent years electrospinning has been used to prepare molecular sieve fibers. Such fibrous molecular sieves, including zeolite fibers, siliceous and nonsiliceous fibers prepared by electrospinning, find promising applications in optics, adsorption, catalysis, electronics, etc.

Porous materials can be classified based on the diameter (d) of the pores as microporous ($d < 2$ nm); mesoporous (2 nm $\leq d \leq$ 50 nm) or macroporous ($d > 50$ nm) (Sing *et al.*, 1985). Microporous materials known as zeolites are crystalline inorganic solids characterized by TO_4 tetrahedra (T = Si, P, Al, etc.) with a regular pore system (Xu, 2007). Depending on their unique pore characteristics, such as size, shape, volume, etc., they show important properties in catalysis, adsorption and ion-exchange. Some of the properties can also be influenced by their morphologies (Perez-Ramirez *et al.*,

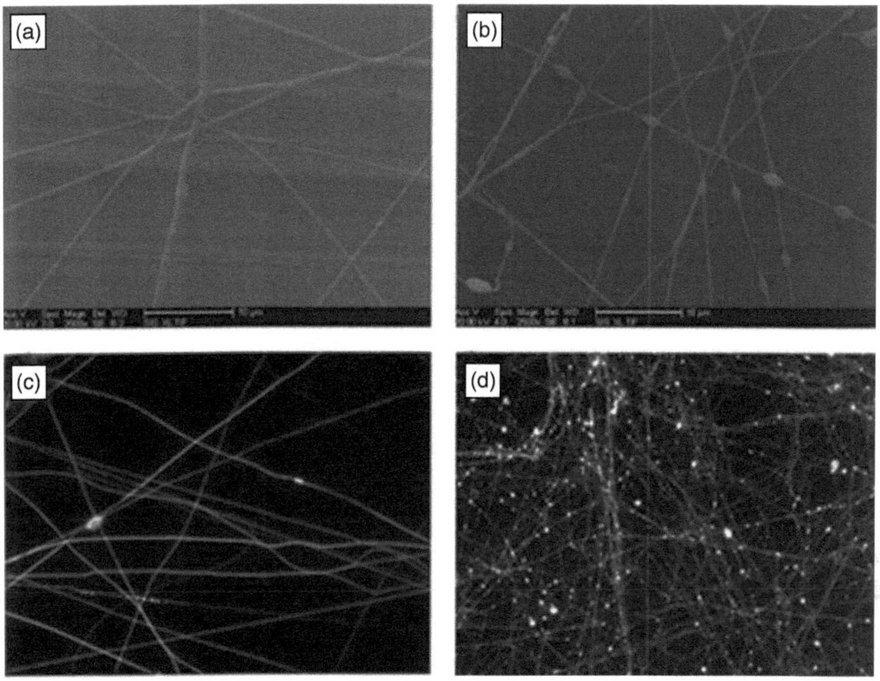

16.15 Top: FE-SEM images of (a) pure PPV fibers and (b) TiO_2/PPV hybrid nanoparticle-polymer fibers. Bottom: Fluorescence photos of (c) PPV fibers and (d) TiO_2/PPV hybrid nanoparticle-polymer fibers (Wang *et al.,* 2007).

2008). Fibrous zeolites have attracted growing attention in recent times as they promise fast diffusion and low pressure drop while maintaining high performance when they are used in catalysis. Such zeolite fibers have been prepared by different methods such as self-assembly, dip-coating and hydrothermal systhesis, etc. (Lee *et al.*, 2000; Ke *et al.*, 2001; Lin *et al.*, 2003; Liu *et al.*, 2005a; Zhang *et al.*, 2005; Song *et al*,. 2006; Wang *et al.*, 2009b). Recently efforts were made to obtain these materials with controllable morphologies by regulating the synthesis conditions and by changing the inorganic precursors. Thus mesoporous thin films (Meng *et al.*, 2005; Dong *et al.*, 2007; Zhao *et al.*, 2007), tubes (Davis *et al.*, 1997), spheres (Schacht *et al.*, 1996; Lu *et al.*, 1999; Trewyn *et al.*, 2007), fibers (Huo *et al.*, 1997; Marlow *et al.*, 2001; Wang *et al.*, 2003a, 2005; Lu *et al.*, 2004; Wu *et al.*, 2004), etc., have been prepared. However, electrospinning has been the most simple and versatile method to fabricate continuous fibers in micro/nanoscale. A recent review on fibrous molecular sieves prepared by electrospinning discusses the different types of zeolites prepared and their applications (Di *et al.*, 2011).

A photoconductor device made using a single electrospun aluminum-doped zinc oxide (AZO) nanowire which is sensitive to illumination with below-gap light and a memory assembled has been reported recently (Lin *et al.*, 2007). There have been a few attempts to prepare spatially oriented electrospun nanowires for device fabrications (Li *et al.*, 2003, 2004; Jalili *et al.*, 2006; Pu *et al.*, 2010). By using a 'Cu bridge' as the cathode, which consists of two conductive copper filaments separated by a gap of 1 cm, the electrospun nanowires could be uniaxially aligned during the process, as shown in Fig. 16.16a. The aligned nanowires were then directly transferred to a silicon substrate with a 500 nm-thick thermally grown SiO_2 layer, and calcined at 550°C at a heating rate of 10°C per minute in air for 3 h to obtain well-aligned polycrystalline AZO nanowires. The optical image of the precursor fibers obtained and the AZO fibers after calcination are shown in Fig. 16.16b and c, respectively. Gold electrodes were then deposited on the AZO nanowires by thermal evaporation using a mask to form channels

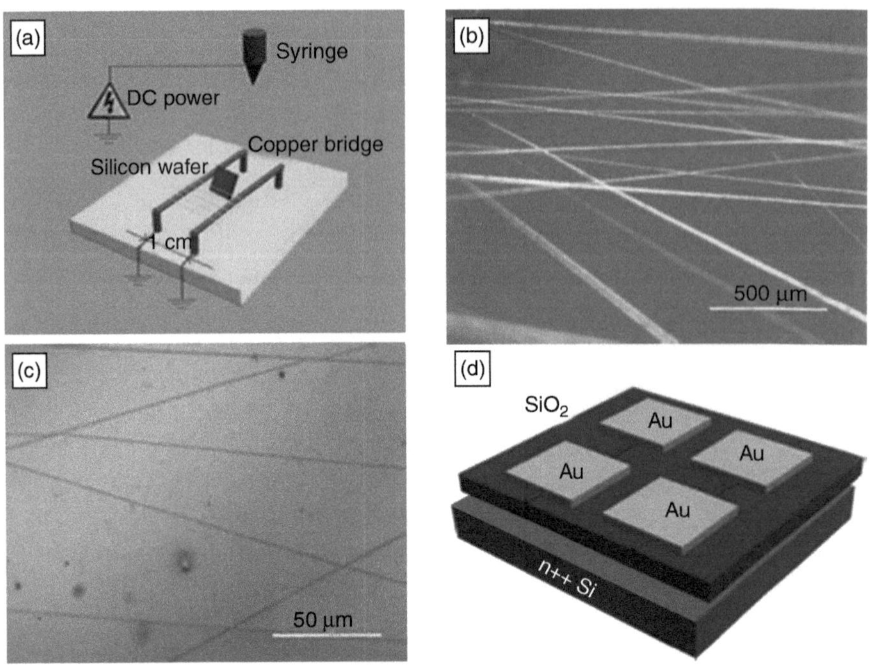

16.16 (a) Schematic illustration of the setup used for electrospinning nanowires as uniaxially aligned arrays, (b) metallographic microscope image of electrospinning precursor nanowires collected across the void gap between the 'Cu bridge', (c) AZO nanowire arrays after calcinations, (d) Schematic illustration of the fabricated device configurations for electrical measurements (Lin *et al.*, 2007).

that were 500 lm in length. Schematic illustration of the fabricated device configurations for electrical measurements is shown in Fig. 16.16d.

The pre-calcined nanowires have a relatively smooth surface and uniform morphology, with an average diameter of 200 nm. The surface of the nanowire exhibits shrinkage and roughness after calcination at 550°C due to the decomposition of the poly(vinyl alcohol) (PVA) solvent and inorganic compounds. There was almost no weight loss above 500°C as seen from the results of thermogravimetry (TG)/differential thermal analysis (DTA), which confirms that the obtained nanowires are pure AZO nanowires. Selected-area electron diffraction (SAED) analysis indicated the formation of the hexagonal wurtzite structure of ZnO.

A typical schematic of the device structure is shown in Fig. 16.16d. A typical current–voltage (*I–V*) curve through a single AZO nanowire with different Al concentrations is shown in Fig. 16.17a, and the corresponding variation of conductivity with concentration is shown in Fig. 16.17b.

Compared to the pure ZnO nanowire with a conductivity of 2.55×10^{-5} S cm^{-1}, the Al-doped ZnO nanowires show a steep increase up to 9.73×10^{-3} S cm^{-1} (Al = 5%), indicating a great enhancement in conductivity. Here, aluminum acts as a cationic dopant in the ZnO lattice – that is, the trivalent $A1^{3+}$ ion will occupy the divalent Zn^{2+} site (Al_Zn) allowing electrons to move to the conduction band easily.

16.3.6 Fiber-based actuators

Poly(vinylidene fluoride) (PVDF) is an inexpensive, flexible, highly nonreactive, polymer with good piezoelectric property (Choi and Jiang, 2006). It might be very useful for conversion of energy between electric and mechanical domains and, hence, as electromechanical actuators, artificial muscles, motion sensors and energy harvesters. There are many reports on actuators based on PVDF films (Lee *et al.*, 2005a; Chen *et al.*, 2008). However, there are few reports on the actuator made of a single PVDF fiber. PVDF is a semicrystalline polymer consisting of four crystalline phases α, β, γ and δ. Among these, nonpolar α phase is the most common and usually found in commercially available powders and films. It has random orientation of dipole moments, and thus the dipole moments cancel each other out. β Phase has all the dipole moments pointing in the same direction, such that it is responsible for the piezoelectric properties of the polymer as the piezoelectric behavior is based upon the dipole orientation within crystalline phase of PVDF. A piezoelectric actuator using a doubly clamped suspended (PVDF) fiber prepared by near-field electrospinning, shown schematically in Fig. 16.18, was demonstrated by Pu *et al.* (2010). The piezoelectric effect was further enhanced by polarizing the fiber under

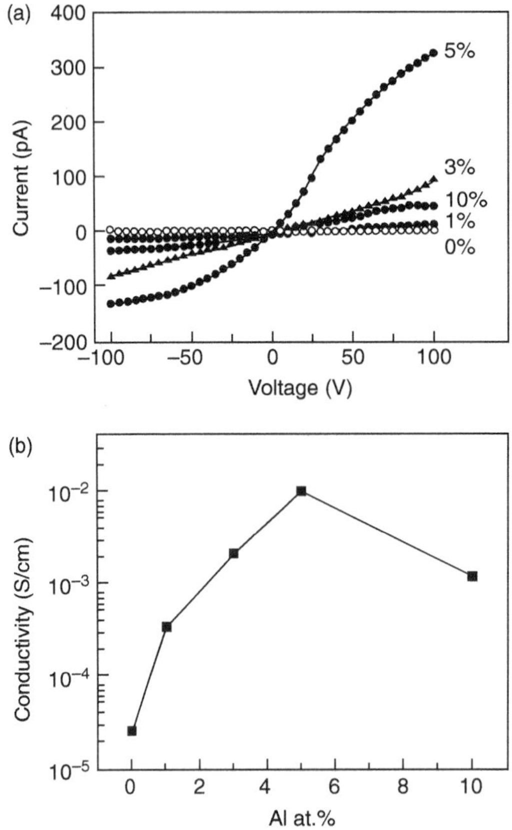

16.17 (a) The measured current–voltage (*I–V*) curves of undoped and Al-doped samples, (b) Al at % dependent conductivity of AZO nanowires (Lin *et al.*, 2007).

a voltage of 2 kV for 2 h at a temperature of 80°C. A displacement of 2 μm at the center of a suspended PVDF fiber has been detected corresponding to a strain of 0.004%, as a result of piezoelectric effect. The applied electric field was 1.6 V/μm. This technology opens up actuation applications for nano/microelectromechanical systems (NEMS/MEMS).

16.3.7 Gas sensors

A gas sensor is a device which detects the presence of different gases in an area, especially those gases which might be harmful to humans or animals. The developments of gas sensor technology have found wide applications in environmental monitoring, protection, etc. It is well known that properties of the sensing materials such as surface area, agglomeration and porosity

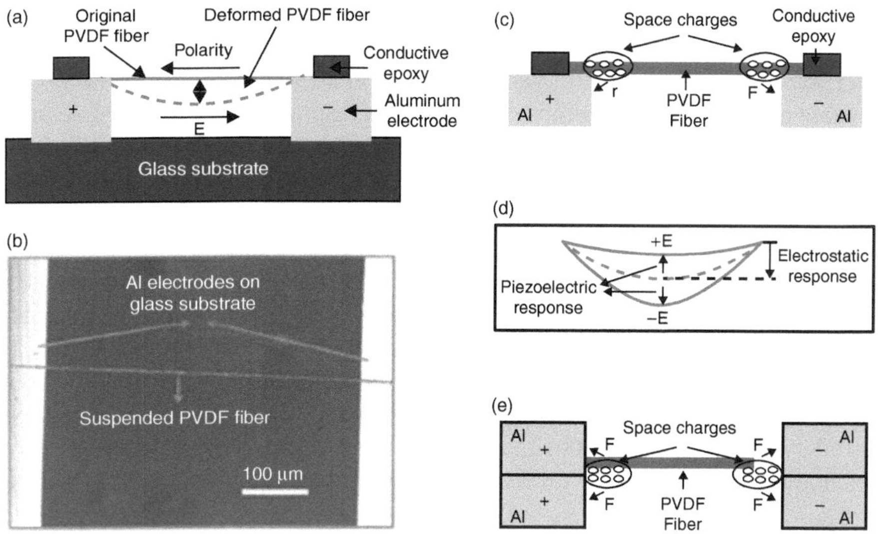

16.18 (a) Cross section view of the piezoelectric PVDF fiber actuator. When applying an electrical field that is opposite to the polarity along the fiber length, the suspended fiber elongates and buckles down. (b) SEM image of a suspended electrospun PVDF fiber (2.6 μm in diameter and 500 μm in length). (c) Side view for illustration of electrostatic effect, the electrostatic force between the trapped space charges in the fiber and electrodes deforms the fiber, thus its central point moves downwards. (d) Deformed PVDF fiber under +E or –E as a result of piezoelectric effect and electrostatic effect. When applying –E, the total deformation is equal to the combined downward electrostatic deformation and downward piezoelectric deformation. (e) View showing balanced electrostatic force by adding aluminum electrodes on top of the original electrodes. The PVDF fiber in (a, c) were enlarged for illustration purpose (Pu *et al.*, 2010).

greatly affect the gas-sensing characteristics, such as sensitivity, selectivity, time response, stability, durability, reproducibility and reversibility. Various types of materials such as semiconductors, polymers and organic/inorganic composites have been used as sensing material to detect the targeted gases based on various sensing techniques and principles. A higher specific surface of a sensing material leads to a higher sensitivity, therefore many techniques have been adopted to increase the specific surface of material, especially to form the nanostructures. Thus, large specific surfaces are expected to be ideal candidates as the structure of sensing materials. Recently electrospun fiber-based gas sensors have attracted much attention as the surface area-to-volume ratio of nanofibers is very high.

Electrospinning provides the ability to produce long, continuous fibers of different polymers and inorganic precursors through the sol-gel route.

During the electrospinning process, reactions such as hydrolysis, condensation and gelation of the precursors are involved in the morphological and microstructural evolution of the fibers. Upon subsequent calcination, the organic components decompose while the inorganic precursors oxidize and crystallize to form metal oxide nanoparticles aligned along what used to be the fibers in the as-spun state. This process results in highly porous structures that often display a bi-modal pore size distribution with relatively large pores (submicrons to a few microns) side by side with nano-sized pores. This unique morphology facilitates effective penetration of the surrounding gas or liquid phase into the layer, which is the key for the exceptionally high sensitivity observed in electrospun gas sensors.

In a typical example of gas sensors based on electrospun inorganic oxide fibers, nanocrystalline fiber mats of unloaded and Pd loaded SnO_2 were prepared by Yang *et al.* (2010) by electrospinning, followed by hot-pressing and calcination at 450°C or 600°C (Fig. 16.19). They observed significantly smaller crystallites in the Pd-loaded SnO_2 fiber mats compared to their unloaded counterparts, indicating grain growth and inhibition of densification due to Pd loading which resulted in higher surface area due to smaller particle size.

However, Pd was not incorporated into the SnO_2 lattice but segregated to their surfaces and formed tiny PdO nanocrystallites decorating the larger SnO_2 crystallites. They found that the Pd-loaded SnO_2 fiber mats showed significantly enhanced sensitivity to H_2 compared to their unloaded counterparts as seen from Fig. 16.20a and b. However, the sensitivity to NO_2 was found to

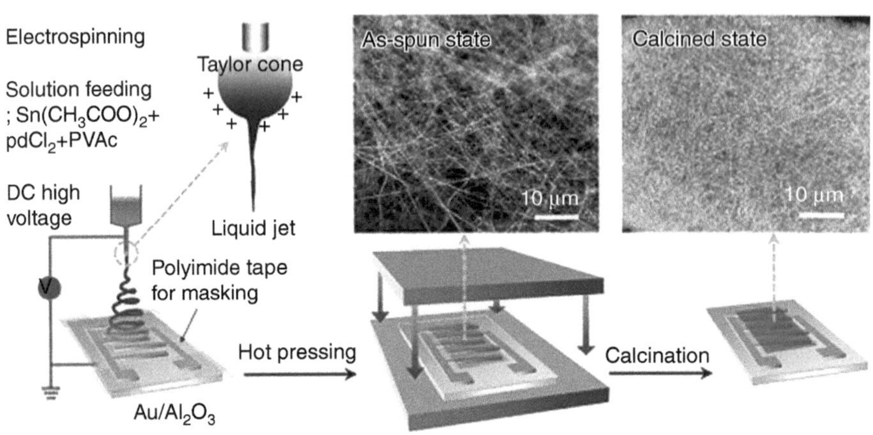

16.19 Illustrates the processing steps involved in the fabrication of SnO_2 sensor prototypes. Non-woven fiber mats were deposited from the liquid precursor solution by electrospinning. Subsequently, hot pressing was applied to achieve good adhesion to the substrates. Finally, the sample was calcined in order to decompose the organic components and crystallize the inorganic ones. The inset shows confocal microscopy images in the as-spun state and after hot-pressing and calcination (Yang *et al.*, 2010).

be reversed with the unloaded SnO_2 sensors demonstrating higher sensitivity than their Pd-loaded counterparts. They attributed their observations to the catalytic effect of the PdO nanocrystallites in promoting the oxidation of H_2 to H_2O and enhancing electron depletion at the surface of the SnO_2 crystallites.

Nonwoven fiber mats were deposited from the liquid precursor solution by electrospinning. Subsequently, hot pressing was applied to achieve good adhesion to the substrates. Finally, the sample was calcined in order to decompose the organic components and crystallize the inorganic ones. The inset shows confocal microscopy images in the as-spun state and after hot-pressing and calcination.

In another example, recently, Zhu *et al.* (2009) reported the fabrication of ultraviolet sensors based on electrospun ZnO. Schematic representation of the UV nanosensors made of the uniaxially aligned electrospun ZnO nanofibers is shown in Fig. 16.21a. The gap between the two gold electrodes was 250 mm, and there were 100 nanofibers over the 3.0 mm device width. Figure 16.21b gives the current–voltage curves of the nanosensor under UV irradiation with different intensities. The response time and recovery behavior of the nanosensors were characterized by measuring the current at a voltage of 40 V as a function of time when the nanosensors were in the dark and exposed to the UV light alternatively. As shown in Fig. 16.21c, the current of the ZnO nanofibers could be reversibly modulated by UV irradiation. The response time of the nanosensors with the UV light was less than 0.5 s, and the recovery

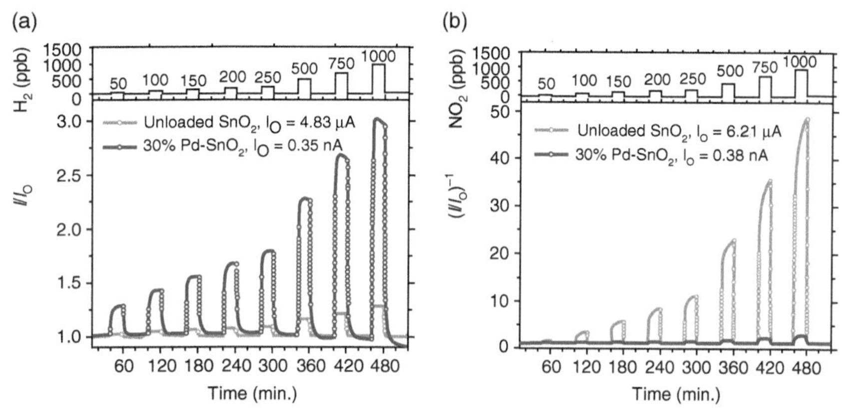

16.20 The electrical response of unloaded and 30 mol% Pd-loaded SnO_2 sensors to H_2 (a) and NO_2 (b). The sensitivity was measured at 350°C during cyclic exposure to 20 min pulses with increasing concentrations of the test gases mixed in dry air. The profiles of the gas concentration (in ppb) versus time are shown on top of the response curves. The response for H_2 (a) is presented as the normalized current I/I_0 and the response to NO_2 (b) is presented as the normalized resistance I_0/I. The baseline currents (I_0) at the beginning of the respective measurements (in clean air) are given in the figure labels (Yang *et al.*, 2010).

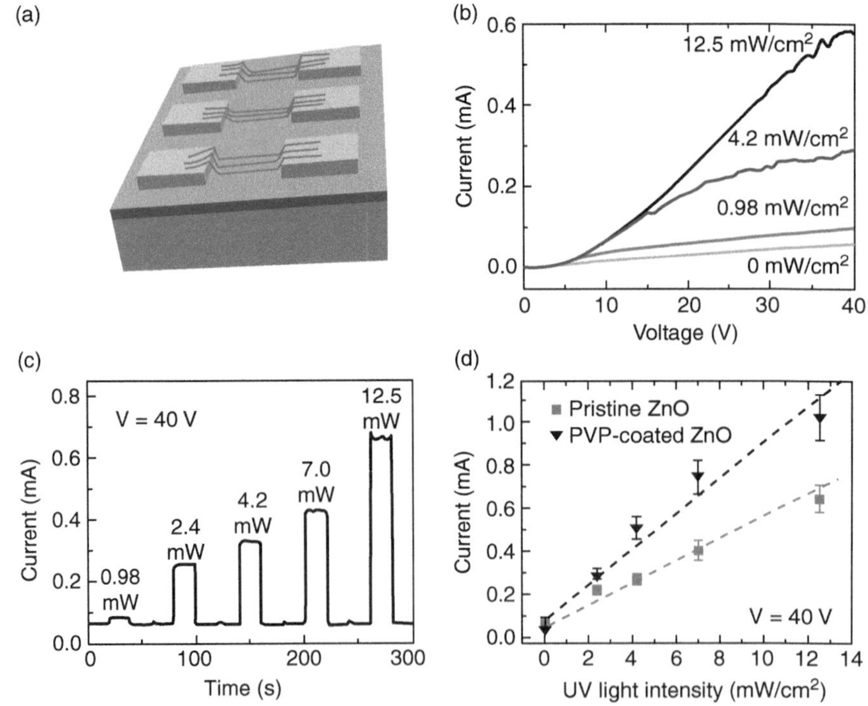

16.21 (a) Schematic representation of the UV nanosensors made of the uniaxially aligned electrospun ZnO nanofibers. The gap between the two gold electrodes was 250 mm, and there were 100 nanofibers over the 3.0 mm device width. (b) The current–voltage curves of the nanosensor under UV irradiation with different intensities. (c) Time dependence of the current measured at the voltage of 40 V. The ZnO nanosensor was irradiated under UV light at 365 nm and in dark alternatively. (d) The current measured at 40 V as a function of the UV light intensity for the two devices made of pristine ZnO and PVP-coated ZnO nanofibers, respectively. The dashed line was the linear fit of the data (Zhu *et al.*, 2009).

time from the UV light exposure to the dark was <1 s. The current increased with increased intensity of the UV light, and the device had a high photoresponse at high bias. Compared with that in the dark, the current increased about an order of magnitude at 40 V with a UV light intensity of 12.5 mW cm^2. However, the current stayed nearly constant at small bias. These results suggested that the charge carriers in the ZnO polycrystalline nanofibers played the major role in determining the current at high voltage, and the charge carriers generated by the UV irradiation led to the increased current.

Figure 16.21d shows the current measured at 40 V as a function of the UV light intensity for the two devices made of pristine ZnO and PVP-coated ZnO nanofibers, respectively. The dashed line was the linear fit of the data.

The photo-conduction in ZnO is determined by the trapping states at the grain (nanocrystallite) boundary and the surface of the material due to adsorption/desorption of O_2 in the air. In the electrospun ZnO nanofibers, the adsorbed O_2 can trap free electrons and become O_2^-. This further leads to the formation of a depletion layer and an energy barrier for conduction, resulting in low conductivity in the dark. When the device is irradiated with the UV light having energy higher than the band gap of ZnO, holes generated near O_2^- discharge the ions and the trapped electrons are released; additionally, the height of the barrier for conduction at the grain boundaries may be reduced. Both effects lead to the increase of the current under UV irradiation. The mechanism can also explain the high photosensitivity of the device made of the PVP-coated ZnO nanofibers. Under UV irradiation, the electrons trapped by PVP at the grain boundary and the surface of the nanofibers could recombine with the photon-generated holes, leaving the photon-generated free electrons for conduction.

16.4 Conclusions

In this chapter we reviewed the use of polymer and other ceramic nanofibers in electronics. A brief overview of electrospinning method and its variations was given, along with a mention of their wide range of applications in biomedical, military, filtering and other related areas. Next, this chapter reviewed the research undertaken all over the world in developing fiber-based electronic devices such as nanowires, Schottky diodes, FETs, LEDs, photoconductors, actuators, gas sensors, etc. The applications of electrospun fibers in electronic and other allied areas seem to be inexhaustible.

16.5 Acknowledgments

The author wishes to thank some of his students, namely, Dr S. Anitha, Dr K. P. Rajesh, Mr Anant Raheja and Dr Chitralekha, who have helped in collating the large collection of reference materials and in reading the draft and coming up with useful suggestions to improve the manuscript.

16.6 References

Ahn, J. H., Kim, D. H. *et al.* (2011), Micro- and nanostructured semiconductor materials for flexible and stretchable electronics, *Comprehensive Semiconductor Science and Technology*, Amsterdam, Elsevier, 499–526.

Ajayan, P. M., Schadler, L. S. *et al.* (2000), Single-walled carbon nanotube–polymer composites: Strength and weakness, *Advanced Materials*, **12**(10), 750–3.

Alam, M. M., Wang, J. *et al.* (2005), Electrolyte-gated transistors based on conducting polymer nanowire junction arrays, *Journal of Physical Chemistry B*, **109**(26), 12777–84.

Babel, A., Li, D. *et al.* (2005), Electrospun nanofibers of blends of conjugated polymers: Morphology, optical properties, and field-effect transistors, *Macromolecules*, **38**(11), 4705–11.

Bae, J., Thompson-Flagg, R. *et al.* (2008), Pattern formation of nanoflowers during the vapor-liquid-solid growth of silicon nanowires, *Physica B: Condensed Matter*, **403**(19–20), 3514–18.

Barink, M., Van den Berg, D. *et al.* (2011), Flexible electronics: Prediction of substrate deformation during different steps of the lithography process, *Microelectronic Engineering*, **88**(6), 999–1005.

Boroumand, F. A., Fry, P. W. *et al.* (2005), Nanoscale conjugated-polymer light-emitting diodes, *Nano letters*, **5**(1), 67–71.

Braun, D. and Heeger, A. J. (1991), Visible light emission from semiconducting polymer diodes, *Applied Physics Letters*, **58**(18), 1982–4.

Burroughes, J. H., Bradley, D. D. C. *et al.* (1990), Light-emitting diodes based on conjugated polymers, *Nature*, **347**(6293), 539–41.

Chang, C. and Zhang, L. (2011), Cellulose-based hydrogels: Present status and application prospects, *Carbohydrate Polymers*, **84**(1), 40–53.

Chang, J. F., Sun, B. *et al.* (2004), Enhanced mobility of poly (3-hexylthiophene) transistors by spin-coating from high-boiling-point solvents, *Chemistry of Materials*, **16**(23), 4772–6.

Chen, Z., Kwon, K. Y. *et al.* (2008), Integrated IPMC/PVDF sensory actuator and its validation in feedback control, *Sensors and Actuators A: Physical*, **144**(2), 231–41.

Cho, C. C., Smith, D. M. *et al.* (1995), Low dielectric-constant insulators for electronics applications, *Materials Chemistry and Physics*, **42**(2), 91–5.

Cho, S. I., Choi, D. H. *et al.* (2005a), Electrochemical synthesis and fast electrochromics of poly (3, 4-ethylenedioxythiophene) nanotubes in flexible substrate, *Chemistry of Materials*, **17**(18), 4564–6.

Cho, S. I., Kwon, W. J. *et al.* (2005b), Nanotube based ultrafast electrochromic display, *Advanced Materials*, **17**(2),171–5.

Cho, S. I., Xiao, R. *et al.* (2007), Electrochemical synthesis of poly (3, 4-ethylenedioxythiophene) nanotubes towards fast window-type electrochromic devices, *Nanotechnology*, **18**, 405705.

Choi, S. and Jiang, Z. (2006), A novel wearable sensor device with conductive fabric and PVDF film for monitoring cardiorespiratory signals, *Sensors and Actuators A: Physical*, **128**(2), 317–26.

Davis, S. A., Burkett, S. L. *et al.* (1997), Bacterial templating of ordered macrostructures in silica and silica-surfactant mesophases, *Nature*, **385**, 420–3.

Deitzel, J. M., Kleinmeyer, J. *et al.* (2001), The effect of processing variables on the morphology of electrospun nanofibers and textiles, *Polymer*, **42**(1), 261–72.

Deitzel, J. M., Kosik, W. *et al.* (2002), Electrospinning of polymer nanofibers with specific surface chemistry, *Polymer*, **43**(3), 1025–9.

Demazeau, G., Millet, J. M. *et al.* (1997), Solvothermal synthesis of microcrystallites of transition metal oxides, *Journal of Alloys and Compounds*, **262**, 271–4.

Dersch, R., Liu, T. *et al.* (2003), Electrospun nanofibers: Internal structure and intrinsic orientation, *Journal of Polymer Science Part A: Polymer Chemistry*, **41**(4), 545–53.

Di, J., Y. Zhao, *et al.* (2011), Fabrication of molecular sieve fibers by electrospinning, *Journal of Materials Chemistry*, **21**(24), 8511–20.

Dodabalapur, A. (2006), Organic and polymer transistors for electronics, *Materials Today*, **9**(4), 24–30.
Dong, W., Sun, Y. *et al.* (2007), Controllable and repeatable synthesis of thermally stable anatase nanocrystal-silica composites with highly ordered hexagonal mesostructures, *Journal of the American Chemical Society*, **129**(45), 13894–904.
Doudou, B. B., Chen, J. *et al.* (2011), Size-dependent properties of amino-functionalized single walled carbon nanotubes, *Computational and Theoretical Chemistry*, **967**(2–3), 231–4.
Dudhe, R. S., Sinha, J. *et al.* (2010), Polymer composite-based OFET sensor with improved sensitivity towards nitro based explosive vapors, *Sensors and Actuators B: Chemical*, **148**(1), 158–65.
Dutta, S. and Narayan, K. S. (2004), Gate voltage control of optically induced charges and memory effects in polymer field effect transistors, *Advanced Materials*, **16**(23–4), 2151–5.
Eom, S. H., Senthilarasu, S. *et al.* (2009), Polymer solar cells based on inkjet-printed PEDOT: PSS layer, *Organic Electronics*, **10**(3), 536–42.
Fong, H., Chun, I. *et al.* (1999), Beaded nanofibers formed during electrospinning, *Polymer-London*, **40**, 4585–92.
Formhals, A. (1934), *Process and Apparatus for Spinning*, US Patent, 1975504.
Formhals, A. (1939), *Method and Apparatus for Spinning*, US Patent, 2160962.
Formhals, A. (1940), *Artificial Thread and Method of Producing Same*, US Patent, 2187306.
Frackowiak, E. and Beguin, F. (2002), Electrochemical storage of energy in carbon nanotubes and nanostructured carbons, *Carbon*, **40**(10), 1775–87.
Fréchet, J. M. J. (2005), Functional polymers: From plastic electronics to polymer-assisted therapeutics, *Progress in Polymer Science*, **30**(8–9), 844–57.
Granström, M., M. Berggren, *et al.* (1995), Micrometer-and nanometer-sized polymeric light-emitting diodes, *Science*, **267**(5203), 1479.
Granström, M., M. Berggren, *et al.* (1996), Polymeric light-emitting diodes of submicron size – structures and developments, *Synthetic Metals*, **76**(1–3), 141–3.
Grimsdale, A. C., Leok Chan, K. *et al.* (2009), Synthesis of light-emitting conjugated polymers for applications in electroluminescent devices, *Chemical Reviews*, **109**(3), 897–1091.
Guo, Y., Tang, Q. *et al.* (2008), Light-controlled organic/inorganic p-n junction nanowires, *Journal of the American Chemical Society*, **130**(29), 9198–9.
Guo, Y., Zhang, Y. *et al.* (2009), Assembled organic/inorganic p-n junction interface and photovoltaic cell on a single nanowire, *Journal of Physical Chemistry Letters*, **1**(1), 327–30.
Gurunathan, K., Murugan, A. V. *et al.* (1999), Electrochemically synthesised conducting polymeric materials for applications towards technology in electronics, optoelectronics and energy storage devices, *Materials Chemistry and Physics*, **61**(3), 173–91.
Hangarter, C. M., Bangar, M. *et al.* (2010), Conducting polymer nanowires for chemiresistive and FET-based bio/chemical sensors, *Journal of Materials Chemistry*, **20**(16), 3131–40.
He, J. H., Wan, Y. Q. *et al.* (2005), Scaling law in electrospinning: Relationship between electric current and solution flow rate, *Polymer*, **46**(8), 2799–801.
He, J. H., Wan, Y. Q. *et al.* (2007), Nano-effects, quantum-like properties in electrospun nanofibers, *Chaos, Solitons & Fractals*, **33**(1), 26–37.

He, Y. (2005), A novel emulsion route to sub-micrometer polyaniline/nano-ZnO composite fibers, *Applied Surface Science*, **249**(1–4), 1–6.

Hongquan, Y., Wang, S. *et al.* (2009), Fabrication and Luminescent Properties of $Eu^{3}\pm$ Doped Lanthanide Oxide Nanowires and Nanotubes by Electrospinning, Power and Energy Engineering Conference, APPEEC, Asia-Pacific.

Howard, W. E. and Prache, O. F. (2001), Microdisplays based upon organic light-emitting diodes, *IBM Journal of Research and Development*, **45**(1), 115–27.

Hsieh, C. T., Chen, J. M. *et al.* (2003), Synthesis of well-ordered CuO nanofibers by a self-catalytic growth mechanism, *Applied Physics Letters*, **82**, 3316.

Huang, Z. H., Yang, J. H. *et al.* (1997), Study on EL emission region of polymer thin film in PPV LED., *Synthetic Metals*, **91**(1–3), 315–16.

Huang, Z. M., Zhang, Y. Z. *et al.* (2003), A review on polymer nanofibers by electrospinning and their applications in nanocomposites, *Composites Science and Technology*, **63**(15), 2223–53.

Huck, J. (1988), Protective clothing systems: A technique for evaluating restriction of wearer mobility, *Applied Ergonomics*, **19**(3), 185–90.

Huo, Q., Zhao, D. *et al.* (1997), Room temperature growth of mesoporous silica fibers: A new high surface area optical waveguide, *Advanced Materials*, **9**(12), 974–8.

Ishii, Y., Sakai, H. *et al.* (2011), Fabrication of a submicron-channel organic field-effect transistor using a controllable electrospun single fibre as a shadow mask, *Nanotechnology*, **22**, 205202.

Jacob, R., Jacob, A. P. *et al.* (2009), Mechanism of the dielectric enhancement in polymer-alumina nano-particle composites, *Journal of Molecular Structure*, **933**(1–3), 77–85.

Jalili, R., Morshed, M. *et al.* (2006), Fundamental parameters affecting electrospinning of PAN nanofibers as uniaxially aligned fibers, *Journal of Applied Polymer Science*, **101**(6), 4350–7.

Jiang, X., Herricks, T. *et al.* (2002), CuO nanowires can be synthesized by heating copper substrates in air, *Nano Letters*, **2**(12), 1333–8.

Jie, J., Zhang, W. *et al.* (2010), One-dimensional II-VI nanostructures: Synthesis, properties and optoelectronic applications, *Nano Today*, **5**(4), 313–36.

Katta, P., Alessandro, M. *et al.* (2004), Continuous electrospinning of aligned polymer nanofibers onto a wire drum collector, *Nano Letters*, **4**(11), 2215–18.

Ke, C., Yang, W. L. *et al.* (2001), Electrophoretic assembly of nanozeolites: Zeolite coated fibers and hollow zeolite fibers, *Chemical Communications*, **8**, 783–4.

Kim, B. H., Kim, M. S. *et al.* (2003), Characteristics and field emission of conducting poly (3, 4-ethylenedioxythiophene) nanowires, *Applied Physics Letters*, **83**, 539.

Kim, B. H., Park, D. H. *et al.* (2005a), Synthesis, characteristics, and field emission of doped and de-doped polypyrrole, polyaniline, poly (3, 4-ethylenedioxythiophene) nanotubes and nanowires, *Synthetic Metals*, **150**(3), 279–84.

Kim, B. K., Kim, Y. H. *et al.* (2005b), Electrical properties of polyaniline nanofibre synthesized with biocatalyst, *Nanotechnology*, **16**, 1177.

Kim, Y., Baek, J. *et al.* (2008), Electrochromic nanostructures grown on a silicon nanowire template, *Ultramicroscopy*, **108**(10), 1224–7.

Kim, Y. S., Ha, S. C. *et al.* (2005c), Portable electronic nose system based on the carbon black-polymer composite sensor array, *Sensors and Actuators B: Chemical*, **108**(1–2), 285–91.

Kobayashi, Y., Tanase, T. *et al.* (2008), Fabrication and dielectric properties of the BaTiO3-polymer nano-composite thin films, *Journal of the European Ceramic Society*, **28**(1), 117–22.
Könenkamp, R., Word, R. C. *et al.* (2004), Vertical nanowire light-emitting diode, *Applied Physics Letters*, **85**, 6004.
Krebs, F. C., Sondergaard, R. *et al.* (2010), Printed metal back electrodes for R2R fabricated polymer solar cells studied using the LBIC technique, *Solar Energy Materials and Solar Cells*, **95**(5), 1348–53.
Kumar, R. V., Elgamiel, R. *et al.* (2001), Sonochemical preparation and characterization of nanocrystalline copper oxide embedded in poly (vinyl alcohol) and its effect on crystal growth of copper oxide, *Langmuir*, **17**(5), 1406–10.
Laudenslager, M. J. and Sigmund, W. M. (2011), Developments in electrohydrodynamic forming: Fabricating, *American Ceramic Society Bulletin*, **90**(2), 22–7.
Lee, B. Y., Cheon, C. I. *et al.* (2002), Low temperature firing of PZT thick films prepared by screen printing method, *Materials Letters*, **56**(4), 518–21.
Lee, C. S., Joo, J. *et al.* (2005a), An approach to durable PVDF cantilevers with highly conducting PEDOT/PSS (DMSO) electrodes, *Sensors and Actuators A: Physical*, **121**(2), 373–81.
Lee, G. S., Lee, Y. J. *et al.* (2000), Self-assembly of -glucosidase and D-glucose-tethering zeolite crystals into fibrous aggregates, *Journal of the American Chemical Society*, **122**(49), 12151–7.
Lee, M. (1980), Properties of CdSe thin film transistors prepared by photolithography, *Solid State Electronics*, **23**, 1087–8.
Lee, S. Y., Choi, G. R. *et al.* (2009), Electronic transport characteristics of electrolyte-gated conducting polyaniline nanowire field-effect transistors, *Applied Physics Letters*, **95**(1), 013113–013113–3.
Lee, S. Y., Shim, E. S. *et al.* (2005b), Fabrication of ZnO thin film diode using laser annealing, *Thin Solid Films*, **473**(1), 31–4.
Li, D., Frey, M. W. *et al.* (2006), Electrospun polylactic acid nanofiber membranes as substrates for biosensor assemblies, *Journal of Membrane Science*, **279**(1–2):,354–63.
Li, D., Wang, Y. *et al.* (2003), Electrospinning of polymeric and ceramic nanofibers as uniaxially aligned arrays, *Nano Letters*, **3**(8), 1167–71.
Li, D., Wang, Y. *et al.* (2004), Electrospinning nanofibers as uniaxially aligned arrays and layer by layer stacked films, *Advanced Materials*, **16**(4), 361–6.
Li, D. and Xia, Y. (2004), Electrospinning of nanofibers: Reinventing the wheel?, *Advanced Materials*, **16**(14), 1151–70.
Lin, D. D., Wu, H. *et al.* (2007), Photoswitches and memories assembled by electrospinning aluminum doped zinc oxide single nanowires, *Advanced Materials*, **19**(22), 3968–72.
Lin, J. C., Dipre, J. T. *et al.* (2003), Microemulsion-directed synthesis of molecular sieve fibers, *Chemistry of Materials*, **15**(14), 2764–73.
Ling, Q. D.,. Liaw, D. J *et al.* (2007), Polymer memories: Bistable electrical switching and device performance, *Polymer*, **48**(18), 5182–201.
Liu, B. S., Tang, D. C. *et al.* (2005a), Fabrication of analcime zeolite fibers by hydrothermal synthesis, *Microporous and Mesoporous Materials*, **86**(1–3), 106–11.
Liu, H., Reccius, C. H. *et al.* (2005b), Single electrospun regioregular poly (3-hexylthiophene) nanofiber field-effect transistor, *Applied Physics Letters*, **87**, 253106.

Liu, L., Zhao, Y. *et al.* (2006), Electrochemical fabrication and electronic behavior of polypyrrole nano-fiber array devices, *Thin Solid Films*, **503**(1–2), 241–5.
Lotus, A. F., Bhargava, S. *et al.* (2009), Electrospinning route for the fabrication of pn junction using nanofiber yarns, *Journal of Applied Physics*, **106**(1), 014303–014303–4.
Lu, Q., Gao, F. *et al.* (2004), Ordered SBA-15 nanorod arrays inside a porous alumina membrane, *Journal of the American Chemical Society*, **126**(28), 8650–1.
Lu, X., Wang, C. *et al.* (2009), One dimensional composite nanomaterials: Synthesis by electrospinning and their applications, *Small*, **5**(21), 2349–70.
Lu, X., Zhang, D. *et al.* (2006), Large scale synthesis of necklace like single crystalline PbTiO3 nanowires, *Macromolecular Rapid Communications*, **27**(1), 76–80.
Lu, Y., Fan, H. *et al.* (1999), Aerosol-assisted self-assembly of mesostructured spherical nanoparticles, *Nature*, **398**(6724), 223–6.
Luo, C. J., Nangrejo, M. *et al.* (2010), A novel method of selecting solvents for polymer electrospinning, *Polymer*, **51**(7), 1654–62.
Luoh, R. and Hahn, H. T. (2006), Electrospun nanocomposite fiber mats as gas sensors, *Composites Science and Technology*, **66**(14), 2436–41.
MacDiarmid, A G., Shirikawa, H. et al. (2000), The Nobel Prize in Chemistry 2000. Retrieved 2009-06-02.
Marlow, F., Leike, I. *et al.* (2001), Mesostructured silica fibers: Ring structures in reciprocal space, *Advanced Materials*, **13**(5), 307–10.
Meng, Y., Gu, D. *et al.* (2005), Ordered mesoporous polymers and homologous carbon frameworks: Amphiphilic surfactant templating and direct transformation, *Angewandte Chemie*, **117**(43), 7215–21.
Merlo, J. A. and Frisbie, C. D. (2004), Field effect transport and trapping in regioregular polythiophene nanofibers, *Journal of Physical Chemistry B*, **108**(50), 19169–79.
Miao, J., Miyauchi, M. *et al.* (2010), Electrospinning of nanomaterials and applications in electronic components and devices, *Journal of Nanoscience and Nanotechnology*, **10**(9), 5507–19.
Möller, S., Forrest, S. R. *et al.* (2003), Electrochromic conductive polymer fuses for hybrid organic/inorganic semiconductor memories, *Journal of Applied Physics*, **94**, 7811.
Moran-Mirabal, J. M., Slinker, J. D. *et al.* (2007), Electrospun light emitting nanofibers, *Nano Letters*, **7**(2), 458–63.
Nakada, T., Hirabayashi, Y. *et al.* (2004), Novel device structure for Cu (In, Ga) Se2 thin film solar cells using transparent conducting oxide back and front contacts, *Solar Energy*, **77**(6), 739–47.
Nalwa, H. S. (2001), *Handbook of advanced electronic and photonic materials and devices: Conducting polymers*, Elsevier; Academic Press, London.
Nomura, K., Ohta, H. *et al.* (2003), Thin-film transistor fabricated in single-crystalline transparent oxide semiconductor, *Science*, **300**(5623), 1269.
O'Connor, B., Pipe, K. P. *et al.* (2008), Fiber based organic photovoltaic devices, *Applied Physics Letters*, **92**, 193306.
Ostermann, R., Li, D. *et al.* (2006), V_2O_5 nanorods on TiO_2 nanofibers: A new class of hierarchical nanostructures enabled by electrospinning and calcination, *Nano Letters*, **6**(6), 1297–302.
Pantano, A., Modica, G. *et al.* (2008), Multiwalled carbon nanotube reinforced polymer composites, *Materials Science and Engineering: A*, **486**(1–2), 222–7.

Park, M. J., Choi, W. M. *et al.* (2006), Patterning polymer light-emitting diodes by micromolding in capillary, *Current Applied Physics*, **6**(4), 627–31.

Pearton, S. J., Heo, W. H. *et al.* (2004), Dilute magnetic semiconducting oxides, *Semiconductor Science and Technology*, **19**, R59.

Pekkanen, V., Mäntysalo, M. *et al.* (2010), Utilizing inkjet printing to fabricate electrical interconnections in a system-in-package, *Microelectronic Engineering*, **87**(11), 2382–90.

Perez-Ramirez, J., Christensen, C. H. *et al.* (2008), Hierarchical zeolites: Enhanced utilisation of microporous crystals in catalysis by advances in materials design, *Chemical Society Reviews*, **37**(11), 2530–42.

Pinto, N. J., González, R. *et al.* (2006), Electrospun hybrid organic/inorganic semiconductor Schottky nanodiode, *Applied Physics Letters*, **89**, 033505.

Pinto, N. J., Johnson Jr., A. T. *et al.* (2003), Electrospun polyaniline/polyethylene oxide nanofiber field-effect transistor, *Applied Physics Letters*, **83**, 4244.

Pinto, N. J., Ramos, I. *et al.* (2008), Electric response of isolated electrospun polyaniline nanofibers to vapors of aliphatic alcohols, *Sensors and Actuators B: Chemical*, **129**(2), 621–7.

Pu, J., Yan, X. *et al.* (2010), Piezoelectric actuation of direct-write electrospun fibers, *Sensors and Actuators A: Physical*, **164**, 131–6.

Qi, P., Javey, A. *et al.* (2004), Miniature organic transistors with carbon nanotubes as quasi-one-dimensional electrodes, *Journal of the American Chemical Society*, **126**(38), 11774–5.

Ratnamala, A., Suresh, G. *et al.* (2008), Template synthesized nano-crystalline natrotantite: Preparation and photocatalytic activity for water decomposition, *Materials Chemistry and Physics*, **110**(1), 176–9.

Reneker, D. H. and Chun, I. (1996), Nanometre diameter fibres of polymer, produced by electrospinning, *Nanotechnology*, **7**, 216.

Riul, A. (2003), An artificial taste sensor based on conducting polymers, *Biosensors and Bioelectronics*, **18**(11), 1365–9.

Rivera, R. and Pinto, N. J. (2009), Schottky diodes based on electrospun polyaniline nanofibers: Effects of varying fiber diameter and doping level on device performance, *Physica E: Low-Dimensional Systems and Nanostructures*, **41**(3), 423–6.

Roth, S., Mahler, G. *et al.* (1989), Molecular electronics of conducting polymers, *Synthetic Metals*, **28**(1–2), 815–22.

Rutledge, G. C. and Fridrikh, S. V. (2007), Formation of fibers by electrospinning, *Advanced Drug Delivery Reviews*, **59**(14), 1384–91.

Ryu, Y. R., Lee, T. S. *et al.* (2005), ZnO devices: Photodiodes and p-type field-effect transistors, *Applied Physics Letters*, **87**, 153504.

Sambaer, W., Zatloukal, M. *et al.* (2010), 3D modeling of filtration process via polyurethane nanofiber based nonwoven filters prepared by electrospinning process, *Chemical Engineering Science*, **66**(4), 613–23.

Sanada, K., Tada, Y. *et al.* (2009), Thermal conductivity of polymer composites with close-packed structure of nano and micro fillers, *Composites Part A: Applied Science and Manufacturing*, **40**(6–7), 724–30.

Schacht, S., Huo, Q. *et al.* (1996), Oil-water interface templating of mesoporous macroscale structures, *Science*, **273**(5276), 768.

Schweitzer, B. and Bassler, H. (2000), Excitons in conjugated polymers, *Synthetic Metals*, **109**(1–3), 1–6.

Shabani, I., Hasani-Sadrabadi, M. M. *et al.* (2010), Nanofiber-based polyelectrolytes as novel membranes for fuel cell applications, *Journal of Membrane Science*, **368**(1–2), 233–40.

Shirakawa, H., Louis, E. J. *et al.* (1977), Synthesis of electrically conducting organic polymers: Halogen derivatives of polyacetylene, $(CH)_x$, *Journal of the Chemical Society, Chemical Communications*, (16), 578–80.

Sing, K. S., Everett, D. H. *et al.* (1985), Reporting physisorption data for gas/solid systems with special reference to the determination of surface area and porosity, *Pure and Applied Chemistry*, **57**, 603–19.

Sirringhaus, H., Brown, P. J. *et al.* (2000), Microstructure-mobility correlation in self-organised, conjugated polymer field-effect transistors, *Synthetic Metals*, **111**, 129–32.

Sirringhaus, H., Brown, P. J. *et al.* (1999), Two-dimensional charge transport in self-organized, high-mobility conjugated polymers, *Nature*, **401**(6754), 685–8.

Sirringhaus, H., Tessler, N. *et al.* (1998), Integrated optoelectronic devices based on conjugated polymers, *Science*, **280**(5370), 1741.

Song, W., Grassian, V. H. *et al.* (2006), Fiber and film formation by self-assembly of colloidal silicalite-1 and copper coated silicalite-1 nanocrystals, *Microporous and Mesoporous Materials*, **88**(1–3), 77–83.

Spivak, A. F., Dzenis, Y. A. *et al.* (2000), A model of steady state jet in the electrospinning process, *Mechanics Research Communications*, **27**, 37–42.

Subbiah, T., Bhat, G. S. *et al.* (2005), Electrospinning of nanofibers, *Journal of Applied Polymer Science*, **96**(2), 557–69.

Sundaray, B., Subramanian, V. *et al.* (2006), Electrical conductivity of a single electrospun fiber of poly (methyl methacrylate) and multiwalled carbon nanotube nanocomposite, *Applied Physics Letters*, **88**, 143114.

Swihart, M. T. (2003), Vapor-phase synthesis of nanoparticles, *Current Opinion in Colloid & Interface Science*, **8**(1), 127–33.

Tang, H., Yan, F. *et al.* (2010), The improvement of glucose bioelectrocatalytic properties of platinum electrodes modified with electrospun TiO2 nanofibers, *Biosensors and Bioelectronics*, **25**(7), 1646–51.

Taylor, G. (1964), Disintegration of water drops in an electric field, *Proceedings of the Royal Society of London. Series A. Mathematical and Physical Sciences*, **280**, 383.

Teo, W. E. and Ramakrishna, S. (2006), A review on electrospinning design and nanofibre assemblies, *Nanotechnology*, **17**, R89.

Tessler, N., Harrison, N. T. *et al.* (1998), High peak brightness polymer light emitting diodes, *Advanced Materials*, **10**(1), 64–8.

Thakur, R. A., Florek, C. A. *et al.* (2008), Electrospun nanofibrous polymeric scaffold with targeted drug release profiles for potential application as wound dressing, *International Journal of Pharmaceutics*, **364**(1), 87–93.

Theron, S. A., Zussman, E. *et al.* (2004), Experimental investigation of the governing parameters in the electrospinning of polymer solutions, *Polymer*, **45**(6), 2017–30.

Thompson, C. J., Chase, G. G. *et al.* (2007), Effects of parameters on nanofiber diameter determined from electrospinning model, *Polymer*, **48**(23), 6913–22.

Toprakci, O., Ji, L. *et al.* (2011), Fabrication and electrochemical characteristics of electrospun $LiFePO_4$/carbon composite fibers for lithium-ion batteries, *Journal of Power Sources*, **196**(18), 7692–9.

Trewyn, B. G., Slowing, I. I. *et al.* (2007), Synthesis and functionalization of a mesoporous silica nanoparticle based on the sol–gel process and applications in controlled release, *Accounts of Chemical Research*, **40**(9), 846–53.

Tsumura, A., Koezuka, H. *et al.* (1986), Macromolecular electronic device: Field effect transistor with a polythiophene thin film, *Applied Physics Letters*, **49**(18), 1210–12.

Uyar, T. and Besenbacher, F. (2008), Electrospinning of uniform polystyrene fibers: The effect of solvent conductivity, *Polymer*, **49**(24), 5336–43.

Vasita, R. and Katti, D. S. (2006), Nanofibers and their applications in tissue engineering, *International Journal of Nanomedicine*, **1**(1), 15.

Vayssieres, L. (2006), Advanced semiconductor nanostructures, *Comptes Rendus Chimie*, **9**(5–6), 691–701.

Wang, C., Wang, Z. *et al.* (2001), Well-aligned polyaniline nano-fibril array membrane and its field emission property, *Chemical Physics Letters*, **341**(5–6), 431–4.

Wang, C., Yan, E. *et al.* (2007), Fabrication of highly photoluminescent TiO_2/PPV hybrid nanoparticle polymer fibers by electrospinning, *Macromolecular Rapid Communications*, **28**(2), 205–9.

Wang, J., Tsung, C. K. *et al.* (2005), Single crystal mesoporous silica ribbons, *Angewandte Chemie International Edition*, **44**(2), 332–6.

Wang, J., Zhang, J. *et al.* (2003a), Structure-selective synthesis of mesostructured/mesoporous silica nanofibers, *Journal of the American Chemical Society*, **125**(46), 13966–7.

Wang, W., Liu, Z. *et al.* (2003b), A simple wet-chemical synthesis and characterization of CuO nanorods, *Applied Physics A: Materials Science & Processing*, **76**(3), 417–20.

Wang, W., Zhan, Y. *et al.* (2002), Synthesis and characterization of CuO nanowhiskers by a novel one-step, solid-state reaction in the presence of a nonionic surfactant, *Materials Research Bulletin*, **37**(6), 1093–100.

Wang, X. L., Oh, I. K. *et al.* (2009a), Enhanced electromechanical performance of carbon nano-fiber reinforced sulfonated poly (styrene-b-[ethylene/butylene]-b-styrene) actuator, *Composites Science and Technology*, **69**(13), 2098–101.

Wang, Z., Ge, Q. *et al.* (2009b), High performance zeolite LTA pervaporation membranes on ceramic hollow fibers by dipcoating-wiping seed deposition, *Journal of the American Chemical Society*, **131**(20), 6910–11.

Wee, G., Soh, H. Z. *et al.* (2010), Synthesis and electrochemical properties of electrospun V2O5 nanofibers as supercapacitor electrodes, *Journal of Materials Chemistry*, **20**(32), 6720–5.

Wu, H., Lin, D. *et al.* (2006), Fabrication, assembly, and electrical characterization of CuO nanofibers, *Applied Physics Letters*, **89**, 133125.

Wu, H., Sun, Y. *et al.* (2009), GaN nanofibers based on electrospinning: Facile synthesis, controlled assembly, precise doping, and application as high performance UV photodetector, *Advanced Materials*, **21**(2), 227–31.

Wu, Y., Cheng, G. *et al.* (2004), Composite mesostructures by nano-confinement, *Nature Materials*, **3**(11), 816–22.

Xie, R., Fu, H. *et al.* (2002), Characterization of the size-dependent optical properties of p-terphenyl nanocrystals, *Journal of Photochemistry and Photobiology A: Chemistry*, **147**(1), 31–7.

Xu, L. (2009), A mathematical model for electrospinning process under coupled field forces, *Chaos, Solitons & Fractals*, **42**(3), 1463–5.
Xu, L., Wu, Y. *et al.* (2010a), Numerical study of magnetic electrospinning processes, *Computers & Mathematics with Applications*, **61**(8), 2116–19.
Xu, R. (2007), *Chemistry of zeolites and related porous materials: Synthesis and structure*. Wiley-Interscience, Singapore.
Xu, Y., Long, G. *et al.* (2010b), Polymer photovoltaic devices with transparent graphene electrodes produced by spin-casting, *Carbon*, **48**(11), 3308–11.
Xu, Y., Zheng, L. *et al.* (2010c), Fluorescence sensors based on chiral polymer for highly enantioselective recognition of phenylglycinol, *Polymer*, **51**(5), 994–7.
Yan, H., Zhang, L. *et al.* (2006), Synthesis, property and field-emission behaviour of amorphous polypyrrole nanowires, *Nanotechnology*, **17**, 3446.
Yang, D. J., Kamienchick, I.*et al.* (2010), Ultrasensitive and highly selective gas sensors based on electrospun SnO_2 nanofibers modified by Pd loading, *Advanced Functional Materials*, **20**(24), 4258–64.
Yao, W. T., Yu, S. H. *et al.* (2005), Formation of uniform CuO nanorods by spontaneous aggregation: selective synthesis of CuO, Cu_2O, and Cu nanoparticles by a solid-liquid phase arc discharge process, *Journal of Physical Chemistry B*, **109**(29), 14011–16.
Yoon, Y. I. and Ko, J. M. (2008), CoNi oxide/carbon-nanofiber composite electrodes for supercapacitors, *International Journal of Electrochemical Science*, **3**, 1340–7.
Yun, K. M., Suryamas, A. B. *et al.* (2010), Morphology optimization of polymer nanofiber for applications in aerosol particle filtration, *Separation and Purification Technology*, **75**(3), 340–5.
Zeng, S., Tang, K. *et al.* (2007), Controlled synthesis of [alpha]-Fe_2O_3 nanorods and its size-dependent optical absorption, electrochemical, and magnetic properties, *Journal of Colloid and Interface Science*, **312**(2), 513–21.
Zhang, J., Dong, J. *et al.* (2005), Zeolite-fiber integrated optical chemical sensors for detection of dissolved organics in water, *Langmuir*, **21**(19), 8609–12.
Zhao, D., Seo, S. J. *et al.* (2007), Full-color mesophase silicate thin film phosphors incorporated with rare earth ions and photosensitizers, *Advanced Materials*, **19**(21), 3473–9.
Zhu, H. X. (2010), Size-dependent elastic properties of micro-and nano-honeycombs, *Journal of the Mechanics and Physics of Solid's*, **58**(5), 696–709.
Zhu, Z., Zhang, L. *et al.* (2009), Aligned electrospun ZnO nanofibers for simple and sensitive ultraviolet nanosensors, *Chemical Communications*, (18), 2568–70.
Zou, C. D., Gao, Y. L. *et al.* (2010), Size-dependent melting properties of Sn nanoparticles by chemical reduction synthesis, *Transactions of Nonferrous Metals Society of China*, **20**(2), 248–53.
Zussman, E., Theron, A. *et al.* (2003), Formation of nanofiber crossbars in electrospinning, *Applied Physics Letters*, **82**, 973.

17

Future trends in the processing of functional nanofibers

Q. WEI,
Jiangnan University, China

Abstract: Electrospinning has been extensively explored to produce various nanofibers for the expanding applications in many industries. This chapter reviews some new materials and techniques used in making electrospun nanofibers. The future development and challenges of electrospinning is also discussed in this chapter.

Key words: electrospinning, nanofibers, mass production.

Electrospinning provides an effective technique by which to produce biopolymer nanofibers for hygiene and medical applications, since biopolymers can be sustainable, carbon neutral and are always renewable. Biopolymers include polysaccharides (cellulose, chitin, chitosan, dextrose), proteins (collagen, gelatin, silk, etc.) and DNA, as well as some biopolymer derivatives and composites that have been electrospun into nanofibers (Schiffman and Schauer, 2008). Attention in this area has been paid to the use of new materials with different functions for specific applications. Wu *et al.* (2011) reported the inclusion of tobacco mosaic virus into polyvinyl alcohol (PVA) by electrospinning to form continuous TMV–PVA composite nanofibers as an extracellular matrix (ECM) mimetic. As a demonstration of its potential as ECM mimetic, a mutant TMV containing RGD peptide was cospun with PVA and the resulting fibrous substrates were used to promote cell growth.

The use of new materials has brought various new functions and potential applications of nanofibers. However, innovations in the use of new techniques for producing nanofibers will further promote the development of this promising technology. Inspired by the geometric structure of ecribellate spider-capture silk and its spinning characteristics, Tian *et al.* (2011) have proposed a one-step electrohydrodynamic method to fabricate bead-on-string heterostructured fibers (BSHFs) by the combination of electrospinning and electrosparying. Hydrophilic poly(ethylene glycol) beads were successfully imprinted on a hydrophobic polystyrene string. It was demonstrated that the BSHFs were capable of intelligently responding to environmental change. With a change in relative humidity, the fibers showed a segmented swelling and shrinking behavior in the 'bead' parts, whereas the 'string' parts remained the same, which may bring about new properties and applications.

Continuous efforts have also been made in the exploration of new approaches to the fabrication of nanofibers. Liu and He (2007) used a bottom-up gas-jet electrospinning process for mass production of polymer nanofibers. The gas jet helped the formation of air bubbles in the polymer solution and bubble-induced cones on the surface of polymer solution performed a similar function to the Taylor cone in conventional electrospinning. It was found that bubbles could form multiple jets during the electrospinning, which facilitated mass production of nanofibers. A simple fabrication of cross-linked hyaluronic acid (HA) hydrogel nanofibers by reactive electrospinning was also reported by Ji *et al.* (2006). In this work, a thiolated HA derivative, 3,3'dithiobis(propanoic dihydrazide)-modified HA (HA-DTPH) and poly(ethylene glycol) diacrylate (PEGDA) were selected as the cross-linking system. It was described that the cross-linking reaction occured simultaneously during the electrospinning process using a dual-syringe mixing technique. The experimental results revealed that the NIH 3T3 fibroblasts migrated into the electrospun HA scaffold through the nanofibrous network, which suggested the application of electrospun HA nanofibrous scaffolds as a potential material for wound healing and tissue regeneration.

A conventional electrospinning setup usually consists of a high-voltage power supply, a spinneret (typically a syringe needle) and a collector, and a pump is normally used to control the flow rate. However, the needle-type electrospinning systems limit productivity. A lot of research work has been focused on increasing the number of needles to improve productivity. The main disadvantage of multineedle electrospinning is the operating space and the spacing between the needles in order to avoid strong charge repulsion between the jets and adjacent needles, and associated uneven fiber deposition. Niu *et al.* (2009) reported the electrospinning of poly(vinyl alcohol) (PVA) nanofibers with a rotating disk and a cylinder. The study revealed that the disk electrospinning needed a relatively low applied voltage to initiate fiber formation, and the fibers were mainly formed on the top disk edge. However, the cylinder electrospinning showed a higher dependence on the applied voltage and polymer concentration. It was found by the finite element analysis of electric field profiles of the nozzles that a concentrated electric field formed on the disk edge, but an uneven distribution of the electric field intensity profile along the nozzle surface was observed for the cylinder nozzle. It was believed that their findings would benefit the design and further development of large-scale electrospinning systems for the mass production of nanofibers for advanced applications. Jirsak *et al.* (2005) described the generation of multiple jets from a liquid uploaded onto a slowly rotating horizontal cylinder, which was subsequently commercialized by Elmarco under the brand name Nanospider™. The multiple jets from a liquid uploaded onto a slowly rotating horizontal cylinder create Taylor cones without nozzles, needles or spinnerets.

However, environmental safety is now a major concern, although solvent-based electrospinning has been used increasingly in many areas. There is always a persistent desire to produce nanofibers by environmentally friendly methods. Inspired by melt spinning of polymer fibers, the use of molten polymers to produce fibers by electrospining has gained a great deal of attention. Larrondo and St. John Manley (1981) pioneered the electrospinning of polypropylene (melt flow indexes 0.5–2.0) and the production of electrospun fibers with diameters greater than 50 mm. The large diameters of the electrospun fibers were attributed to the viscosity of the polymer melt that could be many orders of magnitude greater than that of a polymer solution. Dalton *et al.* (2007) has demonstrated that polypropylene fibers could be reduced from 35 ± 8 μm in diameter, to 840 ± 190 nm with a viscosity-reducing additive. Melt electrospun blends of poly(ethylene glycol)-block-poly(epsilon-caprolactone) (PEG(47)-b-PCL95) and poly(epsilon-caprolactone) (PCL) produced fibers with micron-scale diameters (2.0 ± 0.3 μm), but this could be reduced to 270 100 nm by using the gap method of alignment for collection. A new hybrid methodology that fully combined the processing capabilities of the twin-screw extrusion process (conveying solids, melting, dispersive and distributive mixing, pressurization, temperature profiling, devolatilization) with electrospinning was reported by Erisken *et al.* (2008). The integrated process is especially suited to the dispersion of nanoparticles into polymeric binders and the generation of nanoparticle-incorporated fibers and nanofibers. It was demonstrated that the new hybrid fibers with the dispersion of β-tricalcium phosphate (β-TCP) nanoparticles into poly(ε-caprolactone) (PCL) could be obtained and the fiber diameters in the range of 200–2000 nm could be used as scaffolds for tissue engineering applications.

Technological innovations and commercial demands have been driving research in this areas toward creating simple, clean and mass production technologies for a wide spectrum of applications of nanofibers in many industries.

17.1 References

Dalton, P. D., Grafahrend, D., Klinkhammer, K., Klee, D. and Moeller, M. (2007), Electrospinning of polymer melts: Phenomenological observations, *Polymer*, **48**, 6823–33.

Erisken, C., Kalyon, D. M., Wang, H. J. (2008), A hybrid twin screw extrusion/electrospinning method to process nanoparticle-incorporated electrospun nanofibres, *Nanotechnology*, **19**, 165302–9.

Ji, Y., Ghosh, K., Li, B. Q., Sokolov, J. C., Clark, R. A. F. and Rafailovich, M. H. (2006), Dual-syringe reactive electrospinning of cross-linked hyaluronic acid hydrogel nanofibers for tissue engineering applications, *Macromolecular Bioscience*, **6**, 811–17.

Jirsak, O., Sanetrnik, F., Lukas, D., Kotek, V., Martinova, L. and Chaluopek, J. (2005), International Patent, WO 2005/024101.

Larrondo, L. and St. John Manley, R. (1981), Electrostatic fiber spinning from polymer melts, *Journal of Polymer Science: Polymer Physics*, **19**, 909–40.

Liu, Y. and He, J. H. (2007), Bubble electrospinning for mass production of nanofibers, *International Journal of Nonlinear Sciences and Numerical Simulation*, **8**, 393–6.

Niu, H. T., Lin, T. and Wang, X. G. (2009), Needleless electrospinning, a comparison of cylinder and disk nozzles, *Journal of Applied Polymer Science*, **114**, 3524–30.

Schiffman, J. D. and Schauer, C. L. (2008), A review: Electrospinning of biopolymer nanofibers and their applications, *Polymer Reviews*, **48**(2), 317–52.

Tian, X. L., Bai, H., Zheng, Y. M. and Jiang, L. (2011), Bio-inspired heterostructured bead-on-string fibers that respond to environmental wetting, *Advanced Functional Materials*, **21**, 1398–402.

Wu, L. Y., Zang, J. F., Lee, L. A., Niu, Z. W., Horvatha, G. C., Braxtona, V., Wibowo, A. C., Bruckman, M. A., Ghoshroy, S., Loye, H. C., Li, X. D. and Wang, Q. (2011), Electrospinning fabrication, structural and mechanical characterization of rod-like virus-based composite nanofibers, *Journal of Materials Chemistry*, **21**, 8550–7.

Index

α-chymotrypsin, 276, 318
acetone, 160
acoustic wave sensors, 219–20
  dynamic response of PVP/$LiTaO_3$ SAW device, 222
  gas testing system for TMA detection and sensor response, 221
activated carbon nanofibres, 378
adsorption *see* depth filtration
air filtration, 123
  filtration membranes, 316–17
aligned nanofibres, 27–30
  electrospinning system for various collection apparatus, 28
  electrospun nanofibre alignment mechanism, 29
  normal and modified electrospinning, 29
Allied Forces, 238
allyl isothiocyanate, 288
amidoxime PAN nanofibres, 102
ammonium persulfate (APS), 343
anode, 198–9
  half-reaction, 199
antioxidants, 290
applied voltage, 11
ASTM E 96, 250
ASTM F 2298, 250
ASTM F 2130, 249
atomic deposition, 107–8
attenuation characteristics, 307
Aum Shinrikyo Tokyo subway attack, 239, 241

Bacillus anthracis, 238–9, 241
Bacillus diminuta, 341–3, 348
basis weight, 252
battledress overgarment, 242–3, 249
battlefield filter, 245
bead-on-string heterostructured fibres (BSHFs), 411
bending instability, 244
bilge water, 349–50
bioactive packaging, 262
biological hazards
  functional nanofibres in protection clothing, 236–58
    future trends, 256–8
  warfare background, 237–43
    agents, 239–41
    Joint Service Lightweight Integrated Suit Technology (JSLIST), 243
    protective clothing, 241–3
    warfare agent classification, 240
biological polymers, 258
biological recognition element, 281
biological sensors, 227–9
  biosensor structure and membrane assembly, 228
biomimetic devices, 292
biopharmaceuticals, 124
biopolymers, 411
bioreceptor *see* biological recognition element
bioremediation, 289
biosensors, 280–1
biotin coupling, 270
biotin–streptavidin specific binding, 283
bipolar junction transistor, 381

bone, 177–81
MSCs seeded on electrospun PCL scaffolds, 179
bottom-up gas-jet electrospinning, 412
bovine serum albumin, 270
bovine viral diarrhoea virus, 283
breathability, 249–51
BRUKERS-AXS D 8 FOCUS diffractometer, 162
Brunauer–Emmett–Teller nitrogen adsorption method, 204
BS 7209, 250–1

carbon fibre-reinforced composites, 62
carbon nanofibres, 197, 201–2, 202, 204–5
carbon nanofibres and nanotubes structures, 56
future trends, 67–8
nanofibres composites applications, 61–7
nanofibres composites formation, 57–60
reinforcements in polymer matrix composites, 55–68
strengths and weaknesses, 60–1
carbon nanotubes, 345
cartilage, 184–5
morphology of day 21 cell pellet and PCL nanofibrous scaffold, 184
cathode, 198–9
half-reaction, 199
cellulose acetate membrane, 357
ceramic membranes, 334
CFRP matrix composites, 59
chelation, 101–4
chemical hazards
functional nanofibres in protection clothing, 236–58
future trends, 256–8
warfare background, 237–43
agents, 239–41
Joint Service Lightweight Integrated Suit Technology (JSLIST), 243
protective clothing, 241–3
warfare agent classification, 240
Chemical Weapons Convention, 238
chemically controlled drug release, 159
chemoresistive carbon black–polymer composite sensors, 285–6
chitosan, 288, 322, 355
chlorine gas, 237
Cibacron blue F3GA, 344–5
circulating fluids, 16
clarithromycin, 160
CLM/PLLA composite nanofibre scaffolds, 167
CNF/epoxy composites, 58
CNF/polymer composites, 58
coaxial electrospinning, 39, 325
coaxial electrospun nanofibres, 289
coaxial two-capillary spinneret, 385
coelectrospinning, 324
collector, 5
distance between nozzle, 12
main electrospinning, 7
composite electrospinning, 216–17
composite nanofibres
processing, 38–49
examples and applications, 43–9
future trends, 49
polymer/inorganic salt composite formation, 43
polymer/nanoparticle composite formation, 40–2
polymer/polymer composite formation, 38–40
computational fluid dynamics (CFD), 126–7
conducting *see* electroactive polymers
continuum flows, 125–8
copper oxide (CuO), 385–7
core-shell nanofibres, 23–7
electrospinning parameters, 24–7
properties, 23–4
coaxial electrospinning, 23
solution, 24
cross-linking, 101–4, 325
Cu bridge, 394
Cu(II) chelating scheme, 320

depth filtration, 343–4
dichloromethane, 160

differential scanning calorimetry (DSC), 351
differential thermal analysis (DTA), 395
direct electrospinning, 38
doped electrospinning, 216–17
doping, 82
  Fe^{3+} doped TiO_2 nanofibres, 84
Dow Corning 193, 350
drag force, 138
drug compatibility, 156
drug delivery
  applications, 160–7
    clarithromycin chemical structure, 161
    electrospun fibre diameters distribution, 164
    materials and methods, 160–2
    PLLA basic parameters, 163
    PLLA electrospun nanofibres, 162
    PLLA nanofibres containing clarithromycin, 164
    results and discussion, 162–7
  blood concentration changes, 154
  future trends, 167–8
  mechanisms, 155–9
    chemically controlled drug release, 159
    physically controlled drug release, 157–9
  nanofibres, 153–68
dual-syringe mixing technique, 412
DuPont hybrid membrane technology, 246
dynamic moisture permeation cell (DMPC), 250

electric field shielding effectiveness, 313
electric field superposition, 256
electrical heating, 16
electroactive polymers, 372
electroless deposition, 42, 104–6
  electrospun PA6 fibres, 105
electroluminescence, 388
electromagnetic attenuation
  functional nanofibres, 305–26
    Fe_3O_4, PANI, Fe_3O_4/PANI and Fe_3O_4/PANI -PAN permittivity, 312
    Fe_3O_4 nanoparticles and PANI/Fe_3O_4 nanocomposites image, 309
    Fe_3O_4 nanoparticles and PANI/Fe_3O_4 XRD patterns, 310
    future trends, 325–6
    PANI and PANI/Fe_3O_4 nanofibres SEM images, 311
  functional nanofibres applications, 306–15
electromagnetic interference (EMI), 56, 57, 60, 62
electromagnetic irradiation shielding effect (EMISE), 307
electromagnetic radiation, 306
electromagnetism, 86–7
electron transfer, 323
electrospinning, 5–10, 24–7, 72–3, 161, 172, 214–15, 244, 254, 262–3, 287, 291, 312, 340, 361, 374–6, 411
  core-shell nanofibres formation, 25
  devices, 5–8
    illustration, 8
  fibre formation mechanism, 8–10
  modern technology, 209–15
    comparison of the annual number of scientific publications, 214
    electrospun fibres morphologies, 211
    nanonets, 212–14
  O/W emulsion, 26
  palladium doped alumina nanofibre, 144–5
  rhodium doped alumina nanofibres, 145
electrospun fibres, 8, 14, 17–18
electrospun fibrous mats, 272
  filtration and clarification applications, 266–70
  food related separation applications, 265–71
    nanofibre for air filtration, 267
    three-layer composite membrane and fractured layer fabrication schematics, 268

electrospun fibrous mats (*cont.*)
solid-phase extraction applications, 270–1
Elmarco, 326
emulsion electrospinning, 25
encapsulation, 286–90
enzyme immobilisation, 317–25
encapsulation in nanofibres, 324–5
modified nanofibres, 319–21
pH effect on immobilised and free catalases, 321
PVA- Cu(II) chetalated nanofibrous membrane, 320
storage stability of immobilised and free catalases, 321
temperature effect on immobilised and free catalases, 320
thermal stability of immobilised and free catalases, 322
surface attachment on nanofibres, 318–24
modification for biocompatible surface, 322
modification for electrical conductivity, 323–4
modification for enzyme mobility, 323
untreated nanofibres, 318–19
epigallocatechin- 3-gallate, 288–90
Escherichia coli, 257, 342
Escherichia coli O157:H7, 283
ethanol sensors, 285
extracellular matrix (ECM) mimetic, 411

Fe_3O_4/polyaniline nanofibre, 308
fibre mat, 156
fibre strength, 128–9
field-effect transistor (FET), 381–7
BJT and FET transistors circuit symbols, 383
electrospinning polymers structures and saturation region field-effect mobility, 386
gate-dependent I-V characteristics and gate voltages, 388
nanofibre field effect transistor vs RRP3HT nanofibre FET, 387
NiO/ZnO nanofibre yarns OM and SEM images, 382
figure of merit *see* quality factor
filtration, 84
filter capture efficiency, 122
functional nanofibres, 121–49
future trends, 147–9
nanofibre catalyst support structures, 142–7
nanoparticles from gas streams, 131–42
principles, 125–31
continuum, slip and molecular flows, 125–8
effectiveness factor for a spherical catalyst pellet, 131
fibre and mat strength, 128–9
gas velocities for flow past fibres in continuum and slip flow, 128
mass and heat transfer, 129–31
specific surface area, 125
filtration membranes, 315–17
air filtration, 316–17
water filtration, 316
filtration technology, 333
filtration theory, 247
flow control pump, 5
flow rate, 11–12
fluid jet, 9
food processing
electrospun fibrous mats, 265–71
filtration and clarification applications, 266–70
solid-phase extraction applications, 270–1
encapsulation and bioactive compounds release, 286–90
antimicrobial electrospun nanofibrous mats, 287–9
bioactive compounds encapsulation, 290
microorganism encapsulating electrospun nanofibre, 289–90
enzyme immobilisation, 271–80

dispersed nanofibre and trypsin coating preparation, 277
enzyme functionalised nano- and microfibres, 277–80
immobilised lipases in electrospun polymer fibres, 274–5
immobilised oxireductases and hydrolases in electrospun polymer, 278
lipases, 273–6
proteases, 276–7
food sensors electrospun nanofibres, 280–6
metal oxide gas sensors preparation diagram, 284
role in biosensor improvement, 281–3
role in nonenzymatic sensors, 283–6
functional nanofibres, 262–92
fibre sizes and morphologies, 264–5
future trends, 290–2
future trends, 290–2
forward osmosis, 361
fouling, 338–9
Fourier transform infrared spectrometer, 165–7
blank PLLA fibre and clarithromycin IR spectrum, 165
Fourier transform infrared spectrophotometer, 161–2
fresh water, 331
fructose amperometric sensor, 282
functional nanofibres
advantages of electrospun nanofibres protective clothing, 248–55
DMPC test method illustration, 250
evaporative cooling, 249–51
low basis weight and low pressure drop, 251–2
particulate penetration resistance, 248–9
process simplicity, 254–5
surface activity incorporation to nanofibres, 253
surface functionalisation, 252–4
chemical and biological hazards protection clothing, 236–58
future trends, 256–8
hazards background, 237–43
electrospun fibres development issues, 255–6
electrospun nanofibres, 244–8
filter membrane SEM photo, 245
protective applications use, 246–8
slip flow illustration, 248
spores on nanofibres, 247
food processing, 262–92
electrospun fibrous mats, 265–71
electrospun micro- and nanofibres, 286–90
electrospun microfibres and enzyme immobilisation nanofibres, 271–80
food sensors electrospun nanofibres, 280–6
future trends, 290–2
future trends in processing, 411–13
microelectronics applications, 371–401
conducting polymers for opto-electronics, 372
nanofibres, 377–401
nanofibres electrospinning, 373–7
polymer in electronics, 371–3
sound absorption, electromagnetic wave attenuation and biomedical application, 305–26
electromagnetic wave attenuation, 306–15
enzyme immobilisation, 317–25
filtration membranes, 315–17
future trends, 325–6
sound absorption, 306
thin-film nanofibrous composite (TFNC) membranes, 335–9
hierarchical structures, 336
water purification, 331–62
microfiltration TFNC membranes, 339–48
nanofiltration and reverse osmosis, 357, 360–1
ultrafiltration TFNC membranes, 348–57

gas phase momentum balance, 135
gas sensor, 283–4, 396–401
Pd-loaded SnO_2 sensors electrical response to H_2, 399
SnO_2 sensor processing illustration, 398
UV nanosensors schematic representation, 400
gas streams, 131–42
gelatin, 322
Gibbs-Donnan effect, 360
glucose biosensor, 282
glucosidases, 280
GORE-TEX, 251
gradient nanofibres, 31–2
electrospinning on a tapered mandrel, 32
fabrication, 31
fabrication of model II, 32
green electrospinning approach, 265

heat gun, 16
heat transfer, 129–31
heterocoagulation, 58
high-temperature calcinations, 72–3
high-voltage power supply, 5
horseradish peroxidase, 324–5
hot carrier diode *see* Schottky diode
humidity, 14
sensors, 285
hybrid multilayered hollow nanofibres (HMHNFs), 109
hydrogen cyanide, 241
hydrophobic composite nanofibres, 319
hydrophobic polymers, 338–9

impedance matching ability, 307
in situ emulsion polymerisation, 41
in situ microemulsion polymerisation, 312
in situ polymerisation, 41, 58
in situ reduction, 42
inorganic nanofibres, 71–88
applications, 84–7
electromagnetism, 86–7
filtration, 84
microwave absorption, 87
reaction substrates, 85–6
doping, 82
future trends, 87–8
processing, 72–5
electrospinning techniques, 76–8
fabrication steps by electrospinning, 73
porous and uniaxially aligned arrays TiO_2 nanofibres, 75
PVA/silica composite fibres and calcination, 74
R-Al_2O_3 nanofibres, 75
TiO_2 inorganic nanofibres, preparation, structure and properties, 75, 79–82
inorganic semiconductors, 371
intelligent packaging, 262
interfacial polymerisation, 357, 360
ion beam implantation, 100
ionic liquids, 353–4
ionic transition metal complexes, 390–1
ISO 11092, 250

joint service lightweight integrated suit technology (JSLIST), 242–3, 246

Knudsen number, 247

laser heating, 16
layer-by-layer deposition, 108–10
layer-by-layer technology, 257, 362
ligament, 185–6
light-emitting diode (LED), 387–8
light emitting nanofibres, 387–92
electroluminescent nanofibres, 391
multiple interelectrode gaps light emission, 392
polymer LED configuration, 389
pure PPV fibres and TiO_2 /PPV hybrid nanoparticle-polymer fibres, 393
lipases, 273–6
lithium-ion batteries
applications, 200–2, 202–5

cycle performance of porous carbon nanofibres electrode, 206
initial discharge-charge profile of porous carbon nanofibres electrode, 205
PAN/PMMA composite nanofibres, 203
surface parameters of as-prepared carbon nanofibres, 204
future trends, 205–6
nanofibres, 197–206
overview, 198–200
material components, 198–200
work principle, 199
low-temperature solution processing, 373

magnetic field shielding effectiveness, 313
mass transfer, 129–31
mat strength, 128–9
Maxwell's equations, 313–14
mechanical waves, 306
melt electrospinning, 15–18, 18–19
apparatus, 15
heating, 16–17
processing parameters on fibre diameters, 17–18
melt flow rate (MFR), 17
melt mixing, 58
membrane filtration technology, 266
metal oxide sensors, 284
metal semiconductor field-effect transistor (MESFET), 384
microelectronics
functional nanofibres, 371–401
nanofibres, 377–401
fibre-based actuators, 395–6
field-effect transistor (FET), 381–7
gas sensors, 396–401
light emitting nanofibres, 387–92
nano wires, 378–80
optoelectronic devices, 392–5
piezoelectric actuator with PDVF fibre schematic, 397
Schottky diode, 380–1
Schottky diode symbol and resistor I-V characteristics, 380
single fibre I-V characteristics, 379
nanofibres electrospinning, 373–7
electrospinning principle, 375–6
increasing publications on electrospinning, 374
nanofibres applications, 376
nanofibres applications diagram, 377
setup schematic, 375
polymer in electronics, 371–3
microencapsulation, 289
microfibres, 3398
microfiltration, 315–16, 333
MicroMax, 251
microwave absorption, 87
molecular flows, 125–8
molecular imprinting, 110–12
Molecular Weight Cut-Off's (MWCO) test, 354
molecularly imprinted nanoparticles, 270–1
molecularly imprinted polymer, 111
multienzymatic bioreactors, 291
multifunctional electrospun fibre membranes, 270
multineedle electrospinning, 412
multiple reflection correction term, 314
multisensor devices, 285
multispinneret, 5
different arrangements, 7
multiwalled carbon nanotubes (MWCNTs), 323

nano wires, 378–80
nanofibre catalyst
support structures, 142–7
diameter and particle size of metal doped alumina nanofibres, 147
experimental description, 144–6
manufacture of catalytic filter, 143
PdO/Al_2O_3 and Rh_2O_3/Al_2O_3 fibres, 146
temperature *vs.* concentration for Pd/Al_2O_3, 147

nanofibre catalyst (*cont.*)
temperature *vs.* concentration for Rh/Al_2O_3 catalytic filter, 148
testing result, 146–7
testing set-up used to determine catalytic activity of filter media, 145
nanofibre composite laminates, 267–8
nanofibre composites
applications, 61–7
bacterial cellulose and tapping-mode atomic force microscopy, 65
delamination fracture surface, 64
electrospun PEK-C/CNF nanofibres diameter distribution, 66
epoxies mechanical properties, 67
semi-IPN structure formation, 63
formation, 57–60
fabrication and preparation, 57–9
preparation, 59–60
nanofibre dressing, 258
nanofibre webs, 267
nanofibres, 210–12, 305
drug delivery, 153–68
applications, 160–7
blood concentration changes, 154
future trends, 167–8
mechanisms, 155–9
electrospinning, 5–10
filtration applications, 121–49
catalyst support structures, 142–7
filter capture efficiency, 122
future trends, 147–9
nanoparticles from gas streams, 131–42
principles, 125–31
future trends, 18–19
large scale production, 19
melt electrospinning, 18–19
lithium-ion batteries, 197–206
applications, 200–2, 202–5
future trends, 205–6
overview, 198–200
melt electrospinning, 15–18
principles and manufacture, 3–19
PLA fibres and nanofibres, 4
sensor, 209–29
future trends, 229
modern electrospinning technology, 209–15
nanostructured sensing material formation, 215–18
sensing mechanism, 219–29
solution electrospinning, 10–14
tissue engineering, 171–87
applications, 172–86
electrospinning, 172
nanofiltration, 316, 333
nanofire-based filter media, 145–6
NanoFMG Nanospinner 416, 255
nanonets, 212–14
polymer systems, 213
nanoparticles
filtration from gas streams, 131–42
fibre capture efficiency concept, 133
layer efficiency, 133–4
mechanisms, 131–3
model results, 140–2
particle capture mechanism, 132
pressure drop calculation, 137–9
quality factor calculation, 139–40
RQFs for different sizes microfibres, 141
RQFs for different sizes nanofibres, 141
RQFs for different sizes of particles, 141
species balance and prediction, 134–7
nanoscale fibres, 236
Nanospider, 412
nanotechnology, 262
natural polymers, 176, 186
nerve agents, 238–9
nisin, 288
N,N-dimethylformamide (DMF), 24
nozzle, 12
nutraceuticals, 263

oil-in-water emulsion, 25
one-step electrospinning, 41
electrospinning setup, 42
optical sensors, 225–7
polymer p-doped PS nanofibrous sensing membranes, 226
optoelectronic devices, 392–5
Al I-V curve and variation with conductivity vs concentration, 396
spatially oriented nanowires fabrication illustration, 394
organic electronics *see* plastic electronics
organic light-emitting diode (OLED), 388
organophosphorus nerve agents, 238–9
oxidoreductases, 280

p-n junction diodes, 381
PA6 nanofibres, 97, 100
PAN/PMMA nanofibres, 62, 202, 204
passive collector arrays, 256
PCL scaffolds, 178, 180, 182
peripheral nerve, 181–3
dorsal root ganglia cultured on electrospun PLLA aligned nanofibrous scaffolds, 183
Petryanov filters, 245, 251, 256
pH-controlled assembly method, 257
phase-inversion method, 337, 357
phenolics sensors, 283
photoelectric sensors, 223–5
microreactor and photoelectric response measurements, 224
photon, 306
physical vapour deposition (PVD), 96–100
EDX spectra of the nanofibres before and after sputter coating, 99
PA6 nanofibres, 98
transmittance of PA6 nanofibres, 100
physically controlled drug release, 157–9
plasma treatment, 93–6
dynamic water adsorption of PVDF membrane, 96
electrospinning voltage effect on the fibre diameter, 95
pre-treatment and grafting on PVDF nanofibre membrane, 94
surface contact angles of the PVDF membrane, 95
plastic electronics, 372, 377
PLCL scaffolds, 181
poly(3-hexyl thiophene), 383
poly (phenylene vinylene), 391
poly(acrylonitrile- comaleic acid), 318–19, 322–3
polyacrylonitrile nanofibre, 102, 290
polyacrylonitrile nanofibrous membranes, 276
polyacrylonitrile (PAN), 200
polyamide-based TFNC, 360
polyamide membrane, 357
poly(aniline) nanowire-based transistors, 384
polyanion, 198
polyetherketone cardo nanofibres, 62–3, 65
polyethylene oxide (PEO), 25–7, 39
poly(ethylene terephthalate) membranes, 269
polymer-based electroluminescent displays, 389
polymer/CNT composites, 58
polymer/inorganic salt composite nanofibres, 43, 48–9
polymer light-emitting diode (PLED), 388–90
polymer matrix composites
carbon and polymer nanofibre reinforcements, 55–68
carbon nanofibres and nanotubes structures, 56
future trends, 67–8
nanofibre composites applications, 61–7
nanofibre composites formation, 57–60
strengths and weaknesses, 60–1

polymer membranes, 334
polymer molecular weight, 10
polymer nanofibres
  applications of nanofibres composites, 61–7
  carbon nanofibres and nanotubes structures, 56
  future trends, 67–8
  nanofibres composites formation, 57–60
  reinforcements in polymer matrix composites, 55–68
  strengths and weaknesses, 60–1
  surface functionalisation, 92–116
    biotechnology, 112–15
    chemical technologies, 100–6
    future trends, 115–16
    nanotechnologies, 106–12
    physical technologies, 93–100
polymer/nanoparticle composite nanofibres, 40–2, 44–8
  electroless deposition and *in situ* reduction method, 42
  electrospun PA6, 45
  electrospun PA6 and PA6/Fe-OMT composite nanofibres, 46
  Mechanical properties of PVAc/TiO_2 nanofibres, 48
  one-step electrospinning method, 41
  sol-gel method, 41–2
  TGA curves of PVAc hybrid nanofibres, 48
  two-step electrospinning method, 41
  ZnO coated, preoxidised ZnO coated and carbon/ZnO nanofibres, 47
polymer/polymer composite nanofibres, 38–40, 43–4
  co-axial electrospinning setup, 40
  co-electrospinning setup, 39
poly(methyl methacrylate) (PMMA), 25–7
polyoxometallate, 252–3
polypropylene electrospinning, 413
polystyrene-poly (styrene-co-maleic anhydride) nanofibres, 276–7
polyvinyl alcohol nanofibres, 102
polyvinyl alcohol (PVA), 39, 411
poly(vinyl pyrrolidone)/nanosilver composite fibres, 288
polyvinyl pyrrolidone (PVP), 24
polyvinylidene fluoride, 395
polyvinylidenedifluoride (PVDF), 200
porous nanofibres, 30–1
postelectrospinning assembly method, 257
potable water, 123–4
power law equation, 158
prefabricated electrospun nanofibres, 272–3
pressure drop, 137–9
printed electronics *see* plastic electronics
probiotic bacteria, 286, 289
Pseudomonas nitroreducens, 290
pulmonary agents, 239
PVA-based TFNC, 351–3
PVA-Cu(II) chelated nanofibres, 320
Pyragraf III, 57

quality factor, 139–40
quartz crystal microbalances (QCMs), 219

reaction substrates, 85–6
regioregular poly(3-hexylthiophene) (RRP3HT), 385
relative quality factor (RQF), 140
resistive sensors, 220, 222–3
  fabricated device structure, NiO-SnO_2 composite nanofibres, 223
reverse osmosis, 316, 333
  membranes, 266

S. epidermidis, 257
SARATOGA, 242–3, 249
Sarin gas, 239
scanning electron microscope, 161
Schelkunoff's formula, 314
Schottky diode, 380–1
selected-area electron diffraction (SAED), 395
self-immobilisation techniques, 271

semiconducting polymers, 372
sensor, 214–15
functional nanofibres, 209–29
future trends, 229
modern electrospinning technology, 209–15
nanostructured sensing material formation, 215–18
composite or doped electrospinning, 216–17
single-component electrospinning, 215
surface assembly of sensing materials, 217–18
tungsten oxide/PVAc composite fibres and WO3 nanofibres, 217
sensing mechanism, 219–29
acoustic wave sensors, 219–20
biological sensors, 227–9
optical sensors, 225–7
photoelectric sensors, 223–5
resistive sensors, 220, 222–3
shape memory polymer composites, 308
shielding effectiveness, 313
silica nanofibres, 325
silk fibroin nanofibres, 318
Simon's model, 307–8
single-component electrospinning, 215
single fibre capture efficiency, 136
single-nozzle electrospinning process, 255
single spinneret, 5
different shapes, 6
skin, 172–5
skin wounds, 173
slip flow, 125–8, 247–8, 317
small angle X-ray scattering (SAXS), 353–4
SnO_2 sensor, 398–9
sodium dodecyl sulphate (SDS), 26
sol-gel, 41–2, 72–3, 106–7
solid-phase microextraction (SPME), 271
solution concentration, 10
solution conductivity, 11
solution electrospinning, 10–14
environmental condition, 12–14
process condition, 11–12
properties, 10–11
solution processing, 58
sound absorption
functional nanofibres, 305–26
future trends, 325–6
functional nanofibres applications, 306
sound waves, 306
spacer arms, 323
specific surface area, 125
electrospun Nomex nanofibres, 126
external surface area of fibres as a function of fibre diameter, 125
nylon nanofibres mixed among glass microfibres in a filter medium, 126
spinneret, 5
splaying, 375
Staphylococcus enterotoxin B, 283
streptavidin coupling, 270
structured functional nanofibres
types and processing, 22–36
aligned nanofibres, 27–30
applications, 33–4
C/SnO_2 core-shell nanofibres, 34
core-shell nanofibres, 23–7
future trends, 34–6
gradient nanofibres, 31–2
PMMA/O-MMT composite nanofibres, 35
porous nanofibres, 30–1
surface acoustic wave (SAW), 220
surface assembly
sensing materials, 217–18
assembling fluorescent probes, 218
surface charge density, 11
surface functionalisation, 252–4
biotechnology, 112–15
catalases immobilised onto metal chelated AOPAN nanofibrous membrane, 113
PAN and AOPAN nanofibrous membrane, 114
storage stability of catalases, 115

surface functionalisation (*cont.*)
chemical technologies, 100–6
adsorption isotherm of Cd2 and Cu2 on AOPAN nanofibres, 104
desorption isotherm of Cd2 and Cu2 on metal chelated AOPAN nanofibres, 104
electroless deposition, 104–6
PAN and AOPAN nanofibres, 103
surface grafting, cross-linking and chelation, 101–4
future trends, 115–16
nanotechnologies, 106–12
atomic deposition, 107–8
layer-by-layer deposition, 108–10
molecular imprinting, 110–12
sol-gel method, 106–7
physical technologies, 93–100
physical vapour deposition (PVD), 96–100
plasma treatment, 93–6
polymer nanofibres, 92–116
surface grafting, 101–4
surface tension, 10–11
SympaTex, 251
synthetic polymers, 176, 186

Taylor cone, 244, 263, 375
temperature, 12–14
effect on fibre diameters, 13
effect on surface tension and viscosity of electrospun solution, 14
electrospun PVDF nanofibres preparation, 13
XRD patterns of PVDF nanofibres electrospun, 14
tendon, 185–6
morphological changes of human tendon stem/progenitor cells (hTSPCs), 187
thermogravimetry (TG), 395
thermosetting resins, 62
thin-film nanofibrous composite (TFNC) membranes, 335–9
TiO_2 inorganic nanofibres
crystal forms, 80–2
mats calcined in different temperature, 81
element components, 82
mats calcine in different temperatures, 83
preparation, 75, 79
structures, 79
mats calcined in different temperature, 80
surface morphology, 79–80
PVAc/TiO_2 calcined in different temperatures, 81
tissue engineering
applications, 172–86
bone, 177–81
cartilage, 184–5
cell attachment and spreading of normal human fibroblast, 175
cell attachment and spreading of normal human keratinocytes, 174
ligament and tendon, 185–6
peripheral nerve, 181–3
skin and wound healing, 172–5
vascular tissue, 175–7
functional nanofibres, 171–87
nanofibres electrospinning, 172
tobacco mosaic virus (TMV), 411
transducer, 281
two-phase electrospinning method, 325
two-step electrospinning, 41

Ultra-Web, 245–6
ultrafiltration, 315–16, 333
TFNC membranes, 348–57
cross-linked PVA barrier layer, 350–3
regenerated polysaccharide barrier layer, 353–5
schematic, 349
SEM images, 356
TFNC ultrafiltration membranes, 358–9
ultrafine polysaccharide nanofibres, 355–7
ultrasonication, 58
urea biosensor, 282

US Army Natick Soldier Centre, 246, 250
US Patent 4455187, 243
US Patent 4510193, 243
US Patent 58277963, 243
UV nanosensors, 399

vapour-grown carbon nanofibres (VGCNFs), 307
vascular grafts, 175–7
vascular tissue, 175–7
bilayered electrospun PCL/collagen scaffolds, 177
vessicant agents, 239, 241
viscosity, 10–11
viscous jet, 8
visible spectrum, 306
vitamin encapsulation, 290

water-in-oil emulsion, 25
water purification
functional nanofibres, 331–62
growing population and usable water distribution, 332
membrane technology, 333
membrane technology challenges, 333–5
microfiltration TFNC membranes, 339–48
nanofibres and microfibres role, 340–1
novel fabrication by electrospinning, 340
polysaccharide nanofibres from natural sources, 346
polysaccharide nanofibres properties, 347
pore size and fibre diameter relationship, 341
removal of bacteria, 342–3
removal of viruses, 343–5
ultrafine polysaccharide nanofibres, 345–8
nanofiltration and reverse osmosis membranes, 357, 360–1
ultrafiltration TFNC membranes, 348–57
cross-linked PVA barrier layer, 350–3
regenerated polysaccharide barrier layer, 353–5
ultrafine polysaccharide nanofibres, 355–7
weapons of mass destruction, 238
wide-angle X-ray diffraction (WAXD), 354–5
wound healing, 172–5

X-ray diffraction (XRD), 165–7
CLM, cast films and PLLA nanofibre scaffolds pattern, 166
patterns of PLLA nanofibre scaffolds containing CLM, 166
xanthine determination, 283

zeolites, 392–3

FSC
www.fsc.org
MIX
Papier | Fördert gute Waldnutzung
FSC® C083411

Kolibri 360 GmbH
Ferdinand-Jühlke-Straße 7
99095 Erfurt, Deutschland
produktsicherheit@kolibri360.de